Advanced Mathematics and Mechanics Applications Using

Third Edition

Advanced Mathematics and Mechanics Applications Using

Third Edition

Howard B. Wilson
University of Alabama

Louis H. Turcotte
Rose-Hulman Institute of Technology

David Halpern
University of Alabama

A CRC Press Company
Boca Raton London New York Washington, D.C.

Library of Congress Cataloging-in-Publication Data

Wilson, H.B.
Advanced mathematics and mechanics applications using MATLAB / Howard B. Wilson, Louis H. Turcotte, David Halpern.—3rd ed.
p. cm.
ISBN 1-58488-262-X
1. MATLAB. 2. Engineering mathematics—Data processing. 3. Mechanics, Applied—Data processing. I. Turcotte, Louis H. II. Halpern, David. III. Title.

TA345 . W55 2002
620′.00151—dc21 2002071267

Direct all inquiries to CRC Press LLC, 2000 N.W. Corporate Blvd., Boca Raton, Florida 33431.

Visit the CRC Press Web site at www.crcpress.com

International Standard Book Number 1-58488-262-X
Library of Congress Card Number 2002071267
Printed in the United States of America 1 2 3 4 5 6 7 8 9 0
Printed on acid-free paper

For my dear wife, Emma.

Howard B. Wilson

For my loving wife, Evelyn, our departed cat, Patches, and my parents.

Louis H. Turcotte

Preface

This book uses MATLAB® to analyze various applications in mathematics and mechanics. The authors hope to encourage engineers and scientists to consider this modern programming environment as an excellent alternative to languages such as FORTRAN or C++. MATLAB[1] embodies an interactive environment with a high level programming language supporting both numerical and graphical commands for two- and three-dimensional data analysis and presentation. The wealth of intrinsic mathematical commands to handle matrix algebra, Fourier series, differential equations, and complex-valued functions makes simple calculator operations of many tasks previously requiring subroutine libraries with cumbersome argument lists.

We analyze problems, drawn from our teaching and research interests, emphasizing linear and nonlinear differential equation methods. Linear partial differential equations and linear matrix differential equations are analyzed using eigenfunctions and series solutions. Several types of physical problems are considered. Among these are heat conduction, harmonic response of strings, membranes, beams, and trusses, geometrical properties of areas and volumes, flexure and buckling of indeterminate beams, elastostatic stress analysis, and multi-dimensional optimization.

Numerical integration of matrix differential equations is used in several examples illustrating the utility of such methods as well as essential aspects of numerical approximation. Attention is restricted to the Runge-Kutta method which is adequate to handle most situations. Space limitation led us to omit some interesting MATLAB features concerning predictor-corrector methods, stiff systems, and event locations.

This book is not an introductory numerical analysis text. It is most useful as a reference or a supplementary text in computationally oriented courses emphasizing applications. The authors have previously solved many of the examples in FORTRAN. Our MATLAB solutions consume over three hundred pages (over twelve thousand lines). Although few books published recently present this much code, comparable FORTRAN versions would probably be signifcantly longer. In fact, the conciseness of MATLAB was a primary motivation for writing the book.

The programs contain many comments and are intended for study as separate entities without an additional reference. Consequently, some deliberate redundancy

[1]MATLAB is a registered trademark of The MathWorks, Inc. For additional information contact:
The MathWorks, Inc.
3 Apple Hill Drive
Natick, MA 01760-1500
(508) 647-7000, Fax: (508) 647-7001
Email: info@mathworks.com

exists between program comments and text discussions. We also list programs in a style we feel will be helpful to most readers. The source listings show line numbers adjacent to the MATLAB code. MATLAB code does not use line numbers or permit *goto* statements. We have numbered the lines to aid discussions of particular program segments. To conserve space, we often place multiple MATLAB statements on the same line when this does not interrupt the logical flow.

All of the programs presented are designed to operate under the 6.x version of MATLAB and Microsoft Windows. Both the text and graphics windows should be simultaneously visible. A windowed environment is essential for using capabilities like animation and interactive manipulation of three dimensional figures. The source code for all of the programs in the book is available from the CRC Press website at `http://www.crcpress.com`. The program collection is organized using an independent subdirectory for each of the thirteen chapters.

This third edition incorporates much new material on time dependent solutions of linear partial differential equations. Animation is used whenever seeing the solution evolve in time is helpful. Animation illustrates quite well phenomena like wave propagation in strings and membranes. The interactive zoom and rotation features in MATLAB are also valuable tools for interpreting graphical output.

Most programs in the book are academic examples, but some problem solutions are useful as stand-alone analysis tools. Examples include geometrical property calculation, differentiation or integration of splines, Gauss integration of arbitrary order, and frequency analysis of trusses and membranes.

A chapter on eigenvalue problems presents applications in stress analysis, elastic stability, and linear system dynamics. A chapter on analytic functions shows the efficiency of MATLAB for applying complex valued functions and the Fast Fourier Transform (FFT) to harmonic and biharmonic functions. Finally, the book concludes with a chapter applying multidimensional search to several nonlinear programming problems.

We emphasize that this book is primarily for those concerned with physical applications. A thorough grasp of Euclidean geometry, Newtonian mechanics, and some mathematics beyond calculus is essential to understand most of the topics. Finally, the authors enjoy interacting with students, teachers, and researchers applying advanced mathematics to real world problems.The availability of economical computer hardware and the friendly software interface in MATLAB makes computing increasingly attractive to the entire technical community. If we manage to cultivate interest in MATLAB among engineers who only spend part of their time using computers, our primary goal will have been achieved.

Howard B. Wilson hwilson@bama.ua.edu
Louis H. Turcotte turcotte@rose-hulman.edu
David Halpern david.halpern@ua.edu

Contents

Chapter 1

Introduction

1.1 MATLAB: A Tool for Engineering Analysis

This book presents various MATLAB applications in mechanics and applied mathematics. Our objective is to employ numerical methods in examples emphasizing the appeal of MATLAB as a programming tool. The programs are intended for study as a primary component of the text. The numerical methods used include interpolation, numerical integration, finite differences, linear algebra, Fourier analysis, roots of nonlinear equations, linear differential equations, nonlinear differential equations, linear partial differential equations, analytic functions, and optimization methods. Many intrinsic MATLAB functions are used along with some utility functions developed by the authors. The physical applications vary widely from solution of linear and nonlinear differential equations in mechanical system dynamics to geometrical property calculations for areas and volumes.

For many years FORTRAN has been the favorite programming language for solving mathematical and engineering problems on digital computers. An attractive alternative is MATLAB which facilitates program development with excellent error diagnostics and code tracing capabilities. Matrices are handled efficiently with many intrinsic functions performing familiar linear algebra tasks. Advanced software features such as dynamic memory allocation and interactive error tracing reduce the time to get solutions. The versatile but simple graphics commands in MATLAB also allow easy preparation of publication quality graphs and surface plots for technical papers and books. The authors have found that MATLAB programs are often significantly shorter than corresponding FORTRAN versions. Consequently, more time is available for the primary purpose of computing, namely, to better understand physical system behavior.

The mathematical foundation needed to grasp most topics presented here is covered in an undergraduate engineering curriculum. This should include a grounding in calculus, differential equations, and knowledge of a procedure oriented programming language like FORTRAN. An additional course on advanced engineering mathematics covering linear algebra, matrix differential equations, and eigenfunction solutions of partial differential equations will also be valuable. The MATLAB programs were written primarily to serve as instructional examples in classes traditionally referred to as advanced engineering mathematics and applied numerical methods. The greatest benefit to the reader will probably be derived through study of the programs relating mainly to physics and engineering applications. Furthermore, we believe that several of the MATLAB functions are useful as general utilities. Typical examples include routines for spline interpolation, differentiation, and integration; area and inertial moments for general plane shapes; and volume and inertial properties of arbitrary polyhedra. We have also included examples demonstrating natural frequency analysis and wave propagation in strings and membranes.

MATLAB is now employed in more than two thousand universities and the user community throughout the world numbers in the thousands. Continued growth will be fueled by decreasing hardware costs and more people familiar with advanced analytical methods. The authors hope that our problem solutions will motivate analysts already comfortable with languages like FORTRAN to learn MATLAB. The rewards of such efforts can be considerable.

1.2 MATLAB Commands and Related Reference Materials

MATLAB has a rich command vocabulary covering most mathematical topics encountered in applications. The current section presents instructions on: a) how to learn MATLAB commands, b) how to examine and understand MATLAB's lucidly written and easily accessible "demo" programs, and c) how to expand the command language by writing new functions and programs. A comprehensive online help system is included and provides lengthy documentation of all the operators and commands. Additional capabilities are provided by auxiliary toolboxes. The reader is encouraged to study the command summary to get a feeling for the language structure and to have an awareness of powerful operations such as **null**,**orth**,**eig**, and **fft**.

The manual for The Student Edition of MATLAB should be read thoroughly and kept handy for reference. Other references [47, 97, 103] also provide valuable supplementary information. This book extends the standard MATLAB documentation to include additional examples which we believe are complementary to more basic instructional materials.

Learning to use **help**, **type**, **dbtype**, **demo**, and **diary** is important to understanding MATLAB. **help** function_name (such as **help plot**) lists available documentation on a command or function generically called "function_name." MATLAB responds by printing introductory comments in the relevant function (comments are printed until the first blank line or first MATLAB command after the function heading is encountered). This feature allows users to create online help for their own functions by simply inserting appropriate comments at the top of the function. The instruction **type function_name** lists the entire source code for any function where source code is available (the code for intrinsic functions stored in compiled binary for computational efficiency cannot be listed). Consider the following list of typical examples

Command	Resulting Action
help help	discusses use of the help command
help demos	lists names of various demo programs
type linspace	lists the source code for the function which generates a vector of equidistant data values
type plot	outputs a message indicating that **plot** is a built-in function
intro	executes the source code in a function named **intro** which illustrates various MATLAB functions.
type intro	lists the source code for the **intro** demo program. By studying this example, readers can quickly learn many MATLAB commands
graf2d	demonstrates X-Y graphing
graf3d	demonstrates X-Y-Z graphing
help diary	provides instructions on how results appearing on the command screen can be saved into a file for later printing, editing, or merging with other text
diary *fil_name*	instructs MATLAB to record, into a file called *fil_name,* all text appearing on the command screen until the user types **diary off**. The **diary** command is especially useful for making copies of library programs such as **zerodemo**
demo	initiates access to a lengthy set of programs demonstrating the functionality of MATLAB. It is also helpful to source list some of these programs such as: **zerodemo**, **fitdemo**, **quaddemo**, **odedemo**, **ode45**, **fftdemo**, and **truss**

1.3 Example Problem on Financial Analysis

Let us next analyze a problem showing several language constructs of MATLAB programming. Most of this book is devoted to solving initial value and boundary value problems for physical systems. For sake of variety we study briefly an elementary example useful in business, namely, asset growth resulting from compounded investment return.

The differential equation

$$Q'(t) = R\,Q(t) + S\,\exp(At)$$

describes growth of investment capital earning a rate of investment return R and augmented by a saving rate $S\,\exp(At)$. The general solution of this first order linear equation is

$$Q(t) = \exp(Rt)\left[Q(0) + \int_0^t S\,\exp((A-R)t)dt\right].$$

A realistic formulation should employ inflation adjusted capital defined by

$$q(t) = Q(t)\exp(-It)$$

where I denotes the annual inflation rate. Then a suitable model describing capital accumulation over a saving interval of t_1 years, followed by a payout period of t_2 years, is characterized as

$$q'(t) = r\,q(t) + [s(t \le t_1) - p\exp(-at_1)(t > t_1)]\exp(at), \quad q(0) = q_0.$$

The quantity $(t \le t_1)$ equals one for $t \le t_1$ and is zero otherwise. This equation also uses inflation adjusted parameters $r = R - I$ and $a = A - I$. The parameter s quantifies the initial saving rate and p is the payout rate starting at $t = t_1$.

It is plausible to question whether continuous compounding is a reasonable alternative to a discrete model employing assumptions such as quarterly or yearly compounding. It turns out that results obtained, for example, using discrete monthly compounding over several years differ little from those produced with the continuous model. Since long term rates of investment return and inflation are usually estimated rather than known exactly, the simplified formulas for continuous compounding illustrate reasonably well the benefits of long term investment growth. Integrating the differential equation for the continuous compounding model gives

$$q(t) = q_0\exp(rt) + s[h(t) - (t > t_1)\exp(at_1)h(t - t_1)] - p\,(t > t_1)\,h(t - t_1)$$

where $h(t) = [\exp(rt) - \exp(at)]/(r - a)$. The limiting case for $r = a$ is also dealt with appropriately in the program below. At time $T_2 = t_1 + t_2$ the final capital $q_2 = q(T_2)$ is

$$\begin{aligned} q_2 &= q_0\exp(rT_2) + \frac{s}{r-a}[\exp(rt_1) - \exp(at_1)]\exp(rt_2) \\ &- \frac{p}{r-a}[\exp(rt_2) - \exp(at_2)]. \end{aligned}$$

Therefore, for known r, a, t_1, t_2, the four quantities q_2, q_0, s, p are linearly related and any particular one of these values can be found in terms of the other three. For instance, when $q_0 = q_2 = 0$, the saving factor s needed to provide a desired payout factor p can be computed from the useful equation

$$s = p[1 - \exp((a - r)t_2)]/[\exp(rt_1) - \exp(at_1)]$$

A MATLAB program using the above equations was written to compute and plot $q(t)$ for general combinations of the nine parameters $R, A, I, t_1, t_2, q_0, s, p, q_2$. The program allows data to be passed through the call list of function **finance**, or the interactive input is activated when no call list data is passed. **Finance** calls function **inputv** to read data and the function **savespnd** to evaluate $q(t)$. First we will show some numerical results and then discuss selected parts of the code. Consider a case where someone initially starting with $10,000 of capital expects to save for 40 years and subsequently draw $50,000 annually from savings for 20 years, at which time the remaining capital is to be $100,000. Assume that the investment rate before inflation is $R = 8$ while the inflation rate is $I = 4$. During the 60 year period, annual savings, as well as the pension payout amount, are to be increased to match inflation, so that $A = 4$. The necessary value of s and a plot of the inflation adjusted assets as a function of time are to be determined. The program output shows that when the unknown value of s was input as nan (meaning Not-a-Number in IEEE arithmetic), a corrected value of $6417 was computed. This says that, with the assumed rate of investment return, saving at an initial rate of $6417 per year and continually increasing that amount to match inflation will suffice to provide the desired inflation adjusted payout. Furthermore, the inflation adjusted financial capital accumulated at the end of 40 years is $733,272. The related graph of $q(t)$ duplicates the data listed on the text screen. The reader may find it interesting to repeat the illustrative calculation assuming $R = 11$, in which case the saving coefficient is greatly reduced to only $1060.

1.4 Computer Code and Results

A computer code which analyzes the above equations and presents both numerical and graphical results appears next. First we show the program output, and then discuss particular aspects of the program.

1.4.1 Computer Output

```
>> finance;

   ANALYSIS OF THE SAVE-SPEND PROBLEM BY SOLVING
q'(t)=r*q(t)+[s*(t<=t1)-p*(t>t1)*exp(-a*t1)]*exp(a*t)
where r=R-I, a=A-I, and q(0)=q0

To list parameter definitions enter y
otherwise enter n  ? y
INPUT QUANTITIES:
R    - annual percent earnings on assets
I    - annual percent inflation rate
A    - annual percent increase in savings
       to offset inflation
r,a  - inflation adjusted values of R and I
t1   - saving period (years), 0<t<t1
t2   - payout period (years), t1<t<(t1+t2)
s    - saving rate at t=0, ($K). Saving is
```

```
      expressed as s*exp(a*t),  0<t<t1
p   - payout rate at t=t1, ($K). Payout is
      expressed as
      -p*exp(a*(t-t1)), t1<t<(t1+t2)
q0  - initial savings at t=0, ($K)
q2  - final savings at t=T2=t1+t2, ($K)

OUTPUT QUANTITIES:
q  -  vector of inflation adjusted savings
      values for 0 <= t <= (t1+t2)
t  - vector of times (years) corresponding
     to the components of q
q1 - value of savings at t=t1, when the
     saving period ends

Press return to continue

Input R,A,I (try 11,4,4) ? 8,4,4
Input t1,t2 (try 40,20) ? 40,20
Input q0,s,p,q2 (try 20,5,nan,40) ? 20,nan,50,100

                   PROGRAM RESULTS
    t1          t2          R           A           I
  40.000      20.000      8.000       4.000       4.000

    q0          q1          q2          s           p
  20.000     733.272    100.000       6.417      50.000

>>
```

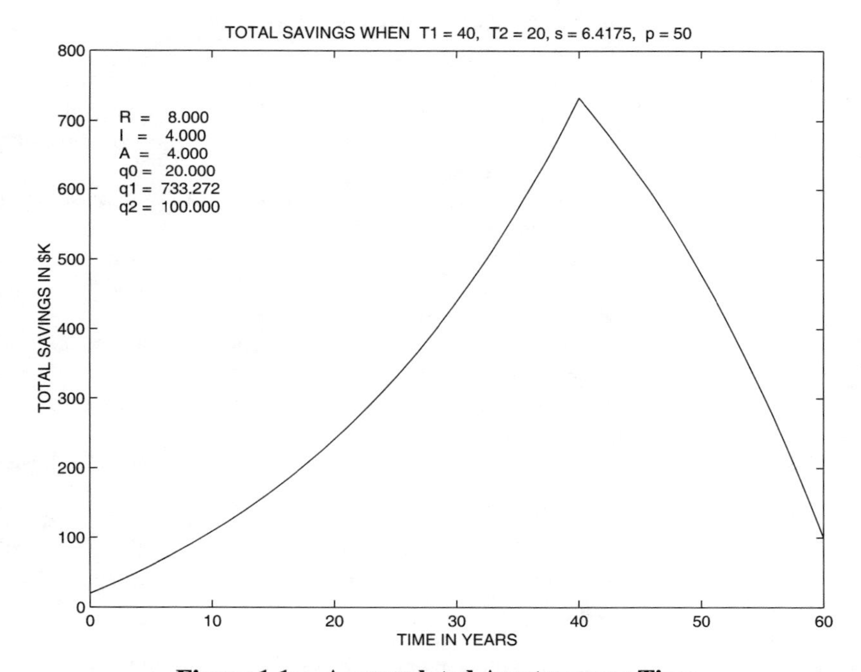

Figure 1.1: Accumulated Assets versus Time

1.4.2 Discussion of the MATLAB Code

Let us examine the following program listing. The line numbers, which are not part of the actual code, are helpful for discussing particular parts of the program. A numbered listing can be obtained with the MATLAB command dbtype.

Line	Comments
1-2	Three dots ... are used to continue function **finance** to handle the long argument list. The output list duplicates some input items to handle cases involving interactive input.
3-16	Comment lines always begin with the % symbol. At the interactive command level in MATLAB, typing help followed by a function name will print documentation in the first unbroken sequence of comments in a function or script file.
20-25	The output heading is printed. Note that q''(t) is used to print q'(t) because special characters such as ' or % must be repeated.
29-50	Intrinsic function **char** is used to store descriptions of program variable in a character matrix.
59	Function **nargin** checks whether the number of input variables is zero. If so, data values are read interactively.
68-69	Function **inputv** reads several variables on the same line.
70-78	While 1,...,end code sequence loops repeatedly to check data input. Break exits to line 80 if data are OK.
85-97	Set multiplier constants to solve for one unknown variable among q0, s, p, q2.
99-105	Determine time vectors to evaluate the solution. Cases where t1 or t2 are zero require special treatment.
108-112	Intrinsic function **isnan** is used to identify the variable which was input as nan.
115-116	User defined function **savespnd** is used to evaluate q(t) and q(t1).
119-127	Program results are printed with a chosen format. The statement b=**inline**('**blanks**(j)','j') just shortens the name for intrinsic function **blanks**.
130-139	Draw the graph along with a title and axis labels.
141-153	Create a label containing data values. Position it on the graph.
154	Turn the grid off and bring the graph to the foreground.
158-176	Function **savespnd** evaluates q(t). The formula for r=a results from the limiting form of q(t) as parameter a tends to r.
180-213	Function **inputv** generalizes the intrinsic function **input** to read several variables on the same line. **Inputv** is used often throughout this text.

1.4.3 Code for Financial Problem

Program finance

```
1: function [q,t,R,A,I,t1,t2,s,p,q0,q1,q2]=finance...
2:                              (R,A,I,t1,t2,s,p,q0,q2)
3: % [q,t,R,A,I,t1,t2,s,p,q0,q1,q2]=finance...
4: %                            (R,A,I,t1,t2,s,p,q0,q2)
5: %~~~~~~~~~~~~~~~~~~~~~~~~~~~~~~~~~~~~~~~~~~~~~~~~~~
6: %
7: % This function solves the SAVE-SPEND PROBLEM
8: % where funds earning interest are accumulated
9: % during one period and paid out in a subsequent
10: % period. The value of assets is adjusted to
11: % account for inflation. This problem is
12: % governed by the differential equation
13: % q'(t)=r*q(t)+[s*(t<=t1)...
14: %       -p*(t>t1)*exp(-a*t1)]*exp(a*t) where
15: % r=R-I, a=A-I and the remaining parameters
16: % are defined below
17:
18: % User m functions required: inputv, savespnd
19:
20: disp(' '), disp(['     ',...
21: 'ANALYSIS OF THE SAVE-SPEND PROBLEM BY SOLVING'])
22: disp(...
23: ['q''(t)=r*q(t)+[s*(t<=t1)-p*(t>t1)*',...
24: 'exp(-a*t1)]*exp(a*t)']), disp(...
25: 'where r=R-I, a=A-I, and q(0)=q0'), disp(' ')
26:
27: % Create a character variable containing
28: % definitions of input and output quantities
29: explain=char('INPUT QUANTITIES:',...
30: 'R   - annual percent earnings on assets',...
31: 'I   - annual percent inflation rate',...
32: 'A   - annual percent increase in savings',...
33: '      to offset inflation',...
34: 'r,a - inflation adjusted values of R and I',...
35: 't1  - saving period (years), 0<t<t1',...
36: 't2  - payout period (years), t1<t<(t1+t2)',...
37: 's   - saving rate at t=0, ($K). Saving is',...
38: '      expressed as s*exp(a*t),  0<t<t1',...
39: 'p   - payout rate at t=t1, ($K). Payout is',...
40: '      expressed as',...
```

```
41: '      -p*exp(a*(t-t1)), t1<t<(t1+t2)',...
42: 'q0  - initial savings at t=0, ($K)',...
43: 'q2  - final savings at t=T2=t1+t2, ($K)',' ',...
44: 'OUTPUT QUANTITIES:',...
45: 'q  -  vector of inflation adjusted savings',...
46: '      values for 0 <= t <= (t1+t2)',...
47: 't  - vector of times (years) corresponding',...
48: '     to the components of q',...
49: 'q1 - value of savings at t=t1, when the',...
50: '     saving period ends',' ');
51:
52: % NOTE: WHEN R,I,A,T1,T2 ARE KNOWN,THEN FIXING
53: % ANY THREE OF THE VALUES q0,s,p,q2 DETERMINES
54: % THE UNKNOWN VALUE WHICH SHOULD BE GIVEN AS
55: % nan IN THE DATA INPUT.
56:
57: % Read data interactively when input data is not
58: % passed through the call list
59: if nargin==0
60:   disp('To list parameter definitions enter y')
61: querry=input('otherwise enter n  ? ','s');
62: if querry=='Y' | querry=='y'
63: disp(explain); disp('Press return to continue')
64: pause, disp(' ')
65: end
66:
67: % Read multiple variables on the same line
68:      [R,A,I]=inputv('Input R,A,I (try 11,4,4) ? ');
69:     [t1,t2]=inputv('Input t1,t2 (try 40,20) ? ');
70: while 1
71:   [q0,s,p,q2]=inputv(...
72:   'Input q0,s,p,q2 (try 20,5,nan,40) ? ');
73:     if sum(isnan([q0,s,p,q2]))==1, break; end
74:   fprintf(['\nDATA ERROR. ONE AND ONLY ',...
75:       'ONE VALUE AMONG\n','THE PARAMETERS ',...
76:       'q0,s,p,q2 CAN EQUAL nan \n\n'])
77:   end
78: end
79:
80: nt=101; T2=t1+t2; r=(R-I)/100; a=(A-I)/100;
81: c0=exp(r*T2);
82:
83: % q0,s,p,q2 are related by q2=c0*q0+c1*s+c2*p
84: % Check special case where t1 or t2 are zero
85: if t1==0
```

```
86: disp(' '), disp('s is set to zero when t1=0')
87:   s=0; c1=0;
88: else
89: c1=savespnd(T2,t1,0,R,A,I,1,0);
90: end
91:
92: if t2==0
93: disp(' '), disp('p is set to zero when t2=0')
94:   p=0; c2=0;
95: else
96: c2=savespnd(T2,t1,0,R,A,I,0,1);
97: end
98:
99: if t1==0 | t2==0
100: t=linspace(0,T2,nt)';
101: else
102:    n1=max(2,fix(t1/T2*nt));
103:    n2=max(2,nt-n1)-1;
104:    t=[t1/n1*(0:n1),t1+t2/n2*(1:n2)]';
105: end
106:
107: % Solve for the unknown parameter
108: if isnan(q0),     q0=(q2-s*c1-p*c2)/c0;
109: elseif isnan(s), s=(q2-q0*c0-p*c2)/c1;
110: elseif isnan(p), p=(q2-q0*c0-s*c1)/c2;
111: else,             q2=q0*c0+s*c1+p*c2;
112: end
113:
114: % Compute results for q(t)
115: q=savespnd(t,t1,q0,R,A,I,s,p);
116: q1=savespnd(t1,t1,q0,R,A,I,s,p);
117:
118: % Print formatted results
119: b=inline('blanks(j)','j'); B=b(3); d='%8.3f';
120: u=[d,B,d,B,d,B,d,B,d,'\n']; disp(' ')
121: disp([b(19),'PROGRAM RESULTS'])
122: disp(['    t1           t2           R',...
123:    '           A            I'])
124: fprintf(u,t1,t2,R,A,I), disp(' ')
125: disp(['    q0           q1           q2',...
126:    '           s            p'])
127: fprintf(u,q0,q1,q2,s,p), disp(' '), pause(1)
128:
129: % Show results graphically
130: plot(t,q,'k')
```

```
131: title(['INFLATION ADJUSTED SAVINGS WHEN ',...
132:  'S = ',num2str(s),' AND P = ',num2str(p)]);
133: titl=...
134: ['TOTAL SAVINGS WHEN  T1 = ',num2str(t1),...
135: ',  T2 = ',num2str(t2),', s = ',num2str(s),...
136: ',  p = ',num2str(p)]; title(titl)
137:
138: xlabel('TIME IN YEARS')
139: ylabel('TOTAL SAVINGS IN $K')
140:
141: % Character label showing data parameters
142: label=char(...
143:  sprintf('R  = %8.3f',R),...
144:  sprintf('I   = %8.3f',I),...
145:  sprintf('A  = %8.3f',A),...
146:  sprintf('q0 = %8.3f',q0),...
147:  sprintf('q1 = %8.3f',q1),...
148:  sprintf('q2 = %8.3f',q2));
149: w=axis; ymin=w(3); dy=w(4)-w(3);
150: xmin=w(1); dx=w(2)-w(1);
151: ytop=ymin+.8*dy; Dy=.065*dy;
152: xlft=xmin+0.04*dx;
153: text(xlft,ytop,label)
154: grid off,  shg
155:
156: %==============================================
157:
158: function q=savespnd(t,t1,q0,R,A,I,s,p)
159: %
160: % q=savespnd(t,t1,q0,R,A,I,s,p)
161: %~~~~~~~~~~~~~~~~~~~~~~~~~~~~~~~~~~~~~~~~~~~~~~
162:
163: % This function determines q(t) satisfying
164: % q'(t)=r*q+[s*(t<=t1)-p*(t>t1)*...
165: % exp(-a*t1)]*exp(a*t), with q(0)=q0,
166: % r=(R-I)/100; a=(A-I)/100
167:
168: r=(R-I)/100; a=(A-I)/100; c=r-a; T=t-t1;
169: if r~=a
170:    q=q0*exp(r*t)+s/c*(exp(r*t)-exp(a*t))...
171:       -(p+s*exp(a*t1))/c*(T>0).*(...
172:       exp(r*T)-exp(a*T));
173: else % limiting case as a=>r
174:    q=q0*exp(r*t)+s*t.*exp(r*t)...
175:       -(p+s*exp(r*t1)).*T.*(T>0).*exp(r*T);
```

```
176: end
177:
178: %=============================================
179:
180: function varargout=inputv(prompt)
181: %
182: % [a1,a2,...,a_nargout]=inputv(prompt)
183: %~~~~~~~~~~~~~~~~~~~~~~~~~~~~~~~~~~~~~~~~~~~~
184: %
185: % This function reads several values on one
186: % line. The items should be separated by
187: % commas or blanks.
188: %
189: % prompt               - A string preceding the
190: %                        data entry.  It is set
191: %                        to ' ? ' if no value of
192: %                        prompt is given.
193: % a1,a2,...,a_nargout - The output variables
194: %                        that are created. If
195: %                        not enough data values
196: %                        are given following the
197: %                        prompt, the remaining
198: %                        undefined values are
199: %                        set equal to NaN
200: %
201: % A typical function call is:
202: % [A,B,C,D]=inputv('Enter values of A,B,C,D: ')
203: %
204: % ---------------------------------------------
205:
206: if nargin==0, prompt=' ? '; end
207: u=input(prompt,'s'); v=eval(['[',u,']']);
208: ni=length(v); no=nargout;
209: varargout=cell(1,no); k=min(ni,no);
210: for j=1:k, varargout{j}=v(j); end
211: if no>ni
212: for j=ni+1:no, varargout{j}=nan; end
213: end
```

Chapter 2

Elementary Aspects of MATLAB Graphics

2.1 Introduction

MATLAB's capabilities for plotting curves and surfaces are versatile and easy to understand. In fact, the effort required to learn MATLAB would be rewarding even if it were only used to construct plots, save graphic images, and output publication quality graphs on a laser printer. Numerous help features and well-written demo programs are included with MATLAB. By executing the demo programs and studying the relevant code, users can quickly understand the techniques necessary to implement graphics within their programs. This chapter discusses a few of the graphics commands. These commands are useful in many applications and do not require extensive time to master. This next section provides a quick overview of the basics of using MATLAB's graphics. The subsequent sections in this chapter present several additional examples (summarized in the table below) involving interesting applications which use these graphics primitives.

Example	Purpose
Polynomial Interpolation	2-D graphics and polynomial interpolation functions
Conformal Mapping	2-D graphics and some aspects of complex numbers
Pendulum Motion	2-D graphics animation and ODE solution
Linear Vibration Model	Animated spring-mass response
String Vibration	2-D and 3-D graphics for a function of form $y(x, t)$
Space Curve Geometry	3-D graphics for a space curve
Intersecting Surfaces	3-D graphics and combined surface plots

2.2 Overview of Graphics

The following commands should be executed since they will accelerate the understanding of graphics functions, and others, included within MATLAB.

Command	Description
help help	discusses use of **help** command.
help	lists categories of help.
help general	lists various utility commands.
help more	describes how to control output paging.
help diary	describes how to save console output to a file.
help plotxy	describes 2D plot functions.
help plotxyz	describes 3D plot functions.
help graphics	describes more general graphics features.
help demos	lists names of various demo programs.
intro	executes the **intro** program showing MATLAB commands including fundamental graphics capabilities.
help funfun	describes several numerical analysis programs contained in MATLAB.
type humps	lists a function employed in several of the MATLAB demos.
fplotdemo	executes program **fplotdemo** which plots the function named **humps**.
help peaks	describes a function **peaks** used to illustrate surface plots.
peaks	executes the function **peaks** to produce an interesting surface plot.
spline2d	executes a demo program to draw a curve through data input interactively.

The example programs can be studied interactively using the **type** command to list programs of interest. Library programs can also be inspected and printed using the MATLAB editor, but care should be taken not to accidentally overwrite the original library files with changes. Furthermore, text output in the command window can be captured in several ways. Some of these are: (1) Use the mouse to highlight material of interest. Then use the "Print Selected" on the file menu to send output to the printer; (2) Use CTRL-C to copy outlined text to the clipboard. Then open a new file and use CTRL-V to paste the text into the new file; and (3) Use a **diary** command such as **diary mysave.doc** to begin printing subsequent command window output into the chosen file. This printing can be turned off using **diary off**. Then the file can be edited, modified, or combined with other text using standard editor commands.

More advanced features of MATLAB graphics, including handle graphics, control of shading and light sources, creation of movies, etc., exceed the scope of the present text. Instead we concentrate on using the basic commands listed below and on producing simple animations. The advanced graphics can be mastered by studying the

MATLAB manuals and relevant demo programs. The principal graphing commands discussed here are

Command	Purpose
plot	draw two-dimensional graphs
xlabel, ylabel, zlabel	define axis labels
title	define graph title
axis	set various axis parameters (min, max, etc.)
legend	show labels for plot lines
shg	bring graphics window to foreground
text	place text at selected locations
grid	turns grid lines on or off
mesh	draw surface using colored lines
surf	draw surface using colored patches
hold	fix the graph limits between successive plots
view	change surface viewing position
drawnow	empty graphics buffer immediately
zoom	magnify graph or surface plot
clf	clear graphics window
contour	draw contour plot
ginput	read coordinates interactively

All of these commands, along with numerous others, are extensively documented by the help facilities in MATLAB. The user can get an introduction to these capabilities by typing "**help plot**" and by running the demo programs. The accompanying code for the demo program should be examined since it provides worthwhile insight into how MATLAB graphics is used.

2.3 Example Comparing Polynomial and Spline Interpolation

Many familiar mathematical functions such as $\arctan(x)$, $\exp(x)$, $\sin(x)$, etc. can be represented well near $x = 0$ by Taylor series expansions. If a series expansion converges rapidly, taking a few terms in the series may produce good polynomial approximations. Assuming such a procedure is plausible, one approach to polynomial approximation is to take some data points, say (x_i, y_i), $1 \le i \le n$ and determine the polynomial of degree $n-1$ passing through those points. It appears reasonable that using evenly spaced data is appropriate and that increasing the number of polynomial terms should improve the accuracy of the approximating function. However, it has actually been shown that a polynomial through points on a function $y(x)$, where the x values are evenly spaced, often gives approximations which are not smooth between the data points and tend to oscillate at the ends of the interpolating interval [20]. Attempting to reduce the oscillation by increasing the polynomial order makes matters worse. Surprisingly, a special set of unevenly spaced points bunching data near the interval ends according to

$$x_j = (a+b)/2 + (a-b)/2 \cos[\pi(j-1/2)/n], \quad 1 \le j \le n$$

for the interval $a \le x \le b$ turns out to be preferable. This formula defines what are called the Chebyshev points optimally chosen in the sense described by Conte and de Boor [20].

The program below employs MATLAB functions **polyfit**, **polyval**, and **spline** to produce interpolated approximations to the known function $1/(1+x^2)$. The example illustrates how strongly the spacing of the data points for polynomial interpolation can influence results, and also shows that a spline interpolation can be a better choice than high order polynomials. A least square fit polynomial of degree n through data points defined by vectors (x_d, y_d) is given by

$$p(x) = \mathbf{polyval}(\mathbf{polyfit}(x_d, y_d, n), x).$$

When the polynomial order is one less than the number of data points, the polynomial passes through the data points exactly, but it may still produce unsatisfactory interpolation because of large oscillations between the data points. A preferable approximation is often provided by function **spline** giving a piecewise cubic curve with continuous first and second derivatives. The program passes polynomials of degree ten through a set of evenly spaced points and a set of Chebyshev points lying in the range $-4 \le x \le 4$. A spline curve passed through the equidistant points is constructed in addition to a least square polynomial fit employing 501 points. Two graphs are created which show results for $x \ge 0$. Only results for positive x were plotted to provide more contrast between different interpolation results. Figure 2.1 plots the exact function, the spline curve, and the polynomial through the equidistant data. The polynomial is clearly an unsatisfactory approximation, whereas the spline appears to deviate imperceptibly from the exact function. By using the interactive **zoom** feature in MATLAB graphics, parts of the graph can be magnified so the difference between the spline and exact results is clearly visible. Figure 2.2 compares the exact function with a polynomial employing the Chebyshev points. This result is much better than what is produced with equidistant data. An approximation generated from a least square fit polynomial and 501 data points is also shown. This curve fits the exact function unpredictably and significantly misses the desired values at $x = 0$ and $x = \pm 4$. While general conclusions about interpolation should not be drawn from this simple example, it certainly implies that high order polynomial interpolation over a large range of the independent variable should be used cautiously.

The graphics functions used in the program include **plot**, **title**, **xlabel**, **ylabel**, and **legend**. Some other features of the program are summarized in the table preceding the code listing.

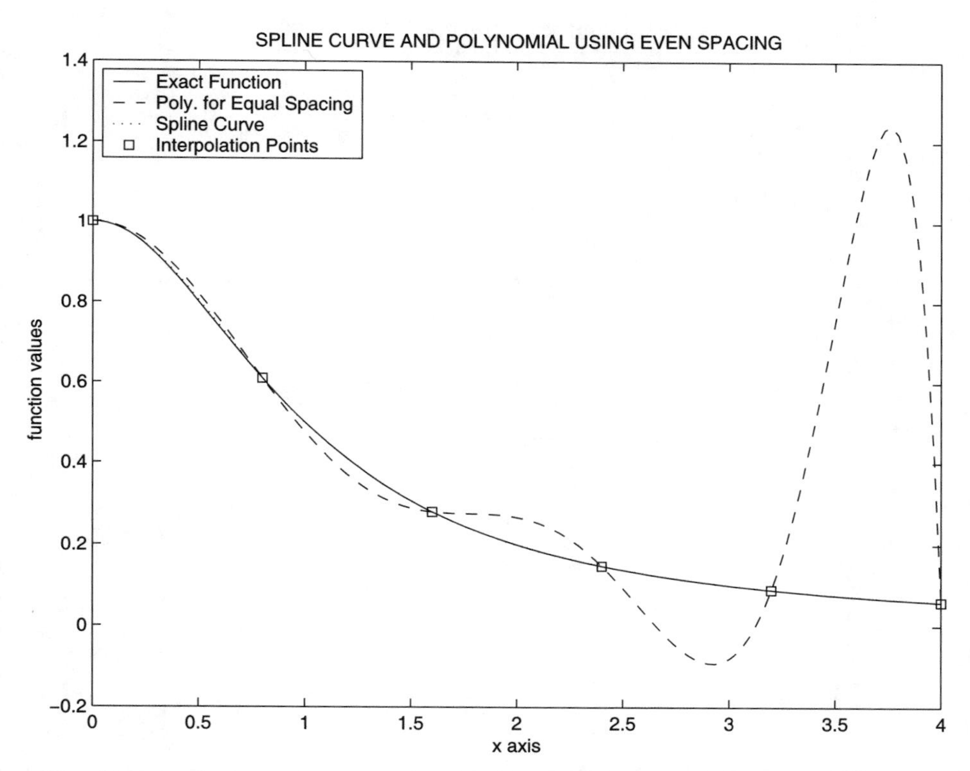

Figure 2.1: Spline and Polynomial Interpolation Using Equidistant Points

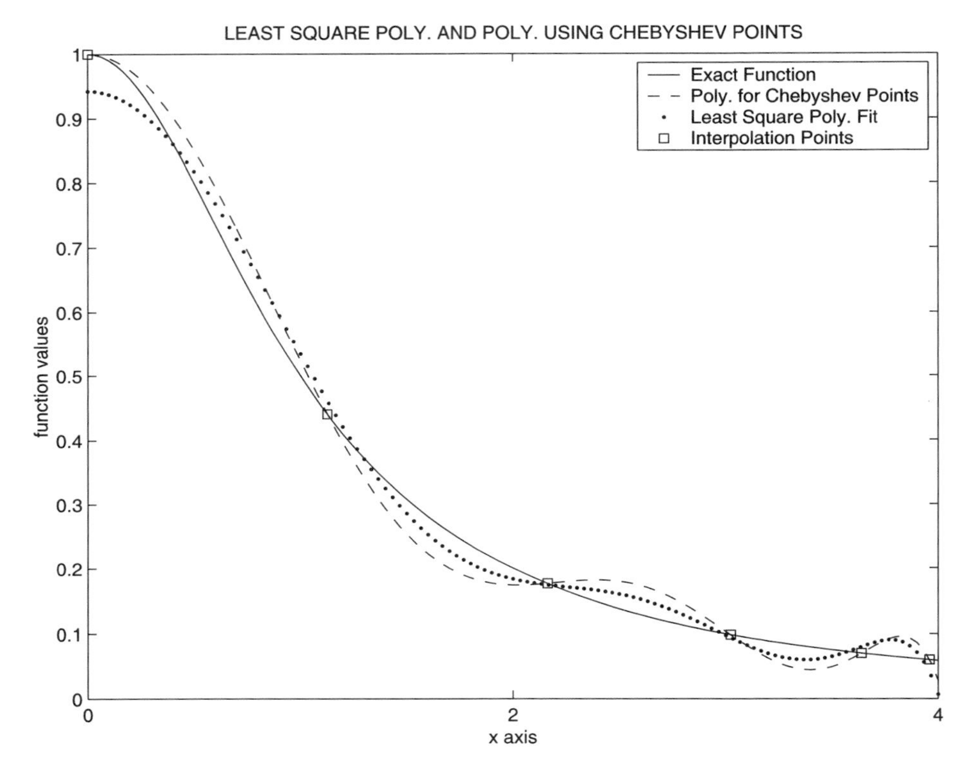

Figure 2.2: Interpolation Using Chebyshev Points and 501 Least Square Points

Line	Operation
12,17,21	several inline functions are defined
27	function linspace generates vector of equidistant points
27,28,34-37	inline functions called
38	intrinsic spline function is used
45,57	graph legends created
52,64	graph images saved to files

Program polyplot

```
1: function polyplot
2: % Example:  polyplot
3: % ~~~~~~~~~~~~~~~~~~~
4: % This program illustrates polynomial and
5: % spline interpolation methods applied to
6: % approximate the function 1/(1+x^2).
7: %
8: % User inline functions used:
9: %   cbp, Ylsq, yexact
10:
11: % Function for Chebyshev data points
12: cbp=inline(['(a+b)/2+(a-b)/2*cos(pi/n*',...
13:     '(1/2:n))'],'a','b','n');
14:
15: % Polynomial of degree n to least square fit
16: % data points in vectors xd,yd
17: Ylsq=inline('polyval(polyfit(xd,yd,n),x)',...
18: 'xd','yd','n','x');
19:
20: % Function to be approximated by polynomials
21: yexact=inline('1./(1+abs(x).^p)','p','x');
22:
23: % Set data parameters. Functions linspace and
24: % cbp generate data with even and Chebyshev
25: % spacing
26: n=10; nd=n+1; a=-4; b=4; p=2;
27: xeven=linspace(a,b,nd); yeven=yexact(p,xeven);
28: xcbp=cbp(a,b,nd); ycbp=yexact(p,xcbp);
29:
30: nlsq=501; % Number of least square points
31: xlsq=linspace(a,b,nlsq); ylsq=yexact(p,xlsq);
32:
33: % Compute interpolated functions for plotting
```

```
34: xplt=linspace(0,b,121); yplt=yexact(p,xplt);
35: yyeven=Ylsq(xeven,yeven,n,xplt);
36: yycbp=Ylsq(xcbp,ycbp,n,xplt);
37: yylsq=Ylsq(xlsq,ylsq,n,xplt);
38: yyspln=spline(xeven,yeven,xplt);
39:
40: % Plot results
41: j=6:nd; % Plot only data points for x>=0
42: plot(xplt,yplt,'-',xplt,yyeven,'--',...
43: xplt,yyspln,'.',xeven(j),yeven(j),...
44:      's','linewidth',2)
45: legend('Exact Function',...
46:     'Poly. for Even Spacing',...
47:         'Spline Curve',...
48:     'Interpolation Points',2)
49: title(['SPLINE CURVE AND POLYNOMIAL ',...
50: 'USING EVEN SPACING'])
51: xlabel('x axis'), ylabel('function values')
52: % print(gcf,'-deps','splpofit')
53: shg, pause
54: plot(xplt,yplt,'-',xplt,yycbp,'--',...
55: xplt,yylsq,'.',xcbp(j),ycbp(j),'s',...
56: 'linewidth',2)
57: legend('Exact Function',...
58:         'Poly. for Chebyshev Points',...
59:         'Least Square Poly. Fit',...
60:         'Interpolation Points',1)
61: title(['LEAST SQUARE POLY. AND POLY. ',...
62: 'USING CHEBYSHEV POINTS'])
63: xlabel('x axis'), ylabel('function values')
64: % print(gcf,'-deps','lsqchfit')
65: shg, disp(' '), disp('All Done')
```

2.4 Conformal Mapping Example

This example involves analytic functions and conformal mapping. The complex function $w(z)$ which maps $|z| \leq 1$ onto the interior of a square of side length 2 can be written in power series form as

$$w(z) = \sum_{k=0}^{\infty} b_k z^{4k+1}$$

where

$$b_k = c\left[\frac{(-1)^k(\frac{1}{2})_k}{k!(4k+1)}\right] \ , \ \sum_{k=0}^{\infty} b_k = 1$$

and c is a scaling coefficient chosen to make $z = 1$ map to $w = 1$ (see reference [75]). Truncating the series after some finite number of terms, say m, produces an approximate square with rounded corners. Increasing m reduces the corner rounding but convergence is rather slow so that using even a thousand terms still gives perceptible inaccuracy. The purpose of the present exercise is to show how a polar coordinate region characterized by

$$z = re^{\imath\theta} \ , \ r_1 \le r \le r_2 \ , \ \theta_1 \le \theta \le \theta_2$$

transforms and to exhibit an undistorted plot of the region produced in the w-plane. The exercise also emphasizes the utility of MATLAB for handling complex arithmetic and complex functions. The program has a short driver **squarrun** and a function **squarmap** which computes points in the w region and coefficients in the series expansion. Salient features of the program are summarized in the table below.

Results produced when $0.5 \le r \le 1$ and $0 \le \theta \le 2\pi$ by a twenty-term series appear in Figure 2.3. The reader may find it interesting to run the program using several hundred terms and take $0 \le \theta \le \pi/2$. The corner rounding remains noticeable even when $m = 1000$ is used. Later in this book we will visit the mapping problem again to show that a better approximation is obtainable using rational functions.

Routine	Line	Operation
squarrun	20-41	functions **input**, **disp**, **fprintf**, and **read** are used to input data interactively. Several different methods of printing were used for purposes of illustration rather than necessity.
	45	function **squarmap** generates results.
	49	function **genprint** is a system dependent routine which is used to create plot files for later printing.
squarmap	31-33	functions **linspace** and **ones** are used to generate points in the z-plane.
	43-45	series coefficients are computed using **cumprod** and the mapping is evaluated using **polyval** with a matrix argument.
	48-51	scale limits are calculated to allow an undistorted plot of the geometry. Use is made of MATLAB functions **real** and **imag**.
	57-73	loops are executed to plot the circumferential lines first and the radial lines second.
cubrange		function which determines limits for a square or cube shaped region.

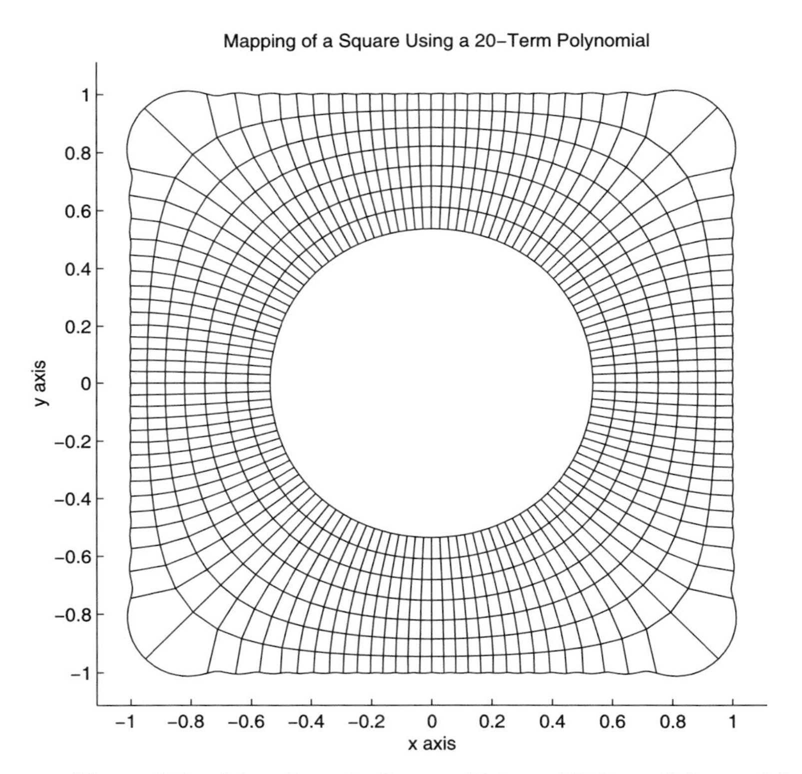

Figure 2.3: Mapping of a Square Using a 20-Term Polynomial

MATLAB Example

Program squarrun

```
1: function squarrun
2: % Example:  squarrun
3: % ~~~~~~~~~~~~~~~~~~~~
4: %
5: % Driver program to plot the mapping of a
6: % circular disk onto the interior of a square
7: % by the Schwarz-Christoffel transformation.
8: %
9: % User m functions required:
10: %    squarmap, inputv, cubrange
11:
12: % Illustrate use of the functions input and
13: % inputv to interactively read one or several
14: % data items on the same line
15:
16: fprintf('\nCONFORMAL MAPPING OF A SQUARE ')
17: fprintf('BY USE OF A\n')
18: fprintf('TRUNCATED SCHWARZ-CHRISTOFFEL ')
19: fprintf('SERIES\n\n')
20:
21: fprintf('Input the number of series ')
22: fprintf('terms used ')
23: m=input('(try 20)? ');
24:
25: % Illustrate use of the function disp
26: disp('')
27: str=['\nInput the inner radius, outer ' ...
28:      'radius and number of increments ' ...
29:      '\n(try .5,1,8)\n'];
30: fprintf(str);
31:
32: % Use function inputv to input several variables
33: [r1,r2,nr]=inputv;
34:
35: % Use function fprintf to print more
36: % complicated heading
37: str=['\nInput the starting value of ' ...
38:      'theta, the final value of theta \n' ...
39:      'and the number of theta increments ' ...
40:      '(the angles are in degrees) ' ...
```

```
41:       '\n(try 0,360,120)\n'];
42: fprintf(str); [t1,t2,nt]=inputv;
43:
44: % Call function squarmap to make the plot
45: hold off; clf;
46: [w,b]=squarmap(m,r1,r2,nr,t1,t2,nt+1);
47:
48: % Save the plot
49: % print -deps squarplt
50:
51: disp(' '); disp('All Done');
52:
53: %==============================================
54:
55: function [w,b]=squarmap(m,r1,r2,nr,t1,t2,nt)
56: %
57: % [w,b]=squarmap(m,r1,r2,nr,t1,t2,nt)
58: % ~~~~~~~~~~~~~~~~~~~~~~~~~~~~~~~~~~~~~
59: % This function evaluates the conformal mapping
60: % produced by the Schwarz-Christoffel
61: % transformation w(z) mapping abs(z)<=1 inside
62: % a square having a side length of two.  The
63: % transformation is approximated in series form
64: % which converges very slowly near the corners.
65: %
66: % m          - number of series terms used
67: % r1,r2,nr - abs(z) varies from r1 to r2 in
68: %              nr steps
69: % t1,t2,nt - arg(z) varies from t1 to t2 in
70: %              nt steps (t1 and t2 are measured
71: %              in degrees)
72: % w          - points approximating the square
73: % b          - coefficients in the truncated
74: %              series expansion which has the
75: %              form
76: %
77: %              w(z)=sum({j=1:m},b(j)*z*(4*j-3))
78: %
79: % User m functions called:  cubrange
80: %----------------------------------------------
81:
82: % Generate polar coordinate grid points for the
83: % map.  Function linspace generates vectors
84: % with equally spaced components.
85: r=linspace(r1,r2,nr)';
```

```
86: t=pi/180*linspace(t1,t2,nt);
87: z=(r*ones(1,nt)).*(ones(nr,1)*exp(i*t));
88:
89: % Use high point resolution for the
90: % outer contour
91: touter=pi/180*linspace(t1,t2,10*nt);
92: zouter=r2*exp(i*touter);
93:
94: % Compute the series coefficients and
95: % evaluate the series
96: k=1:m-1;
97: b=cumprod([1,-(k-.75).*(k-.5)./(k.*(k+.25))]);
98: b=b/sum(b); w=z.*polyval(b(m:-1:1),z.^4);
99: wouter=zouter.*polyval(b(m:-1:1),zouter.^4);
100:
101: % Determine square window limits for plotting
102: uu=real([w(:);wouter(:)]);
103: vv=imag([w(:);wouter(:)]);
104: rng=cubrange([uu,vv],1.1);
105: axis('square'); axis(rng); hold on
106:
107: % Plot orthogonal grid lines which represent
108: % the mapping of circles and radial lines
109: x=real(w); y=imag(w);
110: xo=real(wouter); yo=imag(wouter);
111: plot(x,y,'-k',x(1:end-1,:)',y(1:end-1,:)',...
112: '-k',xo,yo,'-k')
113:
114: % Add a title and axis labels
115: title(['Mapping of a Square Using a ', ...
116:        num2str(m),'-term Polynomial'])
117: xlabel('x axis'); ylabel('y axis')
118: figure(gcf); hold off;
119:
120: %==============================================
121:
122: function range=cubrange(xyz,ovrsiz)
123: %
124: % range=cubrange(xyz,ovrsiz)
125: % ~~~~~~~~~~~~~~~~~~~~~~~~~~~~
126: % This function determines limits for a square
127: % or cube shaped region for plotting data values
128: % in the columns of array xyz to an undistorted
129: % scale
130: %
```

```
131: % xyz    - a matrix of the form [x,y] or [x,y,z]
132: %          where x,y,z are vectors of coordinate
133: %          points
134: % ovrsiz - a scale factor for increasing the
135: %          window size. This parameter is set to
136: %          one if only one input is given.
137: %
138: % range  - a vector used by function axis to set
139: %          window limits to plot x,y,z points
140: %          undistorted. This vector has the form
141: %          [xmin,xmax,ymin,ymax] when xyz has
142: %          only two columns or the form
143: %          [xmin,xmax,ymin,ymax,zmin,zmax]
144: %          when xyz has three columns.
145: %
146: % User m functions called:  none
147: %----------------------------------------------
148:
149: if nargin==1, ovrsiz=1; end
150: pmin=min(xyz); pmax=max(xyz); pm=(pmin+pmax)/2;
151: pd=max(ovrsiz/2*(pmax-pmin));
152: if length(pmin)==2
153:   range=pm([1,1,2,2])+pd*[-1,1,-1,1];
154: else
155:   range=pm([1 1 2 2 3 3])+pd*[-1,1,-1,1,-1,1];
156: end
157:
158: %==============================================
159:
160: % function varargout=inputv(prompt)
161: % See Appendix B
```

2.5 Nonlinear Motion of a Damped Pendulum

Motion of a simple pendulum is one of the most familiar dynamics examples studied in physics. The governing equation of motion can be satisfactorily linearized for small oscillations about the vertical equilibrium position, whereas nonlinear effects become important for large deflections. For small deflections, the analysis leads to a constant coefficient linear differential equation. Solving the general case requires elliptic functions seldom encountered in routine engineering practice. Nevertheless, the pendulum equation can be handled very well for general cases by numerical integration.

Suppose a bar of negligible weight is hinged at one end and has a particle of mass m attached to the other end. The bar has length l and the deflection from the vertical static equilibrium position is called θ. Assuming that the applied forces consist of the particle weight and a viscous drag force proportional to the particle velocity, the equation of motion is found to be

$$\theta''(\tau) + \frac{c}{m}\theta'(t) + \frac{g}{l}\sin(\theta) = 0$$

where τ is time, c is a viscous damping coefficient, and g is the gravity constant. Introducing dimensionless time, t, such that $\tau = \sqrt{l/g}\,t$ gives

$$\theta''(t) + 2\varsigma\theta'(t) + \sin(\theta) = 0$$

where $\varsigma = \sqrt{l/g}\,c/(2m)$ is called the damping factor. When θ is small enough for $\sin(\theta)$ to be approximated well by θ , then a constant coefficient linear equation solvable by elementary means is obtained. In the general situation, a solution can still be obtained numerically without resorting to higher transcendental functions. If we use $\varsigma = 0.10$ for illustrative purposes, and let

$$z = [\theta(t)\,;\ \theta'(t)]$$

then the original differential equation expressed in first order matrix form is

$$z'(t) = [z(2)\,;\ -0.2z(2) - \sin(z(1)].$$

An inline function suitable for use by the **ode45** integrator in MATLAB is simply zdot=inline('[z(2); -0.2*z(2)-sin(z(1))]','t','z').

A program was written to integrate the pendulum equation when the angular velocity ω_0 for $\theta = 0$ is specified. For the undamped case, it is not hard to show that a starting angular velocity exceeding 2 is sufficient to push the pendulum over the top, but the pendulum will fall back for values smaller than two. For the amount of viscous damping chosen here, a value of about $\omega_0 = 2.42$ barely pushes the pendulum over the top, whereas the top is not reached for $\omega_0 = 2.41$. These cases vividly illustrate that, for a nonlinear system, small changes in initial conditions can sometimes produce very large changes in the response of the system.

In the computer program that follows, a driver function **runpen** controls input, calls the differential equation solver **ode45**, as well as a function **animpen** which plots θ versus t, and performs animation by drawing successive positions of the pendulum. Because the animation routine is very simple and requires little knowledge of MATLAB graphics, the images and the titles flicker somewhat. This becomes particularly evident unless the graph axes are left off. A better routine using more detailed graphics commands to eliminate the flicker problem is presented in Article 2.7 on wave motion in a string. The current program permits interactive input repeatedly specifying the initial angular velocity, or two illustrative data cases can be run by executing the command **runpen(1)**. The differential equation for the problem is defined as function **zdot** on lines 26 and 27. This equation is integrated numerically

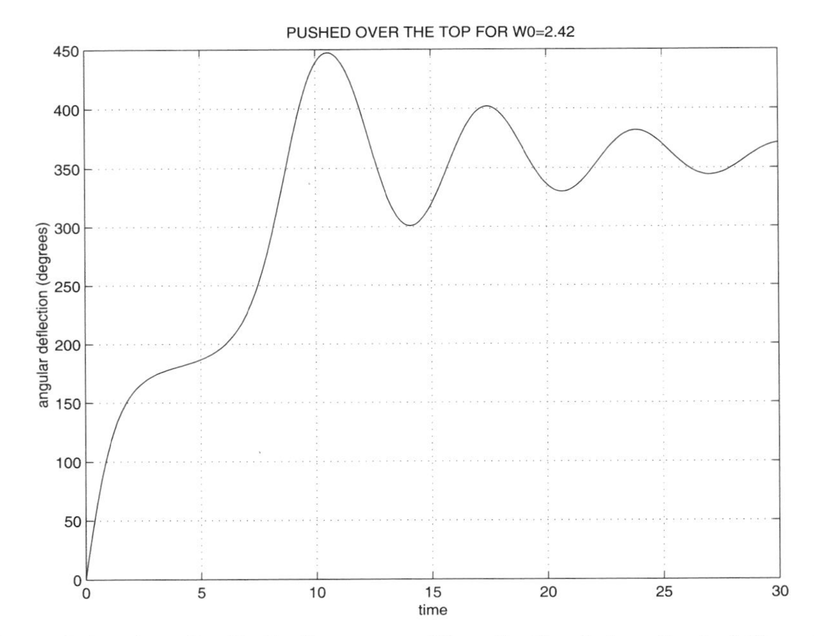

Figure 2.4: Angular Deflection versus Time for Pendulum Pushed Over the Top

by calls to function **ode45** on lines 59, 75, and 80. Integration tolerance values were chosen at line 30, and a time span for the simulation is defined interactively at lines 46 and 47. Function **penanim(t,th,titl,tim)** plots theta versus time and animates the system response by computing the range of (x,y) values, fixing the window size to prevent distortion, and sequentially plotting positions of the pendulum to show the motion history. The output results produced by **runpen(1)** are shown below for reference.

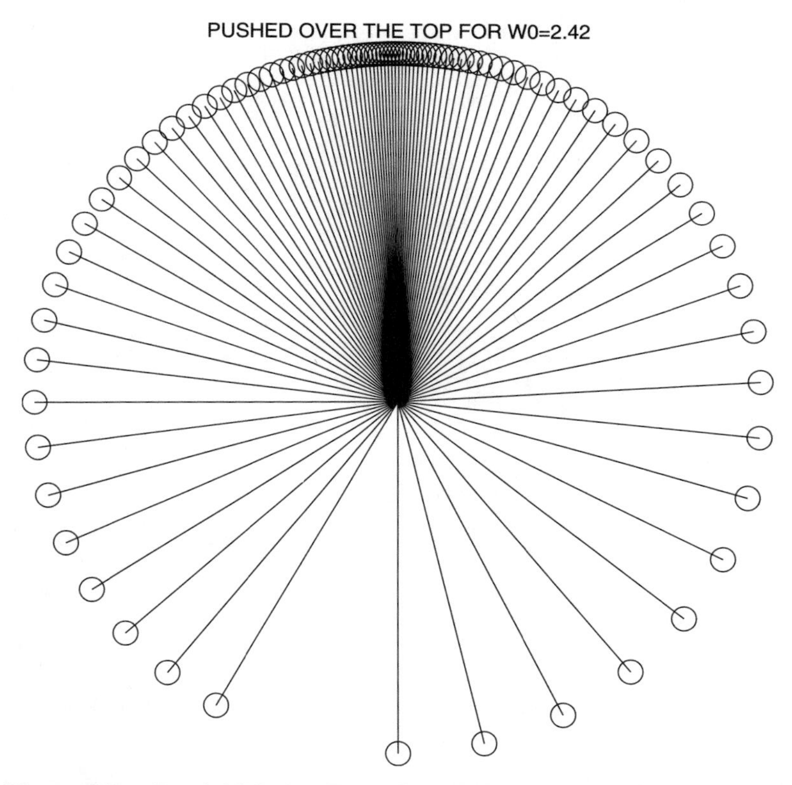

Figure 2.5: **Partial Motion Trace for Pendulum Pushed Over the Top**

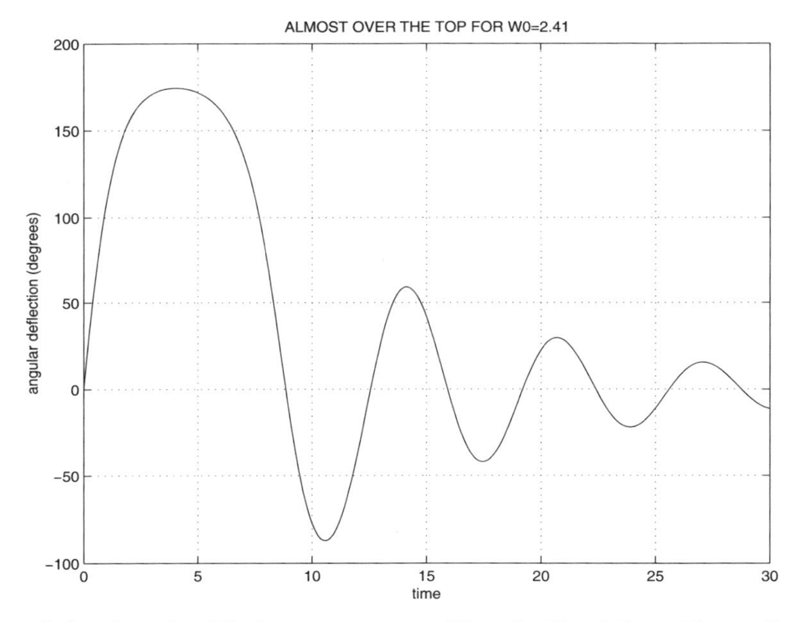

Figure 2.6: Angular Displacement versus Time for Pendulum Almost Pushed Over the Top

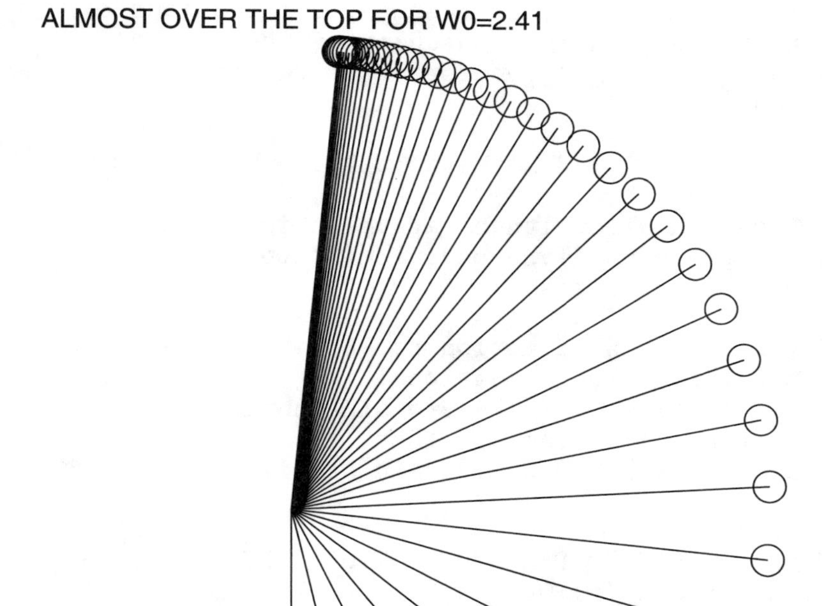

Figure 2.7: Partial Motion Trace for Pendulum Almost Pushed Over the Top

Program pendulum

```
1: function pendulum(rundemo)
2: % pendulum(rundemo)
3: % This example analyzes damped oscillations of
4: % a simple pendulum and animates the motion.
5: % The governing second order differential
6: % equation is
7: %
8: %   theta"(t) + 0.2*theta'(t)+sin(theta) = 0
9:
10: % Type pendulum with no argument for inter-
11: % active input. Type pendulum(1) to run two
12: % example problems
13:
14: % The equation of motion can be written as
15: % two first order equations:
16: % theta'(t)=w; w'(t)=-.2*w-sin(theta)
17: % Letting z=[theta; w], then
18: % z'(t)=[z(2); -0.2*z(2)-sin(z(1))]
19:
20: disp(' ')
21: disp('   DAMPED PENDULUM MOTION DESCRIBED BY')
22: disp(' theta"(t)+0.2*theta''(t)+sin(theta) = 0')
23:
24: % Create an inline function defining the
25: % differential equation in matrix form
26: zdot=inline(...
27:    '[z(2);-0.2*z(2)-sin(z(1))]','t','z');
28:
29: % Set ode45 integration tolerances
30: ops=odeset('reltol',1e-5,'abstol',1e-5);
31:
32: % Interactively input angular velocity repeatedly
33: if nargin==0
34:
35:   while 1, close, disp(' ')
36:     disp('Select the angular velocity at the lowest')
37:     disp('point. Values of 2.42 or greater push the')
38:     disp(...
39:     'the pendulum over the top. Input zero to stop.')
40:     w0=input('w0 = ? > ');
41:
```

```
42:     if isempty(w0) | w0==0
43:       disp(' '), disp('All Done'), disp(' '), return
44:     end
45:     disp(' ')
46:     t=input(['Input a vector of time values ',...
47:             '(Try 0:.1:30) > ? ']);
48:
49:     disp(' ')
50:     titl=input('Input a title for the graphs : ','s');
51:     disp(' '), disp(...
52:     'Input 1 to leave images of all positions shown')
53:     trac=input(...
54:         'in the animation, otherwise input 0 > ? ');
55:
56:     % Specify the initial conditions and solve the
57:     % differential equation using ode45
58:     theta0=0; z0=[theta0;w0];
59:     [t,th]=ode45(zdot,t,z0,ops);
60:
61:     % Animate the motion
62:     animpen(t,th(:,1),titl,.05,trac)
63:   end
64:
65: % Run two typical data cases
66: else
67:
68:   % Choose time limits for the solution
69:   tmax=30; n=351; t=linspace(0,tmax,n);
70:
71:   disp(' ')
72:   disp('Press return to see two examples'), pause
73:
74:   w0=2.42; W0=num2str(w0);
75:   [t,th]=ode45(zdot,t,[0;w0],ops);
76:   titl=['PUSHED OVER THE TOP FOR W0 = ',W0];
77:   animpen(t,th(:,1), titl,.05), pause(2)
78:
79:   w0=2.41; W0=num2str(w0);
80:   [t,th]=ode45(zdot,t,[0;w0],ops);
81:   titl=['NEARLY PUSHED OVER THE TOP FOR W0 = ',W0];
82:   animpen(t,th(:,1),titl,.05)
83:   close, disp(' '), disp('All Done'), disp(' ')
84:
85: end
86:
```

```
87: %==============================================
88:
89: function animpen(t,th,titl,tim,trac)
90: %
91: % animpen(t,th,titl,tim,trac)
92: % ~~~~~~~~~~~~~~~~~~~~~~~~~~~~
93: % This function plots theta versus t and animates
94: % the pendulum motion
95: %
96: % t    - time vector for the solution
97: % th   - angular deflection values defining the
98: %        pendulum positions
99: % titl - a title shown on the graphs
100: % tim  - a time delay between successive steps of
101: %        the animation. This is used to slow down
102: %        the animation on fast computers
103: % trac - 1 if successive positions plotted in the
104: %        animation are retained on the screen, 0
105: %        if each image is erased after it is
106: %        drawn
107:
108: if nargin<5, trac=0; end; if nargin<4, tim=.05; end;
109: if nargin<3, titl=''; end
110:
111: % Plot the angular deflection
112: plot(t,180/pi*th(:,1),'k'), xlabel('time')
113: ylabel('angular deflection (degrees)'), title(titl)
114: grid on, shg, disp(' ')
115: disp('Press return to see the animation'), pause
116: % print -deps penangle
117:
118: nt=length(th); z=zeros(nt,1);
119: x=[z,sin(th)]; y=[z,-cos(th)];
120: hold off, close
121: if trac
122:    axis([-1,1,-1,1]), axis square, axis off, hold on
123: end
124: for j=1:nt
125:    X=x(j,:); Y=y(j,:);
126:    plot(X,Y,'k-',X(2),Y(2),'ko','markersize',12)
127:    if ~trac
128:   axis([-1,1,-1,1]), axis square, axis off
129:    end
130:    title(titl), drawnow, shg
131:    if tim>0, pause(tim), end
```

```
132: end
133: % if trac==1, print -deps pentrace,  end
134: pause(1),hold off
```

2.6 A Linear Vibration Model

Important aspects of linear vibration theory are illustrated by the one-dimensional motion of a mass subjected to an elastic restoring force, a viscous damping force proportional to the velocity, and a harmonically varying forcing function. The related differential equation is

$$m\,x''(t) + c\,x'(t) + k\,x(t) = f_1\cos(\omega\,t) + f_2\sin(\omega\,t) = \mathrm{real}((f_1 - i\,f_2)\exp(i\,\omega\,t))$$

with initial conditions of $x(0) = x_0$ and $x'(0) = v_0$. The general solution is the sum of a particular solution to account for the forcing function, and a homogeneous solution corresponding to a zero right hand side. The initial conditions are applied to the sum of the two solution components. The particular solution is given by

$$X(t) = \mathrm{real}(F\ \exp(i\,\omega\,t))$$

with

$$F = (f_1 - if_2)/(k - m\,\omega^2 + i\,c\,\omega).$$

The initial conditions given by this particular solution are

$$X(0) = \mathrm{real}(F)$$

and

$$X'(0) = \mathrm{real}(i\,\omega\,F).$$

The characteristic equation for the homogeneous equation is

$$m\,s^2 + c\,s + k = 0$$

which has roots

$$s_1 = (-c + r)/(2m),\quad s_2 = (-c - r)/(2m),\quad r = \sqrt{c^2 - 4m\,k}.$$

Then the homogeneous solution has the form

$$u(t) = d(1)\ \exp(s_1 t) + d(2)\exp(s_2 t)$$

where

$$d = [1,\ 1;\ s_1,\ s_2\,]\setminus[x_0\ - X(0);\ v_0\ - X'(0)]$$

and the complete solution is

$$x(t) = u(t) + X(t).$$

A couple of special cases arise. The first corresponds to zero damping and a forcing function matching the undamped natural frequency, i.e.,

$$c = 0, \ \ \omega = \sqrt{k/m}.$$

This case can be avoided by including a tiny amount of damping to make $c = 2\sqrt{mk}/10^6$. The second case happens when the characteristic roots are equal. This is remedied by perturbing the value of c to $(1+10^{-6})$ times c. Such small changes in a system model where realistic physical parameters are only known approximately will not affect the final results significantly.

In practice, enough damping often exists in the system to make the homogeneous solution components decay rapidly so the total solution approaches the particular solution with the displacement having the same frequency as the forcing function but out of phase with that force. To illustrate this effect, a program was written to solve the given differential equation, plot $x(t)$, and show an animation for a block connected to a wall with a spring and sliding on a surface with viscous damping resistance. Applying the oscillating force of varying magnitude on the block helps illustrate how the homogeneous solution dies out and the displacement settles into a constant phase shift relative to the driving force.

The following program either reads data interactively or runs a default data example. The solution procedure described above is implemented in function **smdsolve**. For arbitrary values of the system parameters, $x(t)$ is plotted and a simple animation scheme is used to plot the block, a spring, and the applied force throughout the time history. Figure 2.8 shows $x(t)$ for the default data case. The input data values for this case use

$$[m, \ c, \ k, \ f_1, \ f_2, \ w, \ x_0, \ v_0, \ tmax, \ nt] <=> [1, \ 3, \ 1, \ 1, \ 0, \ 2, \ 0, \ 2, \ 30, \ 250].$$

Note that near $t = 11$, the transient and forced solution components interact so that the block almost pauses momentarily. However, the solution then quickly approaches the steady state. Figure 2.9 shows the final position of the mass and the applied force at the end of the chosen motion cycle.

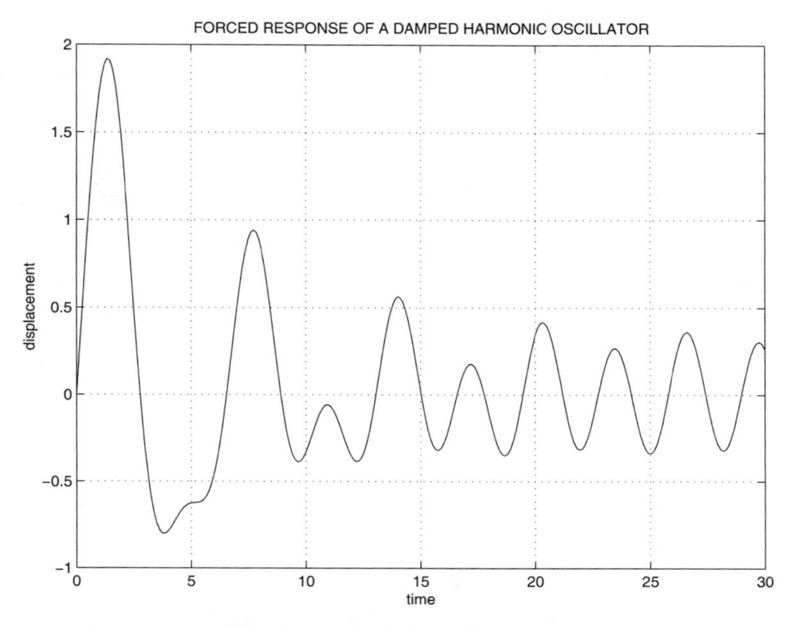

Figure 2.8: Plot of x(t) for a Linear Harmonic Oscillator

FORCED MOTION WITH DAMPING

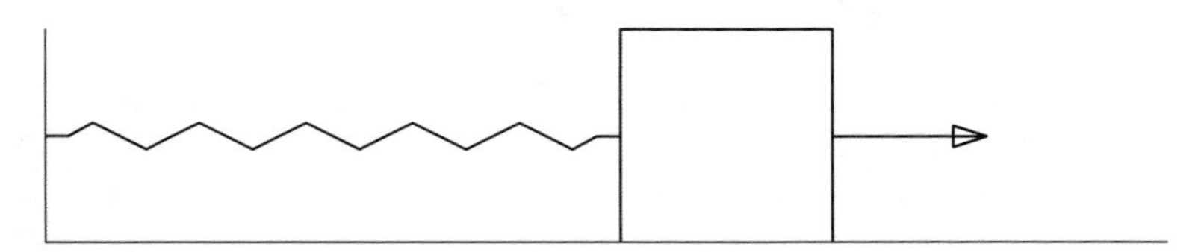

Figure 2.9: Block Sliding On a Plane with Viscous Damping

Program smdplot

```
1: function [t,X,m,c,k,f1,f2,w,x0,v0]= smdplot(example)
2: %
3: % [t,X,m,c,k,f1,f2,w,x0,v0]= smdplot(example)
4: % ~~~~~~~~~~~~~~~~~~~~~~~~~~~~~~~~~~~~~~~~~~~
5: % This function plots the response and animates the
6: % motion of a damped linear harmonic oscillator
7: % characterized by the differential equation
8: % m*x''+c*x'+k*x=f1*cos(w*t)+f2*sin(w*t)
9: % with initial conditions x(0)=x0, x'(0)=v0.
10: % The animation depicts forced motion of a block
11: % attached to a wall by a spring. The block
12: % slides on a horizontal plane which provides
13: % viscous damping.
14:
15: % example - Omit this parameter for interactive input.
16: %           Use smdplot(1) to run a sample problem.
17: % t,X     - time vector and displacement response
18: % m,c,k   - mass, damping coefficient,
19: %           spring stiffness constant
20: % f1,f2,w - force components and forcing frequency
21: % x0,v0   - initial position and velocity
22: %
23: % User m functions called: spring smdsolve inputv
24: % ----------------------------------------------------
25:
26: pltsave=0; disp(' '), disp(...
27: '                SOLUTION OF '), disp(...
28: 'M*X" + C*X'' + K*X = F1*COS(W*T) + F2*SIN(W*T)')
29: disp(...
30: '        WITH ANIMATION OF THE RESPONSE')
31: disp(' ')
32:
33: % Example data used when nargin > 0
34: if nargin > 0
35:   m=1; c=.3; k=1; f1=1; f2=0; w=2; x0=0; v0=2;
36:   tmax=25; nt=250;
37: else % Interactive data input
38:   [m,c,k]=inputv(...
39:           'Input m, c, k (try 1, .3, 1) >> ? ');
40:
41:   [f1,f2,w]=inputv(...
```

```
42:             'Input f1, f2, w (try 1, 0, 2) >> ? ');
43:
44:   [x0,v0]=inputv(...
45:           'Input x0, v0 (try 0, 2) >> ? ');
46:
47:   [tmax,nt]=inputv(...
48:             'Input tmax, nt (try 30, 250) >> ? ');
49:  end
50:
51:  t=linspace(0,tmax,nt);
52:  X=smdsolve(m,c,k,f1,f2,w,x0,v0,t);
53:
54: % Plot the displacement versus time
55: plot(t,X,'k'), xlabel('time')
56: ylabel('displacement'), title(...
57: 'FORCED RESPONSE OF A DAMPED HARMONIC OSCILLATOR')
58: grid on, shg, disp(' ')
59: if pltsave, print -deps smdplotxvst; end
60: disp('Press return for response animation')
61: pause
62:
63: % Add a block and a spring to the displacement
64: xmx=max(abs(X)); X=X/1.1/xmx;
65: xb=[0,0,1,1,0,0]/2; yb=[0,-1,-1,1,1,0]/2;
66:
67: % Make an arrow tip
68: d=.08; h=.05;
69: xtip=[0,-d,-d,0]; ytip=[0,0,0,h,-h,0];
70:
71: % Add a spring and a block to the response
72: [xs,ys]=spring; nm=length(X); ns=length(xs);
73: nb=length(xb); x=zeros(nm,ns+nb);y=[ys,yb];
74: for j=1:nm, x(j,:)=[-1+(1+X(j))*xs,X(j)+xb];end
75: xmin=min(x(:)); xmax=max(x(:)); d=xmax-xmin;
76: xmax=xmin+1.1*d; r=[xmin,xmax,-2,2];
77: rx=r([1 1 2]); ry=[.5,-.5,-.5]; close;
78:
79: % Plot the motion
80: for j=1:nm
81:    % Compute and scale the applied force
82:    f=f1*cos(w*t(j))+f2*sin(w*t(j));
83:    f=.5*f; fa=abs(f); sf=sign(f);
84:    xj=x(j,:); xmaxj=max(xj);
85:    if sf>0
86:       xforc=xmaxj+[0,fa,fa+xtip];
```

```
87:     else
88:        xforc=xmaxj+[fa,0,-xtip];
89:     end
90:
91:     % Plot the spring, block, and force
92:     % plot(xj,y,rx,ry,'k',xforc,ytip,'r')
93:     %plot(xj,y,'k-',rx,ry,'k-',xforc,ytip,'k-')
94:     plot(xj,y,'k-',xforc,ytip,'k-',...
95:                    rx,ry,'k-','linewidth',1)
96:     title('FORCED MOTION WITH DAMPING')
97:     xlabel('FORCED MOTION WITH DAMPING')
98:     axis(r), axis('off'), drawnow
99:     figure(gcf), pause(.05)
100: end
101: if pltsave, print -deps smdplotanim; end
102: disp(' '), disp('All Done')
103:
104: %=====================================
105:
106: function [x,y] = spring(len,ht)
107: % This function generates a set of points
108: % defining a spring
109:
110: if nargin==0, len=1; ht=.125; end
111: x=[0,.5,linspace(1,11,10),11.5,12];
112: y=[ones(1,5);-ones(1,5)];
113: y=[0;0;y(:);0;0]'; y=ht/2/max(y)*y;
114: x=len/max(x)*x;
115:
116: %=====================================
117:
118: function [x,v]=smdsolve(m,c,k,f1,f2,w,x0,v0,t)
119: %
120: % [x,v]=smdsolve(m,c,k,f1,f2,w,x0,v0,t)
121: % ~~~~~~~~~~~~~~~~~~~~~~~~~~~~~~~~~~~~~
122: % This function solves the differential equation
123: % m*x''(t)+c*x'(t)+k*x(t)=f1*cos(w*t)+f2*sin(w*t)
124: % with x(0)=x0 and x'(0)=v0
125: %
126: % m,c,k  - mass, damping and stiffness coefficients
127: % f1,f2  - magnitudes of cosine and sine terms in
128: %          the forcing function
129: % w      - frequency of the forcing function
130: % t      - vector of times to evaluate the solution
131: % x,v    - computed position and velocity vectors
```

```
132:
133: ccrit=2*sqrt(m*k); wn=sqrt(k/m);
134:
135: % If the system is undamped and resonance will
136: % occur, add a little damping
137: if c==0 & w==wn; c=ccrit/1e6; end;
138:
139: % If damping is critical, modify the damping
140: % very slightly to avoid repeated roots
141: if c==ccrit; c=c*(1+1e-6); end
142:
143: % Forced response solution
144: a=(f1-i*f2)/(k-m*w^2+i*c*w);
145: X0=real(a); V0=real(i*w*a);
146: X=real(a*exp(i*w*t)); V=real(i*w*a*exp(i*w*t));
147:
148: % Homogeneous solution
149: r=sqrt(c^2-4*m*k);
150: s1=(-c+r)/(2*m); s2=(-c-r)/(2*m);
151: p=[1,1;s1,s2]\[x0-X0;v0-V0];
152:
153: % Total solution satisfying the initial conditions
154: x=X+real(p(1)*exp(s1*t)+p(2)*exp(s2*t));
155: v=V+real(p(1)*s1*exp(s1*t)+p(2)*s2*exp(s2*t));
156:
157: %====================================
158:
159: % function [a1,a2,...,a_nargout]=inputv(prompt)
160: % See Appendix B
```

2.7 Example of Waves in an Elastic String

One-dimensional wave propagation is illustrated well by the response of a tightly stretched string of finite length released from rest with given initial deflection. The transverse deflection $y(x, t)$ satisfies the wave equation

$$a^2 y_{xx} = y_{tt}$$

and the general solution for an infinite length string, released from rest, is given by

$$y(x, t) = [F(x - at) + F(x + at)]/2$$

where $F(x)$ is the initial deflection for $-\infty < x < \infty$. The physical interpretation for this equation is that the initial deflection splits in two parts translating at speed a,with one part moving to the right and the other moving to the left. The translating wave solution can be adapted to handle a string of finite length l by requiring

$$y(0,t) = y(l,t) = 0.$$

These end conditions, along with initial deflection $f(x)$ (defining $F(x)$ between 0 and l), are sufficient to continue the solution outside the original interval. We write the initial condition for the finite length string as

$$y(x,0) = f(x), \ 0 < x < l.$$

To satisfy the end conditions, $F(x)$ must be an odd-valued function of period $2l$. Introducing a function $g(x)$ such that

$$g(x) = f(x), \ 0 \leq x \leq l$$

and

$$g(x) = -f(2l - x), \ l < x \leq 2l$$

leads to

$$F(x) = \mathbf{sign}(x) g(\mathbf{rem}(\mathbf{abs}(x), 2l))$$

where the desired periodicity is achieved using the MATLAB remainder function, **rem**. This same problem can also be solved using a Fourier sine series (see chapter 9). For the present we concentrate on the solution just obtained.

A program was written to implement the translating wave solution when $f(x)$ is a piecewise linear function computed using **interp1**. The system behavior can be examined from three different aspects. 1) The solution $y(x,t)$ for a range of x and t values describes a surface. 2) The deflection curve at a particular time t_0 is expressed as $y(x,t_0), \ 0 < x < l$. 3) The motion history at a particular point x_0 is $y(x_0,t), \ t \geq 0$. The nature of $F(x)$ implies that the motion has a period of $2l/a$. Waves striking the boundary are reflected in inverted form so that for any time $y(x,t+l/a) = -y(x,t)$. The character of the motion is typified by the default data case the program uses to define a triangular initial deflection pattern where

$$a = 1, \ l = 1, \ xd = [0, \ 0.33, \ 0.5, \ 0.67, \ 1], \ yd = [0, \ 0, \ -1, \ 0, \ 0].$$

The program reads the wave speed, the string length, and data points specifying the initial deflection. The solution is evaluated for a range of x, t values. The function **plot3** was used to create Figure 2.10, which is a three-dimensional plot of traces of the string deflection for a sequence of times. Figure 2.11 shows the string position at $t = 0.33$. Figure 2.12 plots the deflection history at position $x = 0.25$. Finally, a function to animate the solution over two motion cycles illustrates how the initial deflection splits, translates, and reflects from the boundaries. In an attempt to illustrate successive positions assumed in the animation, traces of the motion for a brief period are shown in Figure 2.13

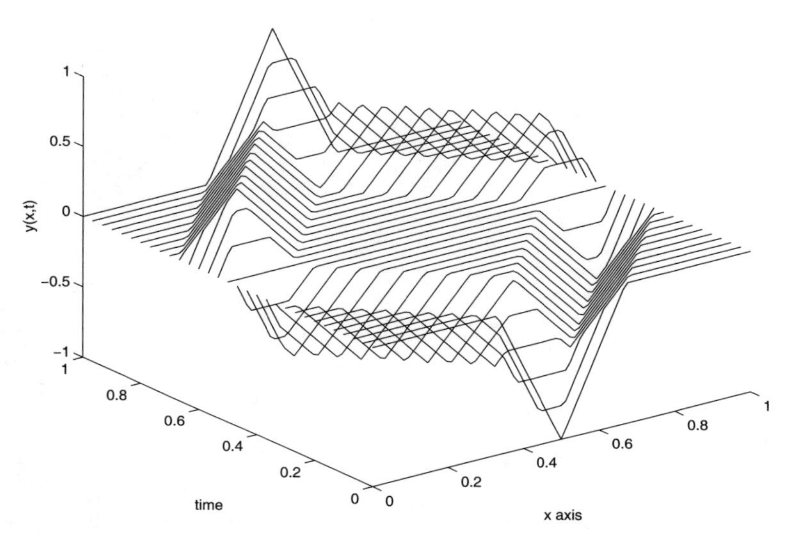

Figure 2.10: String Position as a Function of Position and Time

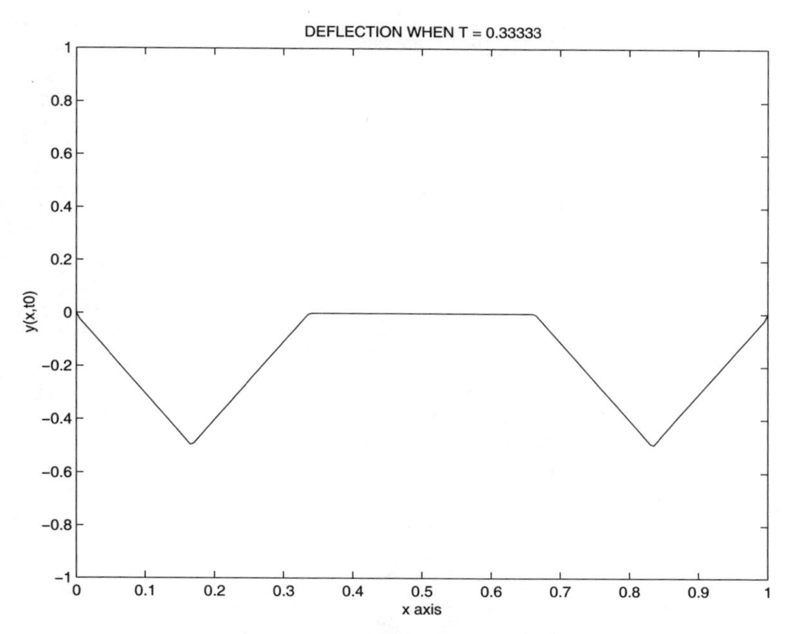

Figure 2.11: String Deflection when $t = 0.33$

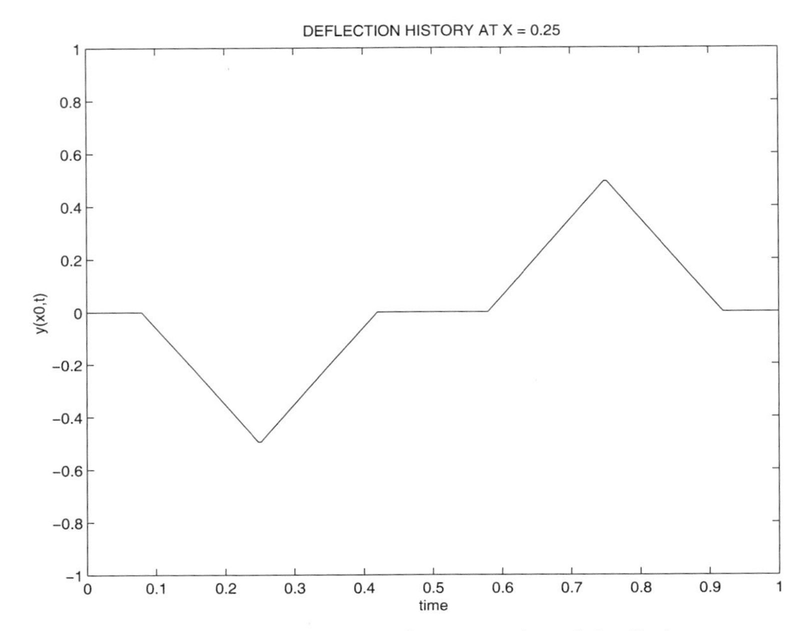

Figure 2.12: Motion at Quarterpoint of the String

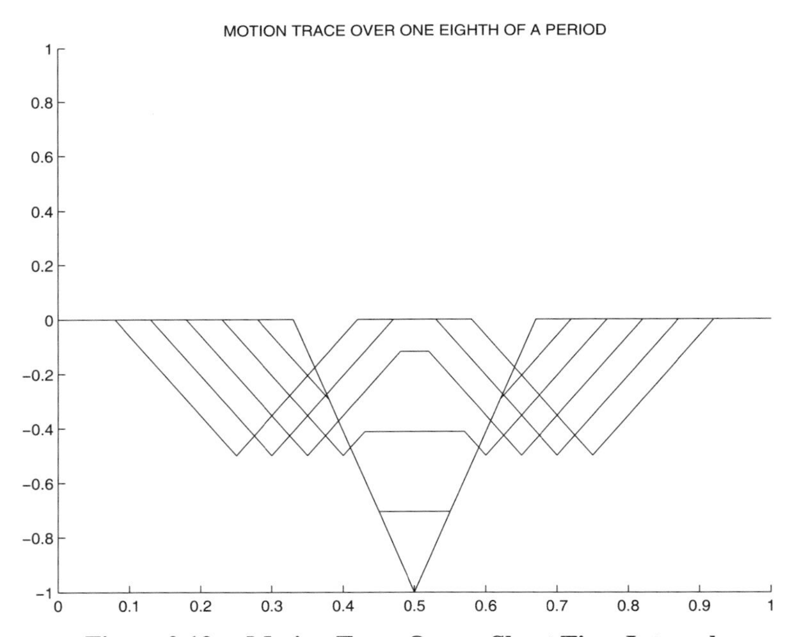

Figure 2.13: Motion Trace Over a Short Time Interval

MATLAB Example

Program strngrun

```
1: function strngrun(rundemo)
2: %
3: % strngrun(rundemo)
4: % ~~~~~~~~~~~~~~~~~
5: % This function illustrates propagation of
6: % waves in a tightly stretched string having
7: % given initial deflection. Calling strngrun
8: % with no input argument causes data to be
9: % read interactively. Otherwise, strngrun(1)
10: % executes a sample data case.
11: %
12: % User m functions called: strngwav animate
13:
14: pltsav=0; % flag to save or not save graphs
15:
16: disp(' ')
17: disp('WAVE PROPAGATION IN A STRING'), disp(' ')
18: if nargin==0 % Input data interactively
19:   [a,len]=inputv(['Input wave speed (a) and ',...
20: 'string length (len) > ? ']);
21:   disp(' ')
22:   disp(['Enter the number of interior ',...
23:         'data points (the fixed'])
24:   disp(['end point coordinates are ',...
25:         'added automatically)'])
26:   n=input('? '); if isempty(n), return, end
27:   xd=zeros(n+2,1); xd(n+2)=len;
28:   yd=zeros(n+2,1); disp(' ')
29:   disp(['The string stretches between ',...
30:         'fixed endpoints at'])
31:   disp(['x=0 and x=',num2str(len),'.']),disp(' ')
32:   disp(['Enter ',num2str(n),...
33:        ' sets of x,y to specify interior'])
34:   disp(['initial deflections ',...
35:       '(one pair per line)'])
36:   for j=2:n+1,[xd(j),yd(j)]=inputv; end;
37:   disp(' ')
38: disp('Input tmax and the number of time steps')
39: [tmax,nt]=inputv('(Try len/a and 40) > ? ');
40:   disp(' ')
```

```
41: disp('Specify position x=x0 where the time')
42: x0=input(...
43:     'history is to be evaluated (try len/4) > ? ');
44: disp(' ')
45: disp('Specify time t=t0 when the deflection')
46: t0=input('curve is to be plotted > ? ');
47: disp(' ')
48: titl=input('Input a graph title > ? ','s');
49:
50: else % Example for triangular initial deflection
51: a=1; len=1; tmax=len/a; nt=40;
52: xd=[0,.33,.5,.67,1]*len; yd=[0,0,-1,0,0];
53:
54: % Different example for a truncated sine curve
55: % xd=linspace(0,len,351); yd=sin(3*pi/len*xd);
56: % k=find(yd<=0); xd=xd(k); yd=yd(k);
57:
58: x0=0.25*len; t0=0.33*len/a;
59: titl='TRANSLATING WAVE OVER HALF A PERIOD';
60: end
61:
62: nx=80; x=0:len/nx:len; t=0:tmax/nt:tmax;
63:
64: h=max(abs(yd)); xplot=linspace(0,len,201);
65: tplot=linspace(0,max(t),251)';
66:
67: [Y,X,T]=strngwav(xd,yd,x,t,len,a);
68: plot3(X',T',Y','k'); xlabel('x axis')
69: ylabel('time'), zlabel('y(x,t)'), title(titl)
70: if pltsav, print(gcf,'-deps','strngplot3'); end
71: drawnow, shg, disp(' ')
72:
73: disp('Press return to see the deflection')
74: disp(['when t = ',num2str(t0)]), pause
75:
76: [yt0,xx,tt]=strngwav(xd,yd,xplot,t0,len,a);
77: close; plot(xx(:),yt0(:),'k')
78: xlabel('x axis'), ylabel('y(x,t0)')
79: title(['DEFLECTION WHEN T = ',num2str(t0)])
80: axis([min(xx),max(xx),-h,h])
81: if pltsav, print(gcf,'-deps','strngyxt0'); end
82: drawnow, shg
83:
84: disp(' ')
85: disp('Press return to see the deflection history')
```

```
86: disp(['at x = ',num2str(x0)]), pause
87:
88: yx0=strngwav(xd,yd,x0,tplot,len,a);
89: plot(tplot,yx0,'k')
90: xlabel('time'), ylabel('y(x0,t)')
91: title(...
92: ['DEFLECTION HISTORY AT X = ',num2str(x0)])
93: axis([0,max(t),-h,h])
94: if pltsav, print(gcf,'-deps','strngyx0t'); end
95: drawnow, shg
96:
97: disp(' ')
98: disp('Press return to see the animation')
99: disp('over two periods of motion'), pause
100: x=linspace(0,len,101); t=linspace(0,4*len/a,121);
101: [Y,X,T]=strngwav(xd,yd,x,t,len,a);
102: titl='MOTION OVER TWO PERIODS';
103: animate(X(1,:),Y',titl,.1), pause(2)
104:
105: if pltsav, print(gcf,'-deps','strnganim'); end
106:
107: disp(' '), disp('All Done')
108:
109: %================================================
110:
111: function [Y,X,T]=strngwav(xd,yd,x,t,len,a)
112: %
113: % [Y,X,T]=strngwav(xd,yd,x,t,len,a)
114: % ~~~~~~~~~~~~~~~~~~~~~~~~~~~~~~~~~
115: % This function computes the dynamic response of
116: % a tightly stretched string released from rest
117: % with a piecewise linear initial deflection. The
118: % string ends are fixed.
119: %
120: % xd,yd - data vectors defining the initial
121: %         deflection as a piecewise linear
122: %         function. xd values should be increasing
123: %         and lie between 0 and len
124: % x,t   - position and time vectors for which the
125: %         solution is evaluated
126: % len,a - string length and wave speed
127:
128: if nargin<6, a=1; end; if nargin <5, len=1; end
129: xd=xd(:); yd=yd(:);  p=2*len;
130:
```

```
131: % If end values are not zero, add these points
132: if xd(end)~=len, xd=[xd;len]; yd=[yd;0]; end
133: if xd(1)~=0, xd=[0;xd]; yd=[0;yd]; end
134: nd=length(xd);
135:
136: % Eliminate any repeated abscissa values
137: k=find(diff(xd)==0); tiny=len/1e6;
138: if length(k)>0, xd(k)=xd(k)+tiny; end
139:
140: % Extend the data definition for len < x < 2*len
141: xd=[xd;p-xd(nd-1:-1:1)]; yd=[yd;-yd(nd-1:-1:1)];
142: [X,T]=meshgrid(x,t); xp=X+a*T; xm=X-a*T;
143: shape=size(xp); xp=xp(:); xm=xm(:);
144:
145: % Compute the general solution for a piecewise
146: % linear initial deflection
147: Y=(sign(xp).*interp1(xd,yd,rem(abs(xp),p),...
148:   'linear','extrap')+sign(xm).*interp1(xd,yd,...
149:    rem(abs(xm),p),'linear','extrap'))/2;
150: Y=reshape(Y,shape);
151:
152: %================================================
153:
154: function animate(x,y,titl,tim,trace)
155: %
156: % animate(x,y,titl,tim,trace)
157: % ~~~~~~~~~~~~~~~~~~~~~~~~~~~~
158: % This function performs animation of a 2D curve
159: % x,y - arrays with columns containing curve positions
160: %       for successive times. x can also be a single
161: %       vector if x values do not change. The animation
162: %       is done by plotting (x(:,j),y(:,j)) for
163: %       j=1:size(y,2).
164: % titl- title for the graph
165: % tim - the time in seconds between successive plots
166:
167: if nargin<5, trace=0; else, trace=1; end;
168: if nargin<4, tim=.05; end
169: if nargin<3, trac=''; end; [np,nt]=size(y);
170: if min(size(x))==1, j=ones(1,nt); x=x(:);
171: else, j=1:nt; end; ax=newplot;
172: if trace, XOR='none'; else, XOR='xor'; end
173: r=[min(x(:)),max(x(:)),min(y(:)),max(y(:))];
174: %axis('equal') % Needed for an undistorted plot
175: axis(r), % axis('off')
```

```
176: curve = line('color','k','linestyle','-',...
177: 'erase',XOR, 'xdata',[],'ydata',[]);
178: xlabel('x axis'), ylabel('y axis'), title(titl)
179: for k = 1:nt
180:     set(curve,'xdata',x(:,j(k)),'ydata',y(:,k))
181:     if tim>0, pause(tim), end, drawnow, shg
182: end
183:
184: %==================================================
185:
186: % function varargout=inputv(prompt)
187: % See Appendix B
```

2.8 Properties of Curves and Surfaces

In this section some properties of space curves and surfaces are studied. Examples illustrating the graphics capabilities of MATLAB to describe three-dimensional geometries are given. Readers should also study the **demo** examples and intrinsic documentation on functions such as **plot3**, **surf**, and **mesh** to appreciate the wealth of plotting options available.

2.8.1 Curve Properties

A space curve is a one-dimensional region representable in parametric form as

$$\boldsymbol{R}(t) = \hat{\imath}\, x(t) + \hat{\jmath}\, y(t) + \hat{\boldsymbol{k}}\, z(t) \ , \ a < t < b$$

where $\hat{\imath}$, $\hat{\jmath}$, $\hat{\boldsymbol{k}}$ are Cartesian base vectors, and t is a scalar parameter such as arc length s or time. At each point on the curve, differential properties naturally lead to a triad of orthonormal base vectors $\hat{\boldsymbol{T}}$, $\hat{\mathbf{N}}$, and $\hat{\mathbf{B}}$ called the tangent, the principal normal, and the binormal. The normal vector points toward the center of curvature and the binormal is defined by $\hat{\boldsymbol{T}} \times \hat{\mathbf{N}}$ to complete the triad. Coordinate planes associated with the triad are the normal plane containing $\hat{\mathbf{N}}$ and $\hat{\mathbf{B}}$, the tangent plane containing $\hat{\boldsymbol{T}}$ and $\hat{\mathbf{B}}$, and the osculating plane containing $\hat{\boldsymbol{T}}$ and $\hat{\mathbf{N}}$. Two other scalar properties of interest are the curvature κ (the reciprocal of the curvature radius) and the torsion τ, which quantifies the rate at which the triad twists about the direction of $\hat{\boldsymbol{T}}$ as a generic point moves along the curve. When a curve is parameterized in terms of arc length s, the five quantities just mentioned are related by the Frenet formulas [91] which are

$$\frac{d\hat{\boldsymbol{T}}}{ds} = \kappa\hat{\mathbf{N}} \ , \ \frac{d\hat{\mathbf{B}}}{ds} = -\tau\hat{\mathbf{N}} \ , \ \frac{d\hat{\mathbf{N}}}{ds} = -\kappa\hat{\boldsymbol{T}} + \tau\hat{\mathbf{B}}.$$

Since most curves are not easily parameterized in terms of arc length, more convenient formulas are needed for computing $\hat{\boldsymbol{T}}$, $\hat{\mathbf{N}}$, $\hat{\mathbf{B}}$, κ, and τ. All the desired quantities can be found in terms of $\boldsymbol{R}'(t)$, $\boldsymbol{R}''(t)$, and $\boldsymbol{R}'''(t)$. Among the five properties, only torsion, τ, depends on $\boldsymbol{R}'''(t)$. The pertinent formulas are

$$\hat{\boldsymbol{T}} = \frac{\boldsymbol{R}'(t)}{|\boldsymbol{R}'(t)|} \ , \ \hat{\mathbf{B}} = \frac{\boldsymbol{R}'(t) \times \boldsymbol{R}''(t)}{|\boldsymbol{R}'(t) \times \boldsymbol{R}''(t)|}$$

$$\hat{\mathbf{N}} = \hat{\mathbf{B}} \times \hat{\boldsymbol{T}} \ , \ \kappa = \frac{|\boldsymbol{R}'(t) \times \boldsymbol{R}''(t)|}{|\boldsymbol{R}'(t)|^3}$$

and

$$\tau = \frac{\hat{\mathbf{B}} \cdot R'''(t)}{|\boldsymbol{R}'(t) \times \boldsymbol{R}''(t)|}.$$

When the independent variable t means time we get

$$\mathbf{V} = \text{velocity} = \frac{d\boldsymbol{R}}{dt} = \frac{ds}{dt}\frac{d\boldsymbol{R}}{ds} = v\hat{\boldsymbol{T}}$$

where v is the magnitude of velocity called speed. Differentiating again leads to

$$\frac{d\mathbf{V}}{dt} = \text{acceleration} = \frac{dv}{dt}\hat{\boldsymbol{T}} + \kappa v^2 \hat{\mathbf{N}}$$

so the acceleration involves a tangential component with magnitude equal to the time rate of change of speed, and a normal component of magnitude κv^2 directed toward the center of curvature. The torsion is only encountered when the time derivative of acceleration is considered. This is seldom of interest in Newtonian mechanics.

A function **crvprp3d** was written to evaluate $\hat{\boldsymbol{T}}$, $\hat{\mathbf{N}}$, $\hat{\mathbf{B}}$, κ, and τ in terms of $\boldsymbol{R}'(t)$, $\boldsymbol{R}''(t)$, and $\boldsymbol{R}'''(t)$. Another function **aspiral** applies **crvprp3d** to the curve described by

$$\boldsymbol{R}(t) = [(r_o + kt)\cos(t);\ (r_o + kt)\sin(t);\ ht]$$

where t is the polar coordinate angle for cylindrical coordinates. Figure 2.14 depicts results generated from the default data set where

$$r_o = 2\pi \ , \ k = 1 \ , \ h = 2 \ , \ 2\pi \le t \le 8\pi,$$

with 101 data points being used. A cross section normal to the surface would produce a right angle describing the directions of the normal and binormal at a typical point. The spiral itself passes along the apex of the right angle. This surface illustrates how the intrinsic triad of base vectors changes position and direction as a point moves along the curve.

An additional function **crvprpsp** was written to test how well cubic spline interpolation approximates curve properties for the spiral. MATLAB provides function **spline** to connect data points by a piecewise cubic interpolation curve having continuous first and second derivatives [27]. This function utilizes other intrinsic functions[1] such as **unmkpp**, **mkpp**, and **ppval**. Although basic MATLAB does not

[1]These functions are included with MATLAB and are a subset of the more comprehensive *Spline Toolbox* also available from The MathWorks.

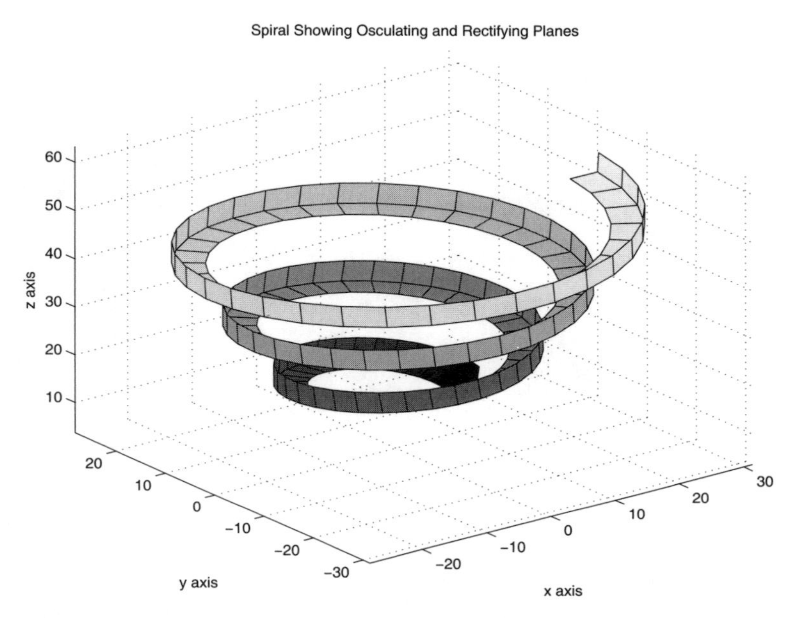

Figure 2.14: Spiral Showing Osculating and Rectifying Planes

include functions for spline differentiation, this can be remedied by the short function **splined** which computes first and second derivatives of the interpolation curve defined by function **spline**. In our example using spline interpolation, approximation of τ was not obtained because a cubic spline only has its first two derivatives continuous. Approximations for $\boldsymbol{R}'''(t)$ could have been generated by interpolating the computed values of $\boldsymbol{R}'(t)$ and differentiating the results twice. That idea was not explored. To assess the accuracy of the spline interpolation, values for $\mathbf{norm}(\hat{\mathbf{B}} - \hat{\mathbf{B}}_{\text{approx}})$ and $|(k - k_{\text{approx}})/k|$ were obtained at 101 sample points along the curve. Results depicted in Figure 2.15 show errors in the third decimal place except near the ends of the interpolation interval where a “not a knot” boundary condition is employed [27].

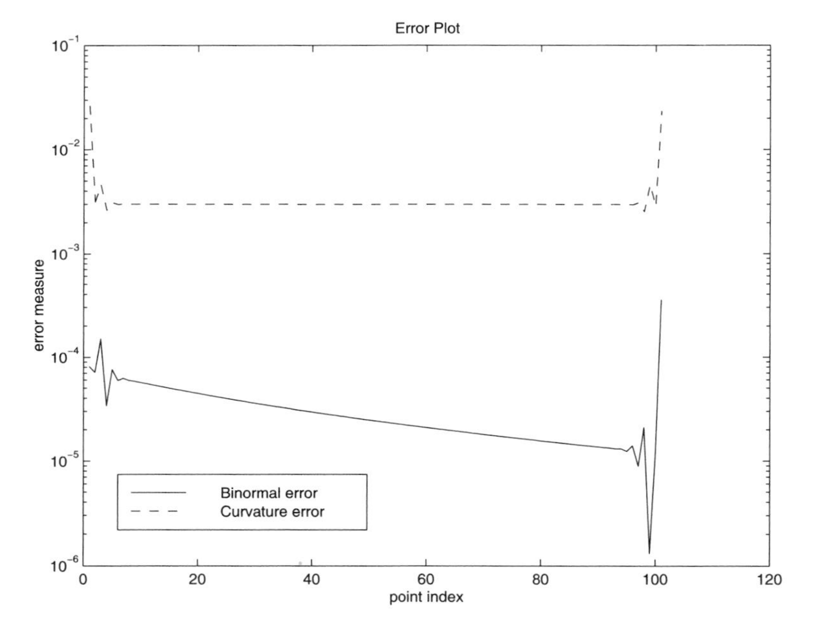

Figure 2.15: Error Plot

Program Output and Code

Program splinerr

```
1: function splinerr
2: % Example: splinerr
3: % ~~~~~~~~~~~~~~~~~~
4: %
5: % This program calculates the binormal and
6: % curvature error for a spiral space curve.
7: %
8: % User m functions called:
9: % aspiral, crvprpsp crvprp3d cubrange splined
10: %----------------------------------------------
11:
12: clear; hold off; clf;
13: [R,T,N,B,KAP]=aspiral; m=size(R,2);
14: [r,t,n,b,k]=crvprpsp(R,m);
15: disp(' '); disp(...
16: 'Press [Enter] to show error curves'); pause
17: errv=sqrt(sum((B-b).^2));
18: errk=abs((KAP-k)./KAP); hold off; clf;
19: semilogy(1:m,errv,'k-',1:m,errk,'k--');
20: xlabel('point index'); ylabel('error measure');
21: title('Error Plot');
22: legend('Binormal error','Curvature error',3);
23: figure(gcf); disp(' ')
24: disp('Press [Enter] to finish'); pause
25: disp(' '), disp('All done'), disp(' ')
26:
27: %================================================
28:
29: function [R,T,N,B,kap,tau,arclen]= ...
30:                           aspiral(r0,k,h,t)
31: %
32: % [R,T,N,B,kap,tau,arclen]=aspiral(r0,k,h,t)
33: % ~~~~~~~~~~~~~~~~~~~~~~~~~~~~~~~~~~~~~~~~~~
34: %
35: % This function computes geometrical properties
36: % of a spiral curve having the parametric
37: % equation
38: %
39: %   R = [(r0+k*t)*cos(t);(r0+k*t)*sin(t);h*t]
40: %
```

```
41: % A figure showing the curve along with the
42: % osculating plane and the rectifying plane
43: % at each point is also drawn.
44: %
45: % r0,k,h - parameters which define the spiral
46: % t      - a vector of parameter values at
47: %          which the curve is evaluated from
48: %          the parametric form.
49: %
50: % R      - matrix with columns containing
51: %          position vectors for points on the
52: %          curve
53: % T,N,B  - matrices with columns containing the
54: %          tangent,normal,and binormal vectors
55: % kap    - vector of curvature values
56: % tau    - vector of torsion values
57: % arclen - value of arc length approximated as
58: %          the sum of chord values between
59: %          successive points
60: %
61: % User m functions called:
62: %          crvprp3d, cubrange
63: %----------------------------------------------
64:
65: if nargin==0
66:   k=1; h=2; r0=2*pi; t=linspace(2*pi,8*pi,101);
67: end
68:
69: % Evaluate R, R'(t), R''(t) and R'''(t) for
70: % the spiral
71: t=t(:)'; s=sin(t); c=cos(t); kc=k*c; ks=k*s;
72: rk=r0+k*t; rks=rk.*s; rkc=rk.*c; n=length(t);
73: R=[rkc;rks;h*t]; R1=[kc-rks;ks+rkc;h*ones(1,n)];
74: R2=[-2*ks-rkc;2*kc-rks;zeros(1,n)];
75: R3=[-3*kc+rks;-3*ks-rkc;zeros(1,n)];
76:
77: % Obtain geometrical properties
78: [T,N,B,kap,tau]=crvprp3d(R1,R2,R3);
79: arclen=sum(sqrt(sum((R(:,2:n)-R(:,1:n-1)).^2)));
80:
81: % Generate points on the osculating plane and
82: % the rectifying plane along the curve.
83: w=arclen/100; Rn=R+w*N;  Rb=R+w*B;
84: X=[Rn(1,:);R(1,:);Rb(1,:)];
85: Y=[Rn(2,:);R(2,:);Rb(2,:)];
```

```
86: Z=[Rn(3,:);R(3,:);Rb(3,:)];
87:
88: % Draw the surface
89: v=cubrange([X(:),Y(:),Z(:)]); hold off; clf; close;
90: surf(X,Y,Z); axis(v); xlabel('x axis');
91: ylabel('y axis'); zlabel('z axis');
92: title(['Spiral Showing Osculating and ', ...
93:        'Rectifying Planes']); grid on; drawnow;
94: figure(gcf);
95:
96: %==============================================
97:
98: function [T,N,B,kap,tau]=crvprp3d(R1,R2,R3)
99: %
100: % [T,N,B,kap,tau]=crvprp3d(R1,R2,R3)
101: % ~~~~~~~~~~~~~~~~~~~~~~~~~~~~~~~~~~
102: %
103: % This function computes the primary
104: % differential properties of a three-dimensional
105: % curve parameterized in the form R(t) where t
106: % can be arc length or any other convenient
107: % parameter such as time.
108: %
109: % R1  - the matrix with columns containing R'(t)
110: % R2  - the matrix with columns containing R''(t)
111: % R3  - the matrix with columns containing
112: %       R'''(t).  This matrix is only needed
113: %       when torsion is to be computed.
114: %
115: % T   - matrix with columns containing the
116: %       unit tangent
117: % N   - matrix with columns containing the
118: %       principal normal vector
119: % B   - matrix with columns containing the
120: %       binormal
121: % kap - vector of curvature values
122: % tau - vector of torsion values. This equals
123: %       [] when R3 is not given
124: %
125: % User m functions called:  none
126: %----------------------------------------------
127:
128: nr1=sqrt(dot(R1,R1)); T=R1./nr1(ones(3,1),:);
129: R12=cross(R1,R2); nr12=sqrt(dot(R12,R12));
130: B=R12./nr12(ones(3,1),:); N=cross(B,T);
```

```
131: kap=nr12./nr1.^3;
132:
133: % Compute the torsion only when R'''(t) is given
134: if nargin==3,   tau=dot(B,R3)./nr12;
135: else, tau=[]; end
136:
137: %==============================================
138:
139: function [R,T,N,B,kappa]=crvprpsp(Rd,n)
140: %
141: % [R,T,N,B,kappa]=crvprpsp(Rd,n)
142: % ~~~~~~~~~~~~~~~~~~~~~~~~~~~~~~~
143: %
144: % This function computes spline interpolated
145: % values for coordinates, base vectors and
146: % curvature obtained by passing a spline curve
147: % through data values given in Rd.
148: %
149: % Rd    - a matrix containing x,y and z values
150: %         in rows 1, 2 and 3.
151: % n     - the number of points at which
152: %         properties are to be evaluated along
153: %         the curve
154: %
155: % R     - a 3 by n matrix with columns
156: %         containing coordinates of interpolated
157: %         points on the curve
158: % T,N,B - matrices of dimension 3 by n with
159: %         columns containing components of the
160: %         unit tangent, unit normal, and unit
161: %         binormal vectors
162: % kappa - a vector of curvature values
163: %
164: % User m functions called:
165: %         splined, crvprp3d
166: %----------------------------------------------
167:
168: % Create a spline curve through the data points,
169: % and evaluate the derivatives of R.
170: nd=size(Rd,2); td=0:nd-1; t=linspace(0,nd-1,n);
171: ud=Rd(1,:)+i*Rd(2,:); u=spline(td,ud,t);
172: u1=splined(td,ud,t); u2=splined(td,ud,t,2);
173: ud3=Rd(3,:); z=spline(td,ud3,t);
174: z1=splined(td,ud3,t); z2=splined(td,ud3,t,2);
175: R=[real(u);imag(u);z]; R1=[real(u1);imag(u1);z1];
```

```
176: R2=[real(u2);imag(u2);z2];
177:
178: % Get curve properties from crvprp3d
179: [T,N,B,kappa]=crvprp3d(R1,R2);
180:
181: %==========================================
182:
183: function val=splined(xd,yd,x,if2)
184: %
185: % val=splined(xd,yd,x,if2)
186: % ~~~~~~~~~~~~~~~~~~~~~~~~
187: %
188: % This function evaluates the first or second
189: % derivative of the piecewise cubic
190: % interpolation curve defined by the intrinsic
191: % function spline provided in MATLAB.If fewer
192: % than four data points are input, then simple
193: % polynomial interpolation is employed
194: %
195: % xd,yd - data vectors determining the spline
196: %         curve produced by function spline
197: % x     - vector of values where the first or
198: %         the second derivative are desired
199: % if2   - a parameter which is input only if
200: %         y''(x) is required. Otherwise, y'(x)
201: %         is returned.
202: %
203: % val   - the first or second derivative values
204: %         for the spline
205: %
206: % User m functions called: none
207:
208: n=length(xd); [b,c]=unmkpp(spline(xd,yd));
209: if n>3 % Use a cubic spline
210:   if nargin==3, c=[3*c(:,1),2*c(:,2),c(:,3)];
211:   else, c=[6*c(:,1),2*c(:,2)]; end
212:   val=ppval(mkpp(b,c),x);
213: else % Use a simple polynomial
214:   c=polyder(polyfit(xd(:),yd(:),n-1));
215:   if nargin==4, c=polyder(c); end
216:   val=polyval(c,x);
217: end
218:
219: %=============================================
220:
```

```
221: % function range=cubrange(xyz,ovrsiz)
222: % See Appendix B
```

2.8.2 Surface Properties

Surfaces are two-dimensional regions described parametrically as

$$\boldsymbol{R}(u,v) = \hat{\boldsymbol{\imath}}x(u,v) + \hat{\boldsymbol{\jmath}}u(u,v) + \hat{\boldsymbol{k}}z(u,v)$$

where u and v are scalar parameters. This parametric form is helpful for generating a grid of points on the surface as well as for computing surface tangents and the surface normal. Holding v fixed while u varies generates a curve in the surface called a u coordinate line. A tangent vector to the u-line is given by

$$g_u = \frac{\partial R}{\partial u} = \hat{\boldsymbol{\imath}}\frac{\partial x}{\partial u} + \hat{\boldsymbol{\jmath}}\frac{\partial y}{\partial u} + \hat{\boldsymbol{k}}\frac{\partial z}{\partial u}.$$

Similarly, holding u fixed and varying v produces a v-line with tangent vector

$$g_v = \frac{\partial R}{\partial v} = \hat{\boldsymbol{\imath}}\frac{\partial x}{\partial v} + \hat{\boldsymbol{\jmath}}\frac{\partial y}{\partial v} + \hat{\boldsymbol{k}}\frac{\partial z}{\partial v}.$$

Consider the following cross product.

$$g_u \times g_v \, du \, dv = \hat{n} \, dS.$$

In this equation $\hat{n}$ is the unit surface normal and dS is the area of a parallelogram shaped surface element having sides defined by $g_u \, du$ and $g_v \, dv$.

The intrinsic functions **surf**(X,Y,Z) and **mesh**(X,Y,Z) depict surfaces by showing a grid network and related surface patches characterized when parameters u and v are varied over constant limits. Thus, values

$$(u_\imath, v_\jmath) \ , \ 1 \le \imath \le n \ , \ 1 \le \jmath \le m$$

lead to matrices

$$X = [x(u_\imath, v_\jmath)] \ , \ Y = [y(u_\imath, v_\jmath)] \ , \ Z = [z(u_\imath, v_\jmath)]$$

from which surface plots are obtained. Function **surf** colors the surface patches whereas **mesh** colors the grid lines.

As a simple example, consider the ellipsoidal surface described parametrically as

$$x = a\cos\theta\cos\phi \ , \ y = b\cos\theta\sin\phi \ , \ z = c\sin\theta$$

where $-\frac{\pi}{2} \le \theta \le \frac{\pi}{2}$, $-\pi \le \phi \le \pi$. The surface equation evidently satisfies the familiar equation

$$\left(\frac{x}{a}\right)^2 + \left(\frac{y}{b}\right)^2 + \left(\frac{z}{c}\right)^2 = 1$$

for an ellipsoid. The function **elipsoid**(a,b,c) called with $a = 2$, $b = 1.5$, $c = 1$ produces the surface plot in Figure 2.18.

Many types of surfaces can be parameterized in a manner similar to the ellipsoid. We will examine two more problems involving a torus and a conical frustum. Consider a circle of radius b lying in the xz-plane with its center at `[a,0,0]`. Rotating the circle about the z-axis produces a torus having the surface equation

$$x = [a + b\cos\theta]\cos\phi \ , \ y = [a + b\cos\theta] \ , \ \sin\phi \ , \ z = b\sin\phi$$

where $-\pi \le \theta \le \pi$, $-\pi \le \phi \le \pi$.

This type of equation is used below in an example involving several bodies. Let us also produce a surface covering the ends and side of a conical frustum (a cone with the top cut off). The frustum has base radius r_b, top radius r_t, and height h, with the symmetry axis along the z-axis. The surface can be parameterized using an azimuthal angle θ and an arc length parameter relating to the axial direction. The lateral side length is

$$r_s = \sqrt{h^2 + (r_b - r_t)^2} \, .$$

Let us take $0 \le s \le (r_b + r_s + r_t)$ and describe the surface $R(s, \theta)$ by coordinate functions

$$x = r(s)\cos\theta \ , \ y = r(s)\sin\theta \ , \ z = z(s)$$

where $0 \le \theta \le 2\pi$ and

$$r(s) = s \ , \ 0 \le s \le r_b$$

$$r(s) = r_b + \frac{(r_t - r_b)(s - r_b)}{r_s} \ , \ z = \frac{h(s - r_b)}{r_s} \ , \ r_b \le s \le (r_b + r_s)$$

$$r(s) = r_b + r_s + r_t - r \ , \ z = h \ , \ (r_b + r_s) \le s \le (r_b + r_s + r_t) \, .$$

The function **frus** produces a grid of points on the surface in terms of r_b, r_t, h, the number of increments on the base, the number of increments on the side, and the number of increments on the top. Figure 2.16 shows the plot generated by **frus**.

An example called **srfex** employs the ideas just discussed and illustrates how MATLAB represents several interesting surfaces. Points on the surface of an annulus symmetric about the z-axis are created, and two more annuli are created by interchanging axes. A pyramid with a square base is also created and the combination of four surfaces is plotted by finding a data range to include all points and then plotting each surface in succession using the `hold` instruction (See Figure 2.16). Although the rendering of surface intersections is not perfect, a useful description of a fairly involved geometry results. Combined plotting of several intersecting surfaces is implemented in a general purpose function **surfmany**. The default data case for **surfmany** produces the six=legged geometry shown in Figure 2.17.

This section is concluded with a discussion of how a set of coordinate points can be moved to a new position by translation and rotation of axes. Suppose a vector

$$r = \hat{\imath}x + \hat{\jmath}y + \hat{k}z$$

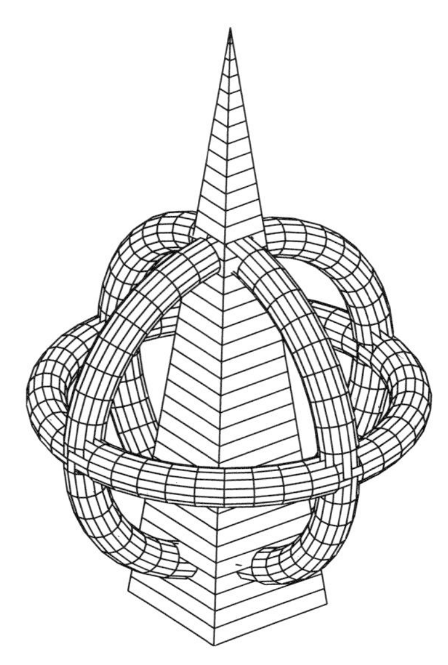

Figure 2.16: Spike and Intersecting Toruses

undergoes a coordinate change which moves the initial coordinate origin to (X_o, Y_o, Z_o) and moves the base vectors $\hat{\imath}$, $\hat{\jmath}$, $\hat{k}$ into $\hat{e}_1$, $\hat{e}_2$, $\hat{e}_3$. Then the endpoint of r passes to

$$\boldsymbol{R} = \hat{\imath}X + \hat{\jmath}Y + \hat{\boldsymbol{k}}Z = R_o + \hat{e}_1 x + \hat{e}_2 y + \hat{e}_3 z$$

where

$$\boldsymbol{R}_o = \hat{\imath}X_o + \hat{\jmath}Y_o + \hat{\boldsymbol{k}}Z_o\,.$$

Let us specify the directions of the new base vectors by employing the columns of a matrix V where we take

$$\hat{e}_3 = \frac{V(:,1)}{\mathbf{norm}[V(:,1)]}\,.$$

If $V(:,2)$ exists we take $V(:,1) \times V(:,2)$ and unitize this vector to produce $\hat{e}_2$. The triad is completed by taking $\hat{e}_1 = \hat{e}_2 \times \hat{e}_3$. In the event that $V(:,2)$ is not provided, we use `[1;0;0]` and proceed as before. The functions **rgdbodmo** and **rotatran** can be used to transform points in the manner described above.

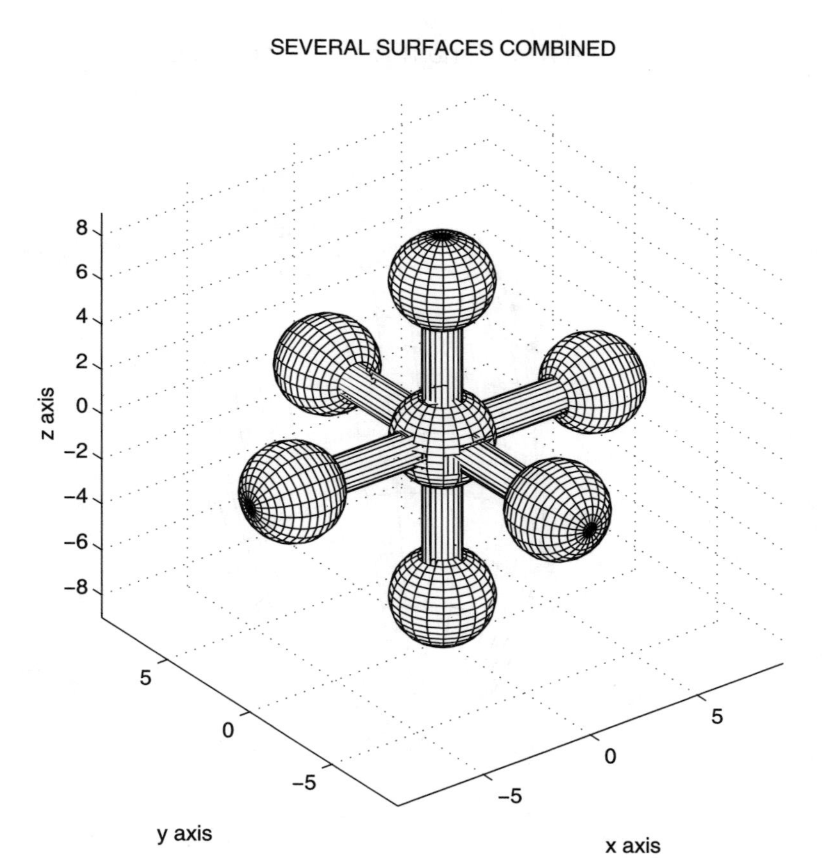

Figure 2.17: **Surface With Six Legs**

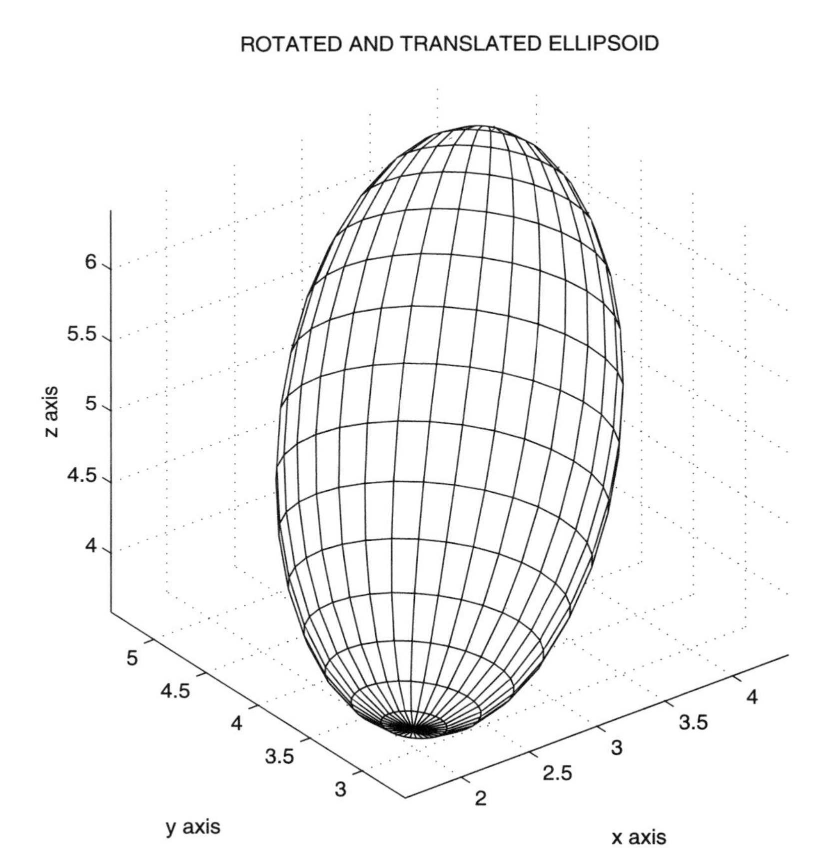

Figure 2.18: Rotated and Translated Ellipsoid Surfaces

2.8.3 Program Output and Code

Function srfex

```
1: function [x1,y1,x2,y2,x3,y3,xf,yf,zf]= ...
2:                                 srfex(da,na,df,nf)
3: % [x1,y1,x2,y2,x3,y3,xf,yf,zf]= ...
4: %                               srfex(da,na,df,nf)
5: % ~~~~~~~~~~~~~~~~~~~~~~~~~~~~~~~~~~~~~~~~~~~~~~~~
6: %
7: % This graphics example draws three toruses
8: % intersecting a spike.
9: %
10: % User m functions called: frus, surfmany
11:
12: if nargin==0
13:   da=[4.0,.45]; na=[42,15];
14:   df=[2.2,0,15]; nf=[43,4];
15: end
16:
17: % Create a torus with polygonal cross section.
18: % Data for the torus is stored in da and na
19:
20: r0=da(1); r1=da(2); nfaces=na(1); nlat=na(2);
21: t=linspace(0,2*pi,nlat)';
22: xz=[r0+r1*cos(t),r1*sin(t)];
23: z1=xz(:,2); z1=z1(:,ones(1,nfaces+1));
24: th=linspace(0,2*pi,nfaces+1);
25: x1=xz(:,1)*cos(th); y1=xz(:,1)*sin(th);
26: y2=x1; z2=y1; x2=z1; y3=x2; z3=y2; x3=z2;
27:
28: % Create a frustum of a pyramid. Data for the
29: % frustum is stored in df and nf
30: rb=df(1); rt=df(2); h=df(3);
31: [xf,yf,zf]=frus(rb,rt,h,nf); zf=zf-.35*h;
32:
33: % Plot four figures combined together
34: hold off; clf; close;
35: surfmany(x1,y1,z1,x2,y2,z2,x3,y3,z3,xf,yf,zf)
36: xlabel('x axis'); ylabel('y axis');
37: zlabel('z axis');
38: title('Spike and Intersecting Toruses');
39: axis equal; axis('off');
40: colormap([1 1 1]); figure(gcf); hold off;
```

```
41: % print -deps srfex
42:
43: %==============================================
44:
45: function [X,Y,Z]=frus(rb,rt,h,n,noplot)
46: %
47: % [X,Y,Z]=frus(rb,rt,h,n,noplot)
48: % ~~~~~~~~~~~~~~~~~~~~~~~~~~~~~~~
49: %
50: % This function computes points on the surface
51: % of a conical frustum which has its axis along
52: % the z axis.
53: %
54: % rb,rt,h - the base radius,top radius and
55: %           height
56: % n       - vector of two integers defining the
57: %           axial and circumferential grid
58: %           increments on the surface
59: % noplot  - parameter input when no plot is
60: %           desired
61: %
62: % X,Y,Z   - points on the surface
63: %
64: % User m functions called: none
65:
66: if nargin==0
67:   rb=2; rt=1; h=3; n=[23, 35];
68: end
69:
70: th=linspace(0,2*pi,n(2)+1)'-pi/n(2);
71: sl=sqrt(h^2+(rb-rt)^2); s=sl+rb+rt;
72: m=ceil(n(1)/s*[rb,sl,rt]);
73: rbot=linspace(0,rb,m(1));
74: rside=linspace(rb,rt,m(2));
75: rtop=linspace(rt,0,m(3));
76: r=[rbot,rside(2:end),rtop(2:end)];
77: hbot=zeros(1,m(1));
78: hside=linspace(0,h,m(2));
79: htop=h*ones(1,m(3));
80: H=[hbot,hside(2:end),htop(2:end)];
81: Z=repmat(H,n(2)+1,1);
82: xy=exp(i*th)*r; X=real(xy); Y=imag(xy);
83: if nargin<5
84:   surf(X,Y,Z); title('Frustum'); xlabel('x axis')
85:   ylabel('y axis'), zlabel('z axis')
```

```
86:   grid on, colormap([1 1 1]);
87:   figure(gcf);
88: end
89:
90: %==============================================
91:
92: function surfmany(varargin)
93: %function surfmany(x1,y1,z1,x2,y2,z2,...
94: %                 x3,y3,z3,..,xn,yn,zn)
95: % This function plots any number of surfaces
96: % on the same set of axes without shape
97: % distortion. When no input is given then a
98: % six-legged solid composed of spheres and
99: % cylinders is shown.
100: %
101: % User m functions called: none
102: %----------------------------------------------
103:
104: if nargin==0
105:   % Default data for a six-legged solid
106:   n=10; rs=.25; d=7; rs=2; rc=.75;
107:   [xs,ys,zs]=sphere;  [xc,yc,zc]=cylinder;
108:   xs=rs*xs; ys=rs*ys; zs=rs*zs;
109:   xc=rc*xc; yc=rc*yc; zc=2*d*zc-d;
110:   x1=xs; y1=ys; z1=zs;
111:   x2=zs+d; y2=ys; z2=xs;
112:   x3=zs-d; y3=ys; z3=xs;
113:   x4=xs; y4=zs-d; z4=ys;
114:   x5=xs; y5=zs+d; z5=ys;
115:   x6=xs; y6=ys; z6=zs+d;
116:   x7=xs; y7=ys; z7=zs-d;
117:   x8=xc; y8=yc; z8=zc;
118:   x9=zc; y9=xc; z9=yc;
119:   x10=yc; y10=zc; z10=xc;
120: varargin={x1,y1,z1,x2,y2,z2,x3,y3,z3,...
121: x4,y4,z4,x5,y5,z5,x6,y6,z6,x7,y7,z7,...
122: x8,y8,z8,x9,y9,z9,x10,y10,z10};
123: end
124:
125: % Find the data range
126: n=length(varargin);
127: r=realmax*[1,-1,1,-1,1,-1];
128: s=inline('min([a;b])','a','b');
129: b=inline('max([a;b])','a','b');
130: for k=1:3:n
```

```
131: x=varargin{k}; y=varargin{k+1};
132: z=varargin{k+2};
133: x=x(:); y=y(:); z=z(:);
134: r(1)=s(r(1),x); r(2)=b(r(2),x);
135: r(3)=s(r(3),y); r(4)=b(r(4),y);
136: r(5)=s(r(5),z); r(6)=b(r(6),z);
137: end
138:
139: % Plot each surface
140: hold off, newplot
141: for k=1:3:n
142: x=varargin{k}; y=varargin{k+1};
143: z=varargin{k+2};
144: surf(x,y,z); axis(r), hold on
145: end
146:
147: % Set axes and display the combined plot
148: axis equal, axis(r), grid on
149: xlabel('x axis'), ylabel('y axis')
150: zlabel('z axis')
151: title('SEVERAL SURFACES COMBINED')
152: % colormap([127/255 1 212/255]); % aquamarine
153: colormap([1 1 1]);, figure(gcf), hold off
```

Function rgdbodmo

```
1: function [X,Y,Z]=rgdbodmo(x,y,z,v,R0)
2: %
3: % [X,Y,Z]=rgdbodmo(x,y,z,v,R0)
4: % ~~~~~~~~~~~~~~~~~~~~~~~~~~~~~~
5: %
6: % This function transforms coordinates x,y,z to
7: % new coordinates X,Y,Z by rotating and
8: % translating the reference frames. When no
9: % input is given, an example involving an
10: % ellipsoid is run.
11: %
12: % x,y,z - initial coordinate matrices referred
13: %         to base vectors [1;0;0], [0;1;0] and
14: %         [0;0;1]. Columns of v are used to
15: %         create new basis vectors i,j,k such
16: %         that a typical point [a;b;c] is
17: %         transformed into [A;B;C] according
18: %         to the equation
```

```
19: %                [A;B;C]=R0(:)+[i,j,k]*[a;b;c]
20: % v      - a matrix having three rows and either
21: %          one or two columns used to construct
22: %          the new basis [i,j,k] according to
23: %          methods employed function rotatran
24: % R0     - a vector which translates the rotated
25: %          coordinates when R0 is input.
26: %          Otherwise no translation is imposed.
27: %
28: % X,Y,Z - matrices containing the transformed
29: %          coordinates
30: %
31: % User m functions called: elipsoid, rotatran
32:
33: if nargin==0
34:   [x,y,z]=elipsoid(1,1,2,[17,33],0);R0=[3;4;5];
35:   v=[[1;1;1],[1;1;0]];
36: end
37: [n,m]=size(x); XYZ=[x(:),y(:),z(:)]*rotatran(v)';
38: X=XYZ(:,1); Y=XYZ(:,2); Z=XYZ(:,3);
39: if ~isempty(R0)
40:   X=X+R0(1); Y=Y+R0(2); Z=Z+R0(3);
41: end
42: X=reshape(X,n,m); Y=reshape(Y,n,m);
43: Z=reshape(Z,n,m);
44: if nargin==0
45:   close; surf(X,Y,Z), axis equal, grid on
46:   title('ROTATED AND TRANSLATED ELLIPSOID')
47:   xlabel('x axis'), ylabel('y axis')
48:   zlabel('z axis'),colormap([1 1 1]); shg
49: end
50:
51: %==============================================
52:
53: function [x,y,z]=elipsoid(a,b,c,n,noplot)
54: %
55: % [x,y,z]=elipsoid(a,b,c,n,noplot)
56: % ~~~~~~~~~~~~~~~~~~~~~~~~~~~~~~~~
57: % This function plots an ellipsoid having semi-
58: % diameters a,b,c
59: % a,b,c  - semidiameters of the ellipsoid defined
60: %          by (x/a)^2+(y/b)^2+(z/c)^2=1
61: % n      - vector [nth,nph] giving the number of
62: %          theta values and phi values used to plot
63: %          the surface
```

```
64: % noplot - omit this parameter if no plot is desired
65: % x,y,z  - matrices of points on the surface
66: %
67: % User m functions called: none
68: %----------------------------------------------------
69:
70: if nargin==0, a=2; b=1.5; c=1; n=[17,33]; end
71: nth=n(1); nph=n(2);
72: th=linspace(-pi/2,pi/2,nth)'; ph=linspace(-pi,pi,nph);
73: x=a*cos(th)*cos(ph); y=b*cos(th)*sin(ph);
74: z=c*sin(th)*ones(size(ph));
75: if nargin<5
76:    surf(x,y,z); axis equal
77:    title('ELLIPSOID'), xlabel('x axis')
78:    ylabel('y axis'), zlabel('z axis')
79:    colormap([1 1 1]); grid on, figure(gcf)
80: end
81:
82: %====================================================
83:
84: function mat=rotatran(v)
85: %
86: % mat=rotatran(v)
87: % ~~~~~~~~~~~~~~~
88: % This function creates a rotation matrix based
89: % on the columns of v.
90: %
91: % v   - a matrix having three rows and either
92: %       one or two columns which are used to
93: %       create an orthonormal triad [i,j,k]
94: %       returned in the columns of mat. The
95: %       third base vector k is defined as
96: %       v(:,1)/norm(v(:,1)). If v has two
97: %       columns then, v(:,1) and v(:,2) define
98: %       the xz plane with the direction of j
99: %       defined by cross(v(:,1),v(:2)). If only
100: %       v(:,1) is input, then v(:,2) is set
101: %       to [1;0;0].
102: %
103: % mat - the matrix having columns containing
104: %       the basis vectors [i,j,k]
105: %
106: % User m functions called: none
107: %----------------------------------------------------
108:
```

```
109: k=v(:,1)/norm(v(:,1));
110: if size(v,2)==2, p=v(:,2); else, p=[1;0;0]; end
111: j=cross(k,p); nj=norm(j);
112: if nj~=0
113:   j=j/nj; mat=[cross(j,k),j,k];
114: else
115:   mat=[[0;1;0],cross(k,[0;1;0]),k];
116: end
```

Chapter 3

Summary of Concepts from Linear Algebra

3.1 Introduction

This chapter briefly reviews important concepts of linear algebra. We assume the reader already has some experience working with matrices, and linear algebra applied to solving simultaneous equations and eigenvalue problems. MATLAB has excellent capabilities to perform matrix operations using the fastest and most accurate algorithms currently available. The books by Strang [96] and Golub and Van Loan [47] give comprehensive treatments of matrix theory and of algorithm developments accounting for effects of finite precision arithmetic. One beautiful aspect of matrix theory is that fairly difficult proofs often lead to remarkably simple results valuable to users not necessarily familiar with all of the theoretical developments. For instance, the property that every real symmetric matrix of order n has real eigenvalues and a set of n orthonormal eigenvectors can be understood and used by someone unfamiliar with the proof. The current chapter summarizes a number of fundamental matrix properties and some of the related MATLAB functions. The intrinsic matrix functions use highly efficient algorithms originally from the LINPACK and EISPACK libraries which have now been superceded by LAPACK. [34, 42, 89]. Dr. Cleve Moler, the Chairman and Chief Scientist at **The MathWorks**, contributed to development of these systems. He also wrote the first version of MATLAB. Readers should simultaneously study the current chapter and the MATLAB demo program on linear algebra.

3.2 Vectors, Norms, Linear Independence, and Rank

Consider an n by m matrix

$$A = [a_{\imath\jmath}] \ , \ 1 \le \imath \le n \ , \ 1 \le \jmath \le m,$$

having real or complex elements. The shape of a matrix is computed by **size**(A) which returns a vector containing n and m. The matrix obtained by conjugating the matrix elements and interchanging columns and rows is called the transpose.

Transposition is accomplished with a $'$ operator, so that

$$A\_\text{transpose} = A'.$$

Transposition without conjugation of the elements can be performed as $A.'$ or as $\mathtt{conj}(A')$. Of course, whenever A is real, A' is simply the traditional transpose.

The structure of a matrix A is characterized by the matrix rank and sets of basis vectors spanning four fundamental subspaces. The rank r is the maximum number of linearly independent rows or columns in the matrix. We discuss these spaces in the context of real matrices. The basic subspaces are:

1. The column space containing all vectors representable as a linear combination of the columns of A. The column space is also referred to as the range or the span.
2. The null space consisting of all vectors perpendicular to every row of A.
3. The row space consisting of all vectors which are linear combinations of the rows of A.
4. The left null space consisting of all vectors perpendicular to every column of A.

MATLAB has intrinsic functions to compute rank and subspace bases

- matrix_rank = **rank**(A)
- column_space = **orth**(A)
- null_space = **null**(A)
- row_space = **orth**$(A')'$
- left_null_space = **null**$(A')'$

The basis vectors produced by **null** and **orth** are orthonormal. They are generated using the singular value decomposition algorithm [47]. The MATLAB function to perform this type of computation is named **svd**.

3.3 Systems of Linear Equations, Consistency, and Least Squares Approximation

Let us discuss the problem of solving systems of simultaneous equations. Representing a vector B as a linear combination of the columns of A requires determination of a vector X to satisfy

$$AX = B \iff \sum_{j=1}^{m} A(:,j)\,x(j) = B$$

where the j'th column of A is scaled by the j'th component of X to form the linear combination. The desired representation is possible if and only if B lies in the column space of A. This implies the consistency requirement that A and $[A, B]$ must have the same rank. Even when a system is consistent, the solution will not be unique unless all columns of A are independent. When matrix A, with n rows and m columns, has rank r less than m, the general solution of $AX = B$ is expressible as any particular solution plus an arbitrary linear combination of $m - r$ vectors forming a basis for the null space. MATLAB gives the solution vector as $X = A\backslash B$. When r is less than m, MATLAB produces a least squares solution having as many components as possible set equal to zero.

In instances where the system is inconsistent, regardless of how X is chosen, the error vector defined by

$$E = AX - B$$

can never be zero. An approximate solution can be obtained by making E normal to the columns of A. We get

$$A'AX = A'B$$

which is known as the system of normal equations. They are also referred to as least squares error equations. It is not difficult to show that the same equations result by requiring E to have minimum length. The normal equations are always consistent and are uniquely solvable when $\mathbf{rank}(A) = m$. A comprehensive discussion of least squares approximation and methods for solving overdetermined systems is presented by Lawson and Hanson [62]. It is instructive to examine the results obtained from the normal equations when A is square and nonsingular. The least squares solution would give

$$X = (A'A)^{-1}A'B = A^{-1}(A')^{-1}A'B = A^{-1}B.$$

Therefore, the least squares solution simply reduces to the exact solution of $AX = B$ for a consistent system. MATLAB handles both consistent and inconsistent systems as $X = A\backslash B$. However, it is only sensible to use the least squares solution of an inconsistent system when AX produces an acceptable approximation to B. This implies

$$\mathbf{norm}(AX - B) < tol * \mathbf{norm}(B)$$

where tol is suitably small.

A simple but important application of overdetermined systems arises in curve fitting. An equation of the form

$$y(x) = \sum_{j=1}^{m} f_j(x)c_j$$

involving known functions $f_j(x)$, such as x^{j-1} for polynomials, must approximately match data values (X_i, Y_i), $1 \le i \le n$, with $n > m$. We simply write an overdetermined system

$$\sum_{j=1}^{n} f_j(X_i)c_j \approx Y_i \ , \ 1 \le i \le n$$

and obtain the least squares solution. The approximation is acceptable if the error components

$$e_i = \sum_{j=1}^{m} f_j(X_i)c_j - Y_i$$

are small enough and the function $y(x)$ is also acceptably smooth between the data points.

Let us illustrate how well MATLAB handles simultaneous equations by constructing the steady-state solution of the matrix differential equation

$$M\ddot{x} + C\dot{x} + Kx = F_1 \cos(\omega t) + F_2 \sin(\omega t)$$

where M, C, and K are constant matrices and F_1 and F_2 are constant vectors. The steady-state solution has the form

$$x = X_1 \cos(\omega t) + X_2 \sin(\omega t)$$

where X_1 and X_2 are chosen so that the differential equation is satisfied. Evidently

$$\dot{x} = -\omega X_1 \sin(\omega t) + \omega X_2 \cos(\omega t)$$

and

$$\ddot{x} = -\omega^2 x.$$

Substituting the assumed form into the differential equation and comparing sine and cosine terms on both sides yields

$$(K - \omega^2 M)X_1 + \omega C X_2 = F_1,$$

$$-\omega C X_1 + (K - \omega^2 M)X_2 = F_2.$$

The equivalent partitioned matrix is

$$\left[\begin{array}{c|c} (K - \omega^2 M) & \omega C \\ \hline -\omega C & (K - \omega^2 M) \end{array}\right] \left[\begin{array}{c} X_1 \\ \hline X_2 \end{array}\right] = \left[\begin{array}{c} F_1 \\ \hline F_2 \end{array}\right].$$

A simple MATLAB function to produce X_1 and X_2 when M, C, K, F_1, F_2, and ω are known is

```
function [x1,x2,xmax]=forcresp(m,c,k,f1,f2,w)
kwm=k-(w*w)*m; wc=w*c;
x=[kwm,wc;-wc,kwm]\[f1;f2]; n=length(f1);
x1=x(1:n); x2=x(n+1:2*n);
xmax=sqrt(x1.*x1+x2.*x2);
```

The vector, xmax, defined in the last line of the function above, has components specifying the maximum amplitude of each component of the steady-state solution.

The main computation in this function occurs in the third line, where matrix concatenation is employed to form a system of $2n$ equations with x being the concatenation of X_1 and X_2. The fourth line uses vector indexing to extract X_1 and X_2 from x. The notational simplicity of MATLAB is elegantly illustrated by these features: a) any required temporary storage is assigned and released dynamically, b) no looping operations are needed, c) matrix concatenation and inversion are accomplished with intrinsic functions using matrices and vectors as sub-elements of other matrices, and d) extraction of sub-vectors is accomplished by use of vector indices. The important differential equation just discussed will be studied further in Article 3.5.3 where eigenvalues and complex arithmetic are used to obtain a general solution satisfying arbitrary initial conditions.

3.4 Applications of Least Squares Approximation

The idea of solving an inconsistent system of equations in the least squares sense, so that some required condition is approximately satisfied, has numerous applications. Typically, we are dealing with a large number of equations (several hundred is common) involving a smaller number of parameters used to closely fit some constraint. Linear boundary value problems often require the solution of a differential equation applicable in the interior of a region while the function values are known on the boundary. This type of problem can sometimes be handled by using a series of functions which satisfy the differential equation exactly. Weighting the component solutions to approximately match the remaining boundary condition may lead to useful results. Below, we examine three instances where least squares approximation is helpful.

3.4.1 A Membrane Deflection Problem

Let us illustrate how least squares approximation can be used to compute the transverse deflection of a membrane subjected to uniform pressure. The transverse deflection u for a membrane which has zero deflection on a boundary L satisfies the differential equation

$$\frac{\partial^2 u}{\partial x^2} + \frac{\partial^2 u}{\partial y^2} = -\gamma \text{ , (x,y) inside L}$$

where γ is a physical constant. Properties of harmonic functions [18] imply that the differential equation is satisfied by a series of the form

$$u = \gamma \left[\frac{-|z|^2}{4} + \sum_{j=1}^{n} c_j \, \textbf{real}(z^{j-1}) \right]$$

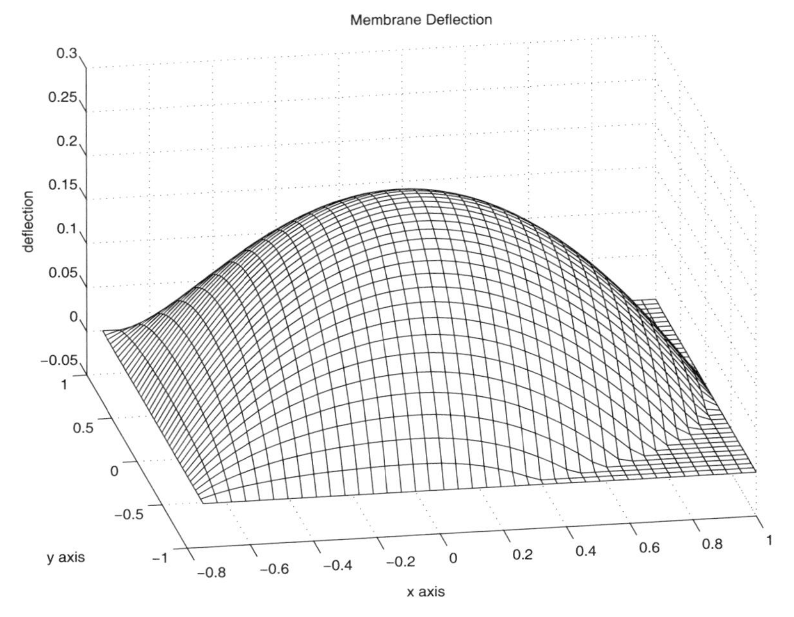

Figure 3.1: **Surface Plot of Membrane**

where $z = x + \imath y$ and constants $c_\jmath$ are chosen to make the boundary deflection as small as possible, in the least squares sense. As a specific example, we analyze a membrane consisting of a rectangular part on the left joined with a semicircular part on the right. The surface plot in Figure 3.1 and the contour plot in Figure 3.2 were produced by the function **membran** listed below. This function generates boundary data, solves for the series coefficients, and constructs plots depicting the deflection pattern. The results obtained using a twenty-term series satisfy the boundary conditions quite well.

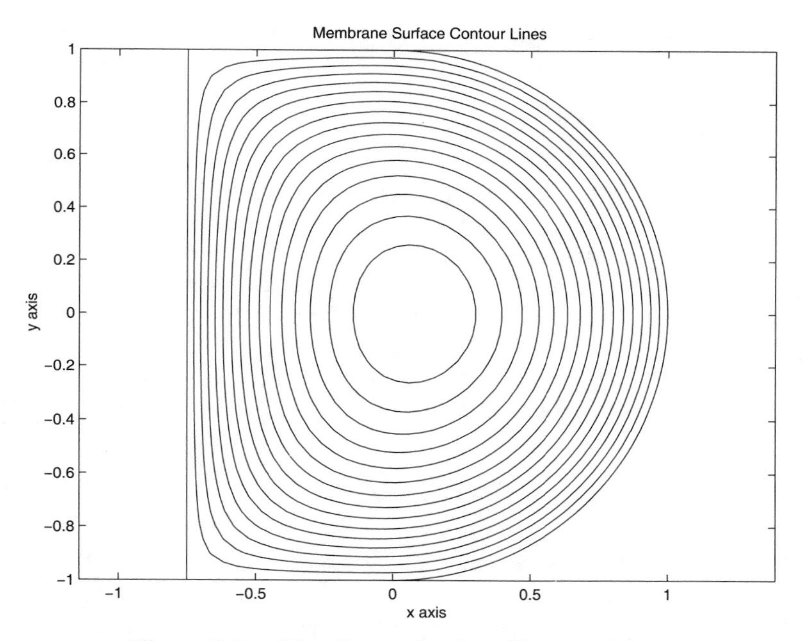

Figure 3.2: Membrane Surface Contour Lines

MATLAB Example

Function membran

```
1: function [dfl,cof]=membran(h,np,ns,nx,ny)
2: % [dfl,cof]=membran(h,np,ns,nx,ny)
3: % ~~~~~~~~~~~~~~~~~~~~~~~~~~~~~~~~~~
4: % This function computes the transverse
5: % deflection of a uniformly tensioned membrane
6: % which is subjected to uniform pressure. The
7: % membrane shape is a rectangle of width h and
8: % height two joined with a semicircle of
9: % diameter two.
10: %
11: % Example use:  membran(0.75,100,50,40,40);
12: %
13: % h       - the width of the rectangular part
14: % np      - the number of least square points
15: %           used to match the boundary
16: %           conditions in the least square
17: %           sense is about 3.5*np
18: % ns      - the number of terms used in the
19: %           approximating series to evaluate
20: %           deflections. The series has the
21: %           form
22: %
23: %           dfl = abs(z)^2/4 +
24: %                 sum({j=1:ns},cof(j)*
25: %                 real(z^(j-1)))
26: %
27: % nx,ny   - the number of x points and y points
28: %           used to compute deflection values
29: %           on a rectangular grid
30: % dfl     - computed array of deflection values
31: % cof     - coefficients in the series
32: %           approximation
33: %
34: % User m functions called:  none
35:
36: if nargin==0
37:   h=.75; np=100; ns=50; nx=40; ny=40;
38: end
39:
40: % Generate boundary points for least square
```

```
41: % approximation
42: z=[exp(i*linspace(0,pi/2,round(1.5*np))),...
43:    linspace(i,-h+i,np),...
44:    linspace(-h+i,-h,round(np/2))];
45: z=z(:); xb=real(z); xb=[xb;xb(end:-1:1)];
46: yb=imag(z); yb=[yb;-yb(end:-1:1)]; nb=length(xb);
47:
48: % Form the least square equations and solve
49: % for series coefficients
50: a=ones(length(z),ns);
51: for j=2:ns, a(:,j)=a(:,j-1).*z; end
52: cof=real(a)\(z.*conj(z))/4;
53:
54: % Generate a rectangular grid for evaluation
55: % of deflections
56: xv=linspace(-h,1,nx); yv=linspace(-1,1,ny);
57: [x,y]=meshgrid(xv,yv); z=x+i*y;
58:
59: % Evaluate the deflection series on the grid
60: dfl=-z.*conj(z)/4+ ...
61:     real(polyval(cof(ns:-1:1),z));
62:
63: % Set values outside the physical region of
64: % interest to zero
65: dfl=real(dfl).*(1-((abs(z)>=1)&(real(z)>=0)));
66:
67: % Make surface and contour plots
68: hold off; close; surf(x,y,dfl);
69: xlabel('x axis'); ylabel('y axis');
70: zlabel('deflection'); view(-10,30);
71: title('Membrane Deflection'); colormap([1 1 1]);
72: shg, disp(...
73: 'Press [Enter] to show a contour plot'), pause
74: % print -deps membdefl;
75: contour(x,y,dfl,15,'k'); hold on
76: plot(xb,yb,'k-'); axis('equal'), hold off
77: xlabel('x axis'); ylabel('y axis');
78: title('Membrane Surface Contour Lines'), shg
79: % print -deps membcntr
```

3.4.2 Mixed Boundary Value Problem for a Function Harmonic Inside a Circular Disk

Problems where a partial differential equation is to be solved inside a region with certain conditions imposed on the boundary occur in many situations. Often the differential equation is solvable exactly in a series form containing arbitrary linear combinations of known functions. An approximation procedure imposing the boundary conditions to compute the series coefficients produces a satisfactory solution if the desired boundary conditions are found to be well satisfied. Consider a mixed boundary value problem in potential theory [73] pertaining to a circular disk of unit radius. We seek $u(r,\theta)$ where function values are specified on one part of the boundary and normal derivative values are specified on the remaining part. The mathematical formulation is

$$\frac{\partial^2 u}{\partial r^2} + \frac{1}{r}\frac{\partial u}{\partial r} + \frac{1}{r^2}\frac{\partial^2 u}{\partial \theta^2} = 0 \;,\; 0 \le r < 1 \;,\; 0 \le \theta \le 2\pi,$$

$$u(1,\theta) = f(\theta) \;,\; -\alpha < \theta < \alpha \;,$$

$$\frac{\partial u}{\partial r}(1,\theta) = g(\theta) \;,\; \alpha < \theta < 2\pi - \alpha.$$

The differential equation has a series solution of the form

$$u(r,\theta) = c_0 + \sum_{n=1}^{\infty} r^n [c_n \cos(n\theta) + d_n \sin(n\theta)]$$

where the boundary conditions require

$$c_0 + \sum_{n=1}^{\infty} [c_n \cos(n\theta) + d_n \sin(n\theta)] = f(\theta) \;,\; -\alpha < \theta < \alpha,$$

and

$$\sum_{n=1}^{\infty} n[c_n \cos(n\theta) + d_n \sin(n\theta)] = g(\theta) \;,\; \alpha < \theta < 2\pi - \alpha.$$

The series coefficients can be obtained by least squares approximation. Let us explore the utility of this approach by considering a particular problem for a field which is symmetric about the x-axis. We want to solve

$$\nabla^2 u = 0 \;,\; r < 1,$$

$$u(1,\theta) = \cos(\theta) \;,\; |\theta| < \pi/2,$$

$$\frac{\partial u}{\partial r}(1,\theta) = 0 \;,\; \pi/2 < |\theta| \le \pi.$$

This problem characterizes steady-state heat conduction in a cylinder with the left half insulated and the right half held at a known temperature. The appropriate series solution is

$$u = \sum_{n=0}^{\infty} c_n r^n \cos(n\theta)$$

subject to

$$\sum_{n=0}^{\infty} c_n \cos(n\theta) = \cos(\theta) \text{ for } |\theta| < \pi/2,$$

and

$$\sum_{n=0}^{\infty} n c_n \cos(n\theta) = 0 \text{ for } \pi/2 < |\theta| \leq \pi.$$

We solve the problem by truncating the series after a hundred or so terms and forming an overdetermined system derived by imposition of both boundary conditions. The success of this procedure depends on the series converging rapidly enough so that a system of least squares equations having reasonable order and satisfactory numerical condition results. It can be shown by complex variable methods (see Muskhelishvili [73]) that the exact solution of our problem is given by

$$u = \textbf{real}\left[z + z^{-1} + (1 - z^{-1})\sqrt{z^2+1}\right]/2\ ,\ |z| \leq 1$$

where the square root is defined for a branch cut along the right half of the unit circle with the chosen branch being that which equals $+1$ at $z = 0$. Readers familiar with analytic function theory can verify that the boundary values of u yield

$$u(1,\theta) = \cos(\theta)\ ,\ |\theta| \leq \pi/2,$$

$$u(1,\theta) = \cos(\theta) + \sin(|\theta|/2)\sqrt{2|\cos(\theta)|}\ ,\ \pi/2 \leq |\theta| \leq \pi.$$

A least squares solution is presented in function **mbvp**. Results from a series of 100 terms are shown in Figure 3.3. The series solution is accurate within about one percent error except for points near $\theta = \pi/2$. Although the results are not shown here, using 300 terms gives a solution error nowhere exceeding 4 percent. Hence the least squares series solution provides a reasonable method to handle the mixed boundary value problem.

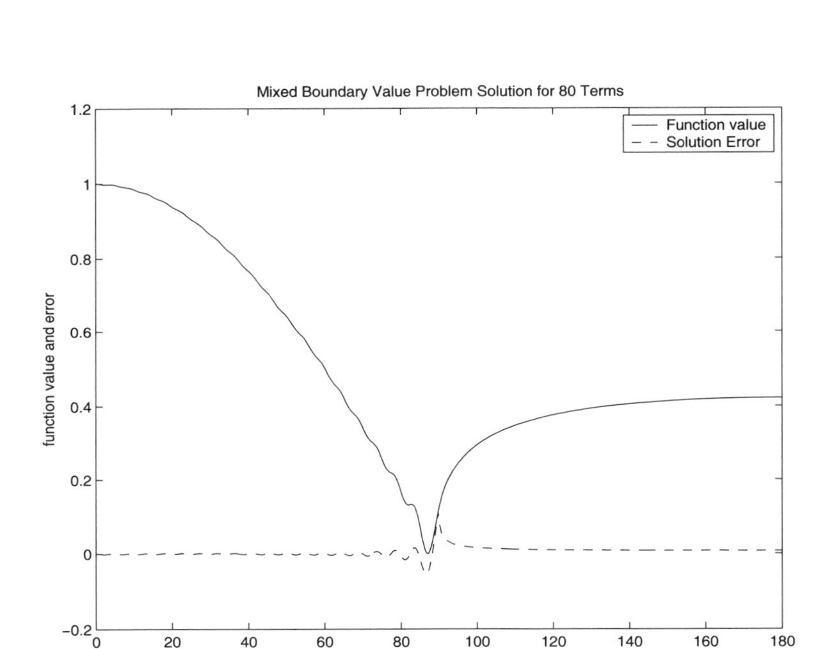

Figure 3.3: Mixed Boundary Value Problem Solution

MATLAB Example

Program mbvprun

```
1: function mbvprun(nser,nf,ng,neval)
2: % Example:  mbvprun(nser,nf,ng,neval)
3: % ~~~~~~~~~~~~~~~~~~
4: % Mixed boundary value problem for a function
5: % harmonic inside a circle.
6:
7: % User m functions required:
8: %   mbvp
9:
10: disp('Calculating');
11:
12: % Set data for series term and boundary
13: % condition points
14: if nargin==0
15:   nser=80; nf=100; ng=100; neval=500;
16: end
17:
18: % Compute the series coefficients
19: [cof,y]=mbvp('cos',pi/2,nser,nf,ng,neval);
20:
21: % Evaluate the exact solution for comparison
22: thp=linspace(0,pi,neval)';
23: y=cos(thp*(0:nser-1))*cof;
24: ye=cos(thp)+sin(thp/2).* ...
25:    sqrt(2*abs(cos(thp))).*(thp>=pi/2);
26:
27: % Plot results showing the accuracy of the
28: % least square solution
29: thp=thp*180/pi; plot(thp,y,'-',thp,y-ye,'--');
30: xlabel('polar angle');
31: ylabel('function value and error')
32: title(['Mixed Boundary Value Problem ', ...
33:       'Solution for ',int2str(nser),' Terms']);
34: legend('Function value','Solution Error');
35: figure(gcf); % print -deps mbvp
36:
37: %============================================
38:
39: function [cof,y]= ...
40:          mbvp(func,alp,nser,nf,ng,neval)
```

```
41: %
42: % [cof,y]=mbvp(func,alp,nser,nf,ng,neval)
43: % ~~~~~~~~~~~~~~~~~~~~~~~~~~~~~~~~~~~~~~~~~
44: % This function solves approximately a mixed
45: % boundary value problem for a function which
46: % is harmonic inside the unit disk, symmetric
47: % about the x axis, and has boundary conditions
48: % involving function values on one part of the
49: % boundary and zero gradient elsewhere.
50: %
51: % func       - function specifying the function
52: %              value between zero and alp
53: %              radians
54: % alp        - angle between zero and pi which
55: %              specifies the point where
56: %              boundary conditions change from
57: %              function value to zero gradient
58: % nser       - number of series terms used
59: % nf         - number of function values
60: %              specified from zero to alp
61: % ng         - number of points from alp to pi
62: %              where zero normal derivative is
63: %              specified
64: % neval      - number of boundary points where
65: %              the solution is evaluated
66: % cof        - coefficients in the series
67: %              solution
68: % y          - function values for the solution
69: %
70: %-----------------------------------------------
71:
72: % Create evenly spaced points to impose
73: % boundary conditions
74: th1=linspace(0,alp,nf);
75: th2=linspace(alp,pi,ng+1); th2(1)=[];
76:
77: % Form an overdetermined system based on the
78: % boundary conditions
79: yv=feval(func,th1);
80: cmat=cos([th1(:);th2(:)]*(0:nser-1));
81: [nr,nc]=size(cmat);
82: cmat(nf+1:nr,:)=...
83:   (ones(ng,1)*(0:nser-1)).*cmat(nf+1:nr,:);
84: cof=cmat\[yv(:);zeros(ng,1)];
85:
```

```
86: % Evaluate the solution on the boundary
87: thp=linspace(0,pi,neval)';
88: y=cos(thp*(0:nser-1))*cof;
```

3.4.3 Using Rational Functions to Conformally Map a Circular Disk onto a Square

Another problem illustrating the value of least squares approximation arises in connection with an example discussed earlier in Section 2.4 where a slowly convergent power series was used to map the interior of a circle onto the interior of a square [75]. It is sometimes possible for slowly convergent power series of the form

$$w = f(z) = \sum_{j=0}^{N} c_j z^j \ , \ |z| \leq 1$$

to be replaceable by a rational function

$$w = \frac{\sum_{j=0}^{n} a_j z^j}{1 + \sum_{j=1}^{m} b_j z^j}.$$

Of course, the polynomial is simply a special rational function form with $m = 0$ and $n = N$. This rational function implies

$$\sum_{j=0}^{n} a_j z^j - w \sum_{j=1}^{m} b_j z^j = w.$$

Coefficients a_j and b_j can be computed by forming least square equations based on boundary data. In some cases, the resulting equations are rank deficient and it is safer to solve a system of the form $UY = V$ as $Y = \textbf{pinv}(U) * V$ rather than using $Y = U \backslash V$. The former solution uses the pseudo inverse function **pinv** which automatically sets to zero any solution components that are undetermined.

Two functions **ratcof** and **raterp** were written to compute rational function coefficients and to evaluate the rational function for general matrix arguments. These functions are useful to examine the conformal mapping of the circular disk $|z| \leq 1$ onto the square defined by $|\,\textbf{real}(w)| \leq 1$, $|\,\textbf{imag}(w)| \leq 1$. A polynomial approximation of the mapping function has the form

$$w/z = \sum_{j=0}^{N} c_j (z^4)^j$$

where N must be quite large in order to avoid excessive corner rounding. If we evaluate w versus z on the boundary for large N (500 or more), and then develop

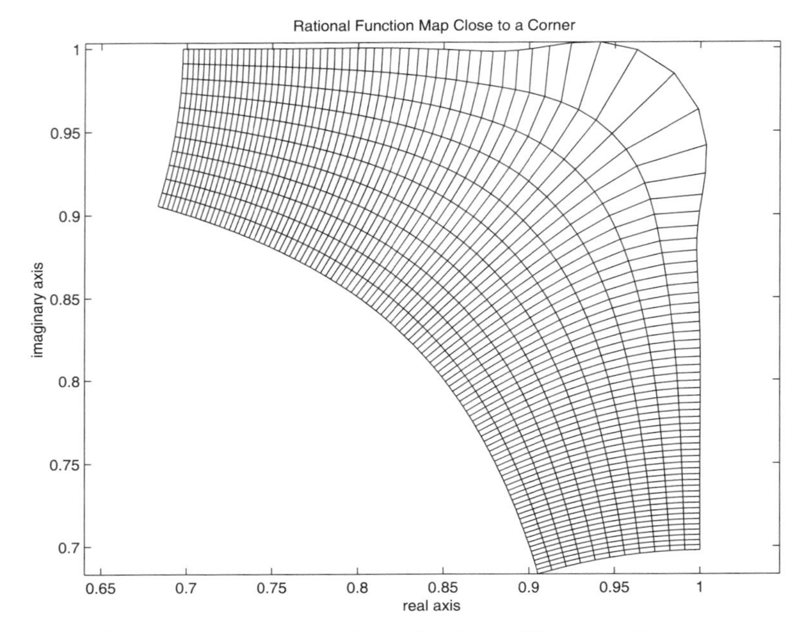

Figure 3.4: **Rational Function Map Close to a Corner**

a rational function fit with $n = m = 10$, a reasonably good representation of the square results without requiring a large number of series terms. The following program illustrates the use of functions **ratcof** and **raterp**. It also includes a function **sqmp** to generate coefficients in the Schwarz-Christoffel series.(See Chapter 11 for further discussion.) Figure 3.4 shows the geometry mapping produced near a corner.

MATLAB Example

Program makratsq

```
1: function [ctop,cbot]=makratsq
2: % Example:   [ctop,cbot]=makratsq
3: % ~~~~~~~~~~~~~~~~~~
4: % Create a rational function map of a unit disk
5: % onto a square.
6: %
7: % User m functions required:
8: %    sqmp, ratcof, raterp
9:
```

```
10: disp(' ');
11: disp('RATIONAL FUNCTION MAPPING OF A CIRCULAR');
12: disp('          DISK ONTO A SQUARE'); disp(' ');
13: disp('Calculating'); disp(' ');
14:
15: % Generate boundary points given by the
16: % Schwarz-Christoffel transformation
17: nsc=501; np=401; ntop=10; nbot=10;
18: z=exp(i*linspace(0,pi/4,np));
19: w=sqmp(nsc,1,1,1,0,45,np);
20: w=mean(real(w))+i*imag(w);
21: z=[z,conj(z)]; w=[w,conj(w)];
22:
23: % Compute the series coefficients for a
24: % rational function fit to the boundary data
25: [ctop,cbot]=ratcof(z.^4,w./z,ntop,nbot);
26: ctop=real(ctop); cbot=real(cbot);
27:
28: % The above calculations produce the following
29: % coefficients
30: % [top,bot]=
31: %           1.0787    1.4948
32: %           1.5045    0.1406
33: %           0.0353   -0.1594
34: %          -0.1458    0.1751
35: %           0.1910   -0.1513
36: %          -0.1797    0.0253
37: %           0.0489    0.2516
38: %           0.2595    0.1069
39: %           0.0945    0.0102
40: %           0.0068    0.0001
41:
42: % Generate a polar coordinate grid to describe
43: % the mapping near the corner of the square.
44: % Then evaluate the mapping function.
45: r1=.95; r2=1; nr=12;
46: t1=.9*pi/4; t2=1.1*pi/4; nt=101;
47: [r,th]=meshgrid(linspace(r1,r2,nr), ...
48:          linspace(t1,t2,nt));
49: z=r.*exp(i*th); w=z.*raterp(ctop,cbot,z.^4);
50:
51: % Plot the mapped geometry
52: close; u=real(w); v=imag(w);
53: plot(u,v,'k',u',v','k'), axis equal
54: title('Rational Function Map Close to a Corner');
```

```
55: xlabel('real axis'); ylabel('imaginary axis');
56: figure(gcf); % print -deps ratsqmap
57:
58: %==============================================
59:
60: function [w,b]=sqmp(m,r1,r2,nr,t1,t2,nt)
61: %
62: % [w,b]=sqmp(m,r1,r2,nr,t1,t2,nt)
63: % ~~~~~~~~~~~~~~~~~~~~~~~~~~~~~~~~
64: % This function evaluates the conformal
65: % mapping produced by the Schwarz-Christoffel
66: % transformation w(z) mapping abs(z)<=1 inside
67: % a square having a side length of two.  The
68: % transformation is approximated in series form
69: % which converges very slowly near the corners.
70: % This function is the same as squarmap of
71: % chapter 2 with no plotting.
72: %
73: % m          - number of series terms used
74: % r1,r2,nr - abs(z) varies from r1 to r2 in
75: %              nr steps
76: % t1,t2,nt - arg(z) varies from t1 to t2 in
77: %              nt steps (t1 and t2 are
78: %              measured in degrees)
79: % w          - points approximating the square
80: % b          - coefficients in the truncated
81: %              series expansion which has
82: %              the form
83: %
84: %              w(z)=sum({j=1:m},b(j)*z*(4*j-3))
85: %
86: % User m functions called:  none.
87: %----------------------------------------------
88:
89: % Generate polar coordinate grid points for the
90: % map. Function linspace generates vectors with
91: % equally spaced components.
92: r=linspace(r1,r2,nr)';
93: t=pi/180*linspace(t1,t2,nt);
94: z=(r*ones(1,nt)).*(ones(nr,1)*exp(i*t));
95:
96: % Compute the series coefficients and evaluate
97: % the series
98: k=1:m-1;
99: b=cumprod([1,-(k-.75).*(k-.5)./(k.*(k+.25))]);
```

```
100: b=b/sum(b); w=z.*polyval(b(m:-1:1),z.^4);
101:
102: %==============================================
103:
104: function [a,b]=ratcof(xdata,ydata,ntop,nbot)
105: %
106: % [a,b]=ratcof(xdata,ydata,ntop,nbot)
107: % ~~~~~~~~~~~~~~~~~~~~~~~~~~~~~~~~~~~~
108: %
109: % Determine a and b to approximate ydata as
110: % a rational function of the variable xdata.
111: % The function has the form:
112: %
113: %    y(x) = sum(1=>ntop) ( a(j)*x^(j-1) ) /
114: %         ( 1 + sum(1=>nbot) ( b(j)*x^(j)) )
115: %
116: % xdata,ydata - input data vectors (real or
117: %               complex)
118: % ntop,nbot   - number of series terms used in
119: %               the numerator and the
120: %               denominator.
121: %
122: % User m functions called: none
123: %----------------------------------------------
124:
125: ydata=ydata(:); xdata=xdata(:);
126: m=length(ydata);
127: if nargin==3, nbot=ntop; end;
128: x=ones(m,ntop+nbot); x(:,ntop+1)=-ydata.*xdata;
129: for i=2:ntop, x(:,i)=xdata.*x(:,i-1); end
130: for i=2:nbot
131:   x(:,i+ntop)=xdata.*x(:,i+ntop-1);
132: end
133: ab=pinv(x)*ydata; %ab=x\ydata;
134: a=ab(1:ntop); b=ab(ntop+1:ntop+nbot);
135:
136: %==============================================
137:
138: function y=raterp(a,b,x)
139: %
140: % y=raterp(a,b,x)
141: % ~~~~~~~~~~~~~~~
142: % This function interpolates using coefficients
143: % from function ratcof.
144: %
```

```
145: % a,b - polynomial coefficients from function
146: %        ratcof
147: % x    - argument at which function is evaluated
148: % y    - computed rational function values
149: %
150: % User m functions called:  none.
151: %----------------------------------------------
152:
153: a=flipud(a(:)); b=flipud(b(:));
154: y=polyval(a,x)./(1+x.*polyval(b,x));
```

3.5 Eigenvalue Problems

3.5.1 Statement of the Problem

Another important linear algebra problem involves the computation of nonzero vectors X and numbers λ such that

$$AX = \lambda X$$

where A is a square matrix of order n having elements which may be real or complex. The number λ, which can also be real or complex, is called the eigenvalue corresponding to the eigenvector X. The eigenvalue equation implies

$$[I\lambda - A]X = 0$$

so that λ values must be selected to make $I\lambda - A$ singular. The polynomial

$$f(\lambda) = \det(I\lambda - A) = \lambda^n + c_1\lambda^{n-1} + \ldots + c_n$$

is called the characteristic equation and its roots are the eigenvalues. It can be factored into

$$f(\lambda) = (\lambda - \lambda_1)(\lambda - \lambda_2)\cdots(\lambda - \lambda_n).$$

The eigenvalues are generally complex numbers and some of the roots may be repeated. In the usual situation, distinct roots $\lambda_1, \cdots, \lambda_n$ yield n linearly independent eigenvectors obtained by solving

$$(A - \lambda_j I)X_j = 0 \ , \ 1 \leq j \leq n.$$

The case involving repeated eigenvalues is more complicated. Suppose a particular eigenvalue such as λ_1 has multiplicity k. Then the general solution of

$$(A - \lambda_1 I)X = 0$$

will yield as few as one, or as many as k, linearly independent vectors. If fewer than k independent eigenvectors are found for any root of multiplicity k, then matrix A is called defective. Occurrence of a defective matrix is not typical. It usually implies special behavior of the related physical system. The combined set of eigenvectors can be written as

$$A[X_1, \cdots, X_n] = [X_1\lambda_1, \cdots, X_n\lambda_n] = [X_1, \cdots, X_n] \ \textbf{diag}(\lambda_1, \cdots, \lambda_n)$$

or

$$AU = U\Lambda$$

where U has the eigenvectors as columns and Λ is a diagonal matrix with eigenvalues on the diagonal. When the eigenvectors are independent, matrix U, known as the modal matrix, is nonsingular. This allows A to be expressed as

$$A = U\Lambda U^{-1}$$

which is convenient for various computational purposes. With repeated eigenvalues, the modal matrix is sometimes singular and the last form of decomposition fails. However, the eigenvectors are always independent when the eigenvalues are distinct. For the important special case of a symmetric matrix, a linearly independent set of eigenvectors always exists, even when some eigenvalues are repeated.

A matrix A is symmetric if $A = A'$ where A' is obtained by interchanging columns and rows, and conjugating all elements. Symmetric matrices always have real eigenvalues and a linearly independent set of eigenvectors which can be orthonormalized. The eigenvectors $X_\jmath$ and X_k for any two unequal eigenvalues automatically satisfy an orthogonality condition

$$X_\jmath' X_k = 0 \ , \ \jmath \neq k.$$

Eigenvectors for the same repeated eigenvalue are not automatically orthogonal. Nevertheless, they can be replaced by an equivalent orthogonal set by applying a process called Gram-Schmidt orthogonalization [47]. In cases we care about here, the symmetric matrix A always has real elements. Therefore the eigenvalues are real with eigenvectors satisfying $X_\imath' X_\jmath = \delta_{\imath\jmath}$, where $\delta_{\imath\jmath}$ is the Kronecker delta symbol. The orthogonality condition is equivalent to the statement that $U'U = I$, so a real symmetric matrix can be expressed as

$$A = U\Lambda U'$$

It is important in MATLAB that the symmetry condition $A' = A$ be satisfied perfectly. This implies a zero value for max(max(abs(A-A'))). Sometimes, results that would be symmetric if roundoff error did not occur may produce unsymmetric results contrary to expectation. For example, $A = BCB'$ should be symmetric if C is symmetric. Replacing A by $(A + A')/2$ finally will assure perfect symmetry. The MATLAB function **eig** computes eigenvalues and eigenvectors. When a matrix is symmetric, **eig** generates real eigenvalues and orthonormalized eigenvectors.

An important property of symmetric matrices and the related orthonormal eigenvector set occurs in connection with quadratic forms expressed as

$$F(Y) = Y'AY$$

where Y is an arbitrary real vector and A is real symmetric. The function $F(Y)$ is a one-by-one matrix; hence, it is a scalar function. The algebraic sign of the form for arbitrary nonzero choices of Y is important in physical applications. Let us use the eigenvector decomposition of A to write

$$F = Y'U\Lambda U'Y = (U'Y)'\Lambda(U'Y).$$

Taking $X = U'Y$ and $Y = UX$ gives

$$F = X'\Lambda X = \lambda_1 x_1^2 + \lambda_2 x_2^2 + \lambda_3 x_3^2 + \ldots + \lambda_n x_n^2.$$

This diagonal form makes the algebraic character of F evident. If all $\lambda_\imath$ are positive, then F is evidently positive whenever X has at least one nonzero component. Then the quadratic form is called positive definite. If the eigenvalues are all positive or zero, the form is called positive semidefinite since the form cannot assume a negative value but can equal zero without having $X = 0$. When both negative and positive eigenvalues occur, the form can change sign and is termed indefinite. When the eigenvalues are all negative, the form is classified as negative definite. Perhaps the most important of these properties is that a necessary and sufficient condition for the form to be positive definite is that all eigenvalues of A be positive.

An important generalization of the standard eigenvalue problem has the form

$$AX = \lambda BX$$

for arbitrary A and nonsingular B. If B is well conditioned, then it is computationally attractive to simply solve

$$B^{-1}AX = \lambda X.$$

In general, it is safer, but much more time consuming, to call **eig** as

```
[EIGVECS,EIGVALS]=eig(A,B)
```

This returns the eigenvectors as columns of EIGVECS and also gives a diagonal matrix EIGVALS containing the eigenvalues.

3.5.2 Application to Solution of Matrix Differential Equations

One of the most familiar applications of eigenvalues concerns the solution of the linear, constant-coefficient matrix differential equation

$$BY'(t) = AY(t)\ ,\ Y(0) = Y_0.$$

Component solutions can be written as

$$Y(t) = Xe^{\lambda t}\ ,\ Y'(t) = \lambda Xe^{\lambda t}$$

where X and λ are constant. Substitution into the differential equation gives

$$(A - \lambda B)Xe^{\lambda t} = 0.$$

Since $e^{\lambda t}$ cannot vanish we need

$$AX = \lambda BX.$$

After the eigenvalues and eigenvectors have been computed, a general solution is constructed as a linear combination of component solutions

$$Y = \sum_{j=1}^{n} X_j e^{\lambda_j t} c_j.$$

The constants c_j are obtained by imposing the initial condition

$$Y(0) = [X_1, X_2, \ldots, X_n]c.$$

Assuming that the eigenvectors are linearly independent we get

$$c = [X_1, \ldots, X_n]^{-1} Y_0.$$

3.5.3 The Structural Dynamics Equation

Eigenvalues are also useful to solve the important second order matrix differential equation for which a particular solution was constructed earlier using real arithmetic. We will now use complex arithmetic and the versatile matrix notation provided in MATLAB. Structural mechanics applications often lead to the second order matrix differential equation

$$M\ddot{X}(t) + C\dot{X}(t) + KX(t) = F_1 \cos(\omega t) + F_2 \sin(\omega t)$$

where M, C, K are constant matrices of order n, and F_1, F_2 are constant vectors of length n, and ω is the forcing function frequency. Initial conditions of the form

$$X(0) = X_0, \quad \dot{X}(0) = V_0$$

also apply. Solving this initial value problem involves combining a particular solution and a homogeneous solution. The solution we present below applies subject to the restriction that 1) the eigenvalues of the homogeneous equation should be nonzero and 2) if matrix C is zero, then $i\omega$ must not coincide with an eigenvalue of the homogeneous differential equation. The particular solution is

$$X_p(t) = \textbf{real}(a\, e^{i\omega t}), \quad a = [K - M\,\omega^2 + iC\omega] \setminus [F_1 - iF_2].$$

where we must assume that the implied matrix inversion exists. The particular solution satisfies initial conditions

$$X_p(0) = \textbf{real}(a), \quad \dot{X}_p(0) = \textbf{real}(i\, a\, \omega).$$

The particular solution plus the homogeneous solution, $X_h(t)$, must satisfy the general initial conditions. Let us introduce

$$Z(t) = [X_h(t)\,;\ \dot{X}_h(t)\,]$$

which obeys the homogeneous first order equation

$$\dot{Z}(t) = AZ(t)\ ,\ A = [\mathbf{eye}(n,n),\ \mathbf{zeros}(n,n)\,;\ -M \setminus [K,\ C]\,]$$

and can be determined using the eigenvectors and eigenvalues of A. Denoting the matrix of eigenvectors as U and the column of eigenvalues as Λ, we find that

$$Z(t) = U\,\mathbf{diag}(D)\ \mathbf{exp}(i\,\Lambda\,t)$$

where

$$D = U \setminus [X_0 - X_p(0)\,;\ V_0 - \dot{X}_p(0)]$$

to satisfy the initial conditions. With t taken as a row of time values, the homogeneous solution is obtained as the first n rows of Z, and the total solution is just

$$X(t) = X_p(t) + X_h(t).$$

A program was written to solve the structural dynamics equation. Error checks are made for the exceptional cases mentioned above. If the system is undamped ($C = 0$) and $i\omega$ matches an eigenvalue of A, then program execution terminates. Occurrence of zero or repeated eigenvalues is also avoided. The program consists of a driver named **strdyneq** which reads data from a function provided by the user. An example function named **threemass** is included as a model for data preparation. Function **fhrcmk** constructs the general solution of the equation. Results of the computation can be plotted one component at a time. In addition to plotting, the program outputs the eigenvalues, a matrix of solution components, and vectors showing the lower and upper limits of motion for each degree for freedom in the system. Function **strdyneq** calls **fhrmck** at lines 25 and 34. The name of a function defining the input data is requested. Users can employ function **threemass** to test the program. **Threemass** models a configuration of three identical masses sliding on a smooth horizontal plane and connected by four identical springs and viscous dampers. The outer two masses are connected to walls and are subjected to forces having equal magnitude but opposite direction. The middle mass has no driving force. The system is initially at rest with zero deflection when forcing functions are applied which nearly resonate with the fourth eigenvalue of the damped homogeneous system. This example was devised to illustrate how the system response grows rapidly when the forcing function is nearly resonant. Function **fhrmck** does most of the computation work which occurs at lines 108-109, 132-134, and 139-140. This example illustrates nicely the power of the intrinsic matrix operators provided in MATLAB. A final caveat about the solution method using eigenvalues is that it is somewhat limited by special cases like repeated eigenvalues or a forcing function resonant with a natural frequency. Numerical integration solvers like **ode45** are not vulnerable to such difficulties.

MATLAB Example

Output Using Function Threemass

```
strdyneq;

SOLUTION OF THE DIFFERENTIAL EQUATION
M*Y''+C*Y'+K*Y=F1*COS(W*T)+F2*SIN(W*T)

Give the name of a function to create data values
(Try threemass as an example)
>? threemass

Input coordinate number, tmin and tmax
(only press return to stop execution)>? 1,0,50

The value of i*w is at distance 0.050001
from the eigenvalue -0.05+1.4133i


Input coordinate number, tmin and tmax
(only press return to stop execution)>? 2,0,50


Input coordinate number, tmin and tmax
(only press return to stop execution)>?

The system eigenvalues are:

lam =

  -0.0146 - 0.7652i
  -0.0146 + 0.7652i
  -0.0500 - 1.4133i
  -0.0500 + 1.4133i
  -0.0854 - 1.8458i
  -0.0854 + 1.8458i

Range of solution values for final times is:

maxy =

    6.4255    0.0000    6.4935


miny =
```

```
   -6.4935    -0.0000    -6.4255

All done
```

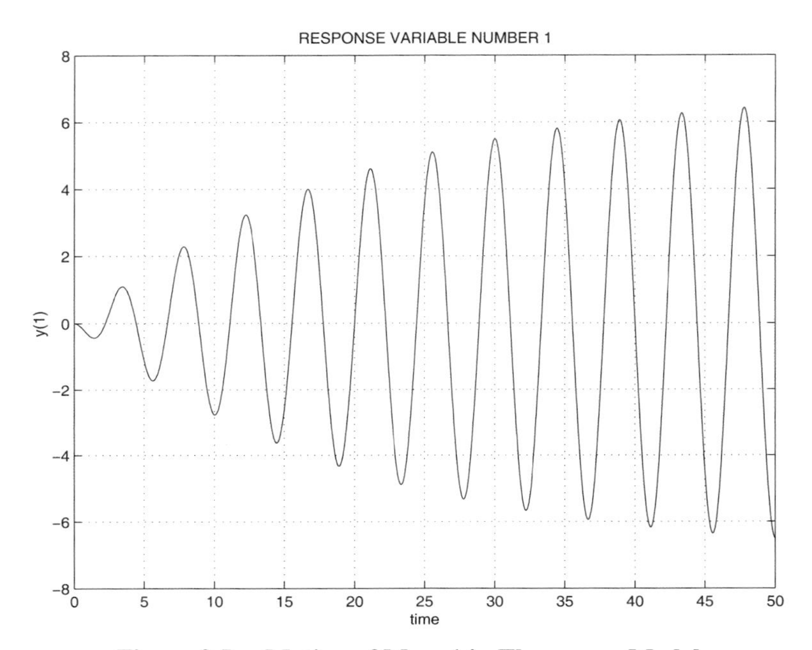

Figure 3.5: Motion of Mass 1 in Threemass Model

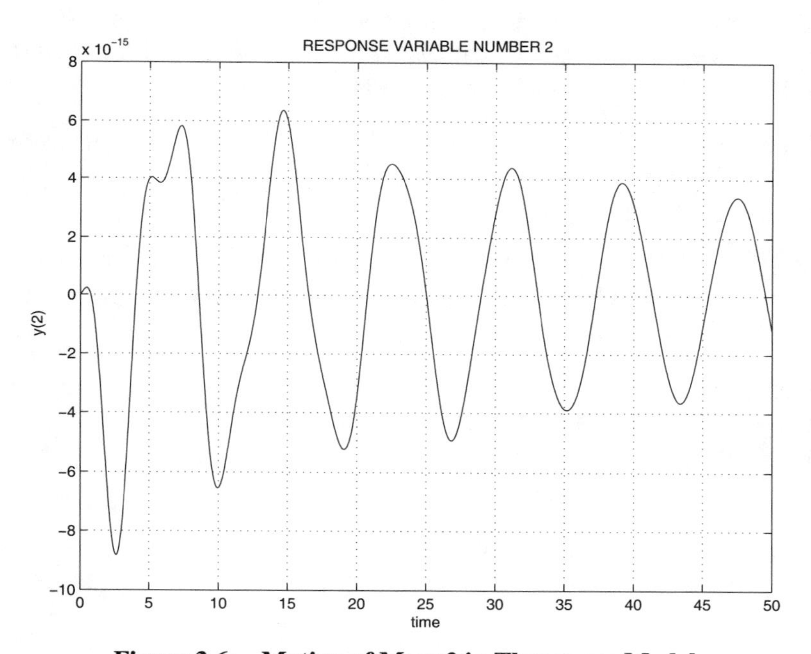

Figure 3.6: Motion of Mass 2 in Threemass Model

Motion of First and Second Mass

MATLAB Code

```
1: function [t,y,lam]=strdyneq
2: %
3: % [t,y,lam]=strdyneq
4: % ~~~~~~~~~~~~~~~~~~
5: % This program integrates the structural dynamics
6: % equation characterized by a general second order
7: % matrix differential equation having a harmonic
8: % forcing function. Input involves mass, stiffness,
9: % and damping matrices as well as force magnitudes,
10: % a forcing frequency, and initial conditions. Data
11: % parameters for the program are created in a user
12: % supplied function provided by the user. (For an
13: % example, see function threemass shown below.)
14:
15: titl=['\nSOLUTION OF THE DIFFERENTIAL EQUATION\n',...
16: 'M*Y''''+C*Y''+K*Y=F1*COS(W*T)+F2*SIN(W*T)\n\n'];
17: fprintf(titl);
18: disp(...
19: 'Give the name of a function to create data values')
20: disp('(Try threemass as an example)')
21: name=input('>? ','s');
22: eval(['[m,c,k,f1,f2,w,nt,y0,v0]=',name,';']); jj=1;
23: while 1
24:   fprintf('\nInput coordinate number, tmin and tmax')
25:   fprintf('\n(only press return to stop execution)')
26:   [j,t1,t2]=inputv('>? ');
27:   if isnan(j), break; end; J=int2str(j);
28:   [t,y,lam]=fhrmck(m,c,k,f1,f2,w,[t1,t2],nt,y0,v0);
29:   if isnan(t), return, end
30:   [dif,h]=min(abs(lam-i*w)); lj=num2str(lam(h));
31:   if jj==1, jj=jj+1; disp(' ')
32:     disp(['The value of i*w is at distance ',...
33:         num2str(dif)])
34:     disp(['from the eigenvalue ',lj])
35:   end
36:   plot(t,y(:,j),'k-'), xlabel('time')
37:   ylabel(['y(',J,')'])
38:   title(['RESPONSE VARIABLE NUMBER ',J])
39:   grid on, shg, dumy=input(' ','s');
40: end
```

```
41: fprintf('\nThe system eigenvalues are:\n')
42: display(lam)
43: fprintf(...
44: 'Range of solution values for final times is:\n')
45: maxy=max(y); miny=min(y); display(maxy)
46: display(miny), fprintf('All done\n')
47:
48: %==================================================
49:
50: function [m,c,k,f1,f2,w,nt,y0,v0]=threemass
51: %
52: % [m,c,k,f1,f2,w,nt,y0,v0]=threemass
53: % ~~~~~~~~~~~~~~~~~~~~~~~~~~~~~~~~~~~~~~
54: % This function creates data for a three mass
55: % system. The name of the function should be
56: % changed to specify different problems. However,
57: % the output variable list should remain unchanged
58: % for compatibility with the data input program.
59:
60: m=eye(3,3); k=[2,-1,0;-1,2,-1;0,-1,2]; c=.05*k;
61:
62: % Data to excite the highest mode
63: f1=[-1;0;1]; f2=[0;0;0]; w=1.413; nt=1000;
64:
65: % Data to excite the lowest mode
66: % f1=[1;1;1]; f2=[0;0;0]; w=.7652; nt=1000;
67:
68: % Homogeneous initial conditions
69: y0=[-.5;0;.5]; v0=zeros(3,1); y0=0*y0;
70:
71: %==================================================
72:
73: function [t,y,lam]=fhrmck(m,c,k,f1,f2,w,tlim,nt,y0,v0)
74: %
75: % [t,y,lam]=fhrmck(m,c,k,f1,f2,w,tlim,nt,y0,v0)
76: % ~~~~~~~~~~~~~~~~~~~~~~~~~~~~~~~~~~~~~~~~~~~~~~~~
77: % This function uses eigenfunction analysis to solve
78: % the matrix differential equation
79: %   m*y''(t)+c*y'(t)+k*y(t)=f1*cos(w*t)+f2*sin(w*t)
80: % with initial conditions of y(0)=y0, y'(0)=v0
81: % The solution is general unless 1) a zero or repeated
82: % eigenvalue occurs or 2) the system is undamped and
83: % the forcing function matches a natural frequency.
84: % If either error condition occurs, program execution
85: % terminates with t and y set to nan.
```

```
86: %
87: % m,c,k    - mass, damping, and stiffness matrices
88: % f1,f2    - amplitude vectors for the sine and cosine
89: %            forcing function components
90: % w        - frequency of the forcing function
91: % tlim     - a vector containing the minimum and
92: %            maximum time limits for evaluation of
93: %            the solution
94: % nt       - the number of times at which the solution
95: %            is evaluated within the chosen limits
96: %            for which y(t) is computed
97: % y0,v0    - initial position and velocity vectors
98: %
99: % t        - vector of time values for the solution
100: % y        - matrix of solution values where y(i,j)
101: %            is the value of component j at time t(i)
102: % lam      - the complex natural frequencies arranged
103: %            in order of increasing absolute value
104:
105: if nargin==0 % Generate default data using 2 masses
106:   m=eye(2,2); k=[2,-1;-1,1]; c=.3*k;
107:   f1=[0;1]; f2=[0;0]; w=0.6; tlim=[0,100]; nt=400;
108: end
109: n=size(m,1); t=linspace(tlim(1),tlim(2),nt);
110: if nargin<10, y0=zeros(n,1); v0=y0; end
111:
112: % Determine eigenvalues and eigenvectors for
113: % the homogeneous solution
114: A=[zeros(n,n), eye(n,n); -m\[k, c]];
115: [U,lam]=eig(A); [lam,j]=sort(diag(lam)); U=U(:,j);
116:
117: % Check for zero or repeated eigenvalues and
118: % for undamped resonance
119: wmin=abs(lam(1)); tol=wmin/1e6;
120: [dif,J]=min(abs(lam-i*w)); lj=num2str(lam(J));
121: if wmin==0, disp(' ')
122:   disp('The homogeneous equation has a zero')
123:   disp('eigenvalue which is not allowed.')
124:   disp('Execution is terminated'), disp(' ')
125:   t=nan; y=nan; return
126: elseif any(abs(diff(lam))<tol)
127:   disp('A repeated eigenvalue occurred.')
128:   disp('Execution is terminated'),disp(' ')
129:   t=nan; y=nan; return
130: elseif dif<tol & sum(abs(c(:)))==0
```

```
131:   disp('The system is undamped and the forcing')
132:   disp(['function resonates with ',...
133:         'eigenvalue ',lj])
134:   disp('Execution is terminated.')
135:   disp(' '), t=nan; y=nan; return
136: else
137:   % Determine the particular solution
138:   a=(-w^2*m+k+i*w*c)\(f1-i*f2);
139:   yp=real(a*exp(i*w*t));
140:   yp0=real(a); vp0=real(i*w*a);
141: end
142:
143: % Scale the homogeneous solution to satisfy the
144: % initial conditions
145: U=U*diag(U\[y0-yp0; v0-vp0]);
146: yh=real(U(1:n,:)*exp(lam*t));
147:
148: % Combine results to obtain the total solution
149: t=t(:); y=[yp+yh]';
150:
151: % Show data graphically only for default case
152: if nargin==0
153:   waterfall(t,(1:n),y'), xlabel('time axis')
154:   ylabel('mass index'), zlabel('Displacements')
155:   title(['DISPLACEMENT HISTORY FOR A ',...
156:          int2str(n),'-MASS SYSTEM'])
157:   colormap([1,0,0]), shg
158: end
```

3.6 Computing Natural Frequencies for a Rectangular Membrane

One of the most useful applications of eigenvalue problems occurs in natural frequency calculations for linear systems. Let us examine finite difference approximation for the natural frequencies of a rectangular membrane and how well the approximate results compare with exact values. Consider a tightly stretched elastic membrane occupying a region R in the (x, y) bounded by a curve L on which the transverse deflection is zero. The differential equation and boundary conditions governing the transverse motion $U(x, y, t)$ are

$$T(U_{xx} + U_{yy}) = \rho U_{tt} \ , \ (x, y)\epsilon R,$$
$$U(x, y, t) = 0 \ , \ (x, y)\epsilon L,$$

where T and ρ denote membrane tension and mass density. The natural vibration modes are motion states where all points of the system simultaneously move with the same frequency, which says $U(x, y, t) = u(x, y)\sin(\Omega t)$. It follows that $u(x, y)$ satisfies

$$u_{xx} + u_{yy} = -\omega^2 u \ , \ (x, y) \,\epsilon R,$$
$$u(x, y) = 0, \ (x, y)\epsilon L,$$

where $\omega = \sqrt{\frac{\rho}{T}}\Omega$. In the simple case of a rectangular membrane lying in the region such that $0 \le x \le a$ and $0 \le y \le b$, the natural frequencies and mode shapes turn out to be

$$\omega_{nm} = \sqrt{\left(\frac{n\pi}{a}\right)^2 + \left(\frac{m\pi}{b}\right)^2} \ , \ u_{nm} = \sin\left(\frac{n\pi x}{a}\right)\sin\left(\frac{m\pi y}{b}\right)$$

where n and m are positive integers. It is interesting to see how closely these values can be reproduced when the partial differential equation is replaced by a second order finite difference approximation defined on a rectangular grid. We introduce grid points expressed as

$$x(i) = (i-1)\Delta_x \ , \ i = 1, \ldots, N,$$
$$y(j) = (j-1)\Delta_y \ , \ j = 1, \ldots, M,$$

where

$$\Delta_x = a\,/(N-1), \ \Delta_y = b\,/(M-1),$$

and we call $u(i, j)$ the value of u at $x(i), y(j)$. Then the Helmholtz equation is replaced by an algebraic eigenvalue problem of the form

$$\Delta_y^2[u(i-1,\, j) - 2u(i,\, j) + u(i+1,\, j)] + \Delta_x^2[u(i,\, j-1)$$
$$- 2u(i,\, j) + u(i,\, j+1)] = \lambda u(i, j)$$

where

$$\lambda = (\Delta_x \Delta_y \omega)^2$$

and associated homogeneous boundary conditions

$$u(1, j) = u(N, j) = u(i, 1) = u(i, M) = 0.$$

This combination of equations can be rearranged into familiar matrix form as

$$A\,u = \lambda\,u, \ B\,u = 0.$$

The MATLAB function **null** can be used to solve the boundary condition equations. We write as $u = Qz$ where $Q = \textbf{null}(B)$ has orthonormal columns. Substituting into the eigenvalue equation and multiplying both sides by Q' then yields a standard eigenvalue problem of the form $Cz = \lambda z$ where $C = Q'AQ$. Denoting the eigenvector matrix of C by V, the eigenvector matrix of the original problem is obtained as $u = QV$, and the desired eigenvalues are simply those of matrix C.

A short function named **recmemfr** was written to form and solve the algebraic equations just discussed. Although the ideas are simple, indexing the double indexed quantities pertaining to the finite difference grid is slightly tedious. Intrinsic functions **ind2sub** and **sub2ind** are helpful to perform the indexing. Lines 32-34 of **rememfr** compute a subset of the lowest frequency values and sort these in ascending order. Lines 37-45 form the homogeneous boundary conditions, and lines 51-56 construct the discretized Helmholtz equation at interior node points. The main computation work is done in lines 59-61 where **null** and **eig** are used. Finally, the results are sorted, the modal arrays are reshaped, and results are plotted to compare the approximate and exact frequencies. In the graph shown below for the case where (a,b)=(2,1), the frequencies obtained using the finite differences are seen to be consistently low. Furthermore, the 50'th frequency is off by about 14 percent, even though 200 grid points were used. Applications leading to eigenvalue problems occur frequently. The ideas touched on in this simple example will be encountered again in Chapters 9 and 10. Readers may find it interesting to modify this example using a higher order difference approximation to see how much the frequency estimates improve.

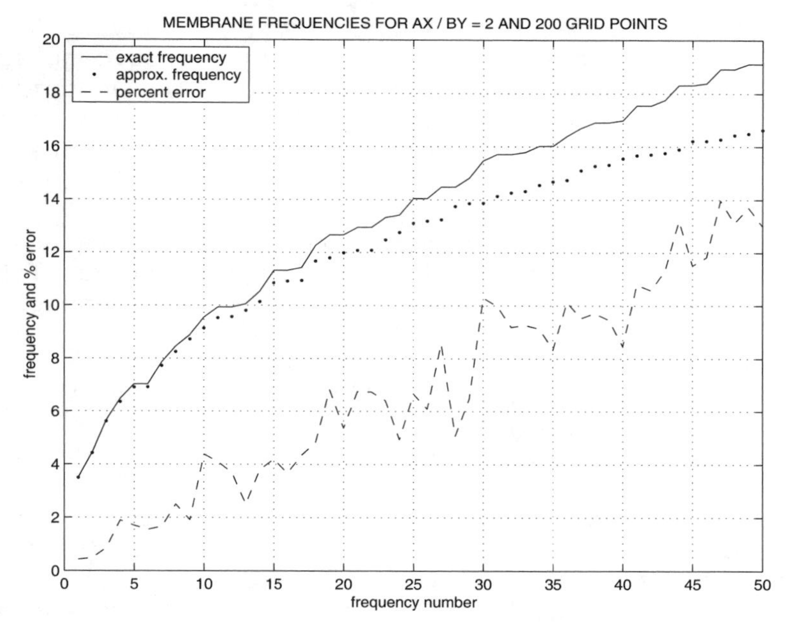

Figure 3.7: Approximate and Exact Frequencies for a Rectangular Membrane

Function recmemfr

```
1: function [w,wex,modes,x,y,nx,ny,ax,by]=recmemfr(...
2:                                        ax,by,nx,ny,noplt)
3: %
4: % [w,wex,modes,x,y,nx,ny,ax,by]=recmemfr(a,b,nx,ny,noplt)
5: % ~~~~~~~~~~~~~~~~~~~~~~~~~~~~~~~~~~~~~~~~~~~~~~~~~~~~~~~~~~
6: % This function employs finite difference methods to
7: % estimate the natural frequencies and mode shapes of
8: % a rectangular membrane having fixed edges.
9: % ax, by   - membrane side lengths along the x and y axes
10: % nx,ny    - number of finite difference points taken in
11: %            the x and y directions including the edges
12: % w        - vector of (nx-2)*(ny-2) frequencies obtained
13: %            by finite difference approximation of the
14: %            wave equation. These are arranged in
15: %            increasing order
16: % wex      - vector of exact frequencies
17: % modes    - three dimensional array containing the mode
18: %            shapes for various frequencies. The array
19: %            size is [nx,ny,(nx-2)*(nx-2)] denoting
20: %            the x direction, y direction, and the
21: %            freqency numbers matching components of the
22: %            w vector. The i'th mode shape is obtained
23: %            as reshape(vecs(:,i),n,m)
24: % x,y      - vectors defining the finite difference grid
25: % noplt    - optional parameter included if no plot of
26: %            the approximate and exact frequencies is to
27: %            be made
28:
29: if nargin==0; ax=2; nx=20; by=1; ny=10; end
30: dx=ax/(nx-1); dy=by/(ny-1);
31: na=(1:nx-1)'/ax; nb=(1:ny-1)/by;
32:
33: % Compute exact frequencies for comparison
34: wex=pi*sqrt(repmat(na.^2,1,ny-1)+repmat(nb.^2,nx-1,1));
35: wex=sort(wex(:)'); x=linspace(0,ax,nx);
36: y=linspace(0,by,ny); neig=(nx-2)*(ny-2); nvar=nx*ny;
37:
38: % Form equations to fix membrane edges
39: k=0; s=[nx,ny]; c=zeros(2*(nx+ny),nvar);
40: for j=1:nx
41:   m=sub2ind(s,[j,j],[1,ny]); k=k+1;
```

```
42:     c(k,m(1))=1; k=k+1; c(k,m(2))=1;
43: end
44: for j=1:ny
45:     m=sub2ind(s,[1,nx],[j,j]); k=k+1;
46:     c(k,m(1))=1; k=k+1; c(k,m(2))=1;
47: end
48:
49: % Form frequency equations at interior points
50: k=0; a=zeros(neig,nvar); b=a;
51: phi=(dx/dy)^2; psi=2*(1+phi);
52: for i=2:nx-1
53:   for j=2:ny-1
54:     m=sub2ind(s,[i-1,i,i+1,i,i],[j,j,j,j-1,j+1]);
55:     k=k+1; a(k,m(1))=-1; a(k,m(2))=psi; a(k,m(3))=-1;
56:     a(k,m(4))=-phi; a(k,m(5))=-phi; b(k,m(2))=1;
57:   end
58: end
59:
60: % Compute frequencies and mode shapes
61: q=null(c); A=a*q; B=b*q; [modes,lam]=eig(B\A);
62: [lam,k]=sort(diag(lam)); w=sqrt(lam)'/dx;
63: modes=q*modes(:,k); modes=reshape(modes(:),nx,ny,neig);
64:
65: % Plot first fifty approximate and exact frequencies
66: if nargin>4, return, end
67: m=1:min([50,length(w),length(wex)]);
68: pcter=100*(wex(m)-w(m))./wex(m);
69:
70: clf; plot(m,wex(m),'k-',m,w(m),'k.',m,pcter,'k--')
71: xlabel('frequency number');
72: ylabel('frequency and % error')
73: legend('exact frequency','approx. frequency',...
74:        'percent error',2)
75: s=['MEMBRANE FREQUENCIES FOR AX / BY = ',...
76:   num2str(ax/by,5),' AND ',num2str(nx*ny),...
77:   ' GRID POINTS'];
78: title(s), grid on, shg
79: % print -deps recmemfr
```

3.7 Column Space, Null Space, Orthonormal Bases, and SVD

One remaining advanced topic discussed in this chapter is the factorization known as singular value decomposition, or SVD. We will briefly explain the structure of

SVD and some of its applications. It is known that any real matrix having n rows, m columns, and rank r can be decomposed into the form

$$A = USV'$$

where

- U is an orthogonal n by n matrix such that $U'U = I$
- V is an orthogonal m by m matrix such that $V'V = I$
- S is an n by m diagonal matrix of the form

$$S = \begin{bmatrix} \sigma_1 & 0 & 0 & 0 & 0 & 0 \\ 0 & \sigma_2 & 0 & 0 & 0 & 0 \\ 0 & 0 & \ddots & 0 & 0 & 0 \\ 0 & 0 & 0 & \sigma_r & 0 & 0 \\ 0 & 0 & 0 & 0 & 0 & 0 \\ 0 & 0 & 0 & 0 & 0 & 0 \end{bmatrix}$$

  where $\sigma_1, \ldots, \sigma_r$ are positive numbers on the main diagonal with $\sigma_\imath \geq \sigma_{\imath+1}$. Constants $\sigma_\jmath$ are called the singular values with the number of nonzero values being equal to the rank r.

To understand the structure of this decomposition, let us study the case where $n \geq m$. Direct multiplication gives

$$A'AV = V\ \mathbf{diag}([\sigma_1^2, \ldots, \sigma_r^2, \mathbf{zeros}(1, m - r)]),$$

and

$$AA'U = U\,\mathbf{diag}([\sigma_1^2, \ldots, \sigma_r^2, \mathbf{zeros}(1, n - r)]).$$

Consequently, the singular values are square roots of the eigenvalues of the symmetric matrix $A'A$. Matrix V contains the orthonormalized eigenvectors arranged so that $\sigma_\imath \geq \sigma_{\imath+1}$. Although the eigenvalues of $A'A$ are obviously real, it may appear that this matrix could have some negative eigenvalues leading to pure imaginary singular values. However, this cannot happen because $A'AY = \lambda Y$ implies $\lambda = (AY)'(AY)/(Y'Y)$, which clearly is nonnegative. Once the eigenvectors and eigenvalues of $A'A$ are computed, columns of matrix U can be found as orthonormalized solutions of

$$[A'A - \sigma_\jmath I]U_\jmath = 0\ ,\ \sigma_\jmath = 0\ ,\ \jmath > r.$$

The arguments just presented show that performing singular value decomposition involves solving a symmetric eigenvalue problem. However, SVD requires additional computation beyond solving a symmetric eigenvalue problem. It can be very time consuming for large matrices. The SVD has various uses, such as solving the normal equations. Suppose an n by m matrix A has $n > m$ and $r = m$. Substituting the SVD into

$$A'AX = A'B$$

gives

$$V\,\mathbf{diag}(\sigma_1^2, \ldots, \sigma_m^2)V'X = VS'U'B.$$

Consequently, the solution of the normal equations is

$$X = V\,\mathbf{diag}(\sigma_1^{-2}, \ldots, \sigma_m^{-2})S'U'B.$$

Another important application of the SVD concerns generation of orthonormal bases for the column space and the row space. The column space has dimension r and the null space has dimension $m - r$. Consider a consistent system

$$AX = B = U(SV'X).$$

Denote $SV'X$ as Y and observe that $y_j = 0$ for $j > r$ since $\sigma_j = 0$. Because B can be any vector in the column space, it follows that the first r columns of U, which are also orthonormal, are a basis for the column space. Furthermore, the decomposition can be written as

$$AV = US.$$

This implies

$$AV_j = U_j\sigma_j = 0\ ,\ j > r$$

which shows that the final $m-r$ columns of V form an orthonormal basis for the null space. The reader can verify that bases for the row space and left null space follow analogously by considering $A' = VS'U'$, which simply interchanges the roles of U and V.

MATLAB provides numerous other useful matrix decompositions such as LU, QR, and Cholesky. Some of these are employed in other sections of this book. The reader will find it instructive to read the built-in help information for MATLAB functions describing these decomposition methods. For instance, the command **help** \ gives extensive documentation on the operation for matrix inversion.

3.8 Computation Time to Run a MATLAB Program

MATLAB is designed to perform matrix computation with maximum speed and accuracy. Consequently, most standard operations like matrix multiplication, Gauss reduction, eigenvalue calculation, SVD, etc. are implemented as highly optimized and compiled intrinsic functions. Efficient program execution requires optimal use of the built-in functions. Executing nested loops can take a lot of time, so using coding with nested loops should be avoided when computation time is important. To illustrate how deeply nested loops can slow down execution speed we will compare slow multiplication of square matrices by a Fortran style triple loop, and fast multiplication using the intrinsic matrix multiply capability. The ratio of the slow time to the fast time is much larger than might initially be expected.

Before proceeding with our example, consider the difficulties of accurately timing a computational process. In the first place, the clock in Intel based systems has a resolution of about 0.06 sec, whereas the time for MATLAB to do a 100 by 100 matrix multiply is about 0.005 secs on a 733 Mhz Pentium 4 computer. This implies that, just to account for the crude clock increment, the matrix multiply has to be repeated at least 1200 times to get a total time accurate within one percent. However, this is not the only timing difficulty. MATLAB continuously performs housekeeping tasks such as memory management. The operating system and other programs running simultaneously in the background also use computer resources and affect recorded times. Hence, any timing of algorithmic processes in MATLAB should be done without having several other programs open. Even then, the authors have found that times recorded for the same computation done repeatedly often vary around five percent.

The following program named **mattimer** was written to compare slow and fast matrix multiplication. The program input includes the matrix order, the number of seconds a loop is performed to improve timing accuracy, and the number of times the basic timing operation is repeated to show how recorded times vary among successive computations. The program also gives the number of floating operations performed per second (Mflops). An n by n matrix multiply involves n^2 dot products each requiring n adds and n multiplies. Hence, the number of floating point operations is $2n^3$. An order 100 matrix multiply done in 0.005 seconds would give 400 megaflops. Function **multimer** does the matrix multiply repeatedly and reads the elapsed time until the specified total number of seconds is reached. Performing loops and reading the clock takes some time, which is subtracted from the time to do the looping, matrix multiplication, and clock reading. We also perform the intrinsic matrix multiplication in a separate function so that both the fast and slow methods have the same computational overhead associated with a function call. Results are shown for matrices of order 100 and 1000. The fast time for an order 100 matrix multiply only took 0.00503 seconds giving 398 megaflops. By comparison, the slow method took more than eighteen hundred times as long as the fast method. This is comparable to making a one hour task take about two and a half months, working twenty-four hours a day, seven days a week. Evidently, intrinsic MATLAB matrix multiply works very well, but nested looping is slow. Something else worth noting is that a dense matrix of order 1000 does not stretch the capabilities of a modern microcomputer. Storing a million word double-precision array only takes 8 megabytes of RAM, which is a small fraction of the 128 megabytes or more typically provided for scientific work. Furthermore, the high order matrix multiply only took 4.6 seconds, which is roughly 1000 times as long as the order 100 time. It turns out that the time needed for most matrix operations increases like the cube of the order, even though a complicated calculation such as singular value decomposition may take around seventeen times as long as a Gauss elimination of the same order.

```
>> mattimer(100,10,60);

MATRIX MULTIPLY TIMING TEST

Get results for a single timer call

The repeated multiplication of matrices
of order 100 may take considerable time.

Fast multiply takes 0.0050238 secs.
Megaflops = 398.1034

Slow multiply takes 9.0714 secs.
Megaflops = 398.1034

tslow/tfast = 1805.6723

Get results for several timer calls
      tfast             tslow             ratio
  5.0473e-003       8.8899e+000       1.7613e+003
  5.0248e-003       8.8271e+000       1.7567e+003
  4.9948e-003       8.9685e+000       1.7956e+003
  5.0075e-003       8.8742e+000       1.7722e+003
  5.3775e-003       8.9599e+000       1.6662e+003
  4.9939e-003       8.8499e+000       1.7721e+003
  5.0013e-003       8.8271e+000       1.7650e+003
  5.0217e-003       8.9842e+000       1.7891e+003
  5.0182e-003       9.0785e+000       1.8091e+003
  4.9905e-003       8.9598e+000       1.7954e+003

Time variation defined by (max(t)-min(t))/mean(t)
Variation for tfast = 0.076656
Variation for tslow = 0.028181

>> mattimer(1000,0,60);

MATRIX MULTIPLY TIMING TEST

Get results for a single timer call

The repeated multiplication of matrices
of order 1000 may take considerable time.

Fast multiply takes 4.5699 secs.
```

```
Megaflops = 437.6421

Slow multiply takes 8882.3899 secs.
Megaflops = 0.22516

tslow/tfast = 1943.654
```

Program mattimer

```
1: function mattimer(norder,ktimes,secs)
2: %
3: % mattimer(norder,ktimes,secs)
4: % ~~~~~~~~~~~~~~~~~~~~~~~~~~~~
5: if nargin==0
6:     norder=100; ktimes=10; secs=30;
7: end
8: fprintf('\nMATRIX MULTIPLY TIMING TEST\n\n')
9:
10: disp('Get results for a single timer call')
11:
12: multimer(norder,secs,1); t=zeros(ktimes,3);
13:
14: secs=max(secs,30); if ktimes==0, return, end
15:
16: disp('Get results for several timer calls')
17: for j=1:ktimes
18:     [t(j,3),t(j,1),t(j,2)]=multimer(norder,secs);
19: end
20: T=(max(t)-min(t))./mean(t);
21:
22: disp(...
23: '      tfast           tslow              ratio')
24: for j=1:ktimes
25:     fprintf('%13.4e   %13.4e   %13.4e\n',t(j,:))
26: end
27: disp(' '), disp(...
28: 'Time variation defined by (max(t)-min(t))/mean(t)')
29: disp(['Variation for tfast = ',num2str(T(1))])
30: disp(['Variation for tslow = ',num2str(T(2))])
31:
32: %=============================================
```

```
33:
34: function [ratio,tfast,tslow]=multimer(...
35:                                norder,secs,doprint)
36: % [ratio,tfast,tslow]=multimer(...
37: %                                  norder,secs,doprint)
38:
39: % This function compares the times to perform a
40: % matrix multiply using the built-in matrix multiply
41: % and the slow method employing scalar triple looping.
42: % The ratio of compute times illustrates how much
43: % faster compiled and vectorized matrix operations
44: % can be compared to similar calculations using
45: % interpreted code with scalar looping.
46: % norder - order of the test matrices used. The
47: %          default for norder is 100.
48: % secs   - number of seconds each computation is run
49: %          to get accurate timing. The default (and
50: %          minimum value) is thirty seconds.
51: % doprint- print intermediate results only if this
52: %          variable is given a value
53: % ratio  - ratio of slow to fast multiply times
54: % tfast  - time in seconds to perform a multiply
55: %          using the built-in precompiled matrix
56: %          multiply
57: % tslow  - time in seconds to perform a multiply
58: %          by triple loop method
59: %
60: % User m functions called: matmultf matmults
61: %
62: % Typical results obtained using a Dell Dimension
63: % XPS B733r computer with 128MB of RAM gave the
64: % following values:
65: %
66: % >> mattimer(100,0,60);
67: %
68: % MATRIX MULTIPLY TIMING TEST
69: %
70: % Fast multiply takes 0.0050238 secs.
71: % Megaflops = 398.1034
72: %
73: % Slow multiply takes 9.0714 secs.
74: % Megaflops = 398.1034
75: %
76: % tslow/tfast = 1805.6723
77:
```

```
78: % >> mattimer(1000,0,60);
79: %
80: % MATRIX MULTIPLY TIMING TEST
81: %
82: % Fast multiply takes 4.5699 secs.
83: % Megaflops = 437.6421
84: %
85: % Slow multiply takes 8882.3899 secs.
86: % Megaflops = 0.22516
87: %
88: % tslow/tfast = 1943.654
89: % >>
90:
91: % Find time to make a loop and call the clock
92: nmax=5e3; nclock=0; tstart=cputime;
93: while nclock<nmax
94:     tclock=cputime-tstart; nclock=nclock+1;
95: end
96: % Time to do one loop and call the timer
97: tclock=tclock/nclock;
98:
99: if nargin<3, doprint=0; else, doprint=1; end
100: if nargin<2, secs=30; end; secs=max(secs,30);
101: if nargin==0, norder=100; end
102: a=rand(norder,norder); b=rand(norder,norder);
103:
104: if doprint
105:    disp(' ')
106:    disp('The repeated multiplication of matrices')
107:    disp(['of order ',num2str(norder),...
108:          ' may take considerable time.'])
109:    disp(' ')
110: end
111:
112: % Time using intrinsic multiply function
113: pack; tfast=0; nfast=0; tstart=cputime;
114: while tfast<secs
115:     cf=matmultf(a,b); nfast=nfast+1;
116:     tfast=cputime-tstart;
117: end
118: tfast=tfast/nfast-tclock;
119:
120: % Time using Fortran style, triple for:next looping
121: pack; tslow=0; nslow=0; tstart=cputime;
122: while tslow<secs
```

```
123:      cs=matmults(a,b); nslow=nslow+1;
124:      tslow=cputime-tstart;
125: end
126:
127: tslow=tslow/nslow-tclock; ratio=tslow/tfast;
128: mflops=inline('num2str(2*n^3/1e6/t)','n','t');
129: if doprint
130:    disp(['Fast multiply takes ',...
131:            num2str(tfast),' secs.'])
132:    disp(['Megaflops = ',...
133:            mflops(norder,tfast)]), disp(' ')
134:    disp(['Slow multiply takes ',...
135:            num2str(tslow),' secs.'])
136:    disp(['Megaflops = ',...
137:            mflops(norder,tslow)]), disp(' ')
138:    disp(['tslow/tfast = ',...
139:            num2str(tslow/tfast)]), disp(' ')
140: end
141:
142: %==============================================
143:
144: function v=matmultf(a,b)
145: % v=matmultf(a,b). Matrix multiply using
146: % precompiled function in MATLAB
147: v=a*b;
148:
149: %==============================================
150:
151: function v=matmults(a,b)
152: % v=matmults(a,b). Matrix multiply using
153: % Fortran like triple loop
154: n=size(a,1); m=size(b,2); K=size(a,2);
155: v=zeros(n,m);
156: for i=1:n
157:    for j=1:m
158:       t=0;
159:       for k=1:K
160:          t=t+a(i,k)*b(k,j);
161:       end
162:       v(i,j)=t;
163:    end
164: end
```

Chapter 4

Methods for Interpolation and Numerical Differentiation

4.1 Concepts of Interpolation

Next we study three types of one-dimensional interpolation: polynomial, piecewise linear, and cubic spline. The MATLAB functions implementing these methods are discussed along with some additional software developed by the authors to differentiate and integrate splines. A simple discussion of cubic spline interpolation formulated from the viewpoint of elastic beam flexure is given. The chapter concludes with a program to compute finite difference formulas for derivatives of general order.

Interpolation is a process whereby a function is approximated using data known at a discrete set of points. Typically we have points (x_i, y_i) arranged such that $x_i < x_{i+1}$. These points are to be connected by a continuous interpolation function influenced by smoothness requirements such as: a) the function should not deviate greatly from the data at points lying between the data values; and b) the function should satisfy a differentiability condition such as continuity of first and second derivatives.

Piecewise linear interpolation simply connects successive points by straight lines. This has the disadvantage of producing a function with piecewise constant slope and finite slope discontinuities. An obvious cure for slope discontinuity is to use a curve such as a polynomial of degree n-1 (through n points) to produce an interpolation function having all derivatives continuous. However, it was seen in Section 2.3 that a polynomial passing exactly through the data points may be highly irregular at intermediate values. Using polynomial interpolations higher than order five or six often produces disappointing results. An excellent alternative to allowing either slope discontinuities or demanding slope continuity of all orders is to use cubic spline interpolation. This method connects successive points by cubic curves joined such that function continuity as well as continuity of the first two function derivatives is achieved.

The MATLAB function **polyfit**(xd,yd,n) can be used to obtain coefficients in a polynomial of degree n which either passes through points in data vectors (xd,yd) or fits the data in the least square sense. Since a polynomial of degree n-1 can pass through n data points, the computation c=**polyfit**(xd,yd,**length**(xd)-1) would produce coefficients in a polynomial passing through the data values. Evaluating the polynomial for an array argument x is accomplished by y=**polyval**(c,x). Combining the two operations gives y=**polyval**(**polyfit**(xd,yd,**length**(xd)-1),x). If the chosen polynomial order is less than **length**(xd)-1, then a polynomial fitting the data in the least square sense is produced. For example, a polynomial of order 4 might be fitted to several hundred points. Of course, how well the least square polynomial actually fits the data should be assessed by examining a plot of the curve and the data. MATLAB also has various utility functions to work with polynomials such as **polyder**, **polyint**, **conv**, and **deconv** which differentiate, integrate, multiply, and divide.

Function **interp1**(xd,yd,x,'method','extrap') is a general purpose interpolation function providing several types of interpolation including linear and spline. The default value for 'method' is 'linear', If the 'extrap' parameter is omitted, then a value of NaN (not a number) is returned for any input argument not lying between min(xd) and max(xd). Otherwise, extrapolation is performed using the interpolation functions for the outermost intervals. Readers should be cautious about extrapolating far outside the known data range, because this often leads to unreasonable results.

Engineering applications often use idealized functions which are piecewise linear and have finite jump discontinuities. Since function **interp1** rejects cases where any successive values in the xd vector are equal, we remedy this situation with function **lintrp**(xd,yd,x) to search xd for any repeated values and separate these values by a small fraction of max(xd)-min(xd). Then **interp1** is used to perform the interpolation as indicated below.

Function lintrp

```
1: function y=lintrp(xd,yd,x)
2: %
3: % y=lintrp(xd,yd,x)
4: % ~~~~~~~~~~~~~~~~~~
5: % This function performs piecewise linear
6: % interpolation through data values stored in
7: % xd, yd, where xd values are arranged in
8: % nondecreasing order. The function can handle
9: % discontinuous functions specified when some
10: % successive values in xd are equal. Then the
11: % repeated xd values are shifted by a small
12: % amount to remove the discontinuities.
13: % Interpolation for any points outside the range
14: % of xd is also performed by continuing the line
15: % segments through the outermost data pairs.
16: %
17: % xd,yd - vectors of interpolation data values
18: % x     - matrix of values where interpolated
19: %         values are required
```

```
20: %
21: % y      - matrix of interpolated values
22:
23: k=find(diff(xd)==0);
24: if length(k)~=0
25:   xd(k+1)=xd(k+1)+(xd(end)-xd(1))*1e3*eps;
26: end
27: y=interp1(xd,yd,x,'linear','extrap');
```

4.2 Interpolation, Differentiation, and Integration by Cubic Splines

Cubic spline interpolation is a versatile method to pass a smooth curve through a sequence of data points. The technique connects the data values with a curve having its third derivative piecewise constant. The curve is piecewise cubic with y(x), y'(x) and y"(x) continuous over the whole data range. The **MathWorks** markets a **Spline Toolbox** providing extensive capabilities to work with spline functions. A few functions from that toolbox are included in standard MATLAB. The intrinsic functions **spline**, **ppval**, **mkpp**, and **unmkpp** are extended here to handle differentiation and integration. Spline interpolation, viewed from Euler beam theory, is also discussed to amplify on the basic ideas. This simple formulation easily accommodates various end conditions. Readers wanting more detail on spline theory will find the books by de Boor [27] and by Ahlberg and Nilson [2] to be helpful.

Cubic spline theory is motivated by a mechanical drafting tool consisting of a flexible strip bent over several supports with heights adjustable to fit given data. Euler beam theory [9] shows that the deflection curve has third derivative values which are constant between successive supports. This implies that the curve is piecewise cubic and the third derivative values (relating to internal shear forces in beam analysis) can be determined to make the support deflections have chosen values. This is the basis of cubic spline interpolation. The method is attractive because the interpolation function $y(x)$ is obtainable analytically as well as $y'(x)$, $y''(x)$ and $\int y(x)dx$.

Let us formulate the problem mathematically by taking a piecewise constant form for $y'''(x)$ and integrating this repeatedly to get $y(x)$. We assume data points (x_i, y_i), $1 \le i \le n$ with $x_i < x_{i+1}$. Each successive data pair can be connected by a cubic curve with $y'(x)$ and $y''(x)$ required to be continuous at all interior data points. If values of $y'(x)$ or $y''(x)$ are known at the curve ends, algebraic conditions to impose those values can be written. Using known values of end slope is appropriate, but specifying good second derivative values when end slopes are not known is usually not obvious. As an alternative, it is customary to apply smoothness conditions requiring continuity of $y'''(x_2)$ and $y'''(x_{n-1})$. Books on spline theory [7, 2] refer to imposition of higher order continuity at interior points as "not-a-knot" conditions.

The piecewise constant third derivative of the interpolation function is described as

$$y'''(x) = \sum_{j=1}^{n-1} c_j < x - x_j >^0$$

where c_j are constants to be determined, and the singularity function

$$< x - a >^n = (x - a)^n (x > a)$$

is used. This formula for $y'''(x)$ is easy to integrate, and making the curve pass through the data points is straightforward. It follows that

$$y''(x) = y_1'' + \sum_{j=1}^{n-1} c_j < x - x_j >^1,$$

$$y'(x) = y_1' + y_1''(x - x_1) + \frac{1}{2} \sum_{j=1}^{n-1} c_j < x - x_j >^2,$$

$$y(x) = y_1 + y_1'(x - x_1) + \frac{1}{2} y_1''(x - x_1)^2 + \frac{1}{6} \sum_{j=1}^{n-1} c_j < x - x_j >^3,$$

$$\int_{x_1}^{x} y(x)dx = y_1(x - x_1) + \frac{1}{2} y_1'(x - x_1)^2 + \frac{1}{6} y_1''(x - x_1)^3$$

$$+ \frac{1}{24} \sum_{j=1}^{n-1} c_j < x - x_j >^4 .$$

The interpolation function automatically goes through the first data point, and the remaining constants are required to satisfy

$$y_i - y_1 = y_1'(x_i - x_1) + \frac{1}{2} y_1''(x_i - x_1)^2 + \frac{1}{6} \sum_{j=1}^{n-1} c_j < x_i - x_j >^3, \quad i = 2, 3, \ldots, n.$$

Since $n + 1$ unknowns are present in the above system, two more end conditions must be included. Five familiar combinations of end conditions include: 1) the "not-a-knot" condition applied at each end; 2) the slope given at each end; 3) the slope given at the left end and the "not-a-knot" condition at the right end; 4) the "not-a-knot" condition at the left end and the slope given at the right end; and 5) a periodic spline is created by making the first and last points have the same values of y, y', and y''.

Spline interpolation involves solution of linear simultaneous equations. A desktop computer solves a system of 200 equations in less than 0.03 seconds; so, the equation solving time is modest unless many points are used. The formulation described above is easy to understand, handles general end conditions, and includes interpolation, differentiation, and integration. It was implemented in two general purpose functions **spterp** and **spcof** used below with function **curvprop** to compute the length and area bounded by a spline curve. Another function **splineg** using intrinsic function **spline** is also discussed at the end of the present article. The spline routines provided here are helpful additions for work with splines since they include spline differentiation and integration which do not come in the standard MATLAB package.

4.2.1 Computing the Length and Area Bounded by a Curve

The ideas just described were implemented in functions **spterp** and **spcof** which are called in the following program **curvprop**. This program computes the length of a spline curve and the area bounded by the curve. The length of a curve parameterized in complex form as

$$z(t) = x(t) + i\,y(t), \;\; a \le t \le b$$

can be computed as

$$length = \int_a^b \mathbf{abs}(z'(t))\,dt.$$

Furthermore, when the the curve is closed and is traversed in a counterclockwise direction, the area is given by

$$area = \frac{1}{2}\int_a^b \mathbf{imag}(\mathbf{conj}(z(t))\,z'(t))\,dt.$$

The curve length is meaningful for an open or closed curve, but the bounded area only makes sense for a closed curve. The next chapter discusses area properties for shapes bounded by several spline curves. Our present example assumes a simple geometry. It is worth mentioning that applying the last integral to an open curve gives the area enclosed within the curve combined with a line from the last point to the origin and a line from the origin to the first point. This fact is clarified in the next chapter which treats general areas bounded by several spline curves.

The following program **curvprop** passes a spline curve through data in vectors x,y. The length, bounded area, and a set of data points on the curve are computed. The curve is assumed to have a smoothly turning tangent. The default data example uses points on an ellipse with semi-diameters of two and one. Readers can verify that approximating the ellipse with a 21 point spline curve gives an area approximation accurate within 0.0055 percent and a boundary length accurate within 0.0068 percent. Of course, better accuracy is achievable with more data points.

4.2.2 Example: Length and Enclosed Area for a Spline Curve

Function curvprop

```
1: function [area,leng,X,Y,closed]=curvprop(x,y,doplot)
```

```
2: %
3: % [area,leng,X,Y,closed]=curvprop(x,y,doplot)
4: % ~~~~~~~~~~~~~~~~~~~~~~~~~~~~~~~~~~~~~~~~~~~
5: % This function passes a cubic spline curve through
6: % a set of data values and computes the enclosed
7: % area, the curve length, and a set of points on
8: % the curve.
9: %
10: % x,y    - data vectors defining the curve.
11: % doplot - plot the curve if a value is input for
12: %          doplot. Otherwise, no plot is made.
13: % area   - the enclosed area is computed. This
14: %          parameter is valid only if the curve
15: %          is closed and the boundary is traversed
16: %          in counterclockwise. For a curve, the
17: %          area agrees with a curve closed using
18: %          a line from the last point to the
19: %          origin, and a line from the origin to
20: %          the first point.
21: % leng   - length of the curve
22: % X,Y    - set of points on the curve. The output
23: %          intersperses three points on each segment
24: %          between the starting data values.
25: % closed - equals one for a closed curve. Equals zero
26: %          for an open curve.
27: %
28:
29: % For default test data, choose an ellipse with
30: % semi-diameters of 2 and 1.
31: if nargin==0
32:   m=21; th=linspace(0,2*pi,m);
33:   x=2*cos(th); y=sin(th); x(m)=2; y(m)=0;
34: end
35:
36: % Use complex data coordinates
37: z=x(:)+i*y(:); n=length(z); t=(1:n)';
38: chord=sum(abs(diff(z))); d=abs(z(n)-z(1));
39:
40: % Use a periodic spline if the curve is closed
41: if d < (chord/1e6)
42:   closed=1; z(n)=z(1); endc=5;
43:   zp=spterp(t,z,1,t,endc);
44:
45: % Use the not-a-knot end condition for open curve
46: else
```

```
47:   closed=0; endc=1; zp=spterp(t,z,1,t,endc);
48: end
49:
50: % Compute length and area
51: % plot(abs(zp)),shg,pause
52: leng=spterp(t,abs(zp),3,n,1);
53: area=spterp(t,1/2*imag(conj(z).*zp),3,n,1);
54: Z=spterp(t,z,0,1:1/4:n,endc);
55: X=real(Z); Y=imag(Z);
56: if nargin>2
57:   plot(X,Y,'-',x,y,'.'), axis equal
58:   xlabel('x axis'), ylabel('y axis')
59:   title('SPLINE CURVE'), shg
60: end
61:
62: %=========================================
63:
64: function [v,c]=spterp(xd,yd,id,x,endv,c)
65: %
66: % [v,c]=spterp(xd,yd,id,x,endv,c)
67: % ~~~~~~~~~~~~~~~~~~~~~~~~~~~~~~~
68: %
69: % This function performs cubic spline interpo-
70: % lation. Values of y(x),y'(x),y''(x) or the
71: % integral(y(t)*dt, xd(1)..x) are obtained.
72: % Five types of end conditions are provided.
73: %
74: % xd, yd - data vectors with xd arranged in
75: %          ascending order.
76: % id     - id equals 0,1,2,3 to compute y(x),
77: %          y'(x), integral(y(t)*dt,t=xd(1)..x),
78: %          respectively.
79: % v      - values of the function, first deriva-
80: %          tive, second derivative, or integral
81: %          from xd(1) to x
82: % c      - the coefficients defining the spline
83: %          curve. If these values are input from
84: %          an earlier computation, then they
85: %          are not recomputed.
86: % endv   - vector giving the end conditions in
87: %          one of the following five forms:
88: %          endv=1 or endv omitted makes
89: %            c(2) and c(n-1) zero
90: %          endv=[2,left_end_slope,...
91: %            right_end_slope] to impose slope
```

```
92: %                values at each end
93: %              endv=[3,left_end_slope] imposes the
94: %                left end slope value and makes
95: %                c(n-1) zero
96: %              endv=[4,right_end_slope] imposes the
97: %                right end slope value and makes
98: %                c(2) zero
99: %              endv=5 defines a periodic spline by
100: %                making y,y',y" match at both ends
101:
102: if nargin<5 | isempty(endv), endv=1; end
103: n=length(xd); sx=size(x); x=x(:); X=x-xd(1);
104:
105: if nargin<6, c=spcof(xd,yd,endv); end
106:
107: C=c(1:n); s1=c(n+1); m1=c(n+2); X=x-xd(1);
108:
109: if id==0      %  y(x)
110: v=yd(1)+s1*X+m1/2*X.*X+...
111:   powermat(x,xd,3)*C/6;
112: elseif id==1  % y'(x)
113:     v=s1+m1*X+powermat(x,xd,2)*C/2;
114: elseif id==2  % y''(x)
115: v=m1+powermat(x,xd,1)*C;
116: else          % integral(y(t)*dt, t=xd(1)..x)
117: v=yd(1)*X+s1/2*X.*X+m1/6*X.^3+...
118: powermat(x,xd,4)*C/24;
119: end
120: v=reshape(v,sx);
121:
122: %===============================================
123:
124: function c=spcof(x,y,endv)
125: %
126: % c=spcof(x,y,endv)
127: % ~~~~~~~~~~~~~~~~
128: % This function determines spline interpolation
129: % coefficients consisting of the support
130: % reactions concatenated with y' and y'' at
131: % the left end.
132: % x,y  - data vectors of interplation points.
133: %        Denote n as the length of x.
134: % endv - vector of data for end conditions
135: %        described in function spterp.
136: %
```

```
137: % c     -  a vector [c(1);...;c(n+2)] where the
138: %          first n components are support
139: %          reactions and the last two are
140: %          values of y'(x(1)) and y''(x(1)).
141:
142: if nargin<3, endv=1; end
143: x=x(:); y=y(:); n=length(x); u=x(2:n)-x(1);
144: a=zeros(n+2,n+2); a(1,1:n)=1;
145: a(2:n,:)=[powermat(x(2:n),x,3)/6,u,u.*u/2];
146: b=zeros(n+2,1); b(2:n)=y(2:n)-y(1);
147: if endv(1)==1      % Force, force condition
148:   a(n+1,2)=1; a(n+2,n-1)=1;
149: elseif endv(1)==2 % Slope, slope condition
150:   b(n+1)=endv(2); a(n+1,n+1)=1;
151:   b(n+2)=endv(3); a(n+2,:)=...
152: [((x(n)-x').^2)/2,1,x(n)-x(1)];
153: elseif endv(1)==3 % Slope, force condition
154:   b(n+1)=endv(2); a(n+1,n+1)=1; a(n+2,n-1)=1;
155: elseif endv(1)==4 % Force, slope condition
156:   a(n+1,2)=1; b(n+2)=endv(2);
157:   a(n+2,:)=[((x(n)-x').^2)/2,1,x(n)-x(1)];
158: elseif endv(1)==5
159:   a(n+1,1:n)=x(n)-x'; b(n)=0;
160:   a(n+2,1:n)=1/2*(x(n)-x').^2;
161:   a(n+2,n+2)=x(n)-x(1);
162: else
163:   error(...
164:   'Invalid value of endv in function spcof')
165: end
166: if endv(1)==1 & n<4, c=pinv(a)*b;
167: else, c=a\b; end
168:
169: %==============================================
170:
171: function a=powermat(x,X,p)
172: %
173: % a=powermat(x,X,p)
174: % ~~~~~~~~~~~~~~~~~
175: % This function evaluates various powers of a
176: % matrix used in cubic spline interpolation.
177: %
178: % x,X   - arbitrary vectors of length n and N
179: % a     - an n by M matrix of elements such that
180: %         a(i,j)=(x(i)>X(j))*abs(x(i)-X(j))^p
181:
```

```
182: x=x(:); n=length(x); X=X(:)'; N=length(X);
183: a=x(:,ones(1,N))-X(ones(n,1),:); a=a.*(a>0);
184: switch p, case 0, a=sign(a); case 1, return;
185: case 2, a=a.*a; case 3; a=a.*a.*a;
186: case 4, a=a.*a; a=a.*a; otherwise, a=a.^p; end
```

4.2.3 Generalizing the Intrinsic Spline Function in MATLAB

The intrinsic MATLAB function **spline** employs an auxiliary function **unmk** to create the piecewise polynomial definitions defining the spline. The polynomials can be differentiated or integrated, and then functions **mkpp** and **ppval** can be used to evaluate results. We have employed the ideas from those routines to develop functions **splineg** and **splincof** extending the minimal spline capabilities of MATLAB. The function **splincof**(xd,yd,endc) computes arrays b and c usable by **mkpp** and **ppval**. The data vector endc defines the first four types of end conditions discussed above. The function **splineg**(xd,yd,x,deriv,endc,b,c) handles the same kind of data as function **spterp**. Sometimes arrays b and c may have been created from a previous call to **splineg** or **spterp**. Whenever these are passed through the call list, they are used by **splineg** without recomputation. Readers wanting more details on spline concepts should consult de Boor's book [7].

Two examples illustrating spline interpolation are presented next. In the first program called, **sinetrp**, a series of equally spaced points between 0 and 2π is used to approximate $y = \sin(x)$ which satisfies

$$y'(x) = \cos(x)\ ,\ y''(x) = -\sin(x)\ ,\ \int_0^x y(x)dx = 1 - \cos(x).$$

The approximations for the function, derivatives, and the integral are evaluated using **splineg**. Results shown in Figure 4.1 are quite satisfactory, except for points outside the data interval $[0,\ 2\pi]$.

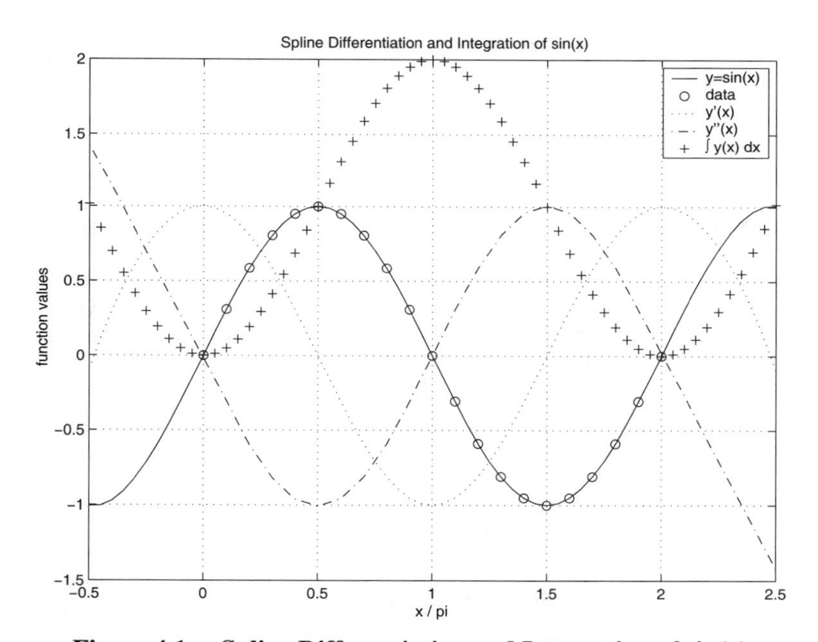

Figure 4.1: Spline Differentiation and Integration of sin(x)

Example: Spline Interpolation Applied to Sin(x)

Program sinetrp

```
1: function sinetrp
2: % Example: sinetrp
3: % ~~~~~~~~~~~~~~~~~~
4: % This example illustrates cubic spline
5: % approximation of sin(x), its first two
6: % derivatives, and its integral.
7: %
8: % User m functions required:
9: %    splineg, splincof
10:
11: % Create data points on the spline curve
12: xd=linspace(0,2*pi,21); yd=sin(xd);
13:
14: % Evaluate function values at a dense
15: % set of points
16: x=linspace(-pi/2,5/2*pi,61);
17: [y,b,c]=splineg(xd,yd,x,0);
18: yp=splineg(xd,yd,x,1,[],b,c);
19: ypp=splineg(xd,yd,x,2,[],b,c);
20: yint=splineg(xd,yd,x,3,[],b,c);
21:
22: % Plot results
23: z=x/pi; zd=xd/pi;
24: plot(z,y,'k-',zd,yd,'ko',z,yp,'k:',...
25:     z,ypp,'k-.',z,yint,'k+');
26: title(['Spline Differentiation and ', ...
27:        'Integration of sin(x)']);
28: xlabel('x / pi'); ylabel('function values');
29: legend('y=sin(x)','data','y''(x)','y''''(x)', ...
30:        '\int y(x) dx',1); grid on
31: figure(gcf); pause;
32: % print -deps sinetrp
33:
34: %=============================================
35:
36: function [val,b,c]=splineg(xd,yd,x,deriv,endc,b,c)
37: %
38: % [val,b,c]=splineg(xd,yd,x,deriv,endc,b,c)
39: % ~~~~~~~~~~~~~~~~~~~~~~~~~~~~~~~~~~~~~~~~~~
40: %
```

```
41: % For a cubic spline curve through data points
42: % xd,yd, this function evaluates y(x), y'(x),
43: % y''(x), or integral(y(x)*dx, xd(1) to x(j) )
44: % for j=1:length(x).The coefficients needed to
45: % evaluate the spline are also computed.
46: %
47: % xd,yd    - data vectors defining the cubic
48: %            spline curve
49: % x        - vector of points where curve
50: %            properties are computed.
51: % deriv    - denoting the spline curve as y(x),
52: %            deriv=0 gives a vector for y(x)
53: %            deriv=1 gives a vector for y'(x)
54: %            deriv=2 gives a vector for y''(x)
55: %            deriv=3 gives a vector of values
56: %               for integral(y(z)*dz) from xd(1)
57: %               to x(j) for j=1:length(x)
58: % endc     - endc=1 makes y'''(x) continuous at
59: %            xd(2) and xd(end-1).
60: %            endc=[2,left_slope,right_slope]
61: %            imposes slope values at both ends.
62: %            endc=[3,left_slope] imposes the left
63: %            end slope and makes the discontinuity
64: %            of y''' at xd(end-1) small.
65: %            endc=[4,right_slope] imposes the right
66: %            end slope and makes the discontinuity
67: %            of y''' at xd(2) small.
68: % b,c        coefficients needed to perform the
69: %            spline interpolation. If these are not
70: %            given, function unmkpp is called to
71: %            generate them.
72: % val        values y(x),y'(x),y''(x) or
73: %            integral(y(z)dz, z=xd(1)..x) for
74: %            deriv=0,1,2, or 3, respectively.
75:
76: if nargin<5 | isempty(endc), endc=1; end
77: if nargin<7, [b,c]=splincof(xd,yd,endc); end
78: n=length(xd); [N,M]=size(c);
79:
80: switch deriv
81:
82: case 0 % Function value
83:   val=ppval(mkpp(b,c),x);
84:
85: case 1 % First derivative
```

```
86:     C=[3*c(:,1),2*c(:,2),c(:,3)];
87:     val=ppval(mkpp(b,C),x);
88:
89: case 2 % Second derivative
90:     C=[6*c(:,1),2*c(:,2)];
91:     val=ppval(mkpp(b,C),x);
92:
93: case 3 % Integral values from xd(1) to x
94:     k=M:-1:1;
95:     C=[c./k(ones(N,1),:),zeros(N,1)];
96:     dx=xd(2:n)-xd(1:n-1); s=zeros(n-2,1);
97:     for j=1:n-2, s(j)=polyval(C(j,:),dx(j)); end
98:     C(:,5)=[0;cumsum(s)]; val=ppval(mkpp(b,C),x);
99:
100: end
101:
102: %==============================================
103:
104: function [b,c]=splincof(xd,yd,endc)
105: %
106: % [b,c]=splincof(xd,yd,endc)
107: % ~~~~~~~~~~~~~~~~~~~~~~~~~~~
108: % This function determines coefficients for
109: % cubic spline interpolation allowing four
110: % different types of end conditions.
111: % xd,yd - data vectors for the interpolation
112: % endc  - endc=1 makes y'''(x) continuous at
113: %         xd(2) and xd(end-1).
114: %         endc=[2,left_slope,right_slope]
115: %         imposes slope values at both ends.
116: %         endc=[3,left_slope] imposes the left
117: %         end slope and makes the discontinuity
118: %         of y''' at xd(end-1) small.
119: %         endc=[4,right_slope] imposes the right
120: %         end slope and makes the discontinuity
121: %         of y''' at xd(2) small.
122: %
123: if nargin<3, endc=1; end;
124: type=endc(1); xd=xd(:); yd=yd(:);
125:
126: switch type
127:
128: case 1
129:     % y'''(x) continuous at the xd(2) and xd(end-1)
130:     [b,c]=unmkpp(spline(xd,yd));
```

```
131:
132: case 2
133:    % Slope given at both ends
134:    [b,c]=unmkpp(spline(xd,[endc(2);yd;endc(3)]));
135:
136: case 3
137:    % Slope at left end given. Compute right end
138:    % slope.
139:    [b,c]=unmkpp(spline(xd,yd));
140:    c=[3*c(:,1),2*c(:,2),c(:,3)];
141:    sright=ppval(mkpp(b,c),xd(end));
142:    [b,c]=unmkpp(spline(xd,[endc(2);yd;sright]));
143:
144: case 4
145:    % Slope at right end known. Compute left end
146:    % slope.
147:    [b,c]=unmkpp(spline(xd,yd));
148:    c=[3*c(:,1),2*c(:,2),c(:,3)];
149:    sleft=ppval(mkpp(b,c),xd(1));
150:    [b,c]=unmkpp(spline(xd,[sleft;yd;endc(2)]));
151:
152: end
```

4.2.4 Example: A Spline Curve with Several Parts and Corners

The final spline example illustrates interpolation of a two-dimensional curve where y cannot be expressed as a single valued function of x. Then we introduce a parameter t_j having its value equal to the index $\jmath$ for each (x_j, y_j) used. Interpolating $x(t)$ and $y(t)$ as continuous functions of t produces a smooth curve through the data. Function **matlbdat** creates data points to define the curve and calls function **spcry2d** to compute points on a general plane curve. We also introduce the idea of 'corner points' where slope discontinuity allows the curve to make sharp turns needed to describe letters such as the 't' in MATLAB. Each curve segment between successive pairs of corner points is parameterized using function **spline**. Results in Figure 4.2 show clearly that spline interpolation can represent a complicated curve. The related code appears after the figure. The same kind of parameterization used for two dimensions also works well for three dimensional curves.

Example: Spline Curve Drawing the Word MATLAB

Program matlbdat

```
1: function matlbdat
```

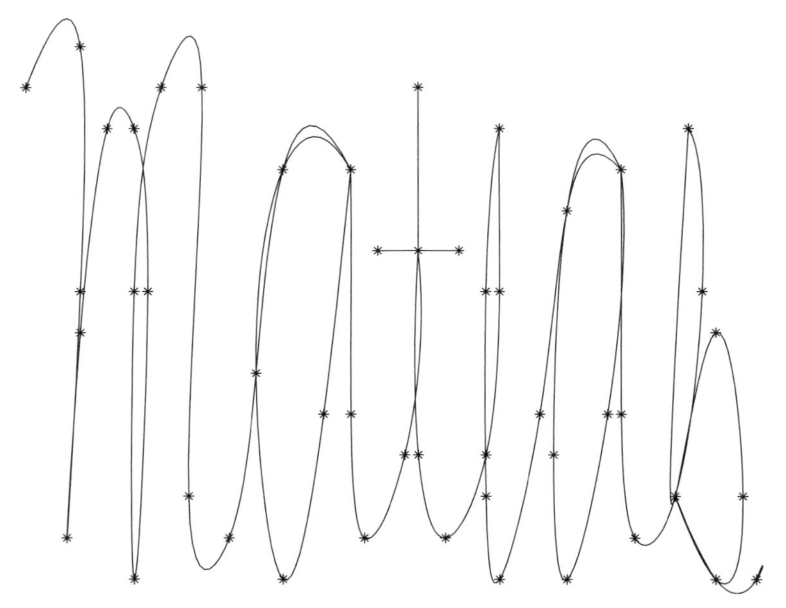

Figure 4.2: Spline Curve Drawing the Word MATLAB

```
 2: % Example: matlbdat
 3: % ~~~~~~~~~~~~~~~~~~
 4: % This example illustrates the use of splines
 5: % to draw the word MATLAB.
 6: %
 7: % User m functions required: spcurv2d
 8:
 9: x=[13 17 17 16 17 19 21 22 21 21 23 26
10:    25 28 30 32 37 32 30 32 35 37 37 38
11:    41 42 42 42 45 39 42 42 44 47 48 48
12:    47 47 48 51 53 57 53 52 53 56 57 57
13:    58 61 63 62 61 64 66 64 61 64 67 67];
14: y=[63 64 58 52 57 62 62 58 51 58 63 63
15:    53 52 56 61 61 61 56 51 55 61 55 52
16:    54 59 63 59 59 59 59 54 52 54 58 62
17:    58 53 51 55 60 61 60 54 51 55 61 55
18:    52 53 58 62 53 57 53 51 53 51 51 51];
19: x=x'; x=x(:); y=y'; y=y(:);
20: ncrnr=[17 22 26 27 28 29 30 31 36 42 47 52];
21: clf; [xs,ys]=curv2d(x,y,10,ncrnr);
22: plot(xs,ys,'k-',x,y,'k*'), axis off;
23: title('A Spline Curve Drawing the Word MATLAB');
24: figure(gcf);
25: % print -deps matlbdat
26:
27: %==============================================
28:
29: function [X,Y]=spcrv2d(xd,yd,nseg,icrnr)
30: %
31: % [X,Y]=spcrv2d(xd,yd,nseg,icrnr)
32: % ~~~~~~~~~~~~~~~~~~~~~~~~~~~~~~~
33: % This function computes points (X,Y) on a
34: % spline curve through (xd,yd) allowing slope
35: % discontinuities at points with corner
36: % indices in icrnr. nseg plot segments are
37: % used between each successive pair of points.
38:
39: if nargin<4, icrnr=[]; end
40: if nargin<3, nseg=10; end
41: zd=xd(:)+i*yd(:); n=length(zd);
42: N=[1;sort(icrnr(:));n]; Z=zd(1);
43: if N(1)==N(2); N(1)=[]; end
44: if N(end)==N(end-1); N(end)=[]; end
45: for k=1:length(N)-1
46:   zk=zd(N(k):N(k+1)); sk=length(zk)-1;
```

```
47:     s=linspace(0,sk,1+sk*nseg)';
48:     Zk=spline(0:sk,zk,s); Z=[Z;Zk(2:end)];
49: end
50: X=real(Z); Y=imag(Z);
```

4.3 Numerical Differentiation Using Finite Differences

Differential equation problems are sometimes solved using difference formulas to approximate the derivatives in terms of function values at adjacent points. Deriving difference formulas by hand can be tedious, particularly when unequal point spacing is used. For this reason, we develop a numerical procedure to construct formulas of arbitrary order and arbitrary truncation error. Of course, as the desired order of derivative and the order of truncation error increases, more points are needed to interpolate the derivative. We will show below that approximating a derivative of order k with a truncation error of order h^m generally requires $(k+m)$ points unless symmetric central differences are used. Consider the Taylor series expansion

$$F(x+\alpha h) = \sum_{k=0}^{\infty} \frac{F^{(k)}(x)}{k!}(\alpha h)^k$$

where $F^{(k)}(x)$ means the k'th derivative of $F(x)$. This relation expresses values of F as linear combinations of the function derivatives at x. Conversely, the derivative values can be cast in terms of function values by solving a system of simultaneous equations. Let us take a series of points defined by

$$x_\imath = x + h\alpha_\imath \ , \ 1 \le \imath \le n$$

where h is a fixed step-size and $\alpha_\imath$ are arbitrary parameters. Separating some leading terms in the series expansion gives

$$F(x_\imath) = \sum_{k=0}^{n-1} \frac{\alpha_\imath^k}{k!}\left[h^k F^{(k)}(x)\right] + \frac{\alpha_\imath^n}{n!}\left[h^n F^{(n)}(x)\right] + \frac{\alpha_\imath^{n+1}}{(n+1)!}\left[h^{(n+1)} F^{(n+1)}(x)\right] + \mathrm{O}(h^{n+2}) \ , \ 1 \le \imath \le n.$$

It is helpful to use the following notation:

α^k	–	a column vector with component $\imath$ being equal to $\alpha_\imath^k$
f	–	a column vector with component $\imath$ being $F(x_\imath)$
fp	–	a column vector with component $\imath$ being $h^\imath F^{(\imath)}(x)$
A	–	$[\alpha^0, \alpha^1, \ldots, \alpha^{n-1}]$, a square matrix with columns which are powers of α.

Then the Taylor series expressed in matrix form is

$$f = A * fp + \frac{h^n F^{(n)}(x)}{n!}\alpha^n + \frac{h^{n+1} F^{(n+1)}(x)}{(n+1)!}\alpha^{n+1} + O(h^{n+2}).$$

Solving this system for the derivative matrix fp yields

$$fp = A^{-1}f - \frac{h^n F^{(n)}(x)}{n!}A^{-1}\alpha^n - \frac{h^{n+1} F^{(n+1)}(x)}{(n+1)!}A^{-1}\alpha^{n+1} + O(h^{n+2}).$$

In the last equation we have retained the first two remainder terms in explicit form to allow the magnitudes of these terms to be examined. Row $k+1$ of the previous equation implies

$$F^{(k)}(x) = h^{-k}(A^{-1}f)_{k+1} - \frac{h^{n-k}}{n!}F^{(n)}(x)(A^{-1}\alpha^n)_{k+1} - \\ \frac{h^{n-k+1}}{(n+1)!}F^{(n+1)}(x)(A^{-1}\alpha^{n+1})_{k+1} + O(h^{n-k+1}).$$

Consequently, the rows of A^{-1} provide coefficients in formulas to interpolate derivatives. For a particular number of interpolation points, say N, the highest derivative approximated will be $F^{(N-1)}(x)$ and the truncation error will normally be of order h^1. Conversely, if we need to compute a derivative formula of order k with the truncation error being m, then it is necessary to use a number of points such that $n - k = m$; therefore $n = m + k$. For the case where interpolation points are symmetrically placed around the point where derivatives are desired, one higher power of accuracy order is achieved than might be expected. We can show, for example, that

$$\frac{d^4F(x)}{dx^4} = \frac{1}{h^4}\left(F(x-2h) - 4F(x-h) + 6F(x) - \\ 4F(x+h) + F(x+2h)\right) + O(h^2)$$

because the truncation error term associated with h^1 is found to be zero. At the same time, we can show that a forward difference formula for $f'''(x)$ employing equidistant point spacing is

$$\frac{d^3F(x)}{dx^3} = \frac{1}{h^3}\left(-2.5F(x) - 9F(x+h) + 12F(x+2h) + \\ 7F(x+3h) - 1.5F(x+4h)\right) + O(h^2).$$

Although the last two formulas contain arithmetically simple interpolation coefficients, due to equal point spacing, the method is certainly not restricted to equal spacing. The following program contains the function **derivtrp** which implements the ideas just developed. Since the program contains documentation that is output when it is executed, no additional example problem is included.

4.3.1 Example: Program to Derive Difference Formulas

Output from Example

```
finitdif;

COMPUTING F(x,k), THE K'TH DERIVATIVE OF
f(x), BY FINITE DIFFERENCE APPROXIMATION

Input the derivative order (give 0 to stop,
or ? for an explanation) > ?

Let f(x) have its k'th derivative denoted by
F(k,x). The finite difference formula for a
stepsize h is given by:

F(x,k)=Sum(c(j)*f(x+a(j)*h), j=1:n)/h^k +...
       TruncationError

with m=n-k being the order of truncation
error which decreases like h^m according to:

TruncationError=-(h^m)*(e(1)*F(x,n)+...
e(2)*F(x,n+1)*h+e(3)*F(x,n+2)*h^2+O(h^3))

Input the derivative order (give 0 to stop,
or ? for an explanation) > 4

Give the required truncation order > 1

To define interpolation points X(j)=x+h*a(j),
input at least 5 components for vector a.

Components of a > -2,-1,0,1,2

The formula for a derivative of order 4 is:
F(x,k)=sum(c(j)*F(X(j),j=1:n)/h^4+order(h^1)
where c is given by:

    1.0000    -4.0000     6.0000    -4.0000     1.0000

and the truncation error coefficients are:

   -0.0000     0.1667    -0.0000     0.0125

Input the derivative order (give 0 to stop,
```

```
or ? for an explanation) > 3

Give the required truncation order > 2

To define interpolation points X(j)=x+h*a(j),
input at least 5 components for vector a.

Components of a > 0,1,2,3,4

The formula for a derivative of order 3 is:
F(x,k)=sum(c(j)*F(X(j),j=1:n)/h^3+order(h^2)
where c is given by:

   -2.5000    9.0000  -12.0000    7.0000   -1.5000

and the truncation error coefficients are:

   -1.7500   -2.5000   -2.1417   -1.3750

Input the derivative order (give 0 to stop,
or ? for an explanation) > 0
```

Program finitdif

```
1: function [c,e,m,crat,k,a]=finitdif(k,a)
2: %
3: % [c,e,m,crat,k,a]=finitdif(k,a)
4: % ~~~~~~~~~~~~~~~~~~~~~~~~~~~~~~~
5: % This program computes finite difference formulas of
6: % general order. For explanation of the input and
7: % output parameters, see the following function
8: % findifco. When the program is executed without input
9: % arguments, then input is read interactively.
10:
11: if nargin==0, disp(' ') % Use interactive input
12:   disp('COMPUTING F(x,k), THE K''TH DERIVATIVE OF')
13:   disp('f(x), BY FINITE DIFFERENCE APPROXIMATION')
14:   disp(' ')
15:   while 1
16:     disp('Input the derivative order (give 0 to stop,')
17:     K=input('or ? for an explanation) > ','s');
18:     k=str2num(K);
19:     if strcmp(K,'') | strcmp(K,'0'); disp(' '),return
```

```
20:     elseif strcmp(K,'?')
21:       disp(' '), disp(...
22:       'Let f(x) have its k''th derivative denoted by')
23:       disp(...
24:       'F(k,x). The finite difference formula for a')
25:       disp('stepsize h is given by:'), disp(' ')
26:       disp(...
27:       'F(x,k)=Sum(c(j)*f(x+a(j)*h), j=1:n)/h^k +...')
28:       disp('        TruncationError'), disp(' ')
29:       disp('with m=n-k being the order of truncation')
30:       disp(...
31:       'error which decreases like h^m according to:')
32:       disp(' ')
33:       disp('TruncationError=-(h^m)*(e(1)*F(x,n)+...')
34:       disp(...
35:       'e(2)*F(x,n+1)*h+e(3)*F(x,n+2)*h^2+O(h^3))')
36:       disp(' ')
37:     else
38:       disp(' ')
39:       m=input('Give the required truncation order > ');
40:       n=m+k; N=num2str(n); disp(' '), disp(...
41:       'To define interpolation points X(j)=x+h*a(j),')
42:       disp(['input at least ',N,...
43:             ' components for vector a.'])
44:       disp(' '), aa=input('Components of a > ','s');
45:       a=eval(['[',aa,']']); n=length(a); m=n-k;
46:       [c,e,m,crat]=findifco(k,a); disp(' '), disp(...
47:       ['The formula for a derivative of order ',...
48:       K,' is:'])
49:       disp(['F(x,k)=sum(c(j)*F(X(j),j=1:n)/h^',K,...
50:             '+order(h^',num2str(m),')'])
51:       disp('where c is given by:')
52:       disp(' '), disp(c), disp(' ')
53:       disp(...
54:       'and the truncation error coefficients are:')
55:       disp(' '), disp(e)
56:     end
57:   end
58: else
59:    [c,e,m,crat]=findifco(k,a);
60: end
61:
62: %==================================================
63:
64: function [c,e,m,crat]=findifco(k,a)
```

```
 65: %
 66: % [c,e,m,crat]=findifco(k,a)
 67: % ~~~~~~~~~~~~~~~~~~~~~~~~~~~
 68: % This function approximates the k'th derivative
 69: % of a function using function values at n
 70: % interpolation points. Let f(x) be a general
 71: % function having its k'th derivative denoted
 72: % by F(x,k). The finite difference approximation
 73: % for the k'th derivative employing a stepsize h
 74: % is given by:
 75: % F(x,k)=Sum(c(j)*f(x+a(j)*h), j=1:n)/h^k +
 76: %        TruncationError
 77: % with m=n-k being the order of truncation
 78: % error which decreases like h^m and
 79: % TruncationError=(h^m)*(e(1)*F(x,n)+...
 80: % e(2)*F(x,n+1)*h+e(3)*F(x,n+2)*h^2+O(h^3))
 81: %
 82: % a    - a vector of length n defining the
 83: %        interpolation points x+a(j)*h where
 84: %        x is an arbitrary parameter point
 85: % k    - order of derivative evaluated at x
 86: % c    - the weighting coeffients in the
 87: %        difference formula above. c(j) is
 88: %        the multiplier for value f(x+a(j)*h)
 89: % e    - error component vector in the above
 90: %        difference formula
 91: % m    - order of truncation order in the
 92: %        formula. The relation m=n-k applies.
 93: % crat - a matrix of integers such that c is
 94: %        approximated by crat(1,:)./crat(2,:)
 95:
 96: a=a(:); n=length(a); m=n-k; mat=ones(n,n+4);
 97: for j=2:n+4; mat(:,j)=a/(j-1).*mat(:,j-1); end
 98: A=pinv(mat(:,1:n)); ec=-A*mat(:,n+1:n+4);
 99: c=A(k+1,:); e=-ec(k+1,:);
100: [ctop,cbot]=rat(c,1e-8); crat=[ctop(:)';cbot(:)'];
```

Chapter 5

Gauss Integration with Geometric Property Applications

5.1 Fundamental Concepts and Intrinsic Integration Tools in MATLAB

Numerical integration methods approximate a definite integral by evaluating the integrand at several points and taking a weighted combination of those integrand values. The weight factors can be obtained by interpolating the integrand at selected points and integrating the interpolating function exactly. For example, the Newton-Cotes formulas result from polynomial interpolation through equidistant base points. This chapter discusses concepts of numerical integration needed in applications.

Let us assume that an integral over limits a to b is to be evaluated. We can write

$$\int_a^b f(x)dx = \sum_{\imath=1}^{n} W_\imath f(x_\imath) + E$$

where E represents the error due to replacement of the integral by a finite sum. This is called an n-point quadrature formula. The points $x_\imath$ where the integrand is evaluated are the base points and the constants $W_\imath$ are the weight factors. Most integration formulas depend on approximating the integrand by a polynomial. Consequently, they give exact results when the integrand is a polynomial of sufficiently low order. Different choices of $x_\imath$ and $W_\imath$ will be discussed below.

It is helpful to express an integral over general limits in terms of some fixed limits, say -1 to 1. This is accomplished by introducing a linear change of variables

$$x = \alpha + \beta t.$$

Requiring that $x = a$ corresponds to $t = -1$ and that $x = b$ corresponds to $t = 1$ gives $\alpha = (a+b)/2$ and $\beta = (b-a)/2$, so that one obtains

$$\int_a^b f(x)dx = \frac{1}{2}(b-a)\int_{-1}^{1} f\left[\frac{a+b}{2} + \frac{b-a}{2}t\right] dt = \int_{-1}^{1} F(t)dt$$

where $F(t) = f[(a+b)/2 + (b-a)t/2](b-a)/2$. Thus, the dependence of the integral on the integration limits can be represented parametrically by modifying the integrand. Consequently, if an integration formula is known for limits -1 to 1, we can write

$$\int_a^b f(x)dx = \beta \sum_{\imath=1}^{n} W_\imath f(\alpha + \beta x_\imath) + E.$$

The idea of shifting integration limits can be exploited further by dividing the interval a to b into several parts and using the same numerical integration formula to evaluate the contribution from each interval. Employing m intervals of length $\ell = (b-a)/m$, we get

$$\int_a^b f(x)dx = \sum_{\jmath=1}^{m} \int_{a+(\jmath-1)\ell}^{a+\jmath\ell} f(x)dx.$$

Each of the integrals in the summation can be transformed to have limits -1 to 1 by taking

$$x = \alpha_\jmath + \beta t$$

with

$$\alpha_\jmath = a + (j - 1/2)\ell \text{ and } \beta = \ell/2.$$

Therefore we obtain the identity

$$\int_a^b f(x)dx = \sum_{\jmath=1}^{m} \cdot \frac{\ell}{2} \int_{-1}^{1} f(\alpha_\jmath + \beta t)dt.$$

Applying the same n-point quadrature formula in each of m equal intervals gives what is termed a composite formula

$$\int_a^b f(x)dx = \frac{\ell}{2} \sum_{\jmath=1}^{m} \sum_{\imath=1}^{n} W_\imath f(\alpha_\jmath + \beta x_\imath) + E.$$

By interchanging the summation order in the previous equation we get

$$\int_a^b f(x)dx = \frac{\ell}{2} \sum_{\imath=1}^{n} W_\imath \sum_{\jmath=1}^{m} f(\alpha_\jmath + \beta x_\imath) + E.$$

Let us now turn to certain choices of weight factors and base points. Two of the most widely used methods approximate the integrand as either piecewise linear or piecewise cubic. Approximating the integrand by a straight line through the integrand end points gives the following formula

$$\int_{-1}^{1} f(x)dx = f(-1) + f(1) + E.$$

A much more accurate formula results by using a cubic approximation matching the integrand at $x = -1, 0, 1$. Let us write

$$f(x) = c_1 + c_2 x + c_3 x^2 + c_4 x^3.$$

Then

$$\int_{-1}^{1} f(x)dx = 2c_1 + \frac{2}{3}c_3.$$

Evidently the linear and cubic terms do not influence the integral value. Also, $c_1 = f(0)$ and $f(-1) + f(1) = 2\, c_1 + 2\, c_3$ so that

$$\int_{-1}^{1} f(x)dx = \frac{1}{3}\left[f(-1) + 4f(0) + f(1)\right] + E.$$

The error E in this formula is zero when the integrand is any polynomial of order 3 or lower. Expressed in terms of more general limits, this result is

$$\int_{a}^{b} f(x)dx = \frac{(b-a)}{6}\left[f(a) + 4f(\frac{a+b}{2}) + f(b)\right] + E$$

which is known as Simpson's rule.

Analyzing the integration error for a particular choice of integrand and quadrature formula can be complex. In practice, the usual procedure taken is to apply a composite formula with m chosen large enough so the integration error is expected to be negligibly small. The value for m is then increased until no further significant change in the integral approximation results. Although this procedure involves some risk of error, adequate results can be obtained in most practical situations.

In the subsequent discussions the integration error that results by replacing an integral by a weighted sum of integrand values will be neglected. It must nevertheless be kept in mind that this error depends on the base points, weight factors, and the particular integrand. Most importantly, the error typically decreases as the number of function values is increased.

It is convenient to summarize the composite formulas obtained by employing a piecewise linear or piecewise cubic integrand approximation. Using m intervals and letting $\ell = (b-a)/m$, it is easy to obtain the composite trapezoidal formula which is

$$\int_{a}^{b} f(x)dx = \ell\left[\frac{f(a)+f(b)}{2} + \sum_{\jmath=1}^{m-1} f(a+\jmath\ell)\right].$$

This formula assumes that the integrand is satisfactorily approximated by piecewise linear functions. The MATLAB function **trapz** implements the trapezoidal rule. A similar but much more accurate result is obtained for the composite integration formula based on cubic approximation. For this case, taking m intervals implies $2m+1$ function evaluations. If we let $g = (b-a)/(2m)$ and $h = 2g$, then

$$f_\jmath = f(x_\jmath) \text{ where } x_\jmath = a + g\jmath\ ,\ \jmath = 0, 1, 2, \ldots, 2m,$$

with $f(x_0) = f(a)$ and $f(x_{2m}) = f(b)$. Combining results for all intervals gives

$$\int_{a}^{b} f(x)dx = \frac{h}{6}\left[f(a) + 4f_1 + f(b) + \sum_{\imath=1}^{m-1}(4f_{2\imath+1} + 2f_{2\imath})\right].$$

This formula, known as the composite Simpson rule, is one of the most commonly used numerical integration methods. The following function **simpson** works for an analytically defined function or a function defined by spline interpolating through discrete data.

Function for Composite Simpson Rule

```
1: function area=simpson(funcname,a,b,n,varargin)
2: %
3: % area=simpson(funcname,a,b,n,varargin)
4: % ------------------------------------
5: % Simpson's rule integration for a general function
6: % defined analytically or by a data array
7: %
8: % funcname  - either the name of a function valid
9: %             for a vector argument x, or an array
10: %            having two columns with x data in the
11: %            first column and y data in the second
12: %            column. If array data is given, then
13: %            the function is determined by piecewise
14: %            cubic spline interpolation.
15: % a,b       - limits of integration
16: % n         - odd number of function evaluations. If
17: %             n is given as even, then the next
18: %             higher odd integer is used.
19: % varargin  - variable number of arguments passed
20: %             for use in funcname
21: % area      - value of the integral when the integrand
22: %             is approximated as a piecewise cubic
23: %             function
24: %
25: % User functions called: function funcname in the
26: %                        argument list
27: %----------------------------------------------------

28: if 2*fix(n/2)==n; n=n+1; end; n=max(n,3);
29: x=linspace(a,b,n);
30: if isstr(funcname)
31:   y=feval(funcname,x,varargin{:});
32: else
33:   y=spline(funcname(:,1),funcname(:,2),x);
34: end
35: area=(b-a)/(n-1)/3*( y(1)-y(n)+...
```

```
36:          4*sum(y(2:2:n))+2*sum(y(3:2:n)));
```

An important goal in numerical integration is to achieve accurate results with only a few function evaluations. It was shown for Simpson's rule that three function evaluations are enough to exactly integrate a cubic polynomial. By choosing the base point locations properly, a much higher accuracy can be achieved for a given number of function evaluations than would be obtained by using evenly spaced base points. Results from orthogonal function theory lead to the following conclusions. If the base points are located at the zeros of the Legendre polynomials (all these zeros are between -1 and 1) and the weight factors are computed as certain functions of the base points, then the formula

$$\int_{-1}^{1} f(x)dx = \sum_{i=1}^{n} W_i f(x_i)$$

is exact for a polynomial integrand of degree $2n-1$. Although the theory proving this property is not elementary, the final results are quite simple. The base points and weight factors for a particular order can be computed once and used repeatedly. Formulas that use the Legendre polynomial roots as base points are called Gauss quadrature formulas. In a typical application, Gauss integration gives much more accurate results than Simpson's rule for an equivalent number of function evaluations. Since it is equally easy to use, the Gauss formula is preferable to Simpson's rule.

MATLAB also has three functions **quad** and **quad8** and **quadl** to numerically integrate by adaptive methods. These functions repeatedly modify approximations for an integral until the estimated error becomes smaller than a specified tolerance. In the current text, the function **quadl** is preferable over the other two functions, and **quadl** is always used when an adaptive quadrature function is needed. Readers should study carefully the system documentation for **quadl** to understand the various combinations of call list parameters allowed.

5.2 Concepts of Gauss Integration

This section summarizes properties of Gauss integration which, for the same number of function evaluations, are typically much more accurate than comparable Newton-Cotes formulas. It can be shown for Gauss integration [20] that

$$\int_{-1}^{1} f(x)\, dx = \sum_{j=1}^{n} w_j f(x_j) + E(f)$$

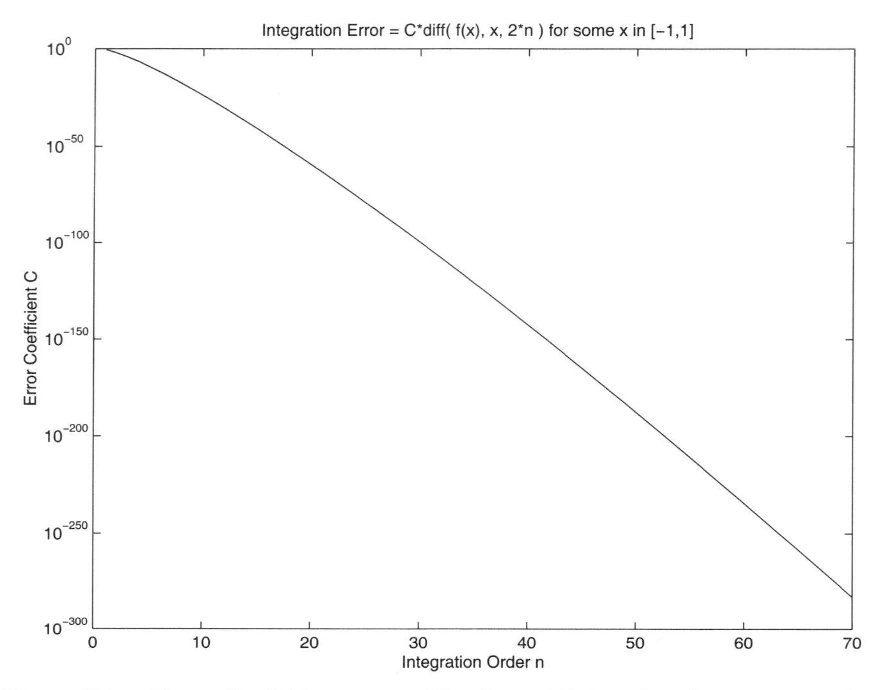

Figure 5.1: Error Coefficient versus Number of Points for Gauss Integration

where the integration error term is

$$E = \frac{2^{2n-1}(n!)^4}{(2n+1)[(2n)!]^3} f^{(2n)}(\xi) \ , \ -1 < \xi < 1.$$

The base points in the Gauss formula of order n are the roots of the Legendre polynomial of order n and the weight factors are expressible concisely in terms of the base points. The quadrature error term for an n-point formula involves the integrand derivative of order $2n$, which implies a zero error for any polynomial of order $2n-1$ or lower. The coefficient of the derivative term in E decreases very rapidly with increasing n, as can be seen in Figure 5.1.

For example, $n = 10$ gives a coefficient of 2.03×10^{-21}. Thus, a function having well behaved high order derivatives can be integrated accurately with a formula of fairly low order. The base points x_j are all distinct, lie between -1 and 1, and are the eigenvalues of a symmetric tridiagonal matrix [26] which can be analyzed very rapidly with the function **eigen**. Furthermore, the weight factors are simply twice the squares of the first components of the orthonormalized eigenvectors. Because **eigen** returns orthonormalized eigenvectors for symmetric matrices, only lines 58-60 in function **gcquad** given below are needed to compute the base points and weight factors.

Function for Composite Gauss Integration

```
1: function [val,bp,wf]=gcquad(func,xlow,...
2:                      xhigh,nquad,mparts,varargin)
3: %
4: % [val,bp,wf]=gcquad(func,xlow,...
5: %     xhigh,nquad,mparts,varargin)
6:
7: % ~~~~~~~~~~~~~~~~~~~~~~~~~~~~~~~~~
8: %
9: % This function integrates a general function using
10: % a composite Gauss formula of arbitrary order. The
11: % integral value is returned along with base points
12: % and weight factors obtained by an eigenvalue based
13: % method. The integration interval is divided into
14: % mparts subintervals of equal length and integration
15: % over each part is performed with a Gauss formula
16: % making nquad function evaluations. Results are
17: % exact for polynomials of degree up to 2*nquad-1.
18: % ~~~~~~~~~~~~~~~~~~~~~~~~~~~~~~~~~~~~~~~~~~~~~~~~~~~
19: % func           - name of a function to be integrated
20: %                  having an argument list of the form
21: %                  func(x,p1,p2,...) where any auxiliary
22: %                  parameters p1,p2,.. are passed through
23: %                  variable varargin. Use [ ] for the
24: %                  function name if only the base points
25: %                  and weight factors are needed.
26: % xlow,xhigh     - integration limits
27: % nquad          - order of Gauss formula chosen
28: % mparts         - number of subintervals selected in
29: %                  the composite integration
30: % varargin       - variable length parameter used to
31: %                  pass additional arguments needed in
32: %                  the integrand func
33: % val            - numerical value of the integral
34: % bp,wf          - vectors containing base points and
35: %                  weight factors in the composite
36: %                  integral formula
37: %
38: % A typical calculation such as:
39: % Fun=inline('(sin(w*t).^2).*exp(c*t)','t','w','c');
40: % A=0; B=12; nquad=21; mparts=10; w=10; c=8;
41: % [value,pcterr]=integrate(Fun,A,B,nquad,mparts,w,c);
```

```
42: % gives value = 1.935685556078172e+040 which is
43: % accurate within an error of 1.9e-13 percent.
44: %
45: % User m functions called:  the function name passed
46: %                           in the argument list
47:
48: %------------------------------------------------
49:
50:  if isempty(nquad),  nquad=10; end
51:  if isempty(mparts), mparts=1; end
52:
53: % Compute base points and weight factors
54: % for the single interval [-1,1]. (Ref:
55: % 'Methods of Numerical Integration' by
56: % P. Davis and P. Rabinowitz, page 93)
57:
58: u=(1:nquad-1)./sqrt((2*(1:nquad-1)).^2-1);
59: [vc,bp]=eig(diag(u,-1)+diag(u,1));
60: [bp,k]=sort(diag(bp)); wf=2*vc(1,k)'.^2;
61:
62: % Modify the base points and weight factors
63: % to apply for a  composite interval
64: d=(xhigh-xlow)/mparts;  d1=d/2;
65: dbp=d1*bp(:); dwf=d1*wf(:);  dr=d*(1:mparts);
66: cbp=dbp(:,ones(1,mparts))+ ...
67: dr(ones(nquad,1),:)+(xlow-d1);
68: cwf=dwf(:,ones(1,mparts)); wf=cwf(:); bp=cbp(:);
69:
70: % Compute the integral
71: if isempty(func)
72:   val=[];
73: else
74:   f=feval(func,bp,varargin{:}); val=wf'*f(:);
75: end
```

5.3 Comparing Results from Gauss Integration and Function QUADL

A program was written to compare the performance of the Gauss quadrature function **gcquad** and the numerical integrator **quadl** provided in MATLAB. **Quadl** is a robust adaptive integration routine which efficiently handles most integrands. It can even deal with special integrals having singularities, like $\log(x)$ or $1/\operatorname{sqrt}(x)$ at the origin. Integrating these functions from zero to one yields correct answers although messages occur warning about integrand singularities at the origin. No capabilities are provided in **quadl** to directly handle vector integrands (except one component at a time), and no options are provided to suppress unwanted warning or error messages. In the timing program given below, warning messages from **quadl** were temporarily turned off for the tests.

Often there are examples involving vector-valued integrands that are to be integrated many times over fixed integration limits. A typical case is evaluation of coefficients in Fourier-Bessel series expansions. Then, computing a set of base points and weight factors once and using these coefficients repeatedly is helpful. To illustrate this kind of situation, let us numerically integrate the vector valued function

$$f(x) = [\sqrt{x}\,;\ \log(x)\ ;\ \text{humps}(x);\ \exp(10x)\cos(10\pi x)\,;\cos(20\pi x - 20\sin(\pi x))]$$

from $x = 0$ to $x = 1$. Several components of this function are hard to integrate numerically because $\sqrt{x}$ has infinite slope at $x = 0$, $\log(x)$ is singular at $x = 0$, the fourth component is highly oscillatory with large magnitude variations, and the last component is highly oscillatory (integrating the last component gives the value of the integer order Bessel function $J_{20}(20)$).

The following function **quadtest** uses functions **quadl** and **gcquad** to integrate $f(x)$ from $x = 0$ to $x = 1$. The Gauss integration employs a formula of order 100 with one subinterval, so integrands are effectively approximated by polynomials of order 199. To achieve accurate timing, it was necessary to evaluate the integrals repeatedly until a chosen number of seconds elapsed. Then average times were computed. The program output shows that **gcquad** was more accurate than **quadl** for all cases except for $\log(x)$ involving a singular integrand. Computations times shown for each component of $f(x)$ are the same when **gcquad** was used because the integration was done for all components at once, and then results were divided by five. The total time used by **quadl** was about 3.5 times as large as the time for **qcquad**. We are not arguing that these results show **gcquad** is superior to **quadl**. However, it does imply that Gauss integration can be attractive in some instances. The geometry problems in the remainder of this chapter include boundary curves defined by cubic splines. Then, using Gauss integration of sufficiently high order produces exact results for the desired geometrical properties.

Output from Program quadtest

```
>> quadtest(10);
PRESS RETURN TO BEGIN COMPUTATION > ?

INTEGRATION TEST COMPARING FUNCTIONS QUADL AND GCQUAD
The functions being integrated are:
sqrt(x)
log(x)
humps(x)
exp(10*x).*cos(10*pi*x)
cos(20*pi*x-20*sin(pi*x))
```

```
              Results Using Function quadl
    Integral      Function       Percent    Computation
     values     evaluations       error      seconds
  6.6667e-001  7.8000e+001 -1.9813e-004  6.9720e-003
 -1.0000e+000  2.2900e+002  2.6064e-005  2.1071e-002
  2.9858e+001  1.9800e+002  6.2164e-010  2.1162e-002
  2.0263e+002  7.0800e+002  2.4425e-013  6.8660e-002
  1.6475e-001  5.2800e+002 -1.0627e-008  5.1370e-002

              Results Using Function gcquad
   Integral      Function       Percent     Computation
    values     evaluations       error        seconds
  6.6667e-001  1.0000e+002  1.5215e-005  9.5628e-003
 -9.9994e-001  1.0000e+002 -6.2513e-003  9.5628e-003
  2.9858e+001  1.0000e+002  8.8818e-014  9.5628e-003
  2.0263e+002  1.0000e+002 -4.1078e-013  9.5628e-003
  1.6475e-001  1.0000e+002 -1.5543e-013  9.5628e-003

(Total time using quadl)/(Total time using gcquad)
equals 3.5395

>>
```

Program Comparing Numerical Integration Methods

```
1: function [L,G,names]=quadtest(secs)
2: %
3: % [L,G,names]=quadtest(secs)
4: % ~~~~~~~~~~~~~~~~~~~~~~~
5: % This program compares the accuracy and
6: % computation times for several integrals
7: % evaluated using quadl and gcquad
8: %
9: % secs  - the number of seconds each integration
10: %         is repeated to get accurate timing. The
11: %         default value is 60 seconds.
12: % L,G   - matrices with columns containing
13: %         results from quadl and from gcquad.
14: %         The matrices are structured as:
15: %         [IntegralValue,PercentError,...
16: %         FunctionEvaluations,ComputationSeconds]
17: % names - character matrix with rows
18: %         describing the functions
```

```
19: %          which were integrated
20: %
21: % User functions called: ftest, gcquad
22: %------------------------------------------
23:
24: global nvals
25:
26: if nargin==0, secs=60; end
27:
28: fprintf('\nPRESS RETURN TO BEGIN COMPUTATION > ')
29: pause
30:
31: % Summary of the five integrands used
32: names=strvcat('sqrt(x)','log(x)','humps(x)',...
33:               'exp(10*x).*cos(10*pi*x)',...
34:               'cos(20*pi*x-20*sin(pi*x))');
35: fprintf(['\n\nINTEGRATION TEST COMPARING',...
36:       ' FUNCTIONS QUADL AND GCQUAD\n'])
37: fprintf('\nThe functions being integrated are:\n')
38: disp(names)
39:
40: % Compute exact values of integrals
41: exact=[2/3; -1; quadl(@humps,0,1,1e-12);
42:       real((exp(10+10*pi*i)-1)/(10+10*pi*i));
43:       besselj(20,20)];
44:
45: % Find time to make a loop and call the clock
46: nmax=5000; nclock=0; t0=clock;
47: while nclock<nmax
48:     nclock=nclock+1; tclock=etime(clock,t0);
49: end
50: tclock=tclock/nclock;
51:
52: % Evaluate each integral individually. Repeat
53: % the integrations for secs seconds to get
54: % accurate timing.  Save results in array L.
55: L=zeros(5,4); tol=1e-6; e=exact; warning off;
56: for k=1:5
57:   nquad=0; tim=0; t0=clock;
58:   while tim<secs
59:     [v,nfuns]=quadl(@ftest,0,1,tol,[],k);
60:     nquad=nquad+1; tim=etime(clock,t0);
61:   end
62:   tim=tim/nquad-tclock; pe=100*(v/e(k)-1);
63:   L(k,:)=[v,nfuns,pe,tim];
```

```
64: end
65: warning on;
66:
67: % Obtain time to compute base points and weight
68: % factors for a Gauss formula of order 100
69: nloop=100; t0=clock;
70: for j=1:nloop
71:   [dumy,bp,wf]=gcquad([],0,1,100,1);
72: end
73: tbpwf=etime(clock,t0)/nloop;
74:
75: % Perform the Gauss integration using a
76: % vector integrand. Save results in array G
77:
78: ngquad=0; tim=0; t0=clock;
79: while tim<secs
80:   v=ftest(bp,6)*wf;
81:   ngquad=ngquad+1; tim=etime(clock,t0);
82: end
83: tim=tim/ngquad+tbpwf-tclock; pe=100*(v./e-1);
84: G=[v,100*ones(5,1),pe,tim/5*ones(5,1)];
85:
86: format short e
87: disp(' ')
88: disp('            Results Using Function quadl')
89: disp(...
90: '    Integral     Function      Percent   Computation')
91: disp(...
92: '     values     evaluations     error     seconds')
93: disp(L)
94: disp('            Results Using Function gcquad')
95: disp(...
96: '   Integral     Function      Percent    Computation')
97: disp(...
98: '    values     evaluations     error        seconds')
99: disp(G)
100: format short
101: disp(['(Total time using quadl)/',...
102:        '(Total time using gcquad)'])
103: disp(['equals ',...
104:        num2str(sum(L(:,end))/sum(G(:,end)))])
105: disp(' ')
106:
107: %=============================================
108:
```

```
109: function y=ftest(x,n)
110: % Integrands used by function quadl
111: global nvals
112: switch n
113: case 1, y=sqrt(x); case 2, y=log(x);
114: case 3, y=humps(x);
115: case 4, y=exp(10*x).*cos(10*pi*x);
116: case 5, y=cos(20*pi*x-20*sin(pi*x));
117: otherwise
118:   x=x(:)'; y=[sqrt(x);log(x);humps(x);
119:               exp(10*x).*cos(10*pi*x);
120:               cos(20*pi*x-20*sin(pi*x))];
121: end
122: if n<6, nvals=nvals+length(x);
123: else, nvals=nvals+5*length(x); end
124:
125: %==============================================
126:
127: % function [val,bp,wf]=gcquad(func,xlow,...
128: %                     xhigh,nquad,mparts,varargin)
129: % See Appendix B
```

5.4 Geometrical Properties of Areas and Volumes

Geometrical properties of areas and volumes are often needed in physical applications such as linear stress analysis and rigid body dynamics. For example, consider a prismatic structural member having a general cross-section area denoted by A with a boundary curve L. Analyzing the stresses occurring when the member undergoes axial compression and bi-axial bending leads to integrals of the form

$$C_{nm} = \iint\limits_A x^n y^m dxdy$$

for integers n and m. The six most important cases and the related property names are:

n	m	Symbol	Geometrical Parameter
0	0	a	Area
1	0	ax	First moment of area about the y-axis
0	1	ay	First moment of area about the x-axis
2	0	axx	Moment of inertia about the y-axis
1	1	axy	Product of inertia with respect to the xy axes
0	2	ayy	Moment of inertia about the x-axis

The integral C_{nm} can be evaluated for very general shapes by converting the area integral to a line integral over the boundary. Then, approximating the boundary curve parametrically (by spline interpolation, for example) and using numerical integration yield the desired values. Green's theorem [119] relates area integrals and line integrals according to

$$\iint_A \left[\frac{\partial U}{\partial x} + \frac{\partial V}{\partial y} \right] dxdy = \oint_L U dy - V dx$$

where $U(x, y)$ and $V(x, y)$ are single-valued and differentiable functions inside and on L. This implies that

$$\iint_A x^n y^m dxdy = \frac{1}{n+m+2} \iint_A \left[\frac{\partial}{\partial x}(x^{n+1} y^m) + \frac{\partial}{\partial y}(x^n y^{m+1}) \right] dxdy$$
$$= \frac{1}{n+m+2} \oint_L x^n y^m (xdy - ydx),$$

provided $n + m + 2 \neq 0$. We can even have negative n provided $x = 0$ is outside L, and negative m provided $y = 0$ is outside L. The case $(n, m) = (0, -1)$ occurring in curved beam theory can also be treated by line integration, but we will confine attention to the six cases listed above.

If the boundary curve, L, is parameterized as $x(t),\ y(t)$, $a \leq t \leq b$, then

$$\oint_L x^n y^m (x\, dy - y\, dx) = \int_a^b x(t)^n y(t)^m [x(t) y'(t) - y(t) x'(t)]\, dt$$

which is a one-dimensional integral amenable to numerical integration. When cubic spline interpolation is used to represent the boundary, then $x(t)y'(t) - y(t)x'(t)$ is a piecewise polynomial function of degree four (not degree five as seems apparent at first glance). Since a Gaussian quadrature formula of order N integrates exactly any polynomial of degree $2N - 1$ or less, the integral of interest can be integrated exactly by taking $2N - 1 \geq 3n + 3m + 4$. For our case, using a composite Gauss formula of order six is appropriate. A program is given below to compute properties for a general geometry that can have several separate parts, and these parts may contain holes. The details of that program are discussed later.

The ideas for plane regions can be extended to three dimensions where volume, gravity center location, and inertia tensor are the quantities being computed. Let

$$\mathbf{R} = [x\,;\ y\,;\ z]$$

be the Cartesian radius vector for points in a three-dimensional region W covered by a surface S. The Gauss divergence theorem [119] implies

$$V = \iiint_W dx\,dy\,dz = \frac{1}{3}\iint_S \boldsymbol{R} \cdot \hat{\boldsymbol{\eta}}\, dS,$$

$$\mathbf{V}_r = \iiint_W \boldsymbol{R}\, dx\,dy\,dz = \frac{1}{4}\iint_S \boldsymbol{R}(\boldsymbol{R} \cdot \hat{\boldsymbol{\eta}})\, dS$$

$$\mathbf{V}_{rr} = \iiint_W \boldsymbol{R}\,\boldsymbol{R}'\, dx\,dy\,dz = \frac{1}{5}\iint_S \boldsymbol{R}\,\boldsymbol{R}'(\boldsymbol{R} \cdot \hat{\boldsymbol{\eta}})\, dS.$$

In the last equation, $\boldsymbol{R}\,\boldsymbol{R}'$ is the matrix product $[x\,;\,y\,;z] * [x\,,\,y\,,\,z]$ and $\hat{\boldsymbol{\eta}}$ is the outward directed unit surface normal. We refer to V, $\mathbf{V}_r$, and $\mathbf{V}_{rr}$ as the volume, the first moment of volume, and the second moment of volume. These quantities can be evaluated exactly for some special cases, such as polyhedra, and volumes of revolution.

The quantity $\mathbf{V}_{rr}$ is useful in rigid body dynamics where the inertia tensor, I_{rr} , is needed to compute the rotational kinetic energy. The inertia tensor for a body having unit mass density can be computed from $\mathbf{V}_{rr}$ as

$$I_{rr} = \mathbf{eye}(3,3)\ \mathbf{sum}(\mathbf{diag}(\mathbf{V}_{rr})) - \mathbf{V}_{rr}.$$

The corresponding inverse is

$$\mathbf{V}_{rr} = \frac{\mathbf{eye}(3,3)}{2}\ \mathbf{sum}(\mathbf{diag}(I_{rr})) - I_{rr}.$$

To illustrate the computation of volume properties, consider the instance where the surface has a parametric equation of the form

$$\boldsymbol{R}(u,v),\quad u_1 \le u \le u_2,\quad v_1 \le v \le v_2.$$

For example, the ellipsoid defined by

$$\left(\frac{x}{a}\right)^2 + \left(\frac{y}{b}\right)^2 + \left(\frac{z}{c}\right)^2 \le 1$$

has a surface equation

$$\boldsymbol{R} = [a\sin(u)\cos(v);\ b\sin(u)\sin(v);\ c\cos(v)],\quad 0 \le u \le \pi,\quad 0 \le v \le 2\pi.$$

The unit surface normal and the differential of surface area can be computed as

$$\hat{\boldsymbol{\eta}}\, dS = \frac{\partial \boldsymbol{R}}{\partial u} \times \frac{\partial \boldsymbol{R}}{\partial v} du dv$$

where the order of the cross product is chosen so that the outward directed normal is produced. Then,

$$(\boldsymbol{R} \cdot \hat{\boldsymbol{\eta}})\, dS = \det([\boldsymbol{R}\,,\ \boldsymbol{R}_u,\ \boldsymbol{R}_v]) = D(u,v)\, du\, dv$$

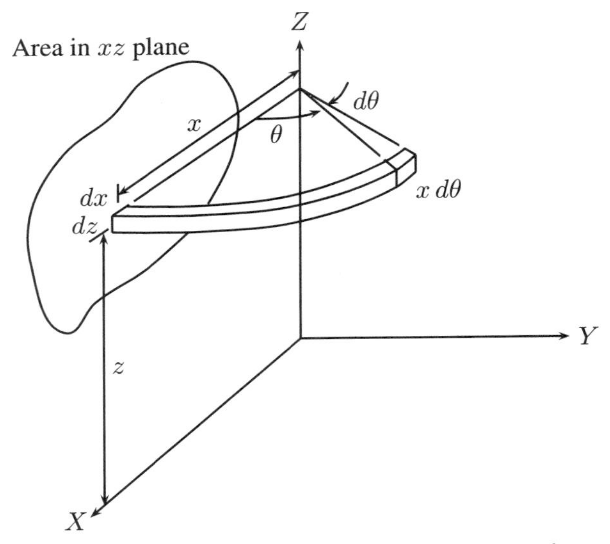

Figure 5.2: Generation of a Volume of Revolution

and the integrals of interest become

$$V = \frac{1}{3} \int_{v_1}^{v_2} \int_{u_1}^{u_2} D(u, v) du dv,$$

$$\mathbf{V}_r = \frac{1}{4} \int_{v_1}^{v_2} \int_{u_1}^{u_2} \boldsymbol{R}\, D(u, v) du dv,$$

$$\mathbf{V}_{rr} = \frac{1}{5} \int_{v_1}^{v_2} \int_{u_1}^{u_2} \boldsymbol{R}\boldsymbol{R}'\ D(u, v) du dv.$$

Note that the function $D(u, v)$ vanishes at points where the radius vector $\boldsymbol{R}$ is perpendicular to the surface normal. A useful instance of the parametric form occurs when a closed curve is rotated to form a volume of revolution as illustrated in Figure 5.2.

Consider a curve defined parametrically in the (x, z) plane as $x(t), z(t), a \leq t \leq b$. If the curve is rotated about the z axis through angular limits $\theta_1 \leq \theta \leq \theta_2$, the lateral surface of the body has a surface equation

$$\boldsymbol{R}(t, \theta) = [\cos(\theta);\ \sin(\theta);\ 1].*[x\,;\ x;\ z]$$

and the volume property integrals reduce to

$$V = \frac{1}{3} \oint_L x[xdz - zdx],$$

$$\mathbf{V}_r = \frac{1}{4} \int_{\theta_1}^{\theta_1} [\cos(\theta);\ \sin(\theta);\ 1]\, d\theta .* \oint_L [x^2;\ x^2;\ xz]\, dxdz,$$

$$\mathbf{V}_{rr} = \frac{1}{5} \int_{\theta_1}^{\theta_1} [\cos(\theta);\ \sin(\theta);\ 1] * [\cos(\theta),\ \sin(\theta),\ 1]\, d\theta.$$

$$* \oint_L [x\,;\ x;\ z] * [x,\ x,\ z]\, x\, dxdz.$$

The line integrals in the last three formulas have similar structure to those for plane area properties, but the highest polynomial power in the integrands of the form $x^{\,n} z^m$ is one higher than that encountered for plane area properties. Taking a composite Gauss formula of order seven produces exact results for the volume properties when a spline interpolated boundary curve is used. A program employing these formulas is developed below.

5.4.1 Area Property Program

A program was written to compute the area, the centroidal coordinates, the inertial moments, and the product of inertia for general plane areas bounded by a series of spline curve segments. Shapes such as polygons, having one or more straight boundary segments, are also handled by allowing slope discontinuities at the ends of the straight segments. The program requires data points $xd(j), yd(j),\ 1 \le j \le nd$ with the boundary traversed in a counterclockwise sense. The first and last points should be identical to make the curve closed. A set of point indices designating any slope discontinuities (such as those at the corners of a square) are also needed.

A typical geometry for the program appears in Figure 5.3. Program data created by the function **makcrcsq** employs 27 data points. A multiply connected geometry is treated as if it were a simply connected region by introducing fictitious cuts connecting outer boundaries and inner holes. Disconnected parts are also joined with zero width strips. Including the cuts and strips has no effect on the area properties because related boundary segments are traversed twice, but in opposite directions. Consequently, the corresponding line integral contributions from the fictitious parts cancel. The complete boundary is parameterized as a spline curve in complex form

$$z(t) = x(t) + i\, y(t),\ \ 1 \le t \le nd$$

with

$$z(j) = xd(j) + i\, yd(j),\ \ j = 1, 2, \ldots, nd.$$

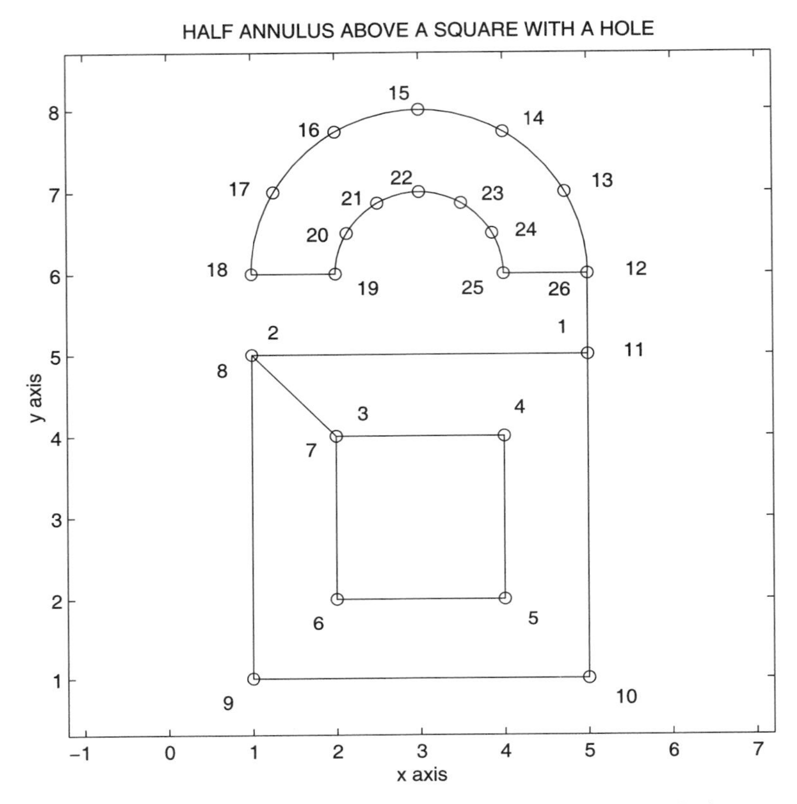

Figure 5.3: Geometry Showing Numbered Boundary Points

The boundary curve and its derivatives are piecewise polynomial functions. Exact results for the geometrical properties are obtained by using the function **gcquad** to generate Gauss base points and weight factors for integration limits from 1 to nd, and a number of integration segments equal to $nd - 1$. The various area properties are accumulated in vector mode for computational efficiency.

The function **runaprop** is the main driver for the program. It accepts boundary data, calls the function **aprop** to compute area properties, prints results, and plots the geometry. If no input data are given, the function **makcrcsq** is called to create data for the illustrative example. The plot produced by the program resembles the one shown, without inclusion of point indices.

The numerical output for our example has the following simple form:

```
>> runaprop;

       GEOMETRICAL PROPERTY ANALYSIS USING FUNCTION APROP

       A          XCG          YCG          AXX          AXY          AYY
    16.7147     3.0000     4.1254   176.3369   206.8620   359.6076

>>
```

Program for Properties of Spline Bounded Areas

```
1: function [p,x,y,xd,yd]=areaprog(xd,yd,icrnr)
2: %
3: % [p,x,y,xd,yd]=areaprog(xd,yd,icrnr)
4: % ~~~~~~~~~~~~~~~~~~~~~~~~~~~~~~~~~~~~
5: % This function calls function aprop which
6: % computes geometrical properties for an area
7: % bounded by a spline curve through data
8: % points in (xd,yd).
9: %
10: % User functions called: aprop
11:
12: if nargin==2,icrnr=[1,length(xd)]; end
13: titl='AREA IN THE XY PLANE';
14:
15: if nargin==0
16:   [xd,yd,icrnr]=makcrcsq;
17:   titl=...
18:   'HALF ANNULUS ABOVE A SQUARE WITH A HOLE';
19: end
20:
21: disp(' ')
22: disp(['        GEOMETRICAL PROPERTY ANALYSIS',...
23:       ' USING FUNCTION APROP'])
24: [p,z]=aprop(xd,yd,icrnr); x=real(z); y=imag(z);
25: disp('  ');
26: disp(['      A          XCG          YCG     ',...
27:       '     AXX          AXY          AYY'])
28: disp(p), close, plot(xd,yd,'ko',x,y,'k-')
29:
30: xlabel('x axis'), ylabel('y axis')
31: title(titl),axis(cubrange([x(:),y(:)],1.2));
```

```
32: axis square; shg
33:
34: %==========================================
35:
36: function [p,zplot]=aprop(xd,yd,kn)
37: %
38: % [p,zplot]=aprop(xd,yd,kn)
39: % ~~~~~~~~~~~~~~~~~~~~~~~~~~~
40: % This function determines geometrical properties
41: % of a general plane area bounded by a spline
42: % curve
43: %
44: % xd,yd - data points for spline interpolation
45: %         with the boundary traversed in counter-
46: %         clockwise direction. The first and last
47: %         points must match for boundary closure.
48: % kn    - vector of indices of points where the
49: %         slope is discontinuous to handle corners
50: %         like those needed for shapes such as a
51: %         rectangle.
52: % p     - the vector [a,xcg,ycg,axx,axy,ayy]
53: %         containing the area, centroid coordinates,
54: %         moment of inertia about the y-axis,
55: %         product of inertia, and moment of inertia
56: %         about the x-axis
57: % zplot - complex vector of boundary points for
58: %         plotting the spline interpolated geometry.
59: %         The points include the numerical quadrature
60: %         points interspersed with data values.
61: %
62: % User functions called: gcquad, curve2d
63: if nargin==0
64:   td=linspace(0,2*pi,13); kn=[1,13];
65:   xd=cos(td)+1; yd=sin(td)+1;
66: end
67: nd=length(xd); nseg=nd-1;
68: [dum,bp,wf]=gcquad([],1,nd,6,nseg);
69: [z,zplot,zp]=curve2d(xd,yd,kn,bp);
70: w=[ones(size(z)), z, z.*conj(z), z.^2].*...
71:    repmat(imag(conj(z).*zp),1,4);
72: v=(wf'*w)./[2,3,8,8]; vr=real(v); vi=imag(v);
73: p=[vr(1:2),vi(2),vr(3)+vr(4),vi(4),vr(3)-vr(4)];
74: p(2)=p(2)/p(1); p(3)=p(3)/p(1);
75:
76: %==========================================
```

```
77:
78: function [z,zplot,zp]=curve2d(xd,yd,kn,t)
79: %
80: % [z,zplot,zp]=curve2d(xd,yd,kn,t)
81: %~~~~~~~~~~~~~~~~~~~~~~~~~~~~~~~~~~~
82: % This function generates a spline curve through
83: % given data points with corners(slope dis-
84: % continuities) allowed as selected points.
85: % xd,yd - real data vectors of length nd
86: %         defining the curve traversed in
87: %         counterclockwise order.
88: % kn    - vectors of point indices, between one
89: %         and nd, where slope discontinuities
90: %         occur
91: % t     - a vector of parameter values at which
92: %         points on the spline curve are
93: %         computed. The components of t normally
94: %         range from one to nd, except when t is
95: %         a negative integer,-m. Then t is
96: %         replaced by a vector of equally spaced
97: %         values using m steps between each
98: %         successive pair of points.
99: % z     - vector of points on the spline curve
100: %         corresponding to the vector t
101: % zplot - a complex vector of points suitable
102: %         for plotting the geometry
103: % zp    - first derivative of z with respect to
104: %         t for the same values of t as is used
105: %         to compute z
106: %
107: % User m functions called:  splined
108: %----------------------------------------------
109:
110: nd=length(xd); zd=xd(:)+i*yd(:); td=(1:nd)';
111: if isempty(kn), kn=[1;nd]; end
112: kn=sort(kn(:)); if kn(1)~=1, kn=[1;kn]; end
113: if kn(end)~=nd, kn=[kn;nd]; end
114: N=length(kn)-1; m=round(abs(t(1)));
115: if -t(1)==m, t=linspace(1,nd,1+N*m)'; end
116: z=[]; zp=[]; zplot=[];
117: for j=1:N
118:   k1=kn(j); k2=kn(j+1); K=k1:k2;
119:   k=find(k1<=t & t<k2);
120:   if j==N, k=find(k1<=t & t<=k2); end
121:   if ~isempty(k)
```

```
122:     zk=spline(K,zd(K),t(k)); z=[z;zk];
123:     zplot=[zplot;zd(k1);zk];
124:     if nargout==3
125:       zp=[zp;splined(K,zd(K),t(k))];
126:     end
127:   end
128: end
129: zplot=[zplot;zd(end)];
130:
131: %=========================================
132:
133: function [x,y,icrnr]=makcrcsq
134: %
135: % [x,y,icrnr]=makcrcsq
136: % ~~~~~~~~~~~~~~~~~~~~~~
137: % This function creates data for a geometry
138: % involving half of an annulus placed above a
139: % square containing a square hole.
140: %
141: % x,y   - data points characterizing the data
142: % icrnr - index vector defining corner points
143: %
144: % User m functions called:  none
145: %----------------------------------------------
146:
147: xshift=3.0; yshift=3.0;
148: a=2; b=1; narc=7; x0=0; y0=2*a-b;
149: xy=[a,-a,-b, b, b,-b,-b,-a,-a, a, a;
150:     a, a, b, b,-b,-b, b, a,-a,-a, a]';
151: theta=linspace(0,pi,narc)';
152: c=cos(theta); s=sin(theta);
153: xy=[xy;[x0+a*c,y0+a*s]];
154: c=flipud(c); s=flipud(s);
155: xy=[xy;[x0+b*c,y0+b*s];[a,y0];[a,a]];
156: x=xy(:,1)+xshift; y=xy(:,2)+yshift;
157: icrnr=[(1:12)';11+narc;12+narc; ...
158:        11+2*narc;12+2*narc;13+2*narc];
159:
160: %=========================================
161:
162: % function [val,bp,wf]=gcquad(func,xlow,...
163: %                        xhigh,nquad,mparts,varargin)
164: % See Appendix B
165:
166: %=========================================
```

```
167:
168: % function range=cubrange(xyz,ovrsiz)
169: % See Appendix B
170:
171: %========================================
172:
173: % function val=splined(xd,yd,x,if2)
174: % See Appendix B
```

5.4.2 Program Analyzing Volumes of Revolution

Since the geometrical property computation for a volume of revolution is quite similar to that for area properties, the same functions **gcquad** and **curve2d** used in the area program are employed below to compute the volume, centroidal coordinates, and inertia tensor of a solid generated by rotating a spline curve through arbitrary angular limits about the z axis. In the following program, the function **volrevol** calls a general purpose function **volrev** which computes the geometrical properties and plots the related volume of revolution. The function **volrev** depends on **gcquad**, **curve2d**, function **rotasurf** to plot the body surface, and function **anglefun** which deals with rotation angle dependence. The function returns volume properties and surface coordinates on the solid. Area properties of the cross section are also obtained.

The geometry in Figure 5.4 was analyzed. The area, rotated through 270 degrees, consists of the bottom half of a semicircle capped on the outer radius by a square which is also capped by a smaller semicircle. Results from **volrev** were confirmed to agree closely with another function **srfv**, which is discussed in the next section. Because the function **srfv** employs triangular surface elements, the two computational models are not identical. This accounts for the slight difference in numerical results.

In conclusion, the volume property program to handle volumes of revolution for spline interpolated cross sections was found to be a useful extension of the methods developed earlier for properties of areas.

Computer Output from Volume of Revolution Program

```
>> volrevol
PROPERTIES OF A VOLUME OF REVOLUTION
Results Using Function VOLREV
Volume = 59.1476
Rg = [0.91028 0.91028 0.13755]
Inertia Tensor =
1.0e+003 *
0.5773 0.1168 -0.0090
0.1168 0.5773 -0.0090
-0.0090 -0.0090 1.1010
Area Properties
```

area xcentr zcentr.
2.9638 4.2350 0.1045
axx axz azz
53.8408 1.7264 1.3101
Results Using Function SRFV
Volume = 59.1056
Rg = [0.91016 0.91016 0.13749]
Inertia Tensor =
1.0e+003 *
0.5768 0.1167 -0.0090
0.1167 0.5768 -0.0090
-0.0090 -0.0090 1.0999
>>

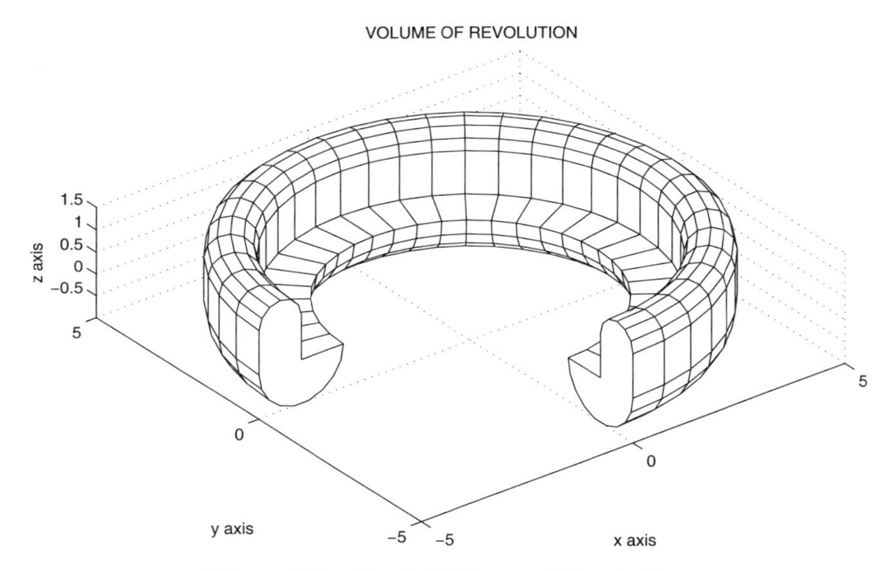

Figure 5.4: Partial Volume of Revolution

Program for Properties of a Volume of Revolution

```
1: function volrevol
2: %
3: % volrevol
4: % ~~~~~~~~
```

```
 5: % This program determines geometrical properties
 6: % for a solid generated by rotating a closed spline
 7: % curve through an arbitrary angle about the z-axis.
 8: % A detailed description of the geometry is given in
 9: % function volrev.
10: %
11: % User m functions called: volrev srfv
12: %----------------------------------------------
13: %
14: % Data for a cross section consisting of the lower
15: % half of a circle plus a square capped by the
16: % upper half of a smaller semicircle. The geometry
17: % is rotated through 270 degrees about the z-axis.
18:
19: n1=9; t1=-pi:pi/n1:0; n2=6; t2=0:pi/n2:pi;
20: Zd=[0,exp(i*t1),1/2+i+exp(i*t2)/2,0];
21: xd=real(Zd)+4; zd=imag(Zd);
22: th=[-pi/2,pi]; nth=31;
23: kn=[1,2,n1+2,n1+3,n1+n2+3,n1+n2+4];
24:
25:
26: % Compute the geometrical properties
27: [v,rg,Irr,x,y,z,aprop]=volrev(...
28:                             xd,zd,kn,th,nth);
29: disp('  ')
30: disp('PROPERTIES OF A VOLUME OF REVOLUTION')
31: disp(' ')
32: disp('Results Using Function VOLREV')
33: disp(['Volume = ',num2str(v)]), %disp(' ')
34: disp(['Rg = [',num2str(rg(:)'),']']), %disp(' ')
35: disp('Inertia Tensor ='), disp(Irr), %disp(' ')
36: disp('Area Properties'), %disp(' ')
37: disp('     area      xcentr    zcentr.')
38: disp(aprop(1:3))
39: disp('      axx        axz         azz')
40: disp(aprop(4:6))
41:
42: % Run a second case to generate a dense set of
43: % surface coordinates to check results using
44: % function srfv.
45:
46: N1=61; T1=-pi:pi/N1:0; N2=41; T2=0:pi/N2:pi;
47: Zd=[0,exp(i*T1),1/2+i+exp(i*T2)/2,0];
48: xxd=real(Zd)+4; zzd=imag(Zd);
49: th=[-pi/2,pi]; Nth=121;
```

```
50: Kn=[1,2,N1+2,N1+3,N1+N2+3,N1+N2+4];
51:
52: [V,Rg,IRR,X,Y,Z]=volrev(...
53:                    xxd,zzd,Kn,th,Nth,1);
54: [vt,rct,vrrt]=srfv(X,Y,Z);
55: disp('Results Using Function SRFV')
56: disp(['Volume = ',num2str(vt)])
57: disp(['Rg = [',num2str(rct(:)'),']'])
58: disp('Inertia Tensor ='), disp(vrrt)
59:
60: %=========================================
61:
62: function [v,rg,Irr,X,Y,Z,aprop,xd,zd,kn]=...
63:                    volrev(xd,zd,kn,th,nth,noplot)
64: %
65: % [v,rg,Irr,X,Y,Z,aprop,xd,zd,kn]=...
66: %                  volrev(xd,zd,kn,th,nth,noplot)
67: %~~~~~~~~~~~~~~~~~~~~~~~~~~~~
68:
69: % This function computes geometrical properties
70: % for a volume of revolution resulting when a
71: % closed curve in the (x,z) plane is rotated,
72: % through given angular limits, about the z axis.
73: % The cross section of the volume is defined by
74: % a spline curve passed through data points
75: % (xd,zd) in the same manner as was done in
76: % function areaprop for plane areas.
77:
78: % xd,zd - data vectors defining the spline
79: %         interpolated boundary, which is
80: %         traversed in a counterclockwise
81: %         direction
82: % kn    - indices of any points where slope
83: %         discontinuity is allowed to turn
84: %         sharp corners
85: % p     - vector of volume properties containing
86: %         [v, xcg, ycg, zcg, vxx, vyy, vzz,...
87: %         vxy, vyz, vzx] where v is the volume,
88: %         (xcg,ycg,zcg) are coordinates of the
89: %         centroid, and the remaining properties
90: %         are volume integrals of the following
91: %         integrand:
92: %         [x.^, y.^2, z.^2, xy, yz, zx]*dxdyxz
93: % X,Y,Z - data arrays containing points on the
94: %         surface of revolution. Plotting these
```

```
95: %           points shows the solid volume with
96: %           the ends left open. Function fill3
97: %           is used to plot the surface with ends
98: %           closed
99: % aprop - a vector containing properties of the
100: %          area in the (x,z) plane which was used
101: %          to generate the volume. aprop=[area,...
102: %          xcentroidal, ycentroidal, axx, axz, azz].
103:
104: % User m functions called: rotasurf, gcquad,
105: %          curve2d, anglefun, splined
106: %----------------------------------------------
107: if nargin==0
108:   t1=-pi:pi/6:0; t2=0:pi/6:pi;
109:   Zd=[0,exp(i*t1),1/2+i+exp(i*t2)/2,0,-1];
110:   xd=real(Zd)+4; zd=imag(Zd);
111:   kn=[1,2,8,9,15,16];
112:   th=[-pi/2,pi]; nth=31;
113: end
114:
115: % Plot a surface of revolution based on the
116: % input data points
117: if nargin==6
118:   [X,Y,Z]=rotasurf(xd,zd,th,nth,1);
119: else
120:   [X,Y,Z]=rotasurf(xd,zd,th,nth); pause
121: end
122:
123: % Obtain base points and weight factors for the
124: % composite Gauss formula of order seven used in
125: % the numerical integration
126: nd=length(xd); nseg=nd-1;
127: [dum,bp,wf]=gcquad([],1,nd,7,nseg);
128:
129: % Evaluate complex points and derivative values
130: % on the spline curve which is rotated to form
131: % the volume of revolution
132: [u,uplot,up]=curve2d(xd,zd,kn,bp);
133: % plot(real(uplot),imag(uplot)), axis equal,shg
134: u=u(:); up=up(:); n=length(bp);
135: x=real(u); dx=real(up); z=imag(u);
136: dz=imag(up); da=x.*dz-z.*dx;
137:
138: % Evaluate line integrals for area properties
139: p=[ones(n,1), x, z, x.^2, x.*z, z.^2, x.^3,...
```

```
140:          (x.^2).*z, x.*(z.^2)].*repmat(da,1,9);
141: p=(wf(:)'*p)./[2 3 3 4 4 4 5 5 5];
142:
143: % Scale area properties by multipliers involving
144: % the rotation angle for the volume
145: f=anglefun(th(2))-anglefun(th(1));
146: v=f(1)*p(2); rg=f([2 3 1]).*p([4 4 5])/v;
147: vrr=[f([4 5 2]); f([5 6 3]); f([2 3 1])].*...
148:     [p([7 7 8]); p([7 7 8]); p([8 8 9])];
149: Irr=eye(3)*sum(diag(vrr))-vrr;
150: aprop=[p(1),p(2:3)/p(1),p(4:6)];
151:
152: %=========================================
153:
154: function f=anglefun(t)
155: % f=anglefun computes multipliers involving
156: % t, the rotation angle of the volume.
157: c=cos(t); s=sin(t);
158: f=[t,s,-c,(t+c*s)/2,s*s/2,(t-c*s)/2];
159:
160: %=========================================
161:
162: function [x,y,z,xd,zd]=rotasurf(xd,zd,th,nth,noplot)
163: % [x,y,z,xd,zd]=rotasurf(xd,zd,th,nth,noplot)
164: % This function generates points on a surface of
165: % revolution generated by rotating an area in
166: % the (x,z) plane about the z-axis
167: %
168: % xd,yz  - coordinate data for the curve in the
169: %          (x,y) which forms the cross section
170: % th     - [ThetaMin,ThetaMax] defining limits of
171: %          rotation angle about the z-axis
172: % nth    - number of theta values used to generate
173: %          surface values
174: % noplot - option given any value if no plot is
175: %          desired. Otherwise omit this value.
176: % x,y,z  - arrays of points on the surface
177: %
178: % User m functions called: none
179: %----------------------------------------------
180:
181: if nargin==0
182:   n1=9; t1=-pi:pi/n1:0; n2=6; t2=0:pi/n2:pi;
183:   Zd=[0,exp(i*t1),1/2+i+exp(i*t2)/2,0];
184:   xd=real(Zd)+4; zd=imag(Zd);
```

```
185:    th=[-pi/2,pi]; nth=31;
186: end
187: xd=xd(:); zd=zd(:); nd=length(xd);
188: t=linspace(th(1),th(2),nth);
189: x=xd*cos(t); y=xd*sin(t); z=repmat(zd,1,nth);
190: if nargin==5, return; end
191: close; surf(x,y,z), title('VOLUME OF REVOLUTION')
192: xlabel('x axis'), ylabel('y axis')
193: zlabel('z axis'), colormap([1 1 1]); hold on
194: fill3(x(:,1),y(:,1),z(:,1),'w')
195: fill3(x(:,end),y(:,end),z(:,end),'w')
196: axis equal, grid on, hold off, shg
197:
198: %==============================================
199:
200: % function [z,zplot,zp]=curve2d(xd,yd,kn,t)
201: % See Appendix B
202:
203: %==============================================
204:
205: % function [val,bp,wf]=gcquad(func,xlow,...
206: %                   xhigh,nquad,mparts,varargin)
207: % See Appendix B
208:
209: %==============================================
210:
211: % function range=cubrange(xyz,ovrsiz)
212: % See Appendix B
213:
214: %==============================================
215:
216: % function val=splined(xd,yd,x,if2)
217: % See Appendix B
218:
219: %==============================================
220:
221: % function [v,rc,vrr]=srfv(x,y,z)
222: % See Appendix B
```

5.5 Computing Solid Properties Using Triangular Surface Elements and Using Symbolic Math

In this section a numerical method is developed to compute properties of a solid covered by triangular surface elements. An example problem is analyzed by a numerical method and also by use of the symbolic math toolbox. Results of the two analyses are compared.

Many familiar solid bodies such as an ellipsoid, a conical frustum, or a torus have surfaces readily parameterized by equations of the form

$$\boldsymbol{R} = \boldsymbol{R}(u, v), \quad U_1 \le u \le U_2, \quad V_1 \le v \le V_2.$$

This is the type of equation implied when MATLAB function **surf** uses rectangular X, Y, Z coordinate arrays to depict a curvilinear coordinate net covering a surface. The surface is approximated by a series of quadrilateral surface patches. Geometrical properties of the related solid can be computed approximately by dividing each quadrilateral into two triangular patches, and accumulating the surface integral contributions of the triangles. This approach is attractive because the surface integral properties of triangles can be computed exactly, and all triangles can be processed in parallel. Although the geometrical properties for a solid covered by triangular patches can be computed exactly, the reader should realize that many surface elements may be required to achieve several digit accuracy for highly curved surfaces.

To fix our ideas, consider the solid in Figure 5.5 which resembles a twisted rope. This body has its outer surface (as distinguished from its ends) described by the following set of equations:

$$\begin{aligned}
&x = x_0 + \rho\cos(p), \quad y = y_0 + \rho\sin(p), \quad z = z_0 - \xi\sin(m\,p) + \eta\cos(m\,p), \\
&\rho = a + \xi\cos(m\,p) + \eta\sin(m\,p), \quad 0 \le t \le 2\pi, \quad 0 \le p \le 3\pi, \\
&\xi = b\cos(t)\ |\cos(t)|\,, \quad \eta = b\sin(t)\ |\cos(t)|\,.
\end{aligned}$$

The cross section of the solid is two circular disks touching tangentially. The solid is swept out as the centroid of the area (where the circles touch) moves along a helical path and twists simultaneously. The parameter choices used in our example are

$$a = 3, \; b = 1, \; m = 6, \; x_0 = y_0 = 0, \; z_0 = 3\pi/2$$

which places the centroid of the solid on the y-axis and makes the ends of the rope lie in the xz plane. Then the geometrical property contributions from both end surfaces are zero because $\hat{\boldsymbol{\eta}} \cdot \boldsymbol{R}$ vanishes on the ends.

Let us next think about a solid with its surface composed of triangular patches. For a generic patch with corners at $\boldsymbol{R}_i, \boldsymbol{R}_j, \boldsymbol{R}_k$, denote the surface area as S_T and the unit surface normal as $\hat{\boldsymbol{\eta}}$. Then

$$\hat{\boldsymbol{\eta}}\, S_T = \frac{1}{2}(\boldsymbol{R}_j - \boldsymbol{R}_i) \times (\boldsymbol{R}_k - \boldsymbol{R}_i),$$

and the triangle centroid is at

$$\boldsymbol{R}_C = \frac{1}{3}(\boldsymbol{R}_i + \boldsymbol{R}_j + \boldsymbol{R}_k).$$

If h is the normal distance from the origin to the plane containing the triangle, then $h = \hat{\boldsymbol{\eta}} \cdot \boldsymbol{R}_i$ and $S_T = |\hat{\boldsymbol{\eta}}\, S_T|$. The first two volume properties are just

$$V_T = \frac{1}{3}\iint\limits_{S_T} \hat{\boldsymbol{\eta}} \cdot \boldsymbol{R}\, dS = \frac{h}{3}\, S_T$$

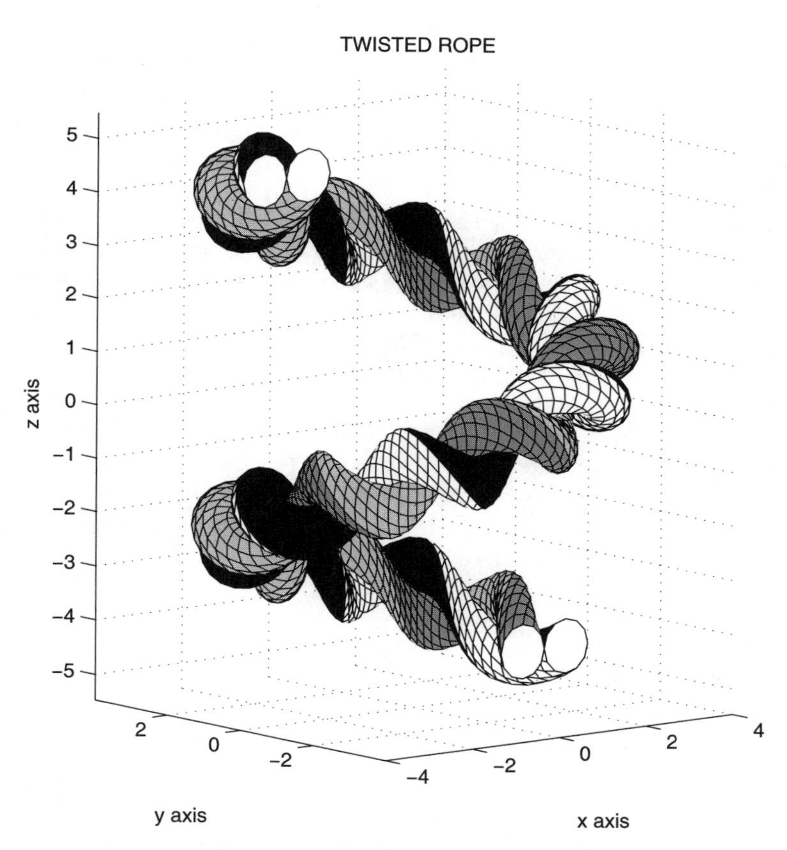

Figure 5.5: Solid Resembling a Twisted Rope

and

$$(\mathbf{V}_{\boldsymbol{R}})_T = \frac{1}{4} \iint_{S_T} (\hat{\boldsymbol{\eta}} \cdot \boldsymbol{R})\, \boldsymbol{R}\; dS = \frac{h}{4}\, S_T\, \boldsymbol{R}_C \,.$$

The remaining inertial property integral is

$$(\mathbf{V}_{\boldsymbol{RR}})_T = \frac{1}{5} \iint_{S_T} (\hat{\boldsymbol{\eta}} \cdot \boldsymbol{R})\, \boldsymbol{R}\, \boldsymbol{R}'\; dS = \frac{h}{60}\, S_T\, (\boldsymbol{R}_i\, \boldsymbol{R}_i' + \boldsymbol{R}_j\, \boldsymbol{R}_j' + \boldsymbol{R}_k\, \boldsymbol{R}_k' + 9\boldsymbol{R}_C\; \boldsymbol{R}_C').$$

These formulas were used to develop the function **srfv** which computes geometrical properties for a surface described by the same type of data arrays as those used by the function **surf**. Each quadrilateral patch is divided into two triangles, and the contributions of all triangles are accumulated in vectorized mode for computational efficiency.

The function **ropesymu** in the following program calls function **twistrope** to perform numerical computation, function **twistprop** to perform symbolic computation, and function **ropedraw** to plot the geometry of the twisted rope. **Twistrope** calls the function **srfv** which is a general routine to compute properties of solid bodies modeled with triangular surface elements. The numerical example employs point arrays of dimension 804 by 100 to obtain the numerical solution. Results for the numerical and symbolic computations are shown next along with the computer code. Note that the numerical and exact solutions agreed within 0.2 percent. The numerical solution took about 1.3 secs compared with 314 seconds for the symbolic solution. Even though the symbolic solution took 238 times as long to compute as the numerical solution, the symbolic coding was simple and might be appealing in specific situations where the related integrals can be evaluated exactly.

5.6 Numerical and Symbolic Results for the Example

```
   COMPARISON OF NUMERICAL AND SYMBOLIC
GEOMETRICAL PROPERTIES FOR A TWISTED ROPE

FOR THE TRIANGULAR SURFACE PATCH MODEL
Volume = 44.3239
Rg = [1.6932e-015  0.64979 3.0068e-015]
Irr =
  548.6015   -0.0000   29.0040
   -0.0000  548.6015   -0.0000
   29.0040   -0.0000  423.7983

Computation Time = 1.3194 Secs.
```

```
FOR THE SYMBOLIC MODEL
Volume = 44.4132
Rg = [0      0.64982          0]
Irr =
  549.7423         0          29.0639
     0         549.7423          0
  29.0639          0         424.7014

Computation Time = 314.28 Secs.

 NUMERICAL APPROXIMATION ERROR USING TRIANGULAR
SURFACE PATCHES. THE ERROR VALUES ARE DEFINED AS
        NORM(APPROX.-EXACT)/NORM(EXACT)
Volume Error = 0.0020102
Centroidal Radius Error = 4.7287e-005
Inertia Tensor Error = 0.0020768

COMPARISON OF SOLUTION TIMES
(Symbolic Time)/(Numerical Time) = 238.1992
```

Program ropesymu

```
1: function [vn,rcn,irrn,vs,rcs,irrs,times,nt,np]=...
2:                       ropesymu(A,B,M,X0,Y0,Z0,nt,np)
3: %
4: % [vn,rcn,irrn,vs,rcs,irrs,times,nt,np]=ropesymu(...
5: %                                   A,B,M,X0,Y0,Z0,nt,np)
6: %~~~~~~~~~~~~~~~~~~~~~~~~~~~~~~~~~~~~~~~~~~~~~~~~~~~~~~
7: %
8: % This program computes geometrical properties of a
9: % twisted rope having a cross section which is two
10: % circles of diameter B touching tangentially. The
11: % tangency point is at distance A from the rotation
12: % axis z. As the area is rotated, it is also twisted
13: % in a helical fashion. For a complete revolution
14: % about the z axis, the area is twisted through m
15: % turns. The resulting surface resembles a rope
16: % composed of two strands. Two results are obtained
17: % 1) by a numerical method where the surface is
18: % modeled with triangular surface patches and
19: % 2) by symbolic math. See functions twistrope and
20: % twistprop for descriptions of the problem parameters.
```

```
21: % Numerical results and computation times for the two
22: % methods are compared, and the related surface
23: % geometry is plotted
24: %
25: % User functions called: twistrope twistprop ropedraw
26: %-------------------------------------------------
27:
28: if nargin==0 % Default data case
29:   A=3; B=1; m=6; np=201; nt=25;
30:   X0=0; Y0=0; Z0=-3*pi/2; M=6;
31: end
32:
33: disp(' ')
34: disp('   COMPARISON OF NUMERICAL AND SYMBOLIC')
35: disp('GEOMETRICAL PROPERTIES FOR A TWISTED ROPE')
36:
37: % Run the first time to get a crude grid for plotting
38: [vn,rcn,irrn,x,y,z,c]=twistrope(A,B,M,X0,Y0,Z0,nt,np);
39:
40: % Numerical solution using a dense point grid to get
41: % close comparison with exact results. Calculations
42: % are run repeatedly for accurate timing.
43: Nt=4*nt; Np=4*np; n=50; tic;
44: for i=1:n
45:  [vn,rcn,irrn]=twistrope(A,B,M,X0,Y0,Z0,Nt,Np);
46: end
47: timn=toc/n;
48:
49: % Perform the symbolic analysis. This takes a long
50: % time.
51: tic;
52: [v,rc,vrr,vs,rcs,irrs]=twistprop(A,B,M,X0,Y0,Z0);
53: tims=toc; times=[timn,tims];
54:
55: disp(' ')
56: disp('FOR THE TRIANGULAR SURFACE PATCH MODEL')
57: disp(['Volume = ',num2str(vn)])
58: disp(['Rg = [',num2str(rcn(:)'),']'])
59: disp('Irr = '), disp(irrn)
60: disp(['Computation Time = ',num2str(timn),' Secs.'])
61:
62: % Print numerical comparisons of results
63: disp(' ')
64: disp('FOR THE SYMBOLIC MODEL')
```

```
65: disp(['Volume = ',num2str(vs)])
66: disp(['Rg = [',num2str(rcs(:)'),']'])
67: disp('Irr = '), disp(irrs)
68: disp(['Computation Time = ',num2str(tims),' Secs.'])
69:
70: disp(' ')
71: disp(' NUMERICAL APPROXIMATION ERROR USING TRIANGULAR')
72: disp('SURFACE PATCHES. THE ERROR VALUES ARE DEFINED AS')
73: disp('          NORM(APPROX.-EXACT)/NORM(EXACT)')
74: evol=abs(vn-vs)/vs; erad=norm(rcs(:)-rcn(:))/norm(rcs);
75: einert=norm(irrn-irrs)/norm(irrs);
76: disp(['Volume Error = ',num2str(evol)])
77: disp(['Centroidal Radius Error = ',num2str(erad)])
78: disp(['Inertia Tensor Error = ',num2str(einert)])
79:
80: disp(' ')
81: disp('COMPARISON OF SOLUTION TIMES')
82: disp(['(Symbolic Time)/(Numerical Time) = ',...
83:       num2str(tims/timn)])
84: disp(' ')
85:
86: % Draw the surface using a crude grid to avoid
87: % crowded grid lines
88: ropedraw(A,B,np,nt,M,X0,Y0,Z0);
89:
90: %==========================================
91:
92: function [x,y,z,t]=ropedraw(a,b,np,nt,m,x0,y0,z0)
93: %
94: % [x,y,z,t]=ropedraw(a,b,np,mp,m,x0,y0,z0)
95: % ~~~~~~~~~~~~~~~~~~~~~~~~~~~~~~~~~~~~~~~~
96: % This function draws the twisted rope.
97: if nargin==0
98:   a=3; b=1; np=200; nt=25; m=6;
99:   x0=0; y0=0; z0=-3*pi/2;
100: end
101:
102: % Draw the surface
103: t=linspace(0,2*pi,nt); p=linspace(0,3*pi,np)';
104: t=repmat(t,np,1); p=repmat(p,1,nt);
105: xi=b*cos(t).*abs(cos(t)); eta=b*sin(t).*abs(cos(t));
106: rho=a+xi.*cos(m*p)+eta.*sin(m*p);
107: x=rho.*cos(p)+x0; y=rho.*sin(p)+y0;
108: z=-xi.*sin(m*p)+eta.*cos(m*p)+p+z0;
109: close; surf(x,y,z,t), title('TWISTED ROPE')
```

```
110: xlabel('x axis'), ylabel('y axis'), zlabel('z axis')
111: colormap('prism(4)'), axis equal, hold on
112:
113: % Fill the ends
114: fill3(x(1,:),y(1,:),z(1,:),'w')
115: fill3(x(end,:),y(end,:),z(end,:),'w')
116: view([-40,10]), hold off, shg
117:
118: %===========================================
119:
120: function [v,rc,vrr,V,Rc,Irr]=twistprop(A,B,M,X0,Y0,Z0)
121: %
122: % [v,rc,vrr,V,Rc,Irr]=twistprop(A,B,M,X0,Y0,Z0)
123: % ~~~~~~~~~~~~~~~~~~~~~~~~~~~~~~~~~~~~~~~~~~~~~
124: % This function computes geometrical properties of
125: % a twisted rope. Exact results are obtained using
126: % symbolic math to evaluate three surface integrals
127: % for the volume, centroidal radius, and inertia
128: % tensor. The symbolic calculations take about five
129: % minutes to run.
130: %
131: % A,B,N      - parameters defining the twisted rope
132: % X0,Y0,Z0 - center coordinates for the centroid of
133: %              the twisted rope
134: % v,rc       - symbolic formulas for the volume and
135: %              centroid radius
136: % vrr        - symbolic formula for integral of
137: %              r*r'*d(vol)
138: % V,Rc       - numerical values for volume and
139: %              centroid radius
140: % Irr        - numerical value for the inertia tensor
141:
142: if nargin==0
143:    A=6; B=1; M=6; X0=1; Y0=2; Z0=3;
144: end
145:
146: syms a b m t p xi eta rho x y z r rt rp real
147: syms x0 y0 z0 real
148: syms n dv dv1 v vr1 vr rg vrr1 vrr real
149: a=sym(A); b=sym(B); Pi=sym('pi');
150: x0=sym(X0); y0=sym(Y0); z0=sym(Z0);
151:
152: % Surface equation for the twisted rope
153: xi=b*cos(t)*abs(cos(t));
154: eta=b*sin(t)*abs(cos(t));
```

```
155: rho=a+xi*cos(m*p)+eta*sin(m*p);
156: x=rho*cos(p)+x0; y=rho*sin(p)+y0;
157: z=-xi*sin(m*p)+eta*cos(m*p)+p+z0;
158: Pi=sym('pi');
159:
160: % Tangent vectors
161: r=[x;y;z]; rt=diff(r,t); rp=diff(r,p);
162:
163: % Integrate to get the volume
164: dv=det([r,rp,rt]); dv1=int(dv,t,0,2*Pi);
165: v=simple(int(dv1,p,0,3*Pi)/3);
166:
167: % First moment of volume
168: vr1=int(r*dv,t,0,2*Pi);
169: vr=simple(int(vr1,p,0,3*Pi)/4);
170:
171: % Radius to the centroid
172: rc=simple(vr/v);
173:
174: % Integral of r*r'*d(vol)
175: vrr1=int(r*r'*dv,t,0,2*Pi);
176: vrr=simple(int(vrr1,p,0,3*Pi)/5);
177:
178: % Obtain numerical values
179: V=double(subs(v,{a,b,m,x0,y0,z0},...
180:    {A,B,M,X0,Y0,Z0}));
181: Rc=double(subs(rc,{a,b,m,x0,y0,z0},...
182:    {A,B,M,X0,Y0,Z0}));
183: Irr=double(subs(vrr,{a,b,m,x0,y0,z0},...
184:    {A,B,M,X0,Y0,Z0}));
185:
186: % Rigid body inertia tensor for a
187: % body of unit mass density
188: Irr=eye(3,3)*sum(diag(Irr))-Irr;
189:
190: %=========================================
191:
192: function [v,rc,vrr,x,y,z,t]=twistrope(...
193:                           a,b,m,x0,y0,z0,nt,np)
194: %
195: % [v,rc,vrr,x,y,z,t]=twistrope(...
196: %                     a,b,m,x0,y0,z0,nt,nm)
197: % ~~~~~~~~~~~~~~~~~~~~~~~~~~~~~~~~~~~~~~~~
198: % Geometrical properties of a twisted rope.
199: % This example takes 1.3 seconds to run
```

```
200: if nargin<8, np=321; end; if nargin<7, nt=161; end
201: if nargin==0
202:    a=6; b=1; m=6; x0=1; y0=2; z0=3;
203: end
204: t=linspace(0,2*pi,nt); p=linspace(0,3*pi,np)';
205: t=repmat(t,np,1); p=repmat(p,1,nt);
206:
207: % Surface equation for the twisted rope
208:
209: xi=b*cos(t).*abs(cos(t));
210: eta=b*sin(t).*abs(cos(t));
211: rho=a+xi.*cos(m*p)+eta.*sin(m*p);
212: x=rho.*cos(p)+x0; y=rho.*sin(p)+y0;
213: z=-xi.*sin(m*p)+eta.*cos(m*p)+p+z0;
214:
215: [v,rc,vrr]=srfv(x,y,z);
216:
217: %=========================================
218:
219: function [v,rc,vrr]=srfv(x,y,z)
220: %
221: % [v,rc,vrr]=srfv(x,y,z)
222: % ~~~~~~~~~~~~~~~~~~~~~~
223: %
224: % This function computes the volume, centroidal
225: % coordinates, and inertial tensor for a volume
226: % covered by surface coordinates contained in
227: % arrays x,y,z
228: %
229: % x,y,z   - matrices containing the coordinates
230: %           of a grid of points covering the
231: %           surface of the solid
232: % v       - volume of the solid
233: % rc      - centroidal coordinate vector of the
234: %           solid
235: % vrr     - inertial tensor for the solid with the
236: %           mass density taken as unity
237: %
238: % User functions called: scatripl proptet
239: %----------------------------------------------
240:
241: % p=inline(...
242: %  'v*(eye(3)*(r(:)''*r(:))-r(:)*r(:)'')','v','r');
243:
244: %d=mean([x(:),y(:),z(:)]);
```

```
245: %x=x-d(1); y=y-d(2); z=z-d(3);
246:
247: [n,m]=size(x); i=1:n-1; I=i+1; j=1:m-1; J=j+1;
248: xij=x(i,j); yij=y(i,j); zij=z(i,j);
249: xIj=x(I,j); yIj=y(I,j); zIj=z(I,j);
250: xIJ=x(I,J); yIJ=y(I,J); zIJ=z(I,J);
251: xiJ=x(i,J); yiJ=y(i,J); ziJ=z(i,J);
252:
253: % Tetrahedron volumes
254: v1=scatripl(xij,yij,zij,xIj,yIj,zIj,xIJ,yIJ,zIJ);
255: v2=scatripl(xij,yij,zij,xIJ,yIJ,zIJ,xiJ,yiJ,ziJ);
256: v=sum(sum(v1+v2));
257:
258: % First moments of volume
259: X1=xij+xIj+xIJ; X2=xij+xIJ+xiJ;
260: Y1=yij+yIj+yIJ; Y2=yij+yIJ+yiJ;
261: Z1=zij+zIj+zIJ; Z2=zij+zIJ+ziJ;
262: vx=sum(sum(v1.*X1+v2.*X2));
263: vy=sum(sum(v1.*Y1+v2.*Y2));
264: vz=sum(sum(v1.*Z1+v2.*Z2));
265:
266: % Second moments of volume
267: vrr=proptet(v1,xij,yij,zij,xIj,yIj,zIj,...
268:      xIJ,yIJ,zIJ,X1,Y1,Z1)+...
269:      proptet(v2,xij,yij,zij,xIJ,yIJ,zIJ,...
270:      xiJ,yiJ,ziJ,X2,Y2,Z2);
271: rc=[vx,vy,vz]/v/4; vs=sign(v);
272: v=abs(v)/6; vrr=vs*vrr/120;
273: vrr=[vrr([1 4 5]), vrr([4 2 6]), vrr([5 6 3])]';
274: vrr=eye(3,3)*sum(diag(vrr))-vrr;
275:
276: %vrr=vrr-p(v,rc)+p(v,rc+d); rc=rc+d;
277:
278: %==========================================
279:
280: function v=scatripl(ax,ay,az,bx,by,bz,cx,cy,cz)
281: %
282: % v=scatripl(ax,ay,az,bx,by,bz,cx,cy,cz)
283: % ~~~~~~~~~~~~~~~~~~~~~~~~~~~~~~~~~~~~~~
284: % Scalar triple product dot(cross(a,b),c) where
285: % the cartesian components of vectors a,b,and c
286: % are given in arrays of the same size.
287: v=ax.*(by.*cz-bz.*cy)+ay.*(bz.*cx-bx.*cz)...
288:   +az.*(bx.*cy-by.*cx);
289:
```

```
290: % ===========================================
291:
292: function vrr=tensprod(v,x,y,z)
293: %
294: % vrr=tensprod(v,x,y,z)
     % ~~~~~~~~~~~~~~~~~~~~~
295: %
296: % This function forms the various components
297: % of v*R*R'. The calculation is vectorized
298: % over arrays of points
299: vxx=sum(sum(v.*x.*x)); vyy=sum(sum(v.*y.*y));
300: vzz=sum(sum(v.*z.*z)); vxy=sum(sum(v.*x.*y));
301: vxz=sum(sum(v.*x.*z)); vyz=sum(sum(v.*y.*z));
302: vrr=[vxx; vyy; vzz; vxy; vxz; vyz];
303:
304: % ===========================================
305:
306: function vrr=proptet(v,x1,y1,z1,x2,y2,z2,...
307:                      x3,y3,z3,xc,yc,zc)
308: %
309: % vrr=proptet(v,x1,y1,z1,x2,y2,z2,x3,y3,z3,...
310: %                                  xc,yc,zc)
     % ~~~~~~~~~~~~~~~~~~~~~~~~~~~~~~~~~~~~~~~~~~~
311: %
312: % This function computes tensor properties of a
313: % tetrahedron with its base being a triangular
314: % surface and its apex at the origin
315: vrr=tensprod(v,x1,y1,z1)+tensprod(v,x2,y2,z2)+...
316:     tensprod(v,x3,y3,z3)+tensprod(v,xc,yc,zc);
```

5.7 Geometrical Properties of a Polyhedron

A polyhedron is a solid covered by polygonal faces. Since polyhedra with sufficiently many faces can approximate volumes of complex shape, computing the volume, centroidal position, and inertia tensor of a polyhedron has useful applications. A polyhedron can be treated as the combination of a number of pyramids with bases which are the polyhedron faces and apexes located at the coordinate origin. Once the geometrical properties of a pyramid are known, results for a polyhedron are found by combining results for all faces [111].

Consider a general volume V covered by surface S. It follows from the divergence theorem of Gauss [59] that

$$\iiint_V X^n Y^m Z^\ell \, dX \, dY \, dZ = \frac{1}{n+m+\ell+3} \iint_S X^n Y^m Z^\ell (\hat{\boldsymbol{\eta}} \cdot \boldsymbol{R}) dS$$

where $\hat{\boldsymbol{\eta}}$ is the outward directed surface normal and $\boldsymbol{R}$ is the column vector $[X;\ Y;\ Z]$.

This formula implies

$$V = \iiint_V dX\, dY\, dZ = \frac{1}{3} \iint_S \hat{\boldsymbol{\eta}} \cdot \boldsymbol{R}\, dS,$$

$$\mathbf{V}_{\boldsymbol{R}} = \iiint_V \boldsymbol{R}\, dX\, dY\, dZ = \frac{1}{4} \iint_S \boldsymbol{R}\, (\hat{\boldsymbol{\eta}} \cdot \boldsymbol{R})\, dS,$$

and

$$\mathbf{V}_{\boldsymbol{RR}} = \iiint_V \boldsymbol{R}\boldsymbol{R}'\, dX\, dY\, dZ = \frac{1}{5} \iint_S \boldsymbol{R}\boldsymbol{R}'\, (\hat{\boldsymbol{\eta}} \cdot \boldsymbol{R})\, dS$$

where $\boldsymbol{R}'$ means the transpose of $\boldsymbol{R}$. Let us apply these formulas to a pyramid with the apex at $\boldsymbol{R} = 0$ and the base being a planar region S_b of area A. For points on the side of the pyramid of height h we find that $\hat{\boldsymbol{\eta}} \cdot \boldsymbol{R} = 0$, and for points on the base $\hat{\boldsymbol{\eta}} \cdot \boldsymbol{R} = h$. Consequently

$$V = \frac{1}{3} \iint_{S_b} h\, dS = \frac{h}{3} A,$$

$$\mathbf{V}_{\boldsymbol{R}} = \frac{1}{4} \iint_{S_b} \boldsymbol{R} h\, dS = \frac{h}{4} \iint_{S_b} \boldsymbol{R}\, dS,$$

$$\mathbf{V}_{\boldsymbol{RR}} = \frac{1}{5} \iint_{S_b} \boldsymbol{R}\boldsymbol{R}' h\, dS = \frac{h}{5} \iint_{S_b} \boldsymbol{R}\boldsymbol{R}'\, dS.$$

The volume is equal to one third of the height times the base area, regardless of the base shape. If $\boldsymbol{R}_b$ and $\boldsymbol{R}_p$ signify the centroidal radii of the base and the pyramid volume, respectively, we get

$$\boldsymbol{R}_p = \frac{\mathbf{V}_{\boldsymbol{R}}}{V} = \frac{\frac{h}{4} \boldsymbol{R}_b A}{\frac{h}{3} A} = \frac{3}{4} \boldsymbol{R}_b.$$

Therefore, the centroid of the volume lies $\frac{3}{4}$ of the way along a line from the apex to the centroid of the base. For any planar area it is not hard to show that the area A and unit surface normal $\hat{\boldsymbol{\eta}}$ can be computed using the line integral

$$\hat{\boldsymbol{\eta}}\, A = \frac{1}{2} \oint_L \boldsymbol{R} \times d\boldsymbol{R}.$$

The last formula simplifies for a polygon having corners at $\boldsymbol{R}_1, \boldsymbol{R}_2, \ldots, \boldsymbol{R}_n$ to yield

$$\hat{\boldsymbol{\eta}}\, A = \frac{1}{2} \sum_{j=1}^{n} \boldsymbol{R}_j \times \boldsymbol{R}_{j+1} \text{ where } \boldsymbol{R}_{n+1} = \boldsymbol{R}_1.$$

To compute the first and second area moments for a general planar area, it is helpful to introduce coordinates centered anywhere in the plane containing the base. We let

$$\boldsymbol{R} = \boldsymbol{R}_0 + \hat{\boldsymbol{\imath}}\, x + \hat{\boldsymbol{\jmath}}\, y$$

where $\boldsymbol{R}_0$ is a vector to a point in the plane of the base, and $\hat{\boldsymbol{\imath}}$ and $\hat{\boldsymbol{\jmath}}$ are orthonormal unit vectors which are tangent to the plane and are chosen such that $\hat{\boldsymbol{\imath}}$, $\hat{\boldsymbol{\jmath}}$, $\hat{\boldsymbol{\eta}}$ form a right-handed system. The local coordinates (x, y) can be computed using

$$x = (\boldsymbol{R} - \boldsymbol{R}_0)^{'}\hat{\boldsymbol{\imath}} \text{ and } y = (\boldsymbol{R} - \boldsymbol{R}_0)^{'}\hat{\boldsymbol{\jmath}}.$$

Then we get

$$\begin{aligned}\mathbf{V}_{\boldsymbol{R}} &= \frac{h}{4}\iint\limits_{S_b} (\boldsymbol{R}_0 + \hat{\boldsymbol{\imath}}\, x + \hat{\boldsymbol{\jmath}}\, y)\, dx\, dy \\ &= \frac{h}{4}(\boldsymbol{R}_0 + \hat{\boldsymbol{\imath}}\, \bar{x} + \hat{\boldsymbol{\jmath}}\, \bar{y})A \\ &= \frac{h}{4}\boldsymbol{R}_b A\end{aligned}$$

where $(\bar{x}, \bar{y})$ are the centroidal coordinates of the area measured relative to the local axes. Similarly we have

$$\begin{aligned}\mathbf{V}_{\boldsymbol{RR}} = \frac{h}{5}\iint\limits_{S_b} [\, &\boldsymbol{R}_0\boldsymbol{R}_0^{'} + (\boldsymbol{R}_0\hat{\boldsymbol{\imath}}^{'} + \hat{\boldsymbol{\imath}}\boldsymbol{R}_0^{'})x + (\boldsymbol{R}_0\hat{\boldsymbol{\jmath}}^{'} + \hat{\boldsymbol{\jmath}}\boldsymbol{R}_0^{'})y + \\ &(\hat{\boldsymbol{\imath}}\,\hat{\boldsymbol{\jmath}}^{'} + \hat{\boldsymbol{\jmath}}\,\hat{\boldsymbol{\imath}}^{'})xy + \hat{\boldsymbol{\imath}}\,\hat{\boldsymbol{\imath}}^{'}x^2 + \hat{\boldsymbol{\jmath}}\,\hat{\boldsymbol{\jmath}}^{'}y^2\,]dx\, dy \\ = \frac{h}{5}&\left[\boldsymbol{R}_0\boldsymbol{R}_0^{'} + (\boldsymbol{R}_0\hat{\boldsymbol{\imath}}^{'} + \hat{\boldsymbol{\imath}}\,\boldsymbol{R}_0^{'})\bar{x} + (\boldsymbol{R}_0\hat{\boldsymbol{\jmath}}^{'} + \hat{\boldsymbol{\jmath}}\,\boldsymbol{R}_0^{'})\bar{y}\right] A + \\ \frac{h}{5}&\left[\hat{\boldsymbol{\imath}}\hat{\boldsymbol{\imath}}^{'}A_{xx} + \hat{\boldsymbol{\jmath}}\,\hat{\boldsymbol{\jmath}}^{'}A_{yy} + (\hat{\boldsymbol{\imath}}\,\hat{\boldsymbol{\jmath}}^{'} + \hat{\boldsymbol{\jmath}}\,\hat{\boldsymbol{\imath}}^{'})A_{xy}\right]\end{aligned}$$

where

$$A_{xx} = \iint\limits_{S_b} x^2\, dx\, dy\ ,\ A_{xy} = \iint\limits_{S_b} xy\, dx\, dy\ ,\ A_{yy} = \iint\limits_{S_b} y^2\, dx\, dy.$$

The formula for $\mathbf{V}_{\boldsymbol{RR}}$ simplifies when $\boldsymbol{R}_0$ is chosen as the centroidal radius $\boldsymbol{R}_b$. Then $\bar{x} = \bar{y} = 0$ so that

$$\mathbf{V}_{\boldsymbol{RR}} = \frac{h}{5}[\boldsymbol{R}_b\,\boldsymbol{R}_b^{\prime}A + \hat{\boldsymbol{\imath}}\,\hat{\boldsymbol{\imath}}^{'}A_{xx}^b + \hat{\boldsymbol{\jmath}}\,\hat{\boldsymbol{\jmath}}^{'}A_{yy}^b + (\hat{\boldsymbol{\imath}}\,\hat{\boldsymbol{\jmath}}^{'} + \hat{\boldsymbol{\jmath}}\,\hat{\boldsymbol{\imath}}^{'})A_{xy}^b]$$

with the quantities A_{xx}^b, A_{yy}^b, A_{xy}^b denoting reference to the centroidal axes.

The analysis to compute polyhedron properties can now be completed using vector algebra along with area property calculations of the type introduced earlier. To define data for a particular polyhedron we provide vectors x, y, z containing global coordinates of all corners. We also employ a matrix named `idface` having a row dimension equal to the number of faces on the polyhedron and a column dimension equal to the largest number of corners on any face. Row $\imath$ of `idface` consists of corner indices of the $\imath$'th face with the row being padded with zeros on the right if necessary. Each face is traversed in the counterclockwise sense relative to the outward normal. Consider a figure showing a triangular block with a hole, having twelve corners and eight faces as shown in Figure 5.6. The required geometry descriptions are defined in example **polhdrun**. The results produced for this example are

```
>> polhdrun;

v = 15

rc =
     0.0000
     2.6667
     1.3333

vrr =
     5.0000     0.0000     0.0000
     0.0000   120.8333    60.4167
     0.0000    60.4167    40.8333

irr =
   161.6667    -0.0000    -0.0000
    -0.0000    45.8333   -60.4167
    -0.0000   -60.4167   125.8333
```

These values can be easily verified by manual calculations.

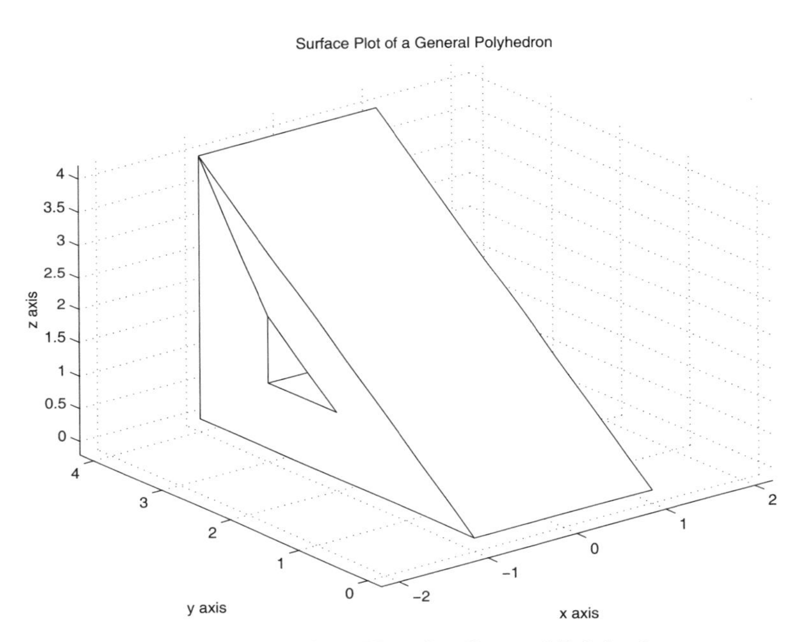

Figure 5.6: Surface Plot of a General Polyhedron

Program polhdrun

```
1: function polhdrun
2: % Example: polhdrun
3: % ~~~~~~~~~~~~~~~~~
4: %
5: % This program illustrates the use of routine
6: % polhedrn to calculate the geometrical
7: % properties of a polyhedron.
8: %
9: % User m functions called:
10: %       crosmat, polyxy, cubrange, pyramid,
11: %       polhdplt, polhedrn
12:
13: x=[2 2 2 2 2 2 0 0 0 0 0 0]-1;
14: y=[0 4 4 2 3 3 0 4 4 2 3 3];
15: z=[0 0 4 1 1 2 0 0 4 1 1 2];
16: idface=[1  2  3  6  5  4  6  3; ...
17:         1  3  9  7  0  0  0  0; ...
18:         1  7  8  2  0  0  0  0; ...
19:         2  8  9  3  0  0  0  0; ...
20:         7  9 12 10 11 12  9  8; ...
21:         4 10 12  6  0  0  0  0; ...
22:         4  5 11 10  0  0  0  0; ...
23:         5  6 12 11  0  0  0  0];
24: polhdplt(x,y,z,idface,[1,1,1]);
25: [v,rc,vrr,irr]=polhedrn(x,y,z,idface)
26:
27: %=============================================
28:
29: function [v,rc,vrr,irr]=polhedrn(x,y,z,idface)
30: %
31: % [v,rc,vrr,irr]=polhedrn(x,y,z,idface)
32: % ~~~~~~~~~~~~~~~~~~~~~~~~~~~~~~~~~~~~~
33: %
34: % This function determines the volume,
35: % centroidal coordinates and inertial moments
36: % for an arbitrary polyhedron.
37: %
38: % x,y,z  - vectors containing the corner
39: %          indices of the polyhedron
40: % idface - a matrix in which row j defines the
41: %          corner indices of the j'th face.
```

```
42: %           Each face is traversed in a
43: %           counterclockwise sense relative to
44: %           the outward normal. The column
45: %           dimension equals the largest number
46: %           of indices needed to define a face.
47: %           Rows requiring fewer than the
48: %           maximum number of corner indices are
49: %           padded with zeros on the right.
50: %
51: % v      - the volume of the polyhedron
52: % rc     - the centroidal radius
53: % vrr    - the integral of R*R'*d(vol)
54: % irr    - the inertia tensor for a rigid body
55: %          of unit mass obtained from vrr as
56: %          eye(3,3)*sum(diag(vrr))-vrr
57: %
58: % User m functions called: pyramid
59: %----------------------------------------------
60:
61: r=[x(:),y(:),z(:)]; nf=size(idface,1);
62: v=0; vr=0; vrr=0;
63: for k=1:nf
64:   i=idface(k,:); i=i(find(i>0));
65:   [u,ur,urr]=pyramid(r(i,:));
66:   v=v+u; vr=vr+ur; vrr=vrr+urr;
67: end
68: rc=vr/v; irr=eye(3,3)*sum(diag(vrr))-vrr;
69:
70: %=============================================
71:
72: function [area,xbar,ybar,axx,axy,ayy]=polyxy(x,y)
73: %
74: % [area,xbar,ybar,axx,axy,ayy]=polyxy(x,y)
75: % ~~~~~~~~~~~~~~~~~~~~~~~~~~~~~~~~~~~~~~~~~
76: %
77: % This function computes the area, centroidal
78: % coordinates, and inertial moments of an
79: % arbitrary polygon.
80: %
81: % x,y       - vectors containing the corner
82: %             coordinates. The boundary is
83: %             traversed in a counterclockwise
84: %             direction
85: %
86: % area      - the polygon area
```

```
87: % xbar,ybar - the centroidal coordinates
88: % axx       - integral of x^2*dxdy
89: % axy       - integral of xy*dxdy
90: % ayy       - integral of y^2*dxdy
91: %
92: % User m functions called: none
93: %----------------------------------------------
94:
95: n=1:length(x); n1=n+1;
96: x=[x(:);x(1)]; y=[y(:);y(1)];
97: a=(x(n).*y(n1)-y(n).*x(n1))';
98: area=sum(a)/2; a6=6*area;
99: xbar=a*(x(n)+x(n1))/a6; ybar=a*(y(n)+y(n1))/a6;
100: ayy=a*(y(n).^2+y(n).*y(n1)+y(n1).^2)/12;
101: axy=a*(x(n).*(2*y(n)+y(n1))+x(n1).* ...
102:     (2*y(n1)+y(n)))/24;
103: axx=a*(x(n).^2+x(n).*x(n1)+x(n1).^2)/12;
104:
105: %==============================================
106:
107: function [v,vr,vrr,h,area,n]=pyramid(r)
108: %
109: % [v,vr,vrr,h,area,n]=pyramid(r)
110: % ~~~~~~~~~~~~~~~~~~~~~~~~~~~~~~
111: %
112: % This function determines geometrical
113: % properties of a pyramid with the apex at the
114: % origin and corner coordinates of the base
115: % stored in the rows of r.
116: %
117: % r    - matrix containing the corner
118: %        coordinates of a polygonal base stored
119: %        in the rows of matrix r.
120: %
121: % v    - the volume of the pyramid
122: % vr   - the first moment of volume relative to
123: %        the origin
124: % vrr  - the second moment of volume relative
125: %        to the origin
126: % h    - the pyramid height
127: % area - the base area
128: % n    - the outward directed unit normal to
129: %        the base
130: %
131: % User m functions called: crosmat, polyxy
```

```
132: %-----------------------------------------------
133:
134: ns=size(r,1);
135: na=sum(crosmat(r,r([2:ns,1],:)))'/2;
136: area=norm(na); n=na/area; p=null(n');
137: i=p(:,1); j=p(:,2);
138: if det([p,n])<0, j=-j; end;
139: r1=r(1,:); rr=r-r1(ones(ns,1),:);
140: x=rr*i; y=rr*j;
141: [areat,xc,yc,axx,axy,ayy]=polyxy(x,y);
142: rc=r1'+xc*i+yc*j; h=r1*n;
143: v=h*area/3; vr=v*3/4*rc;
144: axx=axx-area*xc^2; ayy=ayy-area*yc^2;
145: axy=axy-area*xc*yc;
146: vrr=h/5*(area*rc*rc'+axx*i*i'+ayy*j*j'+ ...
147:     axy*(i*j'+j*i'));
148:
149: %===============================================
150:
151: function polhdplt(x,y,z,idface,colr)
152: %
153: % polhdplt(x,y,z,idface,colr)
154: % ~~~~~~~~~~~~~~~~~~~~~~~~~~~~
155: %
156: % This function makes a surface plot of an
157: % arbitrary polyhedron.
158: %
159: % x,y,z  - vectors containing the corner
160: %          indices of the polyhedron
161: % idface - a matrix in which row j defines the
162: %          corner indices of the j'th face.
163: %          Each face is traversed in a
164: %          counterclockwise sense relative to
165: %          the outward normal. The column
166: %          dimension equals the largest number
167: %          of indices needed to define a face.
168: %          Rows requiring fewer than the
169: %          maximum number of corner indices are
170: %          padded with zeros on the right.
171: % colr   - character string or a vector
172: %          defining the surface color
173: %
174: % User m functions called: cubrange
175: %-----------------------------------------------
176:
```

```
177: if nargin<5, colr=[1 0 1]; end
178: hold off, close; nf=size(idface,1);
179: v=cubrange([x(:),y(:),z(:)],1.1);
180: for k=1:nf
181:   i=idface(k,:); i=i(find(i>0));
182:   xi=x(i); yi=y(i); zi=z(i);
183:   fill3(xi,yi,zi,colr); hold on;
184: end
185: axis(v); grid on;
186: xlabel('x axis'); ylabel('y axis');
187: zlabel('z axis');
188: title('Surface Plot of a General Polyhedron');
189: figure(gcf); hold off;
190:
191: %==============================================
192:
193: function c=crosmat(a,b)
194: %
195: % c=crosmat(a,b)
196: % ~~~~~~~~~~~~~~
197: %
198: % This function computes the vector cross
199: % product for vectors stored in the rows
200: % of matrices a and b, and returns the
201: % results in the rows of c.
202: %
203: % User m functions called: none
204: %----------------------------------------------
205:
206: c=[a(:,2).*b(:,3)-a(:,3).*b(:,2),...
207:    a(:,3).*b(:,1)-a(:,1).*b(:,3),...
208:    a(:,1).*b(:,2)-a(:,2).*b(:,1)];
209:
210: %==============================================
211:
212: % function range=cubrange(xyz,ovrsiz)
213: % See Appendix B
```

5.8 Evaluating Integrals Having Square Root Type Singularities

Consider the problem of evaluating the following three integrals having square root type singularities at one or both ends of the integration interval:

$$I_1 = \int_a^b \frac{f(x)}{\sqrt{x-a}} dx \quad , \quad I_2 = \int_a^b \frac{f(x)}{\sqrt{b-x}} dx \quad , \quad I_3 = \int_a^b \frac{f(x)}{\sqrt{(x-a)(b-x)}} dx.$$

The singularities in these integrals can be removed using substitutions $x - a = t^2$, $b - x = t^2$, and $(x-a)(b-x) = (b+a)/2 + (b-a)/2\cos(t)$ which lead to

$$I_1 = 2 \int_0^{\sqrt{b-a}} f(a+t^2)\, dt \quad , \quad I_2 = 2 \int_0^{\sqrt{b-a}} f(b-t^2)\, dt$$

$$I_3 = \int_0^{\pi} f(\frac{b+a}{2} + \frac{b-a}{2}\cos(t)\,)\, dt.$$

These modified integrals can be evaluated using **gcquad** or **quadl** by creating integrands with appropriate argument shifts. Two integration functions **quadgsqrt** and **quadlsqrt** were written to handle each of the three integral types. Shown below is a program called **sqrtquadtest** which computes results for the case where $f(x) = e^{ux}\cos(vx)$ with constants u and v being parameters passed to the integrators using the **varargin** construct in MATLAB. Function **quadgsqrt** uses Gauss quadrature to evaluate I_1 and I_2, and uses Chebyshev quadrature [1] to evaluate I_3. When $f(x)$ is a polynomial, then taking parameter norder in function **quadgsqrt** equal to the polynomial order gives exact results. With norder taken sufficiently high, more complicated functions can also be integrated accurately. Function **quadlsqrt** evaluates the three integral types using the adaptive integrator **quadl**, which accommodates $f(x)$ of quite general form. The program shown below integrates the test function for parameter choices corresponding to $[a, b, u, v] = [1, 4, 3, 10]$ with norder=10 in **quadgsqrt** and tol=1e-12 in **quadlsqrt** . Output from the program for this data case appears as comments at lines 14 thru 35 of **sqrtquadtest**. The integrators apparently work well and give results agreeing to fifteen digits. However, **quadlsqrt** took more than four hundred times as long to run as **quadgsqrt**. Furthermore, the structure of **quadgsqrt** is such that it could easily be modified to accommodate a form of $f(x)$ which returns a vector.

5.8.1 Program Listing

Singular Integral Program

```
1: function [vg,tg,vL,tL,pctdiff]=sqrtquadtest
```

```
2: %
3: % [vg,tg,vL,tL,pctdiff]=sqrtquadtest
4: %~~~~~~~~~~~~~~~~~~~~~~~~~~~~~~~~~~~~~
5: % This function compares the accuracy and
6: % computation time for functions quadgsqrt
7: % and quadlsqrt to evaluate:
8: % integral(exp(u*x)*cos(v*x)/radical(x), a<x<b)
9: % where radical(x) is sqrt(x-a), sqrt(b-x), or
10: % sqrt((x-a)*(b-x))
11:
12: %-----------------------------------
13: % Program Output
14:
15: % >> sqrtquadtest;
16:
17: % EVALUATING INTEGRALS WITH SQUARE ROOT TYPE
18: %     SINGULARITIES AT THE END POINTS
19:
20: % Function integrated:
21: % ftest(x,u,v)=exp(u*x).*cos(v*x)
22:
23: % a = 1   b = 4
24: % u = 3   v = 10
25:
26: % Results from function gquadsqrt
27: % 4.836504484e+003 -8.060993912e+003 -4.264510048e+003
28: % Computation time = 0.0159 sec.
29:
30: % Results from function quadlsqrt
31: % 4.836504484e+003 -8.060993912e+003 -4.264510048e+003
32: % Computation time = 7.03 sec.
33:
34: % Percent difference for the two methods
35: % -3.6669e-012 -1.5344e-012 1.4929e-012
36: % >>
37:
38: %------------------------------------
39:
40: % The test function
41: ftest=inline('exp(u*x).*cos(v*x)','x','u','v');
42:
43: % Limits and function parameters
44: a=1; b=4; u=3; v=10;
45:
46: nloop=100; tic;
```

```
47: for j=1:nloop
48:   v1g=quadgsqrt(ftest,1,a,b,40,1,u,v);
49:   v2g=quadgsqrt(ftest,2,a,b,40,1,u,v);
50:   v3g=quadgsqrt(ftest,3,a,b,40,1,u,v);
51: end
52: vg=[v1g,v2g,v3g]; tg=toc/nloop;
53: disp(' ')
54: disp('EVALUATING INTEGRALS WITH SQUARE ROOT TYPE')
55: disp('     SINGULARITIES AT THE END POINTS')
56: disp(' ')
57: disp('Function integrated:')
58: disp('ftest(x,u,v)=exp(u*x).*cos(v*x)')
59: disp(' ')
60: disp(['a = ',num2str(a),'   b = ',num2str(b)])
61: disp(['u = ',num2str(u),'   v = ',num2str(v)])
62: disp(' ')
63: disp('Results from function gquadsqrt')
64: fprintf('%17.9e %17.9e %17.9e\n',vg)
65: disp(['Computation time = ',num2str(tg),' sec.'])
66:
67: tol=1e-12; tic;
68: v1L=quadlsqrt(ftest,1,a,b,tol,[],u,v);
69: v2L=quadlsqrt(ftest,2,a,b,tol,[],u,v);
70: v3L=quadlsqrt(ftest,3,a,b,tol,[],u,v);
71: vL=[v1L,v2L,v3L]; tL=toc;
72:
73: disp(' ')
74: disp('Results from function quadlsqrt')
75: fprintf('%17.9e %17.9e %17.9e\n',vL)
76: disp(['Computation time = ',num2str(tL),' sec.'])
77:
78: pctdiff=100*(vg-vL)./vL; disp(' ')
79: disp('Percent difference for the two methods')
80: fprintf('%13.4e %12.4e %12.4e\n',pctdiff)
81:
82: %========================================
83:
84: function v=quadgsqrt(...
85:              func,type,a,b,norder,nsegs,varargin)
86: %
87: % v=quadgsqrt(func,type,a,b,norder,nsegs,varargin)
88: %
89: %~~~~~~~~~~~~~~~~~~~~~~~~~~~~~~~~~~~~~~~~
90: %
91: % This function evaluates an integral having a
```

```
 92: % square root type singularity at one or both ends
 93: % of the integration interval a<x<b. Composite
 94: % Gauss integration is used with func(x) treated
 95: % as a polynomial of degree norder.
 96: % The integrand has the form:
 97: % func(x)/sqrt(x-a) if type==1.
 98: % func(x)/sqrt(b-x) if type==2.
 99: % func(x)/sqrt((x-a)*(b-x)) if type==3.
100: % The integration interval is subdivided into
101: % nsegs subintervals of equal length.
102: %
103: % func   - a character string or function handle
104: %          naming a function continuous in the
105: %          interval from x=a to x=b
106: % type   - 1 if the integrand is singular at x=a
107: %          2 if the integrand is singular at x=b
108: %          3 if the integrand is singular at both
109: %            x=a and x=b.
110: % a,b    - integration limits with b>a
111: % norder - polynomial interpolation order within
112: %          each interval. Lowest norder is 20.
113: % nsegs  - number of integration subintervals
114: %
115: % User m functions called: gcquad
116: %
117: % Reference: Abromowitz and Stegun, 'Handbook of
118: %            Mathematical Functions', Chapter 25
119: % ----------------------------------------------
120:
121: if nargin<6, nsegs=1; end;
122: if nargin<5, norder=50; end
123: switch type
124:   case 1  % Singularity at the left end.
125:           % Use Gauss quadrature
126:     [dumy,bp,wf]=gcquad(...
127:       '',0,sqrt(b-a),norder+1,nsegs);
128:     t=a+bp.^2; y=feval(func,t,varargin{:});
129:     v=wf(:)'*y(:)*2;
130:   case 2  % Singularity at the right end.
131:           % Use Gauss quadrature
132:     [dumy,bp,wf]=gcquad(...
133:       '',0,sqrt(b-a),norder+1,nsegs);
134:     t=b-bp.^2; y=feval(func,t,varargin{:});
135:     v=wf(:)'*y(:)*2;
136:   case 3  % Singularity at both ends.
```

```
137:            % Use Chebyshev integration
138:      n=norder; bp=cos(pi/(2*n+2)*(1:2:2*n+1));
139:      c1=(b+a)/2; c2=(b-a)/2; t=c1+c2*bp;
140:      y=feval(func,t,varargin{:});
141:      v=pi/(n+1)*sum(y);
142: end
143:
144: %=========================================
145:
146: function v=quadlsqrt(fname,type,a,b,tol,trace,varargin)
147: %
148: % v=quadlsqrt(fname,type,a,b,tol,trace,varargin)
149: % ~~~~~~~~~~~~~~~~~~~~~~~~~~~~~~~~~~~~~~~~~~~~~~
150: %
151: % This function uses the MATLAB integrator quadl
152: % to evaluate integrals having square root type
153: % singularities at one or both ends of the
154: % integration interval a < x < b.
155: % The integrand has the form:
156: % func(x)/sqrt(x-a) if type==1.
157: % func(x)/sqrt(b-x) if type==2.
158: % func(x)/sqrt((x-a)*(b-x)) if type==3.
159: %
160: % func   - the handle for a function continuous
161: %          from x=a to x=b
162: % type   - 1 if the integrand is singular at x=a
163: %          2 if the integrand is singular at x=b
164: %          3 if the integrand is singular at both
165: %            x=a and x=b.
166: % a,b    - integration limits with b > a
167:
168: if nargin<6 | isempty(trace), trace=0; end
169: if nargin<5 | isempty(tol), tol=1e-8; end
170: if nargin<7
171:   varargin{1}=type; varargin{2}=[a,b];
172:   varargin{3}=fname;
173: else
174:   n=length(varargin); c=[a,b]; varargin{n+1}=type;
175:   varargin{n+2}=c; varargin{n+3}=fname;
176: end
177:
178: if type==1 | type==2
179:   v=2*quadl(@fshift,0,sqrt(b-a),...
180:     tol,trace,varargin{:});
181: else
```

```
182:     v=quadl(@fshift,0,pi,tol,trace,varargin{:});
183: end
184:
185: %==========================================
186:
187: function u=fshift(x,varargin)
188: % u=fshift(x,varargin)
189: % This function shifts arguments to produce
190: % a nonsingular integrand called by quadl
191: N=length(varargin); fname=varargin{N};
192: c=varargin{N-1}; type=varargin{N-2};
193: a=c(1); b=c(2); c1=(b+a)/2; c2=(b-a)/2;
194:
195: switch type
196:   case 1, t=a+x.^2; case 2, t=b-x.^2;
197:   case 3, t=c1+c2*cos(x);
198: end
199:
200: if N>3, u=feval(fname,t,varargin{1:N-3});
201: else, u=feval(fname,t); end
202:
203: %==========================================
204:
205: % function [val,bp,wf]=gcquad(func,xlow,...
206: %             xhigh,nquad,mparts,varargin)
207: % See Appendix B
```

5.9 Gauss Integration of a Multiple Integral

Gauss integration can be used to evaluate multiple integrals having variable limits. Consider the instance typified by the following triple integral

$$I = \int_{c_1}^{c_2} \int_{b_1(z)}^{b_2(z)} \int_{a_1(y,z)}^{a_2(y,z)} F(x, y, z)\, dx\, dy\, dz.$$

This integral can be changed into one with constant limits by the substitutions

$$z = c_p + c_m u \ , \ -1 \le u \le 1,$$
$$y = b_p + b_m t \ , \ -1 \le t \le 1,$$
$$x = a_p + a_m s \ , \ -1 \le s \le 1$$

where

$$c_p = \frac{c_2 + c_1}{2}, \quad c_m = \frac{c_2 - c_1}{2},$$
$$b_p = \frac{b_2 + b_1}{2}, \quad b_m = \frac{b_2 - b_1}{2},$$
$$a_p = \frac{a_2 + a_1}{2}, \quad a_m = \frac{a_2 - a_1}{2}.$$

The above integral becomes

$$I = \int_{-1}^{1} \int_{-1}^{1} \int_{-1}^{1} c_m b_m a_m f(s, t, u) \, ds \, dt \, du$$

where

$$f(s, t, u) = F(a_p + a_m s, b_p + b_m t, c_p + c_m u),$$
$$a_m = a_m(y, z) = a_m(b_p + b_m t, c_p + c_m u),$$
$$b_m = b_m(z) = b_m(c_p + c_m u).$$

Thus, the integral has the form

$$I = \int_{-1}^{1} \int_{-1}^{1} \int_{-1}^{1} G(s, t, u) \, ds \, dt \, du$$

where

$$G = c_m b_m a_m f.$$

Performing the integration over each limit using an n-point quadrature formula with weight factors $w_\imath$ and base points $x_\imath$ yields

$$I = \sum_{k=1}^{n} \sum_{\jmath=1}^{n} \sum_{\imath=1}^{n} w_k w_\jmath w_\imath G(x_\imath, x_\jmath, x_k).$$

A function allowing an integrand and integration limits of general form was developed. An example is considered where the inertial moment of a sphere having unit radius, unit mass density, and centered at $(0, 0, 0)$ is to be obtained about an axis through $x = 2, y = 0$, parallel to the z-axis. The related integral

$$I = \int_{-1}^{1} \int_{-\sqrt{1-z^2}}^{\sqrt{1-z^2}} \int_{-\sqrt{1-y^2-z^2}}^{\sqrt{1-y^2-z^2}} \left[(x-2)^2 + y^2\right] dx \, dy \, dz$$

has a value of $88\pi/15$. Shown below is a function **quadit3d** and related limit and integrand functions. The function **triplint(n)** computes the ratio of the numerically integrated function to the exact result. The function specification **triplint(20)** yields a value of 1.000067. Even though the triple integration procedure is not computationally very fast, it is nevertheless robust enough to produce accurate results when a sufficiently high integration order is chosen.

5.9.1 Example: Evaluating a Multiple Integral

Triple Integration Program

```
1: function val=triplint(n)
2: %
3: % val=triplint(n)
4: % ~~~~~~~~~~~~~~~
5: % Triple integration example on inertial
6: % moment of a sphere.
7: %
8: % User m functions called:  fsphere, bs1, bs2,
9: %                            as1, as2
10:
11: if nargin==0, n=20; end
12: val=quadit3d('fsphere',[-1,1],'bs1','bs2',...
13:              'as1','as2',n)/(88*pi/15);
14:
15: %==============================================
16:
17: function s = quadit3d(f,c,b1,b2,a1,a2,w)
18: %
19: % s = quadit3d(f,c,b1,b2,a1,a2,w)
20: % ~~~~~~~~~~~~~~~~~~~~~~~~~~~~~~~~
21: % This function computes the iterated integral
22: %
23: % s = integral(...
24: %     f(x,y,z), x=a1..a2, y=b1..b2, z=c1..c2)
25: %
26: % where a1 and a2 are functions of y and z, b1
27: % and b2 are functions of z, and c is a vector
28: % containing constant limits on the z variable.
29: % Hence, as many as five external functions may
30: % be involved in the call list. For example,
31: % when the integrand and limits are:
32: %
33: % f  = x.^2+y^2+z^2
34: % a2 = sqrt(4-y^2-z^2)
35: % a1 = -a2
36: % b2 = sqrt(4-z^2)
37: % b1 = -b2
38: % c = [-2,2]
39: %
40: % Then the exact value is 128*pi/5.
```

```
41: % The approximation produced from a 20 point
42: % Gauss formula is accurate within .007 percent.
43: %
44: % f      - a function f(x,y,z) which must return
45: %          a vector value when x is a vector,
46: %          and y and z are scalar.
47: % a1,a2 - integration limits on the x variable
48: %          which may specify names of functions
49: %          or have constant values. If a1 is a
50: %          function it should have a call list
51: %          of the form a1(y,z). A similar form
52: %          applies to a2.
53: % b1,b2 - integration limits on the y variable
54: %          which may specify functions of z or
55: %          have constant values.
56: % c      - a vector defined by c=[c1,c2] where
57: %          c1 and c2 are fixed integration
58: %          limits for the z direction.
59: % w      - this argument defines the quadrature
60: %          formula used. It has the following
61: %          three possible forms. If w is omitted,
62: %          a Gauss formula of order 12 is used.
63: %          If w is a positive integer n, a Gauss
64: %          formula of order n is used. If w is an
65: %          n by 2 matrix, w(:,1) contains the base
66: %          points and w(:,2) contains the weight
67: %          factors for a quadrature formula over
68: %          limits -1 to 1.
69: %
70: % s      - the numerically evaluated integral
71: %
72: % User m functions called:  gcquad
73: %----------------------------------------------
74:
75: if nargin<7
76: % function gcquad generates base points
77: % and weight factors
78:   n=12; [dummy,x,W]=gcquad('',-1,1,n,1);
79: elseif size(w,1)==1 & size(w,2)==1
80:   n=w; [dummy,x,W]=gcquad('',-1,1,n,1);
81: else
82:  n=size(w,1); x=w(:,1); W=w(:,2);
83: end
84: s=0; cp=(c(1)+c(2))/2; cm=(c(2)-c(1))/2;
85:
```

```
86: for k=1:n
87:   zk=cp+cm*x(k);
88:   if ischar(b1), B1=feval(b1,zk);
89:   else, B1=b1; end
90:
91:   if ischar(b2), B2=feval(b2,zk);
92:   else, B2=b2; end
93:
94:   Bp=(B2+B1)/2; Bm=(B2-B1)/2; sj=0;
95:
96:   for j=1:n
97:     yj=Bp+Bm*x(j);
98:     if ischar(a1), A1=feval(a1,yj,zk);
99:     else, A1=a1; end
100:
101:    if ischar(a2), A2=feval(a2,yj,zk);
102:    else, A2=a2; end
103:
104:    Ap=(A2+A1)/2; Am=(A2-A1)/2;
105:    fval=feval(f, Ap+Am*x, yj, zk);
106:    si=fval(:).'*W(:); sj=sj+W(j)*Am*si;
107:  end
108:  s=s+W(k)*Bm*sj;
109: end
110: s=cm*s;
111:
112: %==============================================
113:
114: function v=fsphere(x,y,z)
115: %
116: % v=fsphere(x,y,z)
117: % ~~~~~~~~~~~~~~~~
118: % Integrand.
119: %----------------------------------------------
120:
121: v=(x-2).^2+y.^2;
122:
123: %==============================================
124:
125: function x=as1(y,z)
126: %
127: % x=as1(y,z)
128: % ~~~~~~~~~~
129: % Lower x integration limit.
130: %----------------------------------------------
```

```
131:
132: x=-sqrt(1-y.^2-z.^2);
133:
134: %==============================================
135:
136: function x=as2(y,z)
137: %
138: % x=as2(y,z)
139: % ~~~~~~~~~~
140: % Upper x integration limit.
141: %----------------------------------------------
142:
143: x=sqrt(1-y.^2-z.^2);
144:
145: %==============================================
146:
147: function y=bs1(z)
148: %
149: % y=bs1(z)
150: % ~~~~~~~~
151: % Lower y integration limit.
152: %----------------------------------------------
153:
154: y=-sqrt(1-z.^2);
155:
156: %==============================================
157:
158: function y=bs2(z)
159: %
160: % y=bs2(z)
161: % ~~~~~~~~~~
162: % Upper y integration limit.
163: %----------------------------------------------
164:
165: y=sqrt(1-z.^2);
166:
167: %==============================================
168:
169: % function [val,bp,wf]=gcquad(func,xlow,...
170: %                  xhigh,nquad,mparts,varargin)
171: % See Appendix B
```

Chapter 6

Fourier Series and the Fast Fourier Transform

6.1 Definitions and Computation of Fourier Coefficients

Trigonometric series are useful to represent periodic functions. A function defined for $-\infty < x < \infty$ has a period of 2π if $f(x+2\pi) = f(x)$ for all x. In most practical situations, such a function can be expressed as a complex Fourier series

$$f(x) = \sum_{j=-\infty}^{\infty} c_j e^{ijx} \text{ where } i = \sqrt{-1}.$$

The numbers c_j, called complex Fourier coefficients, are computed by integration as

$$c_j = \frac{1}{2\pi} \int_0^{2\pi} f(x) e^{-ijx} dx.$$

The Fourier series can also be rewritten using sines and cosines as

$$f(x) = c_0 + \sum_{j=1}^{\infty} (c_j + c_{-j}) \cos(jx) + i(c_j - c_{-j}) \sin(jx).$$

Denoting

$$a_j = c_j + c_{-j} \text{ and } b_j = i(c_j - c_{-j})$$

yields

$$f(x) = \frac{1}{2} a_0 + \sum_{j=1}^{\infty} a_j \cos(jx) + b_j \sin(jx)$$

which is called a Fourier sine-cosine expansion. This series is especially appealing when $f(x)$ is real valued. For that case $c_{-j} = \bar{c}_j$ for all j, which implies that c_0 must be real and

$$a_j = 2 \ \mathbf{real}(c_j) \ , \ b_j = -2 \ \mathbf{imag}(c_j) \text{ for } j > 0.$$

Suppose we want a Fourier series expansion for a more general function $f(x)$ having period p instead of 2π. If we introduce a new function $g(x)$ defined by

$$g(x) = f\left(\frac{px}{2\pi}\right)$$

then $g(x)$ has a period of 2π. Consequently, $g(x)$ can be represented as

$$g(x) = \sum_{\jmath=-\infty}^{\infty} c_\jmath e^{\imath \jmath x}.$$

From the fact that $f(x) = g(2\pi x/p)$ we deduce that

$$f(x) = \sum_{\jmath=-\infty}^{\infty} c_\jmath e^{2\pi \imath \jmath x/p}.$$

A need sometimes occurs to expand a function as a series of sine terms only, or as a series of cosine terms only. If the function is originally defined for $0 < x < \frac{p}{2}$, then making $f(x) = -f(p-x)$ for $\frac{p}{2} < x < p$ gives a series involving only sine terms. Similarly, if $f(x) = +f(p-x)$ for $\frac{p}{2} < x < p$, only cosine terms arise. Thus we get

$$f(x) = c_0 + \sum_{\jmath=1}^{\infty} (c_\jmath + c_{-\jmath}) \cos(2\pi \jmath x/p) \text{ if } f(x) = f(p-x),$$

or

$$f(x) = \sum_{\jmath=1}^{\infty} \imath(c_\jmath - c_{-\jmath}) \sin(2\pi \jmath x/p) \text{ if } f(x) = -f(p-x).$$

When the Fourier series of a function is approximated using a finite number of terms, the resulting approximating function may oscillate in regions where the actual function is discontinuous or changes rapidly. This undesirable behavior can be reduced by using a smoothing procedure described by Lanczos [60]. Use is made of Fourier series of a closely related function $\hat{f}(x)$ defined by a local averaging process according to

$$\hat{f}(x) = \frac{1}{\Delta} \int_{x-\frac{\Delta}{2}}^{x+\frac{\Delta}{2}} f(\zeta) d\zeta$$

where the averaging interval Δ should be a small fraction of the period p. Hence we write $\Delta = \alpha p$ with $\alpha < 1$. The functions $\hat{f}(x)$ and $f(x)$ are identical as $\alpha \to 0$. Even for $\alpha > 0$, these functions also match exactly at any point x where $f(x)$ varies linearly between $x - \frac{\Delta}{2}$ and $x + \frac{\Delta}{2}$. An important property of $\hat{f}(x)$ is that it agrees closely with $f(x)$ for small α but has a Fourier series which converges more rapidly than the series for $f(x)$. Furthermore, from its definition,

$$\hat{f}(x) = \sum_{\jmath=-\infty}^{\infty} c_\jmath \frac{1}{p\alpha} \int_{x-\frac{\alpha p}{2}}^{x+\frac{\alpha p}{2}} e^{2\pi \imath \jmath x/p} \, dx = \sum_{\jmath=-\infty}^{\infty} \hat{c}_\jmath e^{2\pi \imath \jmath x/p}$$

where $\hat{c}_0 = c_0$ and $\hat{c}_\jmath = c_\jmath \sin(\pi \jmath \alpha)/(\pi \jmath \alpha)$ for $\jmath \neq 0$. Evidently the Fourier coefficients of $\hat{f}(x)$ are easily obtainable from those of $f(x)$. When the series for $f(x)$ converges slowly, using the same number of terms in the series for $\hat{f}(x)$ often gives an approximation preferable to that provided by the series for $f(x)$. This process is called smoothing.

6.1.1 Trigonometric Interpolation and the Fast Fourier Transform

Computing Fourier coefficients by numerical integration is very time consuming. Consequently, we are led to investigate alternative methods employing trigonometric polynomial interpolation through evenly spaced data. The resulting formulas are the basis of an important algorithm called the Fast Fourier Transform (FFT) . Although the Fourier coefficients obtained by interpolation are approximate, these coefficients can be computed very rapidly when the number of sample points is an integer power of 2 or a product of small primes. We will discuss next the ideas behind trigonometric polynomial interpolation among evenly spaced data values.

Suppose we truncate the Fourier series and only use harmonics up to some order N. We assume $f(x)$ has period 2π so that

$$f(x) = \sum_{\jmath=-N}^{N} c_\jmath e^{\imath \jmath x}.$$

This trigonometric polynomial satisfies $f(0) = f(2\pi)$ even though the original function might actually have a finite discontinuity at 0 and 2π. Consequently, we may choose to use, in place of $f(0)$, the limit as $\epsilon \to 0$ of $[f(\epsilon) + f(2\pi - \epsilon)]/2$.

It is well known that the functions $e^{\imath \jmath x}$ satisfy an orthogonality condition for integration over the interval 0 to 2π. They also satisfy an orthogonality condition regarding summation over equally spaced data. The latter condition is useful for deriving a discretized approximation of the integral formula for the exact Fourier coefficients. Let us choose data points

$$x_\jmath = \left(\frac{\pi}{N}\right) \jmath\,,\ 0 \le \jmath \le (2N-1),$$

and write the simultaneous equations to make the trigonometric polynomial match the original function at the equally spaced data points. To shorten the notation we let

$$t = e^{\imath\pi/N},$$

and write

$$f_k = \sum_{\jmath=-N}^{N} c_\jmath t^{k\jmath}.$$

Suppose we pick an arbitrary integer n in the range $-N < n < N$. Multiplying the last equation by t^{-kn} and summing from $k = 0$ to $2N-1$ gives

$$\sum_{k=0}^{2N-1} f_k t^{-kn} = \sum_{k=0}^{2N-1} t^{-kn} \sum_{\jmath=-N}^{N} c_\jmath t^{k\jmath}.$$

Interchanging the summation order in the last equation yields

$$\sum_{k=0}^{2N-1} f_k t^{-kn} = \sum_{\jmath=-N}^{N} c_\jmath \sum_{k=0}^{2N-1} \zeta^k$$

where $\zeta = e^{\imath(\jmath-n)\pi/N}$. Summing the inner geometric series gives

$$\sum_{k=0}^{2N-1} \zeta^k = \begin{cases} \frac{1-\zeta^{2N}}{1-\zeta} & \text{for } \zeta \neq 1, \\ 2N & \text{for } \zeta = 1. \end{cases}$$

We find, for all k and n in the stated range, that

$$\zeta^{2N} = e^{\imath 2\pi(k-n)} = 1.$$

Therefore we get

$$\sum_{k=0}^{2N-1} f_k t^{-kn} = 2Nc_n \ , \ -N < n < N.$$

In the cases where $n = \pm N$, the procedure just outlined only gives a relationship governing $c_N + c_{-N}$. Since the first and last terms cannot be computed uniquely, we customarily take N large enough to discard these last two terms and write simply

$$c_n = \frac{1}{2N} \sum_{k=0}^{2N-1} f_k t^{-kn} \ , \ -N < n < N.$$

This formula is the basis for fast algorithms (called FFT for Fast Fourier Transform) to compute approximate Fourier coefficients. The periodicity of the terms depending on various powers of $e^{\imath\pi/N}$ can be utilized to greatly reduce the number of trigonometric function evaluations. The case where N equals a power of 2 is especially attractive. The mathematical development is not provided here. However, the related theory was presented by Cooley and Tukey in 1965 [21] and has been expounded in many textbooks [53, 96]. The result is a remarkably concise algorithm which can be comprehended without studying the details of the mathematical derivation. For our present interests it is important to understand how to use MATLAB's intrinsic function for the FFT (**fft**).

Suppose a periodic function is evaluated at a number of equidistant points ranging over one period. It is preferable for computational speed that the number of sample points should equal an integer power of two ($n = 2^m$). Let the function values for argument vector

$$x = p/n * (0 : n-1)$$

be an array f denoted by

$$f \Longleftrightarrow [f_1, f_2, \cdots, f_n].$$

The function evaluation **fft**(f) produces an array of complex Fourier coefficients multiplied by n and arranged in a peculiar fashion. Let us illustrate this result for $n = 8$. If

$$f = [f_1, f_2, \cdots, f_8]$$

then **fft**$(f)/8$ produces

$$c = [c_0, c_1, c_2, c_3, c_*, c_{-3}, c_{-2}, c_{-1}].$$

The term denoted by c_* actually turns out to equal $c_4 + c_{-4}$, so it would not be used in subsequent calculations. We generalize this procedure for arbitrary n as follows. Let $N = n/2 - 1$. In the transformed array, elements with indices of $1, \cdots, N+1$ correspond to $c_0, \cdots, c_N$ and elements with indices of $n, n-1, n-2, \cdots, N+3$ correspond to $c_{-1}, c_{-2}, c_{-3}, \cdots, c_{-N}$. It is also useful to remember that a real valued function has $c_{-n} = \text{conj}(c_n)$. To fix our ideas about how to evaluate a Fourier series, suppose we want to sum an approximation involving harmonics from order zero to order $(nsum - 1)$. We are dealing with a real valued function defined by **func** with a real argument vector x. The following code expands **func** and sums the series for argument x using $nsum$ terms.

```
function fouval=fftaprox(func,period,nfft,nsum,x)
fc=feval(func,period/nfft*(0:nfft-1));
fc=fft(fc)/nfft; fc(1)=fc(1)/2;
w=2*pi/period*(0:nsum-1);
fouval=2*real(exp(i*x(:)*w)*fc(:));
```

6.2 Some Applications

Applications of Fourier series arise in numerous practical situations such as structural dynamics, signal analysis, solution of boundary value problems, and image processing. Three examples are given below that illustrate use of the FFT. The first example calculates Bessel functions and the second problem studies forced dynamic response of a lumped mass system. The final example presents a program for constructing Fourier expansions and displaying graphical results for linearly interpolated or analytically defined functions.

6.2.1 Using the FFT to Compute Integer Order Bessel Functions

The FFT provides an efficient way to compute integer order Bessel functions $J_n(x)$ which are important in various physical applications [119]. Function $J_n(x)$ can be obtained as the complex Fourier coefficient of $e^{\imath n\theta}$ in the generating function described by

$$e^{\imath x \sin(\theta)} = \sum_{n=-\infty}^{\infty} J_n(x) e^{\imath n\theta}.$$

Orthogonality conditions imply

$$J_n(x) = \frac{1}{2\pi} \int_0^{2\pi} e^{\imath(x\sin(\theta) - n\theta)}\, d\theta.$$

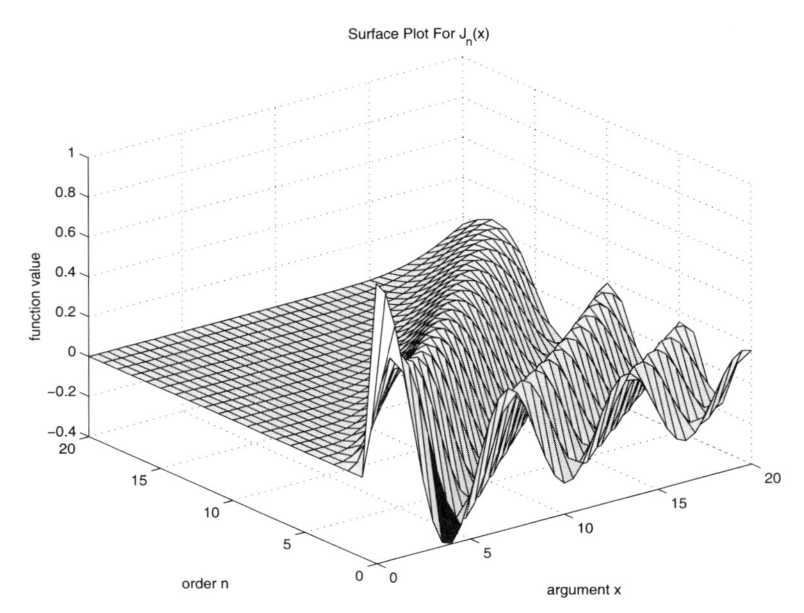

Figure 6.1: **Surface Plot for $J_n(x)$**

The Fourier coefficients represented by $J_n(x)$ can be computed approximately with the FFT. The infinite series converges very rapidly because the function it represents has continuous derivatives of all finite orders. Of course, $e^{\imath x \sin(\theta)}$ is highly oscillatory for large $|x|$, thereby requiring a large number of sample points in the FFT to obtain accurate results. For $n < 30$ and $|x| < 30$, a 128-point transform is adequate to give about ten digit accuracy for values of $J_n(x)$. The following code implements the above ideas and plots a surface showing how J_n changes in terms of n and x.

MATLAB Example

Bessel Function Program plotjrun

```
1: function plotjrun
2: % Example: plotjrun
3: % ~~~~~~~~~~~~~~~~~~
4: % This program computes integer order Bessel
5: % functions of the first kind by using the FFT.
6: %
7: % User m functions required: jnft
8:
9: x=0:.5:20; n=0:20; J=jnft(n,x); surf(x,n,J');
10: title('Surface Plot For J_{n}(x)');
11: ylabel('order n'), xlabel('argument x')
12: zlabel('function value'), figure(gcf);
13: print -deps plotjrun
14:
15: %=================================================
16:
17: function J=jnft(n,z,nft)
18: %
19: % J=jnft(n,z,nft)
20: % ~~~~~~~~~~~~~~~~~~~~~~
21: % Integer order Bessel functions of the
22: % first kind computed by use of the Fast
23: % Fourier Transform (FFT).
24: %
25: % n   - integer vector defining the function
26: %       orders
27: % z   - a vector of values defining the
28: %       arguments
29: % nft - number of function evaluations used
30: %       in the FFT calculation. This value
31: %       should be an integer power of 2 and
32: %       should exceed twice the largest
33: %       component of n. When nft is omitted
34: %       from the argument list, then a value
35: %       equal to 512 is used. More accurate
36: %       values of J are computed as nft is
37: %       increased. For max(n) < 30 and
38: %       max(z) < 30, nft=256 gives about
39: %       ten digit accuracy.
40: % J   - a matrix of values for the integer
```

```
41: %          order Bessel function of the first
42: %          kind. Row position matches orders
43: %          defined by n, and column position
44: %          corresponds to arguments defined by
45: %          components of z.
46: %
47: % User m functions called:  none.
48: %----------------------------------------------
49:
50: if nargin<3, nft=512; end;
51: J=exp(sin((0:nft-1)'* ...
52:    (2*pi/nft))*(i*z(:).'))/nft;
53: J=fft(J); J=J(1+n,:).';
54: if sum(abs(imag(z)))<max(abs(z))/1e10
55:    J=real(J);
56: end
```

6.2.2 Dynamic Response of a Mass on an Oscillating Foundation

Fourier series are often used to describe time dependent phenomena such as earthquake ground motion. Understanding the effects of foundation motions on an elastic structure is important in design. The model in Figure 6.2 embodies rudimentary aspects of this type of system and consists of a concentrated mass connected by a spring and viscous damper to a base which oscillates with known displacement $Y(t)$. The system is assumed to have arbitrary initial conditions $y(0) = y_0$ and $\dot{y}(0) = v_0$ when the base starts moving. The resulting displacement and acceleration of the mass are to be computed.

We assume that $Y(t)$ can be represented well over some time interval p by a Fourier series of the form

$$Y(t) = \sum_{n=-\infty}^{\infty} c_n e^{\imath\omega_n t} \;, \; \omega_n = \frac{2n\pi}{p}$$

where $c_{-n} = \text{conj}(c_n)$ because Y is real valued. The differential equation governing this problem is

$$m\ddot{y} + c\dot{y} + ky = kY(t) + c\dot{Y}(t) = F(t)$$

where the forcing function can be expressed as

$$F(t) = \sum_{n=-\infty}^{\infty} c_n[k + \imath c\omega_n]e^{\imath\omega_n t} = kc_0 + 2\,\textbf{real}\left(\sum_{n=1}^{\infty} f_n e^{\imath\omega_n t}\right)$$

and

$$f_n = c_n(k + \imath c\omega_n).$$

The corresponding steady-state solution of the differential equation is representable

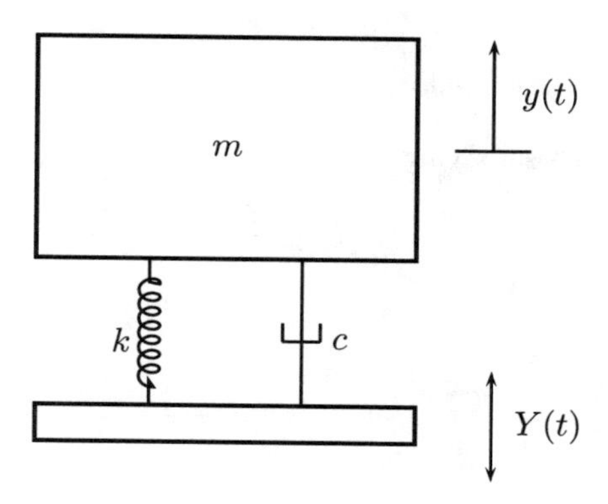

Figure 6.2: **Mass System**

as

$$y_s(t) = \sum_{n=-\infty}^{\infty} y_n e^{\imath\omega_n t}$$

where $y_{-n} = \text{conj}(y_n)$ since $y_s(t)$ is real valued. Substituting the series solution into the differential equation and comparing coefficients of $e^{\imath\omega_n t}$ on both sides leads to

$$y_n = \frac{c_n(k + \imath c\omega_n)}{k - m\omega_n^2 + \imath c\omega_n}.$$

The displacement, velocity, and acceleration corresponding to the steady-state (also called particular) solution are

$$y_s(t) = c_0 + 2\,\textbf{real}\left(\sum_{n=1}^{\infty} y_n e^{\imath\omega_n t}\right),$$

$$\dot{y}_s(t) = 2\,\textbf{real}\left(\sum_{n=1}^{\infty} \imath\omega_n y_n e^{\imath\omega_n t}\right),$$

$$\ddot{y}_s(t) = -2\,\textbf{real}\left(\sum_{n=1}^{\infty} \omega_n^2 y_n e^{\imath\omega_n t}\right).$$

The initial conditions satisfied by y_s are

$$y_s(0) = c_0 + 2\,\textbf{real}\left(\sum_{n=1}^{\infty} y_n\right), \quad \dot{y}_s(0) = 2\,\textbf{real}\left(\sum_{n=1}^{\infty} \imath\omega_n y_n\right).$$

Because these values usually will not match the desired initial conditions, the total solution consists of $y_s(t)$ and $y_h(t)$ which satisfies the homogeneous differential equation

$$m\ddot{y}_h + c\dot{y}_h + ky_h = 0.$$

The solution is

$$y_h = g_1 e^{s_1 t} + g_2 e^{s_2 t}$$

where s_1 and s_2 are roots satisfying

$$ms^2 + cs + k = 0.$$

The roots are

$$s_1 = \frac{-c + \sqrt{c^2 - 4mk}}{2m} \ , \ s_2 = \frac{-c - \sqrt{c^2 - 4mk}}{2m}.$$

Since the total solution is

$$y(t) = y_s(t) + y_h(t)$$

the constants g_1 and g_2 are obtained by solving the two simultaneous equations

$$g_1 + g_2 = y(0) - y_s(0) \ , \ s_1 g_1 + s_2 g_2 = \dot{y}(0) - \dot{y}_s(0).$$

The roots s_1 and s_2 are equal when $c = 2\sqrt{mk}$. Then the homogeneous solution assumes an alternate form given by $(g_1 + g_2 t)e^{st}$ with $s = -c/(2m)$. In this special case we find that

$$g_1 = y(0) - y_s(0) \ , \ g_2 = \dot{y}(0) - \dot{y}_s(0) - s g_1.$$

It should be noted that even though roots s_1 and s_2 will often be complex numbers, this causes no difficulty since MATLAB handles the complex arithmetic automatically (just as it does when the FFT transforms real function values into complex Fourier coefficients).

The harmonic response solution works satisfactorily for a general forcing function as long as the damping coefficient c is nonzero. A special situation can occur when $c = 0$, because the forcing function may resonate with the natural frequency of the undamped system. If c is zero, and for some n we have $\sqrt{k/m} = 2\pi n/p$, a condition of harmonic resonance is produced and a value of zero in the denominator occurs when the corresponding y_n is computed. In the undamped resonant case the particular solution grows like $[te^{\imath\omega_n t}]$, quickly becoming large. Even when c is small and $\sqrt{k/m} \approx 2\pi n/p$, undesirably large values of y_n can result. Readers interested in the important phenomenon of resonance can find more detail in Meirovitch [68].

This example concludes by using a base motion resembling an actual earthquake excitation. Seismograph output employing about 2700 points recorded during the Imperial Valley, California, earthquake of 1940 provided the displacement history for Figure 6.3. The period used to describe the motion is 53.8 seconds. A program was written to analyze system response due to a simulated earthquake base excitation. The following program modules are used:

runimpv	sets data values and generates graphical results
fouaprox	generates Fourier series approximations for a general function
imptp	piecewise linear function approximating the Imperial Valley earthquake data
shkbftss	computes steady-state displacement and acceleration for a spring-mass-dashpot system subjected to base motion expandable in a Fourier series
hsmck	computes the homogeneous solution for the spring-mass-dashpot system subjected to general initial conditions

Numerical results were obtained for a system having a natural period close to one second $(2\pi/6 \approx 1.047)$ and a damping factor of 5 percent. The function **imptp** was employed as an alternative to the actual seismograph data to provide a concisely expressible function which still embodies characteristics of a realistic base motion. Figure 6.4 shows a plot of function **imptp** along with its approximation by a twenty-term Fourier series. The series representation is surprisingly good considering the fact that such a small number of terms is used. The use of two-hundred terms gives an approximation which graphically does not deviate perceptibly from the actual function. Results showing how rapidly the Fourier coefficients diminish in magnitude with increasing order appear in Figure 6.5. The dynamical analysis produced displacement and acceleration values for the mass. Figure 6.6 shows both the total displacement as well as the displacement contributed from the homogeneous solution alone. Evidently, the steady-state harmonic response function captures well most of the motion, and the homogeneous part could probably be neglected without serious error. Figure 6.7 also shows the total acceleration of the mass which is, of course, proportional to the resultant force on the mass due to the base motion.

Before proceeding to the next example, the reader should be sure to appreciate the following important fact. Once a truncated Fourier series expansion of the forcing function using some appropriate number of terms is chosen, the truncated series defines an input function for which the response is computed exactly. If the user takes enough terms in the truncated series so that he/she is well satisfied with the function it approximates, then the computed response value for $y(t)$ will also be acceptable. This situation is distinctly different from the more complicated type of approximations occurring when finite difference or finite element methods produce discrete approximations for continuous field problems. Understanding the effects of grid size discretization error is more complex than understanding the effects of series truncation in the example given here.

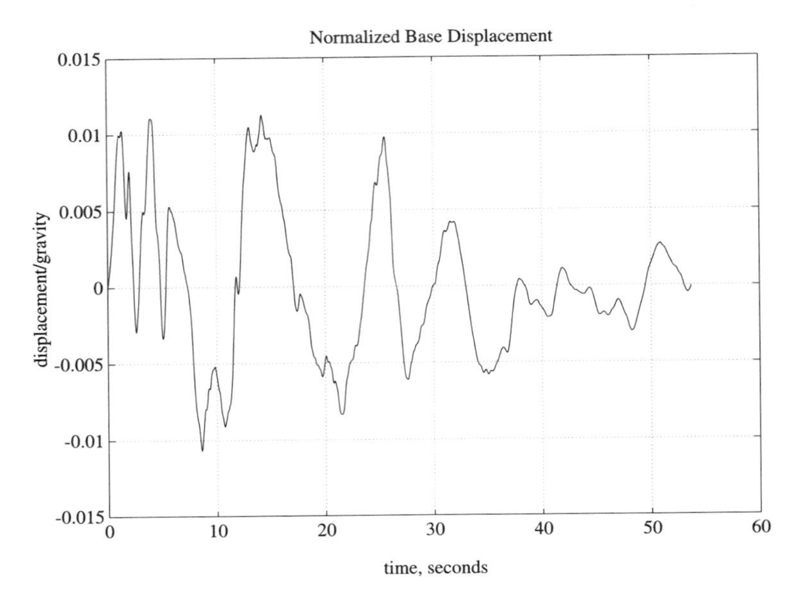

Figure 6.3: Normalized Base Displacement

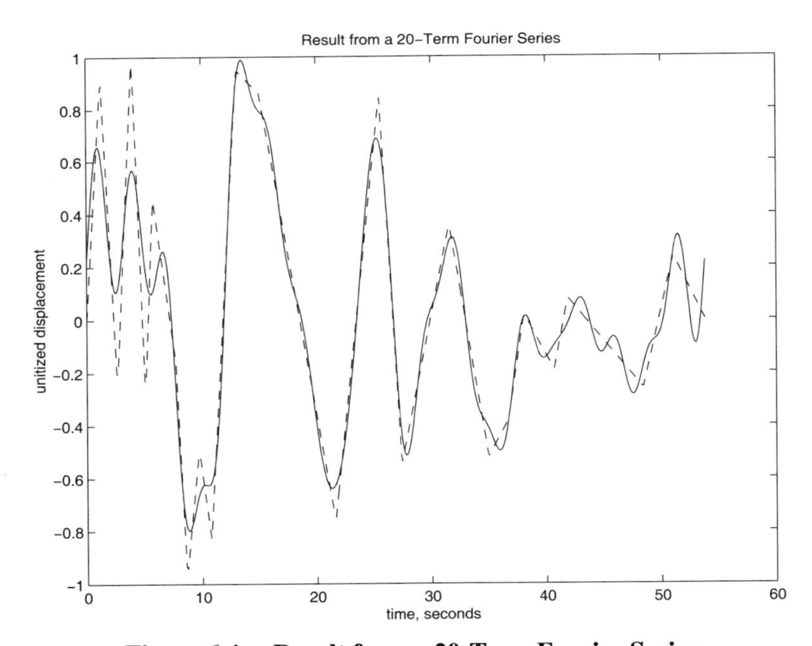

Figure 6.4: Result from a 20-Term Fourier Series

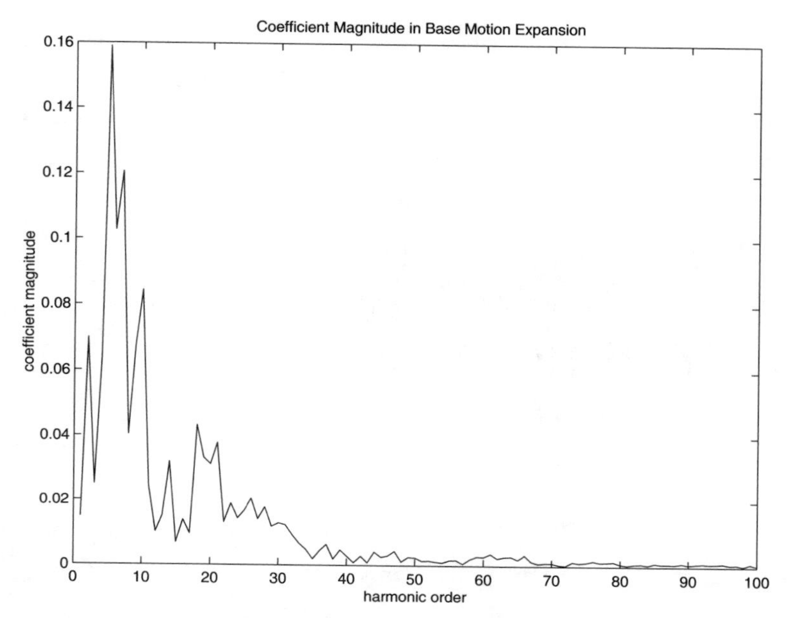

Figure 6.5: Coefficient Magnitude in Base Motion Expansion

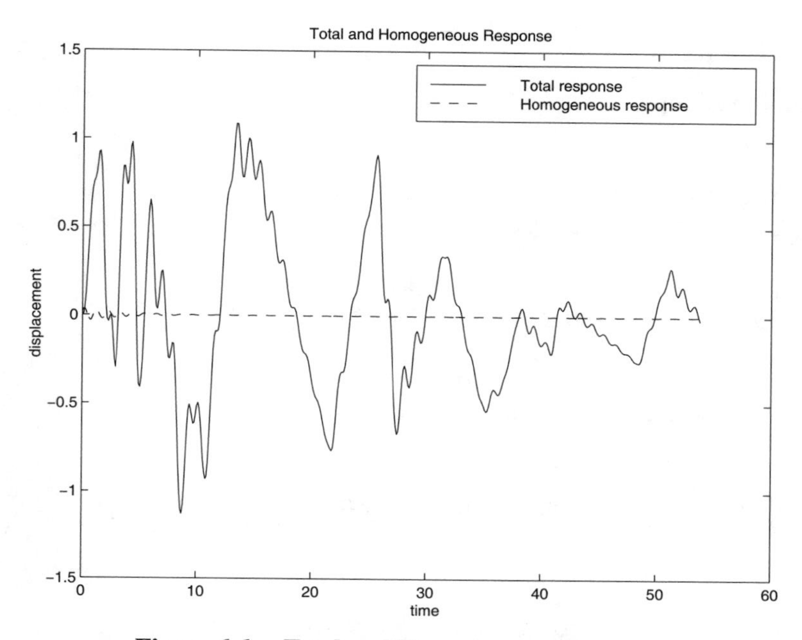

Figure 6.6: Total and Homogeneous Response

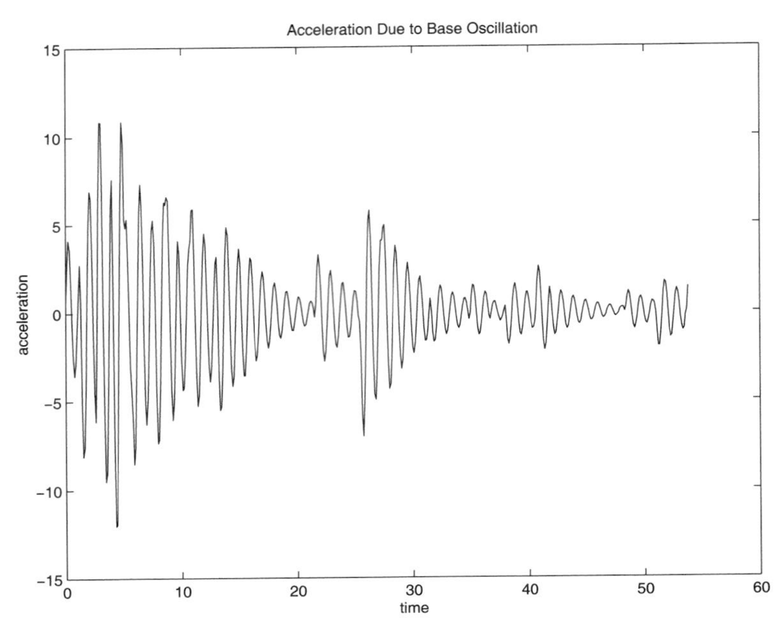

Figure 6.7: **Acceleration Due to Base Oscillation**

MATLAB Example

Program runimpv

```
1: function runimpv
2: % Example:  runimpv
3: % ~~~~~~~~~~~~~~~~~~
4: % This is a driver program for the
5: % earthquake example.
6: %
7: % User m functions required:
8: %    fouaprox, imptp, hsmck,
9: %    shkbftss, lintrp
10:
11: % Make the undamped period about one
12: % second long
13: m=1; k=36;
14:
15: % Use damping equal to 5 percent of critical
16: c=.05*(2*sqrt(m*k));
17:
```

```
18: % Choose a period equal to length of
19: % Imperial Valley earthquake data
20: prd=53.8;
21:
22: nft=1024; tmin=0; tmax=prd;
23: ntimes=200; nsum=80; % ntimes=501; nsum=200;
24: tplt=linspace(0,prd,ntimes);
25: y20trm=fouaprox('imptp',prd,tplt,20);
26: plot(tplt,y20trm,'-',tplt,imptp(tplt),'--');
27: xlabel('time, seconds');
28: ylabel('unitized displacement');
29: title('Result from a 20-Term Fourier Series')
30: figure(gcf);
31: disp('Press [Enter] to continue');
32: dumy=input('','s');
33: % print -deps 20trmplt
34:
35: % Show how magnitudes of Fourier coefficients
36: % decrease with increasing harmonic order
37:
38: fcof=fft(imptp((0:1023)/1024,1))/1024;
39: clf; plot(abs(fcof(1:100)));
40: xlabel('harmonic order');
41: ylabel('coefficient magnitude');
42: title(['Coefficient Magnitude in Base ' ...
43:        'Motion Expansion']); figure(gcf);
44: disp('Press [Enter] to continue');
45: dumy=input('','s');
46: % print -deps coefsize
47:
48: % Compute forced response
49: [t,ys,ys0,vs0,as]= ...
50:   shkbftss(m,c,k,'imptp',prd,nft,nsum, ...
51:            tmin,tmax,ntimes);
52:
53: % Compute homogeneous solution
54: [t,yh,ah]= ...
55:   hsmck(m,c,k,-ys0,-vs0,tmin,tmax,ntimes);
56:
57: % Obtain the combined solution
58: y=ys(:)+yh(:); a=as(:)+ah(:);
59: clf; plot(t,y,'-',t,yh,'--');
60: xlabel('time'); ylabel('displacement');
61: title('Total and Homogeneous Response');
62: legend('Total response','Homogeneous response');
```

```
63: figure(gcf);
64: disp('Press [Enter] to continue');
65: dumy=input('','s');
66: print -deps displac;
67:
68: clf; plot(t,a,'-');
69: xlabel('time'); ylabel('acceleration')
70: title('Acceleration Due to Base Oscillation')
71: figure(gcf); print -deps accel
72:
73: %=============================================
74:
75: function y=fouaprox(func,per,t,nsum,nft)
76: %
77: % y=fouaprox(func,per,t,nsum,nft)
78: % ~~~~~~~~~~~~~~~~~~~~~~~~~~~~~~~~
79: % Approximation of a function by a Fourier
80: % series.
81: %
82: % func   - function being expanded
83: % per    - period of the function
84: % t      - vector of times at which the series
85: %          is to be evaluated
86: % nsum   - number of terms summed in the series
87: % nft    - number of function values used to
88: %          compute Fourier coefficients. This
89: %          should be an integer power of 2.
90: %          The default is 1024
91: %
92: % User m functions called:  none.
93: %----------------------------------------------
94:
95: if nargin<5, nft=1024; end;
96: nsum=min(nsum,fix(nft/2));
97: c=fft(feval(func,per/nft*(0:nft-1)))/nft;
98: c(1)=c(1)/2; c=c(:); c=c(1:nsum);
99: w=2*pi/per*(0:nsum-1);
100: y=2*real(exp(i*t(:)*w)*c);
101:
102: %=============================================
103:
104: function ybase=imptp(t,period)
105: %
106: % ybase=imptp(t,period)
107: % ~~~~~~~~~~~~~~~~~~~~~~
```

```
108: % This function defines a piecewise linear
109: % function resembling the ground motion of
110: % the earthquake which occurred in 1940 in
111: % the Imperial Valley of California. The
112: % maximum amplitude of base motion is
113: % normalized to equal unity.
114: %
115: % period - period of the motion
116: %          (optional argument)
117: % t      - vector of times between
118: %          tmin and tmax
119: % ybase  - piecewise linearly interpolated
120: %          base motion
121: %
122: % User m functions called:  lintrp
123: %----------------------------------------------
124:
125: tft=[ ...
126:    0.00    1.26    2.64    4.01    5.10 ...
127:    5.79    7.74;   8.65    9.74   10.77 ...
128:   13.06   15.07   21.60   25.49;  27.38 ...
129:   31.56   34.94   36.66   38.03   40.67 ...
130:   41.87;  48.40   51.04   53.80    0    ...
131:    0       0       0 ]';
132: yft=[ ...
133:    0       0.92   -0.25    1.00   -0.29 ...
134:    0.46   -0.16;  -0.97   -0.49   -0.83 ...
135:    0.95    0.86   -0.76    0.85;  -0.55 ...
136:    0.36   -0.52   -0.38    0.02   -0.19 ...
137:    0.08;  -0.26    0.24    0.00    0    ...
138:    0       0       0 ]';
139: tft=tft(:); yft=yft(:);
140: tft=tft(1:24); yft=yft(1:24);
141: if nargin == 2
142:   tft=tft*period/max(tft);
143: end
144: ybase=lintrp(tft,yft,t);
145:
146: %=============================================
147:
148: function [t,ys,ys0,vs0,as]=...
149:      shkbftss(m,c,k,ybase,prd,nft,nsum, ...
150:               tmin,tmax,ntimes)
151: %
152: % [t,ys,ys0,vs0,as]=...
```

```
153: %   shkbftss(m,c,k,ybase,prd,nft,nsum, ...
154: %             tmin,tmax,ntimes)
155: % ~~~~~~~~~~~~~~~~~~~~~~~~~~~~~~~~~~~~~~~~~
156: % This function determines the steady-state
157: % solution of the scalar differential equation
158: %
159: %   m*y''(t) + c*y'(t) + k*y(t) =
160: %                  k*ybase(t) + c*ybase'(t)
161: %
162: % where ybase is a function of period prd
163: % which is expandable in a Fourier series
164: %
165: % m,c,k     - Mass, damping coefficient, and
166: %             spring stiffness
167: % ybase     - Function or vector of
168: %             displacements equally spaced in
169: %             time which describes the base
170: %             motion over a period
171: % prd       - Period used to expand xbase in a
172: %             Fourier series
173: % nft       - The number of components used
174: %             in the FFT (should be a power
175: %             of two). If nft is input as
176: %             zero, then ybase must be a
177: %             vector and nft is set to
178: %             length(ybase)
179: % nsum      - The number of terms to be used
180: %             to sum the Fourier series
181: %             expansion of ybase. This should
182: %             not exceed nft/2.
183: % tmin,tmax - The minimum and maximum times
184: %             for which the solution is to
185: %             be computed
186: % t         - A vector of times at which
187: %             the solution is computed
188: % ys        - Vector of steady-state solution
189: %             values
190: % ys0,vs0   - Position and velocity at t=0
191: % as        - Acceleration ys''(t), if this
192: %             quantity is required
193: %
194: % User m functions called:  none.
195: %----------------------------------------------
196:
197: if nft==0
```

```
198:     nft=length(ybase); ybft=ybase(:)
199: else
200:     tbft=prd/nft*(0:nft-1);
201:     ybft=fft(feval(ybase,tbft))/nft;
202:     ybft=ybft(:);
203: end
204: nsum=min(nsum,fix(nft/2)); ybft=ybft(1:nsum);
205: w=2*pi/prd*(0:nsum-1);
206: t=tmin+(tmax-tmin)/(ntimes-1)*(0:ntimes-1)';
207: etw=exp(i*t*w); w=w(:);
208: ysft=ybft.*(k+i*c*w)./(k+w.*(i*c-m*w));
209: ysft(1)=ysft(1)/2;
210: ys=2*real(etw*ysft); ys0=2*real(sum(ysft));
211: vs0=2*real(sum(i*w.*ysft));
212: if nargout > 4
213:   ysft=-ysft.*w.^2; as=2*real(etw*ysft);
214: end
215:
216: %==============================================
217:
218: function [t,yh,ah]= ...
219:          hsmck(m,c,k,y0,v0,tmin,tmax,ntimes)
220: %
221: % [t,yh,ah]=hsmck(m,c,k,y0,v0,tmin,tmax,ntimes)
222: % ~~~~~~~~~~~~~~~~~~~~~~~~~~~~~~~~~~~~~~~~~~~~~
223: % Solution of
224: %     m*yh''(t) + c*yh'(t) + k*yh(t) = 0
225: % subject to initial conditions of
226: %     yh(0) = y0 and yh'(0) = v0
227: %
228: % m,c,k       -  mass, damping and spring
229: %                constants
230: % y0,v0       -  initial position and velocity
231: % tmin,tmax   -  minimum and maximum times
232: % ntimes      -  number of times to evaluate
233: %                solution
234: % t           -  vector of times
235: % yh          -  displacements for the
236: %                homogeneous solution
237: % ah          -  accelerations for the
238: %                homogeneous solution
239: %
240: % User m functions called:  none.
241: %----------------------------------------------
242:
```

```
243: t=tmin+(tmax-tmin)/(ntimes-1)*(0:ntimes-1);
244: r=sqrt(c*c-4*m*k);
245: if r~=0
246:   s1=(-c+r)/(2*m); s2=(-c-r)/(2*m);
247:   g=[1,1;s1,s2]\[y0;v0];
248:   yh=real(g(1)*exp(s1*t)+g(2)*exp(s2*t));
249:   if nargout > 2
250:     ah=real(s1*s1*g(1)*exp(s1*t)+ ...
251:         s2*s2*g(2)*exp(s2*t));
252:   end
253: else
254:   s=-c/(2*m);
255:   g1=y0; g2=v0-s*g1; yh=(g1+g2*t).*exp(s*t);
256:   if nargout > 2
257:     ah=real(s*(2*g2+s*g1+s*g2*t).*exp(s*t));
258:   end
259: end
260:
261: %=============================================
262:
263: % function y=lintrp(xd,yd,x)
264: % See Appendix B
```

6.2.3 General Program to Plot Fourier Expansions

The final example in this chapter is a program to compute Fourier coefficients of general real valued functions and to display series with varying numbers of terms so that a user can see how rapidly such series converge. Since a truncated Fourier series is a continuous differentiable function, it cannot perfectly represent a discontinuous function such as a square wave. Near points where jump discontinuities occur, Fourier series approximations oscillate [18]. The same kind of behavior occurs less seriously near points of slope discontinuity. Adding more terms does not cure the problem at jump discontinuities. The behavior, known as Gibbs phenomenon, produces approximations which overshoot the function on either side of the discontinuity. Illustrations of this behavior appear below.

A program was written to expand real functions of arbitrary period using Fourier series approximations computed with the FFT. A piecewise linear function can be specified interactively by giving data points over a period. Alternatively, a function which is user defined can be employed. For instance, a function varying like a sine curve with the bottom half cut off would be

```
function y=chopsine(x,period)
y=sin(pi*x/period).*(x<period)
```

The program consists of the following functions.

fouseris	main driver
sine	example for exact function input
lintrp	function for piecewise linear interpolation
fousum	sum a real valued Fourier series
read	reads several data items on one line

Comments within the program illustrate how to input data interactively. Details of different input options can be found by executing the program.

Let us see how well the FFT approximates a function of period 3 defined by piecewise linear interpolation through (x, y) values of (0,1), (1,1), (1,–1), (2,–1), (3,1), and (4,0). The function has jump discontinuities at $x = 0$, $x = 1$, and $x = 4$. A slope discontinuity also occurs at $x = 3$. Program results using a twenty-term approximation appear in Figure 6.8. Results produced by 100- and 250-term series plotted near $x = 1$ are shown in Figures 6.9 and 6.10. Clearly, adding more terms does not eliminate the oscillation. However, the oscillation at a jump discontinuity can be reduced with the Lanczos smoothing procedure. Results for a series of 250 terms smoothed over an interval equal to the period times 0.01 appear in Figure 6.11. The oscillation is reduced at the cost of replacing the infinite slope at a discontinuity point by a steep slope of fifty-to-one for this case. Figure 6.12 shows a plot produced using an exact function definition as indicated in the second program execution. The reader may find it instructive to investigate how well Fourier series converge by running the program for other function choices.

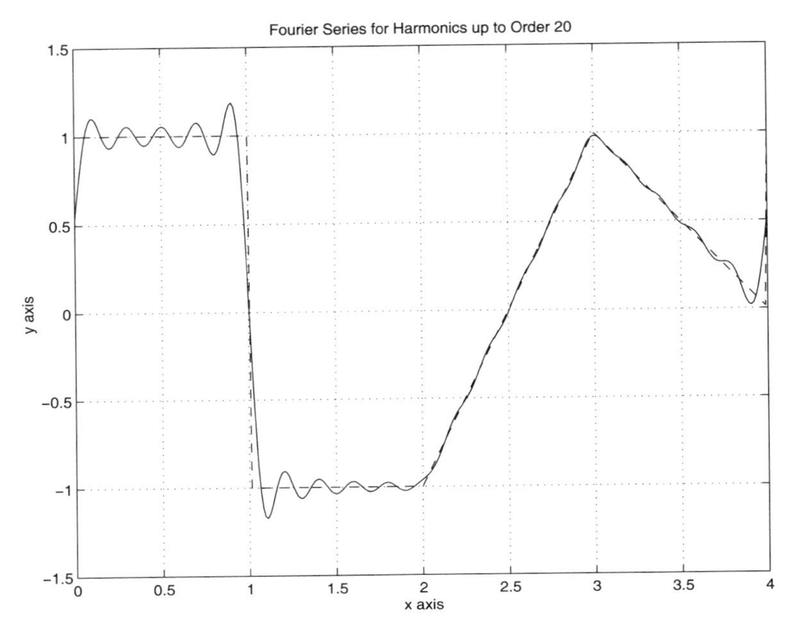

Figure 6.8: **Fourier Series for Harmonics up to Order 20**

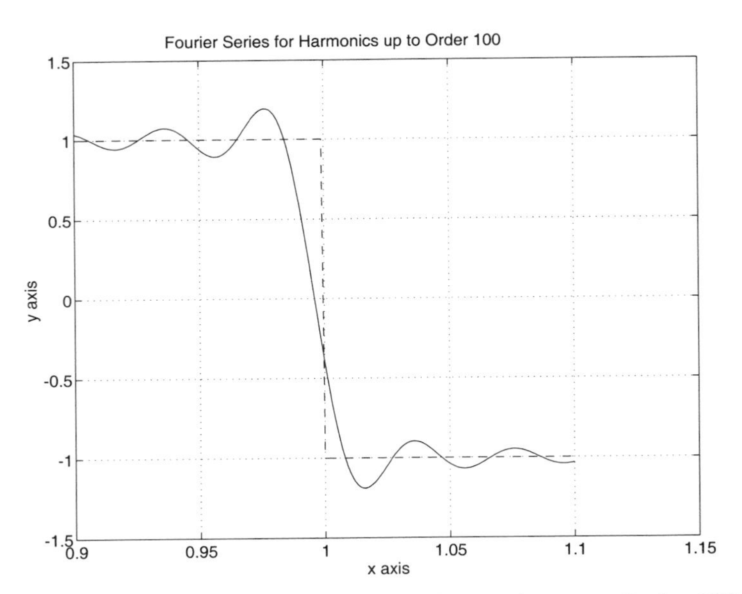

Figure 6.9: **Fourier Series for Harmonics up to Order 100**

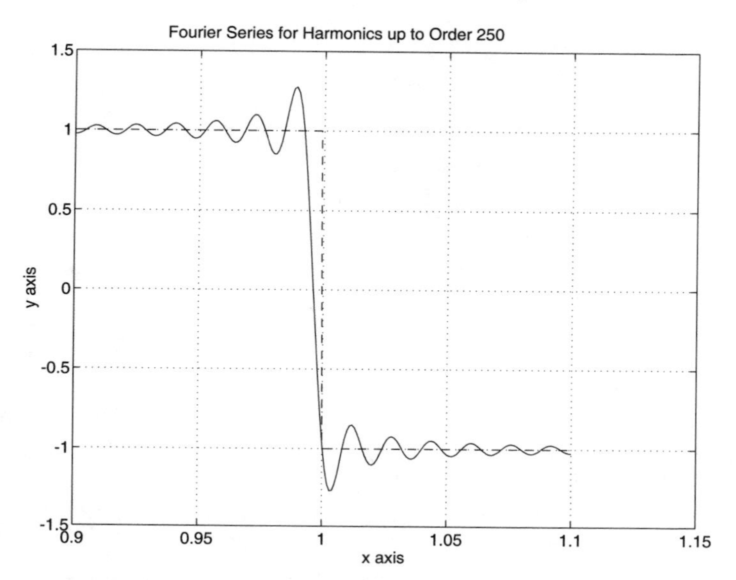

Figure 6.10: **Fourier Series for Harmonics up to Order 250**

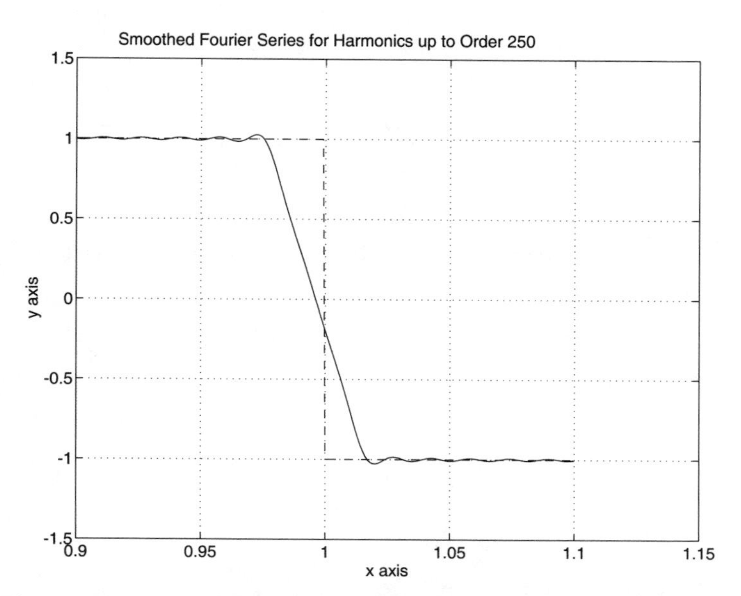

Figure 6.11: **Smoothed Fourier Series for Harmonics up to Order 250**

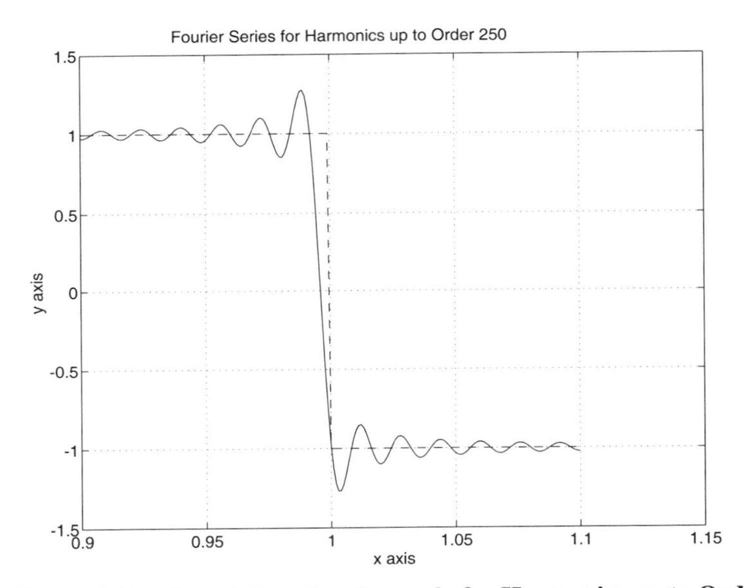

Figure 6.12: Exact Function Example for Harmonics up to Order 20

Examples of Fourier Series Expansions

Output for Piecewise Linear Example

```
>> fouseris

FOURIER SERIES EXPANSION FOR A PIECEWISE LINEAR OR
        ANALYTICALLY DEFINED FUNCTION

Input the period of the function
 ? > 4

Input the number of data points to define the function
by piecewise linear interpolation (input a zero if the
function is defined analytically by the user).
 ? > 6

Input the x,y values one pair per line
 ? > 0,1
 ? > 1,1
 ? > 1,-1
 ? > 2,-1
 ? > 3,1
 ? > 4,0

To plot the series input xmin, xmax, and the highest
harmonic not exceeding 255 (input 0,0,0 to stop)
(Use a negative harmonic number to save your graph)
 ? > 0,4,20

To plot the series smoothed over a fraction of the
period, input the smoothing fraction
(give 0.0 for no smoothing).
 ? > 0

Press RETURN to continue

To plot the series input xmin, xmax, and the highest
harmonic not exceeding 255 (input 0,0,0 to stop)
(Use a negative harmonic number to save your graph)
 ? > 0,0,0
```

Output for Analytically Defined Example

```
>> fouseris
```

```
FOURIER SERIES EXPANSION FOR A PIECEWISE LINEAR OR
        ANALYTICALLY DEFINED FUNCTION

Input the period of the function
 ? > pi/2

Input the number of data points to define the function
by piecewise linear interpolation (input a zero if the
function is defined analytically by the user).
 ? > 0

Select the method used for exact function definition:

1 <=> Use an existing function with syntax defined by
the following example:

function y=sine(x,period)
%
% y=sine(x,period)
% ~~~~~~~~~~~~~~~~~
% This function specifies all or part of
% a sine wave.
%
%   x       - vector of argument values
%   period  - period of the function
%   y       - vector of function values
%
% User m functions called:  none
%----------------------------------------------
y=sin(rem(x,period));

or

2 <=> Use a one-line character string definition
involving argument x and period p. For example a sine
wave with the bottom cut off would be defined by:
sin(x*2*pi/p).*(x<p/2)

1 or 2 ? > 1

Enter the name of your function
 ? > sine

To plot the series input xmin, xmax, and the highest
harmonic not exceeding 255 (input 0,0,0 to stop)
```

```
(Use a negative harmonic number to save your graph)
 ? > 0,pi,-20

To plot the series smoothed over a fraction of the
period, input the smoothing fraction
(give 0.0 for no smoothing).
 ? > 0

Give a file name to save the current graph >
exactplt

Press RETURN to continue

To plot the series input xmin, xmax, and the highest
harmonic not exceeding 255 (input 0,0,0 to stop)
(Use a negative harmonic number to save your graph)
 ? > 0,0,0
```

Fourier Series Program fouseris

```
1: function fouseris
2: % Example: fouseris
3: % ~~~~~~~~~~~~~~~~~
4: % This program illustrates the convergence rate
5: % of Fourier series approximations derived by
6: % applying the FFT to a general function which
7: % may be specified either by piecewise linear
8: % interpolation in a data table or by
9: % analytical definition in a function given by
10: % the user. The linear interpolation model
11: % permits inclusion of jump discontinuities.
12: % Series having varying numbers of terms can
13: % be graphed to demonstrate Gibbs phenomenon
14: % and to show how well the truncated Fourier
15: % series represents the original function.
16: % Provision is made to plot the Fourier series
17: % of the original function or a smoothed
18: % function derived by averaging the original
19: % function over an arbitrary fraction of the
20: % total period.
21: %
22: % User m functions required:
23: %    fousum, lintrp, inputv, sine
```

```
24:
25: % The following parameters control the number
26: % of fft points used and the number of points
27: % used for graphing.
28: nft=1024; ngph=1001; nmax=int2str(nft/2-1);
29:
30: fprintf('\nFOURIER SERIES EXPANSION FOR');
31: fprintf(' A PIECEWISE LINEAR OR');
32: fprintf('\n          ANALYTICALLY DEFINED ');
33: fprintf('FUNCTION\n');
34:
35: fprintf('\nInput the period of the function\n');
36: period=input('? > ');
37: xfc=(period/nft)*(0:nft-1)';
38: fprintf('\nHow many points define the function');
39: fprintf('\nby piecewise linear interpolation?');
40: fprintf('\n(Give a zero for analytical definition)\n')
41: nd=input('> ? ');
42: if nd > 0, xd=zeros(nd,1); yd=xd;
43:   fprintf('\nInput the x,y values one ');
44:   fprintf('pair per line\n');
45:   for j=1:nd
46:     [xd(j),yd(j)]=inputv('> ? ');
47:   end
48:
49: % Use nft interpolated data points to
50: % compute the fft
51:   yfc=lintrp(xd,yd,xfc); c=fft(yfc);
52: else
53:   fprintf('\nSelect the method for ');
54:   fprintf('analytical function definition:\n');
55:   fprintf('\n1 <=> Use an existing function ');
56:   fprintf('with syntax of the form:');
57:   fprintf('\nfunction y=funct(x,period), or \n');
58:   fprintf(['\n2 <=> Give a character string ',...
59:            'in argument x and period p.'])
60:   fprintf(['\n(Such as: sign(sin(2*pi*x/p)) '...
61:            'to make a square wave)\n'])
62:   nopt=input('Enter 1 or 2 ? > ');
63:   if nopt == 1
64:     fprintf('\nEnter the name of your ');
65:     fprintf('function\n');
66:     fnam=input('> ? ','s');
67:     yfc=feval(fnam,xfc,period); c=fft(yfc);
68:   else
```

```
 69:     fprintf('\nInput the one-line definition');
 70:     fprintf(' in terms of x and p\n');
 71:     strng=input('> ? ','s');
 72:     x=xfc; p=period;
 73:     yfc=eval(strng); c=fft(yfc);
 74:   end
 75: end
 76:
 77: while 1
 78:   fprintf('\nTo plot the series input xmin,');
 79:   fprintf(' xmax, and the highest');
 80:   fprintf(['\nharmonic not exceeding ', ...
 81:           nmax,' (press [Enter] to stop)']);
 82:   fprintf('\n(Use a negative harmonic number');
 83:   fprintf(' to save your graph)\n');
 84:   [xl,xu,nh]=inputv('> ? ');
 85:   if isnan(xl), break; end
 86:   pltsav=(nh < 0); nh=abs(nh);
 87:   xtmp=xl+((xu-xl)/ngph)*(0:ngph);
 88:   fprintf('\nTo plot the series smoothed ');
 89:   fprintf('over a fraction of the');
 90:   fprintf('\nperiod, input the smoothing ');
 91:   fprintf('fraction');
 92:   fprintf('\n(give 0.0 for no smoothing).\n');
 93:   alpha=input('> ? ');
 94:   yfou=fousum(c,xtmp,period,nh,alpha);
 95:   xxtmp=xtmp; idneg=find(xtmp<0);
 96:   xng=abs(xtmp(idneg));
 97:   xxtmp(idneg)=xxtmp(idneg)+ ...
 98:                 period*ceil(xng/period);
 99:   if nd>0
100:     yexac=lintrp(xd,yd,rem(xxtmp,period));
101:   else
102:     if nopt == 1
103:       yexac=feval(fnam,xtmp,period);
104:     else
105:       x=xxtmp; yexac=eval(strng);
106:     end
107:   end
108:   in=int2str(nh);
109:   if alpha == 0
110:     titl=['Fourier Series for Harmonics ' ...
111:           'up to Order ',in];
112:   else
113:     titl=['Smoothed Fourier Series for ' ...
```

```
114:            'Harmonics up to Order ',in];
115:   end
116:   clf; plot(xtmp,yfou,'-',xtmp,yexac,'--');
117:   ylabel('y axis'); xlabel('x axis'); zoom on
118:   title(titl); grid on; figure(gcf); disp(' ');
119:   disp('You can zoom in with the mouse button.')
120:   input('You can press [Enter] to continue. ','s');
121:   if pltsav
122:     disp(' ')
123:     filnam=input(['Give a file name to ' ...
124:            'save the current graph > ? '],'s');
125:     if length(filnam) > 0
126:       eval(['print -deps ',filnam]);
127:     end
128:   end
129: end
130: 
131: %==============================================
132: 
133: function y=sine(x,period)
134: % y=sine(x,period)
135: % ~~~~~~~~~~~~~~~~
136: % Function for all or part of a sine wave.
137: %   x,period -  vector argument and period
138: %   y        - function value
139: %
140: y=sin(rem(x,period));
141: 
142: %==============================================
143: 
144: function yreal=fousum(c,x,period,k,alpha)
145: %
146: % yreal = fousum(c,x,period,k,alpha)
147: % ~~~~~~~~~~~~~~~~~~~~~~~~~~~~~~~~~~
148: % Sum the Fourier series of a real
149: % valued function.
150: %
151: %   x      - The vector of real values at
152: %            which the series is evaluated.
153: %   c      - A vector of length n containing
154: %            Fourier coefficients output by
155: %            the fft function
156: %   period - The period of the function
157: %   k      - The highest harmonic used in
158: %            the Fourier sum.  This must
```

```
159: %              not exceed n/2-1
160: %    alpha  - If this parameter is nonzero,
161: %              the Fourier coefficients are
162: %              replaced by those of a function
163: %              obtained by averaging the
164: %              original function over alpha
165: %              times the period
166: %    yreal  - The real valued Fourier sum
167: %              for argument x
168: %
169: % The Fourier coefficients c must have been
170: % computed using the fft function which
171: % transforms the vector [y(1),...,y(n)] into
172: % an array of complex Fourier coefficients
173: % which have been multiplied by n and are
174: % arranged in the order:
175: %
176: %    [c(0),c(1),...,c(n/2-1),c(n/2),
177: %                     c(-n/2+1),...,c(-1)].
178: %
179: % The coefficient c(n/2) cannot be used
180: % since it is actually the sum of c(n/2) and
181: % c(-n/2). For a particular value of n, the
182: % highest usable harmonic is n/2-1.
183: %
184: % User m functions called:  none
185: %----------------------------------------------
186:
187: x=x(:); n=length(c);
188: if nargin <4, k=n/2-1; alpha=0; end
189: if nargin <5, alpha=0; end
190: if nargin <3, period=2*pi; end
191: L=period/2; k=min(k,n/2-1); th=(pi/L)*x;
192: i=sqrt(-1); z=exp(i*th);
193: y=c(k+1)*ones(size(th)); pa=pi*alpha;
194: if alpha > 0
195:   jj=(1:k)';
196:   c(jj+1)=c(jj+1).*sin(jj*pa)./(jj*pa);
197: end
198: for j=k:-1:2, y=c(j)+y.*z; end
199: yreal=real(c(1)+2*y.*z)/n;
200:
201: %=============================================
202:
203: % function y=lintrp(xd,yd,x)
```

```
204: % See Appendix B
205:
206: %==============================================
207:
208: % function varargout=inputv(prompt)
209: % See Appendix B
```

Chapter 7

Dynamic Response of Linear Second Order Systems

7.1 Solving the Structural Dynamics Equations for Periodic Applied Forces

The dynamics of a linear structure subjected to periodic forces obeys the matrix differential equation

$$M\ddot{X} + C\dot{X} + KX = F(t),$$

with initial conditions

$$X(0) = D_0\ ,\ \dot{X}(0) = V_0.$$

The solution vector $X(t)$ has dimension n and M, C, and K are real square matrices of order n. The mass matrix, M, the damping matrix, C, and the stiffness matrix, K, are all real. The forcing function $F(t)$, assumed to be real and having period L, can be approximated by a finite trigonometric series as

$$F(t) = \sum_{k=-N}^{N} c_k e^{\imath \omega_k t} \text{ where } \omega_k = 2\pi k/L$$

and $\imath = \sqrt{-1}$. The Fourier coefficients c_k are vectors that can be computed using the FFT. The fact that $F(t)$ is real also implies that $c_{-k} = \textbf{conj}(c_k)$ and, therefore,

$$F(t) = c_0 + 2\ \textbf{real}\left(\sum_{k=1}^{n} c_k e^{\imath \omega_k t}\right).$$

The solution of the differential equation is naturally resolvable into two distinct parts. The first is the so called particular or forced response which is periodic and has the same general mathematical form as the forcing function. Hence, we write

$$X_p = \sum_{k=-n}^{n} X_k e^{\imath \omega_k t} = X_0 + 2\ \textbf{real}\left(\sum_{k=1}^{n} X_k e^{\imath \omega_k t}\right).$$

Substituting this series into the differential equation and matching coefficients of $e^{\imath \omega_k t}$ on both sides yields

$$X_k = (K - \omega_k^2 M + \imath \omega_k C)^{-1} c_k.$$

The particular solution satisfies initial conditions given by

$$X_p(0) = X_0 + 2\ \mathbf{real}\left(\sum_{k=1}^{n} c_k\right) \text{ and } \dot{X}_p(0) = 2\ \mathbf{real}\left(\sum_{k=1}^{n} \imath\omega_k c_k\right).$$

Since these conditions usually will not equal the desired values, the particular solution must be combined with what is called the homogeneous or transient solution X_h, where

$$M\ddot{X}_h + c\dot{X}_h + KX_h = 0,$$

with

$$X_h(0) = D_0 - X_p(0)\ ,\ \dot{X}_h(0) = V_0 - \dot{X}_p(0).$$

The homogeneous solution can be constructed by reducing the original differential equation to first order form. Let Z be the vector of dimension $2n$ which is the concatenation of X and $\dot{X} = V$. Hence, $Z = [X; V]$ and the original equation of motion is

$$\frac{dZ}{dt} = AZ + P(t)$$

where

$$A = \begin{bmatrix} 0 & I \\ -M^{-1}K & -M^{-1}C \end{bmatrix} \text{ and } P = \begin{bmatrix} 0 \\ m^{-1}F \end{bmatrix}.$$

The homogeneous differential equation resulting when $P = 0$ can be solved in terms of the eigenvalues and eigenvectors of matrix A. If we know the eigenvalues λ_j and eigenvectors U_j satisfying

$$AU_\jmath = \lambda_\jmath U_\jmath\ ,\ 1 \leq \jmath \leq 2n,$$

then the homogeneous solution can be written as

$$Z = \sum_{\jmath=1}^{2n} z_\jmath U_\jmath e^{\imath\omega_\jmath t}.$$

The weighting coefficients $z_\jmath$ are computed to satisfy the desired initial conditions which require

$$\begin{bmatrix} U_1, U_2, \cdots, U_{2n} \end{bmatrix} \begin{bmatrix} z_1 \\ \vdots \\ z_{2n} \end{bmatrix} = \begin{bmatrix} X_0 - X_p(0) \\ V_0 - \dot{X}_p(0) \end{bmatrix}.$$

We solve this system of equations for $z_1, \cdots, z_{2n}$ and replace each $U_\jmath$ by $z_\jmath U_\jmath$. Then the homogeneous solution is

$$X_h = \sum_{\jmath=1}^{n} U_\jmath(1:n)e^{\lambda_\jmath t}$$

where $U_j(1:n)$ means we take only the first n elements of column j.

In most practical situations, the matrix C is nonzero and the eigenvalues $\lambda_1, \cdots, \lambda_{2n}$ have negative real parts. Then the exponential terms $e^{\lambda_j t}$ all decay with increasing time, which is why X_h is often known as the transient solution . In other cases, where the damping matrix C is zero, the eigenvalues λ_j are typically purely imaginary, and the homogeneous solution does not die out. In either instance, it is often customary in practical situations to ignore the homogeneous solution because it is usually small when compared to the contribution of the particular solution.

7.1.1 Application to Oscillations of a Vertically Suspended Cable

Let us solve the problem of small transverse vibrations of a vertically suspended cable. This system illustrates how the natural frequencies and mode shapes of a linear system can be combined to satisfy general initial conditions on position and velocity.

The cable in Figure 7.1 is idealized as a series of n rigid links connected at frictionless joints. Two vectors, consisting of link lengths $[\ell_1, \ell_2, \cdots, \ell_n]$ and masses $[m_1, m_2, \cdots, m_n]$ lumped at the joints, characterize the system properties. The accelerations in the vertical direction will be negligibly small compared to transverse accelerations, because the transverse displacements are small. Consequently, the tension in the chain will remain close to the static equilibrium value. This means the tension in link i is

$$T_i = g b_i \text{ where } b_i = \sum_{j=i}^{n} m_j.$$

We assume that the transverse displacement y_i for mass m_i is small compared to the total length of the cable. A free body diagram for mass i is shown in Figure 7.2. The small deflection angles are related to the transverse deflections by $\theta_{i+1} = (y_{i+1} - y_i)\,\ell_{i+1}$ and $\theta_i = (y_i - y_{i-1})/\ell_i$. Summation of forces shows that the horizontal acceleration is governed by

$$\begin{aligned} m_i \ddot{y}_i &= g(b_{i+1}/\ell_{i+1})\,(y_{i+1} - y_i) - g(b_i/\ell_i)\,(y_i - y_{i-1}) \\ &= g(b_i/\ell_i) y_{i-1} - g(b_i/\ell_i + b_{i+1}/\ell_{i+1}) y_i + g(b_{i+1}/\ell_{i+1}) y_{i+1}. \end{aligned}$$

In matrix form this equation is

$$M\ddot{Y} + KY = 0$$

where M is a diagonal matrix of mass coefficients and K is a symmetric tridiagonal matrix. The natural modes of free vibration are dynamical states where each element of the system simultaneously moves with harmonic motion of the same frequency. This means we seek motions of the form $Y = U\cos(\omega t)$, or equivalently $Y = U\sin(\omega t)$, which implies

$$KU_j = \lambda_j M U_j \text{ where } \lambda_j = \omega_j^2 \text{ for } 1 \le j \le n.$$

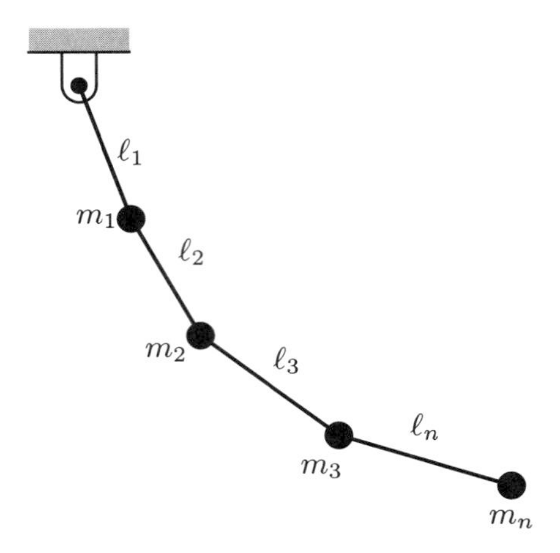

Figure 7.1: Transverse Cable Vibration

Solving the eigenvalue problem $(M^{-1}K)U = \lambda U$ gives the natural frequencies $\omega_1, \cdots, \omega_n$ and the modal vectors $U_1, \cdots, U_n$. The response to general initial conditions is then obtained by superposition of the component modes. We write

$$Y = \sum_{j=1}^{n} \cos(\omega_j t) U_j c_j + \sin(\omega_j t) U_j d_j / \omega_j$$

where the coefficients $c_1, \cdots, c_n$ and $d_1, \cdots, d_n$ (not to be confused with Fourier coefficients) are determined from the initial conditions as

$$\begin{bmatrix} U_1, \cdots, U_n \end{bmatrix} \begin{bmatrix} c_1 \\ \vdots \\ c_n \end{bmatrix} = Y(0) \ , \ c = U^{-1} Y(0),$$

$$\begin{bmatrix} U_1, \cdots, U_n \end{bmatrix} \begin{bmatrix} d_1 \\ \vdots \\ d_n \end{bmatrix} = \dot{Y}(0) \ , \ d = U^{-1} \dot{Y}(0).$$

The following program determines the cable response for general initial conditions. The natural frequencies and mode shapes are computed along with an animation of the motion.

The cable motion produced when an initially vertical system is given the same initial transverse velocity for all masses was studied. Graphical results of the analysis appear in Figures 7.3 through 7.6. The surface plot in Figure 7.3 shows the cable deflection pattern in terms of longitudinal position and time. Figure 7.4 shows the deflection pattern at two times. Figure 7.5 traces the motion of the middle and the

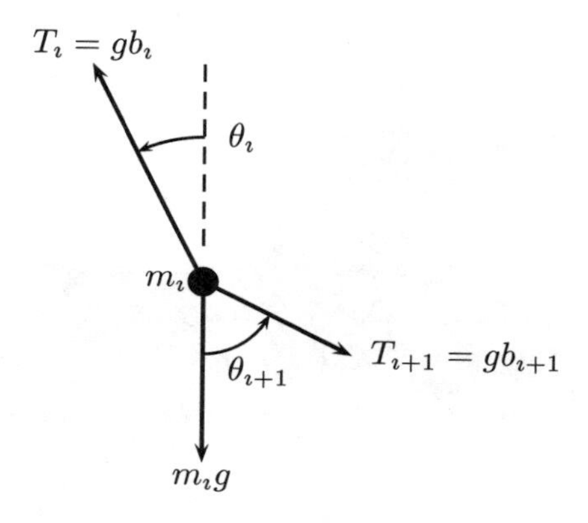

Figure 7.2: Forces on $\imath$'th Mass

free end. At $t = 1$, the wave propagating downward from the support point is about halfway down the cable. By $t = 2$, the wave has reached the free end and the cable is about to swing back. Finally, traces of cable positions during successive stages of motion appear in Figure 7.6.

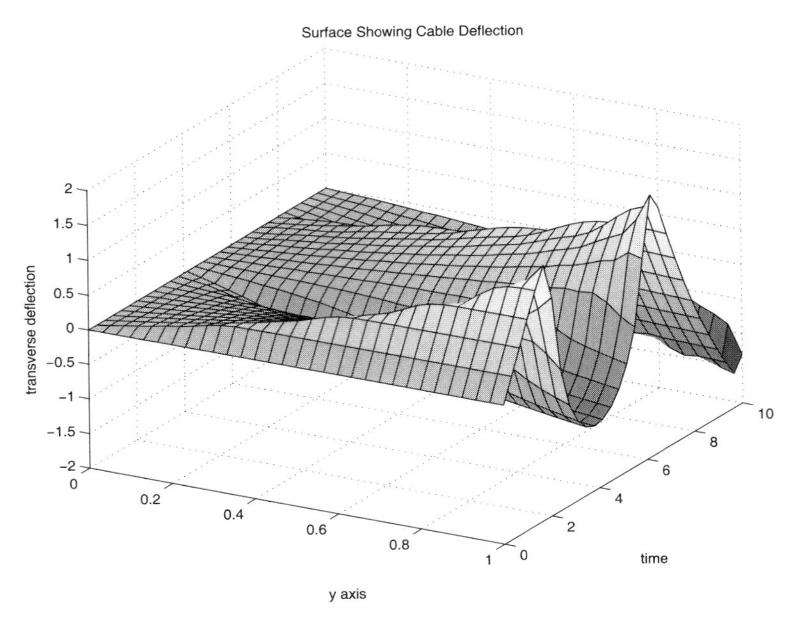

Figure 7.3: Surface Showing Cable Deflection

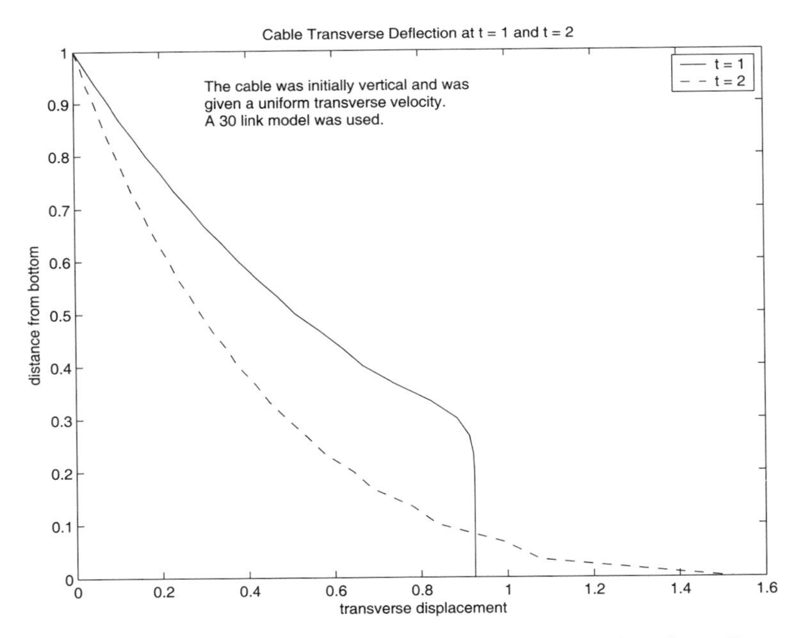

Figure 7.4: Cable Transverse Deflection at $t = 1$ and $t = 2$

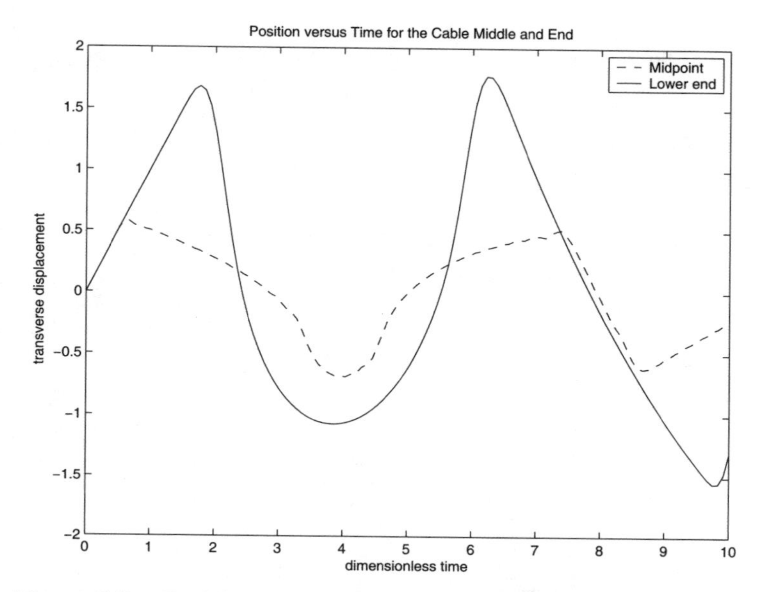

Figure 7.5: Position Versus Time for the Cable Middle and End

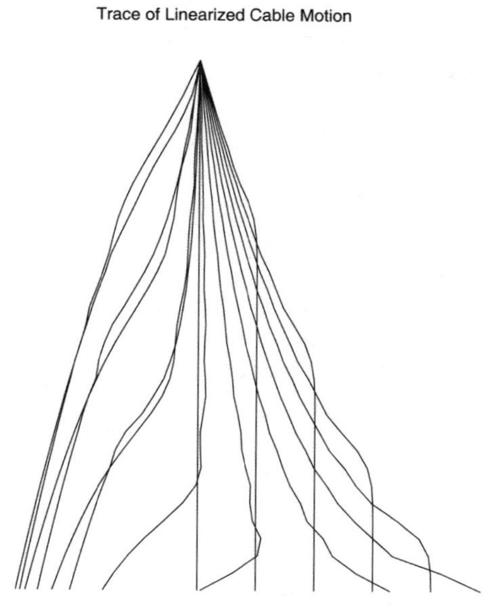

Figure 7.6: Trace of Cable Motion

MATLAB Example

Program cablinea

```
1: function cablinea
2: % Example: cablinea
3: % ~~~~~~~~~~~~~~~~~~
4: % This program uses modal superposition to
5: % compute the dynamic response of a cable
6: % suspended at one end and free at the other.
7: % The cable is given a uniform initial
8: % velocity. Time history plots and animation
9: % of the motion are provided.
10: %
11: % User m functions required:
12: %    cablemk, udfrevib, canimate
13:
14: % Initialize graphics
15: hold off; axis('normal'); close;
16:
17: % Set physical parameters
18: n=30; gravty=1.; masses=ones(n,1)/n;
19: lengths=ones(n,1)/n;
20:
21: % Obtain mass and stiffness matrices
22: [m,k]=cablemk(masses,lengths,gravty);
23:
24: % Assign initial conditions & time limit
25: % for solution
26: dsp=zeros(n,1); vel=ones(n,1);
27: tmin=0; tmax=10; ntim=30;
28:
29: % Compute the solution by modal superposition
30: [t,u,modvc,natfrq]=...
31:   udfrevib(m,k,dsp,vel,tmin,tmax,ntim);
32:
33: % Interpret results graphically
34: nt1=sum(t<=tmin); nt2=sum(t<=tmax);
35: u=[zeros(ntim,1),u];
36: y=cumsum(lengths); y=[0;y(:)];
37:
38: % Plot deflection surface
39: disp(' '), disp('TRANSVERSE MOTION OF A CABLE')
40: surf(y,t,u); xlabel('y axis'); ylabel('time');
```

```
41: zlabel('transverse deflection');
42: title('Surface Showing Cable Deflection');
43: colormap('default'), view([30,30]); figure(gcf);
44: disp(['Press [Enter] to see the cable ',...
45:       'position at two times'])
46: pause,  %print -deps surface
47:
48: % Show deflection configuration at two times
49: % Use closer time increment than was used
50: % for the surface plots.
51: mtim=4*ntim;
52: [tt,uu,modvc,natfrq]=...
53:   udfrevib(m,k,dsp,vel,tmin,tmax,mtim);
54: uu=[zeros(mtim,1),uu];
55: tp1=.1*tmax; tp2=.2*tmax;
56: s1=num2str(tp1); s2=num2str(tp2);
57: np1=sum(tt<=tp1); np2=sum(tt<=tp2);
58: u1=uu(np1,:); u2=uu(np2,:);
59: yp=flipud(y(:)); ym=max(yp);
60: plot(u1,yp,'-',u2,yp,'--');
61: ylabel('distance from bottom');
62: xlabel('transverse displacement');
63: title(['Cable Transverse Deflection ' ...
64:        'at t = ',s1,' and t = ',s2]);
65: legend('t = 1', 't = 2');
66:
67: xm=.2*max([u1(:);u2(:)]);
68: ntxt=int2str(n); n2=1+fix(n/2);
69: str=strvcat(...
70: 'The cable was initially vertical and was',...
71: 'given a uniform transverse velocity.',...
72: ['A ',ntxt,' link model was used.']);
73: text(xm,.9*ym,str), figure(gcf);
74: disp(['Press [Enter] to show the time ',...
75: 'response at the middle and free end'])
76: pause, %print -deps twoposn
77:
78: % Plot time history for the middle and the end
79: clf; plot(tt,uu(:,n2),'--',tt,uu(:,n+1),'-');
80: xlabel('dimensionless time');
81: ylabel('transverse displacement');
82: title(['Position versus Time for the ' ...
83:        'Cable Middle and End'])
84: legend('Midpoint','Lower end');
85: figure(gcf);
```

```
86: disp('Press [Enter] for a motion trace')
87: pause, %print -deps 2timhist
88:
89: % Plot animation of motion history
90: clf; canimate(y,u,t,0,.5*max(t),1);
91: %print -deps motntrac
92: disp('Press [Enter] to finish'), pause, close;
93:
94: %==============================================
95:
96: function [m,k]=cablemk(masses,lngths,gravty)
97: %
98: % [m,k]=cablemk(masses,lngths,gravty)
99: % ~~~~~~~~~~~~~~~~~~~~~~~~~~~~~~~~~~~~~
100: % Form the mass and stiffness matrices for
101: % the cable.
102: %
103: % masses     - vector of masses
104: % lngths     - vector of link lengths
105: % gravty     - gravity constant
106: % m,k        - mass and stiffness matrices
107: %
108: % User m functions called:  none.
109: %----------------------------------------------
110:
111: m=diag(masses);
112: b=flipud(cumsum(flipud(masses(:))))* ...
113:   gravty./lngths;
114: n=length(masses); k=zeros(n,n); k(n,n)=b(n);
115: for i=1:n-1
116:   k(i,i)=b(i)+b(i+1); k(i,i+1)=-b(i+1);
117:   k(i+1,i)=k(i,i+1);
118: end
119:
120: %==============================================
121:
122: function [t,u,mdvc,natfrq]=...
123:                udfrevib(m,k,u0,v0,tmin,tmax,nt)
124: %
125: % [t,u,mdvc,natfrq]= ...
126: %              udfrevib(m,k,u0,v0,tmin,tmax,nt)
127: % ~~~~~~~~~~~~~~~~~~~~~~~~~~~~~~~~~~~~~~~~~~~~~
128: % This function computes undamped natural
129: % frequencies, modal vectors, and time response
130: % by modal superposition.  The matrix
```

```
131: % differential equation and initial conditions
132: % are
133: %
134: %    m u'' + k u = 0,  u(0) = u0, u'(0) = v0
135: %
136: % m,k        - mass and stiffness matrices
137: % u0,v0      - initial position and velocity
138: %              vectors
139: % tmin,tmax - time limits for solution
140: %              evaluation
141: % nt         - number of times for solution
142: % t          - vector of solution times
143: % u          - matrix with row j giving the
144: %              system response at time t(j)
145: % mdvc       - matrix with columns which are
146: %              modal vectors
147: % natfrq     - vector of natural frequencies
148: %
149: % User m functions called:  none.
150: %----------------------------------------------
151:
152: % Call function eig to compute modal vectors
153: % and frequencies
154: [mdvc,w]=eig(m\k);
155: [w,id]=sort(diag(w)); w=sqrt(w);
156:
157: % Arrange frequencies in ascending order
158: mdvc=mdvc(:,id); z=mdvc\[u0(:),v0(:)];
159:
160: % Generate vector of equidistant times
161: t=linspace(tmin,tmax,nt);
162:
163: % Evaluate the displacement as a
164: % function of time
165: u=(mdvc*diag(z(:,1)))*cos(w*t)+...
166:   (mdvc*diag(z(:,2)./w))*sin(w*t);
167: t=t(:); u=u'; natfrq=w;
168:
169: %==============================================
170:
171: function canimate(y,u,t,tmin,tmax,norub)
172: %
173: % canimate(y,u,t,tmin,tmax,norub)
174: % ~~~~~~~~~~~~~~~~~~~~~~~~~~~~~~~~~
175: % This function draws an animated plot of
```

```
176: % data values stored in array u. The
177: % different columns of u correspond to position
178: % values in vector y. The successive rows of u
179: % correspond to different times. Parameter
180: % tpause controls the speed of the animation.
181: %
182: % u          - matrix of values for which
183: %              animated plots of u versus y
184: %              are required
185: % y          - spatial positions for different
186: %              columns of u
187: % t          - time vector at which positions
188: %              are known
189: % tmin,tmax - time limits for graphing of the
190: %              solution
191: % norub      - parameter which makes all
192: %              position images remain on the
193: %              screen. Only one image at a
194: %              time shows if norub is left out.
195: %              A new cable position appears each
196: %              time the user presses any key
197: %
198: % User m functions called:  none.
199: %----------------------------------------------
200:
201: % If norub is input,
202: %   all images are left on the screen
203: if nargin < 6
204:   rubout = 1;
205: else
206:   rubout = 0;
207: end
208:
209: % Determine window limits
210: umin=min(u(:)); umax=max(u(:)); udif=umax-umin;
211: uavg=.5*(umin+umax);
212: ymin=min(y); ymax=max(y); ydif=ymax-ymin;
213: yavg=.5*(ymin+ymax);
214: ywmin=yavg-.55*ydif; ywmax=yavg+.55*ydif;
215: uwmin=uavg-.55*udif; uwmax=uavg+.55*udif;
216: n1=sum(t<=tmin); n2=sum(t<=tmax);
217: t=t(n1:n2); u=u(n1:n2,:);
218: u=fliplr (u); [ntime,nxpts]=size(u);
219:
220: hold off; cla; ey=0; eu=0; axis('square');
```

```
221: axis([uwmin,uwmax,ywmin,ywmax]);
222: axis off; hold on;
223: title('Trace of Linearized Cable Motion');
224:
225: % Plot successive positions
226: for j=1:ntime
227:   ut=u(j,:); plot(ut,y,'-');
228:   figure(gcf); pause(.5);
229:
230:   % Erase image before next one appears
231:   if rubout & j < ntime, cla, end
232: end
```

7.2 Direct Integration Methods

Using stepwise integration methods to solve the structural dynamics equation provides an alternative to frequency analysis methods. If we invert the mass matrix and save the result for later use, the n degree-of-freedom system can be expressed concisely as a first order system in $2n$ unknowns for a vector $z = [x; v]$, where v is the time derivative of x. The system can be solved by applying the variable step-size differential equation integrator **ode45** as indicated in the following function:

```
function [t,x]=strdynrk(t,x0,v0,m,c,k,functim)
% [t,x]=strdynrk(t,x0,v0,m,c,k,functim)
global Mi C K F n n1 n2
Mi=inv(m); C=c; K=k; F=functim;
n=size(m,1); n1=1:n; n2=n+1:2*n;
[t,z]=ode45(@sde,t,[x0(:);v0(:)]); x=z(:,n1);
%==============================
function zp=sde(t,z)
global Mi C K F n n1 n2
zp=[z(n2); Mi*(feval(F,t)-C*z(n2)-K*z(n1))];
%==============================
function f=func(t)
% m=eye(3,3); k=[2,-1,0;-1,2,-1;0,-1,2];
% c=.05*k;
f=[-1;0;1]*sin(1.413*t);
```

In this function, the inverted mass matrix is stored in a global variable Mi, the damping and stiffness matrices are in C and K, and the forcing function name is stored in a character string called functim. Although this approach is easy to implement, the resulting analysis can be very time consuming for systems involving several hundred degrees of freedom. Variable step integrators make adjustments to control stability and accuracy which can require very small integration steps. Consequently, less sophisticated formulations employing fixed step-size are often employed in finite element programs. We will investigate two such algorithms derived from trapezoidal integration rules [7, 113]. The two fundamental integration formulas [26] needed are:

$$\int_a^b f(t)dt = \frac{h}{2}[f(a) + f(b)] - \frac{h^3}{12} f''(\epsilon_1)$$

and

$$\int_a^b f(t)dt = \frac{h}{2}[f(a) + f(b)] + \frac{h^2}{12}[f\,'(a) - f\,'(b)] \; + \frac{h^5}{720} f^{(4)}(\epsilon_2)$$

where $a < \epsilon_i < b$ and $h = b - a$. The first formula, called the trapezoidal rule , gives a zero truncation error term when applied to a linear function. Similarly, the second formula, called the trapezoidal rule with end correction , has a zero final term for a cubic integrand.

The idea is to multiply the differential equation by dt, integrate from t to $(t+h)$, and employ numerical integration formulas while observing that M, C, and K are constant matrices, or

$$M \int_t^{t+h} \dot{V}\; dt + C \int_t^{t+h} \dot{X}\; dt + K \int_t^{t+h} X\; dt = \int_t^{t+h} P(t)\; dt$$

and

$$\int_t^{t+h} \dot{X}\; dt = \int_t^{t+h} V\; dt.$$

For brevity we utilize a notation characterized by $X(t) = X_0$, $X(t+h) = X_1$, $\tilde{X} = X_1 - X_0$. The trapezoidal rule immediately leads to

$$\left[M + \frac{h}{2}C + \frac{h^2}{4}K\right] \tilde{V} = \int_t^{t+h} P(t)dt - h\left[CV_0 + K(X_0 + \frac{h}{2}V_0)\right] + O(h^3).$$

The last equation is a balance of impulse and momentum change involving the effective mass matrix

$$M_e = \left[M + \frac{h}{2}C + \frac{h^2}{4}K\right]$$

which can be inverted once and used repeatedly if the step-size is not changed.

To integrate the forcing function we can use the midpoint rule [26] which states that

$$\int_a^b P(t)\; dt = hP\left(\frac{a+b}{2}\right) + O(h^3).$$

Solving for $\tilde{V}$ yields

$$\tilde{V} = \left[M + \frac{h}{2}C + \frac{h^2}{4}K\right]^{-1} \left[P\left(t + \frac{h}{2}\right) - CV_0 - K\left(X_0 + \frac{h}{2}V_0\right)h\right] + O(h^3).$$

The velocity and position at $(t + h)$ are then computed as

$$V_1 = V_0 + \tilde{V}\ ,\ X_1 = X_0 + \frac{h}{2}[V_0 + V_1] + O(h^3).$$

A more accurate formula with truncation error of order h^5 can be developed from the extended trapezoidal rule. This leads to

$$M\tilde{V} + C\tilde{X} + K\left[\frac{h}{2}(\tilde{X} + 2X_0) - \frac{h^2}{12}\tilde{V}\right] = \int_t^{t+h} P(t)dt + O(h^5)$$

and

$$\tilde{X} = \frac{h}{2}[\tilde{V} + 2V_0] + \frac{h^2}{12}[\dot{V_0} - \dot{V_1}] + O(h^5).$$

Multiplying the last equation by M and employing the differential equation to reduce the $\dot{V_0} - \dot{V_1}$ terms gives

$$M\tilde{X} = \frac{h}{2}M[\tilde{V} + 2V_0] + \frac{h^2}{12}[-\tilde{P} + C\tilde{V} + K\tilde{X}] + O(h^5).$$

These results can be arranged into a single matrix equation to be solved for $\tilde{X}$ and $\tilde{V}$:

$$\begin{bmatrix} -(\frac{h}{2}M + \frac{h^2}{12}C) & (M - \frac{h^2}{12}K) \\ (M - \frac{h^2}{12}K) & (C + \frac{h}{2}K) \end{bmatrix} \begin{bmatrix} \tilde{V} \\ \tilde{X} \end{bmatrix} = \begin{bmatrix} hMV_0 + \frac{h^2}{12}(P_0 - P_1) \\ \int Pdt - hKX_0 \end{bmatrix} + O(h^5).$$

A Gauss two-point formula [26] evaluates the force integral consistent with the desired error order so that

$$\int_t^{t+h} P(t)dt = \frac{h}{2}\left[P(t + \alpha h) + P(t + \beta h)\right] + O(h^5)$$

where $\alpha = \frac{3-\sqrt{3}}{6}$ and $\beta = \frac{3+\sqrt{3}}{6}$.

7.2.1 Example on Cable Response by Direct Integration

Functions implementing the last two algorithms appear in the following program which solves the previously considered cable dynamics example by direct integration. Questions of computational efficiency and numerical accuracy are examined for two different step-sizes. Figures 7.7 and 7.8 present solution times as multiples of the times needed for a modal response solution. The accuracy measures employed

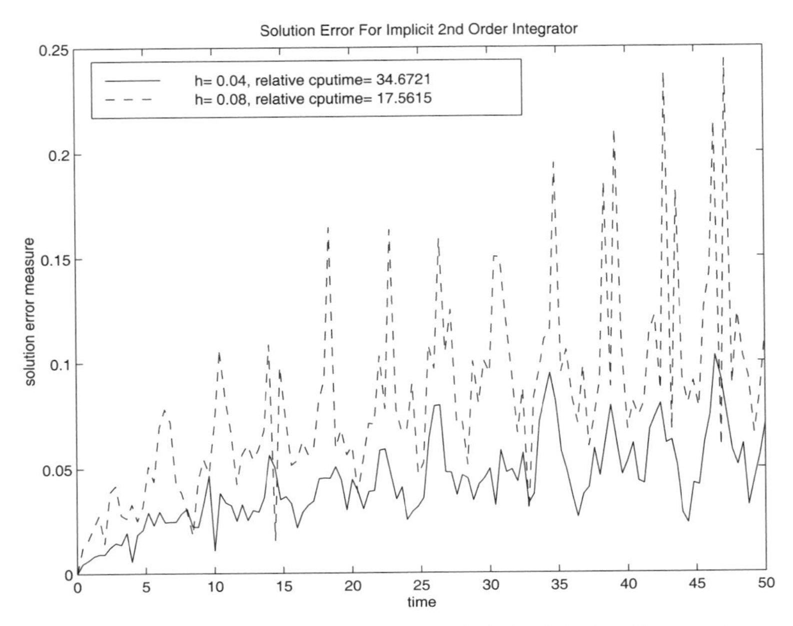

Figure 7.7: Solution Error for Implicit 2nd Order Integrator

are described next. Note that the displacement response matrix has rows describing system positions at successive times. Consequently, a measure of the difference between approximate and exact solutions is given by the vector

```
error_vector = \bsqrt(\bsum(((x_aprox-x_exact).^2)'));
```

Typically this vector has small initial components (near $t = 0$) and larger components (near the final time). The error measure is compared for different integrators and time steps in the figures. Note that the fourth order integrator is more efficient than the second order integrator because a larger integration step can be taken without excessive loss in accuracy. Using $h = 0.4$ for **mckde4i** achieved nearly the same accuracy as that given by **mckde2i** with $h = 0.067$. However, the computation time for **mckde2i** was several times as large as that for **mckde4i**.

In the past it has been traditional to use only second order methods for solving the structural dynamics equation. This may have been dictated by considerations on computer memory. Since workstations widely available today have relatively large memories and can invert a matrix of order two hundred in about half a second, it appears that use of high order integrators may gain in popularity.

The following computer program concludes our chapter on the solution of linear,

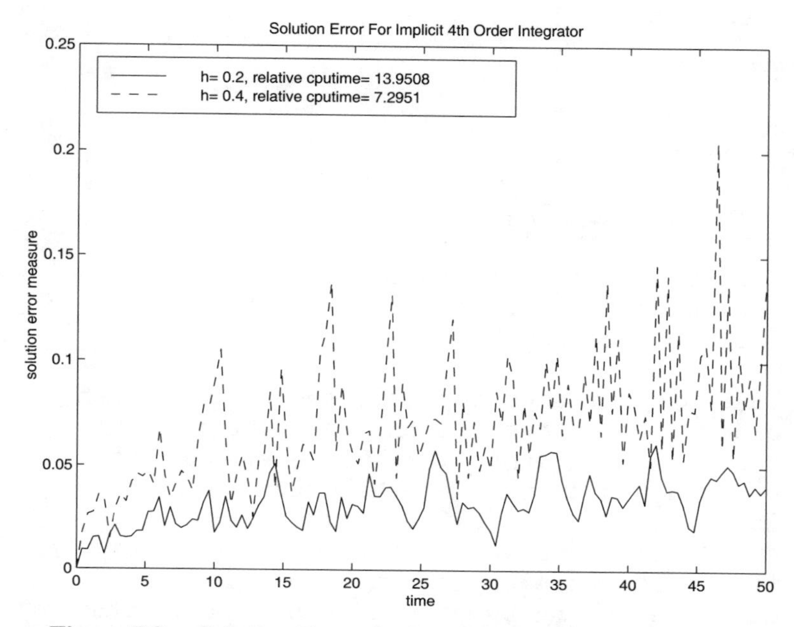

Figure 7.8: Solution Error for Implicit 4th Order Integrator

constant-coefficient matrix differential equations. Then we will study, in the next chapter, the Runge-Kutta method for integrating nonlinear problems.

MATLAB Example

Program deislner

```
1: function deislner
2: %
3: % Example:  deislner
4: % ~~~~~~~~~~~~~~~~~~~~
5: % Solution error for simulation of cable
6: % motion using a second or a fourth order
7: % implicit integrator.
8: %
9: % This program uses implicit second or fourth
10: % order integrators to compute the dynamical
11: % response of a cable which is suspended at
12: % one end and is free at the other end. The
13: % cable is given a uniform initial velocity.
14: % A plot of the solution error is given for
15: % two cases where approximate solutions are
16: % generated using numerical integration rather
17: % than modal response which is exact.
18: %
19: % User m functions required:
20: %    mckde2i, mckde4i, cablemk, udfrevib,
21: %    plterror
22:
23: % Choose a model having twenty links of
24: % equal length
25:
26: fprintf(...
27: '\nPlease wait: solution takes a while\n')
28: clear all
29: n=20; gravty=1.; n2=1+fix(n/2);
30: masses=ones(n,1)/n; lengths=ones(n,1)/n;
31:
32: % First generate the exact solution by
33: % modal superposition
34: [m,k]=cablemk(masses,lengths,gravty);
35: c=zeros(size(m));
36: dsp=zeros(n,1); vel=ones(n,1);
37: t0=0; tfin=50; ntim=126; h=(tfin-t0)/(ntim-1);
38:
39: % Numbers of repetitions each solution is
40: % performed to get accurate cpu times for
```

```
41: % the chosen step sizes are shown below.
42: % Parameter jmr may need to be increased to
43: % give reliable cpu times on fast computers
44:
45: jmr=500;
46: j2=fix(jmr/50); J2=fix(jmr/25);
47: j4=fix(jmr/20); J4=fix(jmr/10);
48:
49: % Loop through all solutions repeatedly to
50: % obtain more reliable timing values on fast
51: % computers
52: tic;
53: for j=1:jmr;
54:   [tmr,xmr]=udfrevib(m,k,dsp,vel,t0,tfin,ntim);
55: end
56: tcpmr=toc/jmr;
57:
58: % Second order implicit results
59: i2=10; h2=h/i2; tic;
60: for j=1:j2
61:   [t2,x2]=mckde2i(m,c,k,t0,dsp,vel,tfin,h2,i2);
62: end
63: tcp2=toc/j2; tr2=tcp2/tcpmr;
64:
65: I2=5; H2=h/I2; tic;
66: for j=1:J2
67:   [T2,X2]=mckde2i(m,c,k,t0,dsp,vel,tfin,H2,I2);
68: end
69: Tcp2=toc/J2; Tr2=Tcp2/tcpmr;
70:
71: % Fourth order implicit results
72: i4=2; h4=h/i4; tic;
73: for j=1:j4
74:   [t4,x4]=mckde4i(m,c,k,t0,dsp,vel,tfin,h4,i4);
75: end
76: tcp4=toc/j4; tr4=tcp4/tcpmr;
77:
78: I4=1; H4=h/I4; tic;
79: for j=1:J4
80:   [T4,X4]=mckde4i(m,c,k,t0,dsp,vel,tfin,H4,I4);
81: end
82: Tcp4=toc/J4; Tr4=Tcp4/tcpmr;
83:
84: % Plot error measures for each solution
85: plterror(xmr,t2,h2,x2,T2,H2,X2,...
```

```
86:         t4,h4,x4,T4,H4,X4,tr2,Tr2,tr4,Tr4)
87:
88: %==============================================
89:
90: function [t,x,tcp] = ...
91:      mckde2i(m,c,k,t0,x0,v0,tmax,h,incout,forc)
92: %
93: % [t,x,tcp]= ...
94: %    mckde2i(m,c,k,t0,x0,v0,tmax,h,incout,forc)
95: % ~~~~~~~~~~~~~~~~~~~~~~~~~~~~~~~~~~~~~~~~~~~~~~~
96: % This function uses a second order implicit
97: % integrator % to solve the matrix differential
98: % equation
99: %             m x'' + c x' + k x = forc(t)
100: % where m,c, and k are constant matrices and
101: % forc is an externally defined function.
102: %
103: % Input:
104: % ------
105: % m,c,k   mass, damping and stiffness matrices
106: % t0      starting time
107: % x0,v0   initial displacement and velocity
108: % tmax    maximum time for solution evaluation
109: % h       integration stepsize
110: % incout  number of integration steps between
111: %         successive values of output
112: % forc    externally defined time dependent
113: %         forcing function. This parameter
114: %         should be omitted if no forcing
115: %         function is used.
116: %
117: % Output:
118: % -------
119: % t       time vector going from t0 to tmax
120: %         in steps of
121: % x       h*incout to yield a matrix of
122: %         solution values such that row j
123: %         is the solution vector at time t(j)
124: % tcp     computer time for the computation
125: %
126: % User m functions called:  none.
127: %----------------------------------------------
128:
129: if (nargin > 9); force=1; else, force=0; end
130: if nargout ==3, tcp=clock; end
```

```
131: hbig=h*incout;
132: t=(t0:hbig:tmax)'; n=length(t);
133: ns=(n-1)*incout; ts=t0+h*(0:ns)';
134: xnow=x0(:); vnow=v0(:);
135: nvar=length(x0);
136: jrow=1; jstep=0; h2=h/2;
137:
138: % Form the inverse of the effective
139: % stiffness matrix
140: mnv=h*inv(m+h2*(c+h2*k));
141:
142: % Initialize the output matrix for x
143: x=zeros(n,nvar); x(1,:)=xnow';
144: zroforc=zeros(length(x0),1);
145:
146: % Main integration loop
147: for j=1:ns
148:   tj=ts(j);tjh=tj+h2;
149:   if force
150:     dv=feval(forc,tjh);
151:   else
152:     dv=zroforc;
153:   end
154:   dv=mnv*(dv-c*vnow-k*(xnow+h2*vnow));
155:   vnext=vnow+dv;xnext=xnow+h2*(vnow+vnext);
156:   jstep=jstep+1;
157:   if jstep == incout
158:     jstep=0; jrow=jrow+1; x(jrow,:)=xnext';
159:   end
160:   xnow=xnext; vnow=vnext;
161: end
162: if nargout ==3
163:   tcp=etime(clock,tcp);
164: else
165:   tcp=[];
166: end
167:
168: %==============================================
169:
170: function [t,x,tcp] = ...
171:        mckde4i(m,c,k,t0,x0,v0,tmax,h,incout,forc)
172: %
173: % [t,x,tcp]= ...
174: %      mckde4i(m,c,k,t0,x0,v0,tmax,h,incout,forc)
175: % ~~~~~~~~~~~~~~~~~~~~~~~~~~~~~~~~~~~~~~~~~~~~~~
```

```
176: % This function uses a fourth order implicit
177: % integrator with fixed stepsize to solve the
178: % matrix differential equation
179: %           m x'' + c x' + k x = forc(t)
180: % where m,c, and k are constant matrices and
181: % forc is an externally defined function.
182: %
183: % Input:
184: % ------
185: % m,c,k   mass, damping and stiffness matrices
186: % t0      starting time
187: % x0,v0   initial displacement and velocity
188: % tmax    maximum time for solution evaluation
189: % h       integration stepsize
190: % incout  number of integration steps between
191: %         successive values of output
192: % forc    externally defined time dependent
193: %         forcing function. This parameter
194: %         should be omitted if no forcing
195: %         function is used.
196: %
197: % Output:
198: % -------
199: % t       time vector going from t0 to tmax
200: %         in steps of h*incout
201: % x       matrix of solution values such
202: %         that row j is the solution vector
203: %         at time t(j)
204: % tcp     computer time for the computation
205: %
206: % User m functions called:  none.
207: %----------------------------------------------
208:
209: if nargin > 9, force=1; else, force=0; end
210: if nargout ==3, tcp=clock; end
211: hbig=h*incout; t=(t0:hbig:tmax)';
212: n=length(t); ns=(n-1)*incout; nvar=length(x0);
213: jrow=1; jstep=0; h2=h/2; h12=h*h/12;
214:
215: % Form the inverse of the effective stiffness
216: % matrix for later use.
217:
218: m12=m-h12*k;
219: mnv=inv([[(-h2*m-h12*c),m12];
220:         [m12,(c+h2*k)]]);
```

```
221:
222: % The forcing function is integrated using a
223: % 2 point Gauss rule
224: r3=sqrt(3); b1=h*(3-r3)/6; b2=h*(3+r3)/6;
225:
226: % Initialize output matrix for x and other
227: % variables
228: xnow=x0(:); vnow=v0(:);
229: tnow=t0; zroforc=zeros(length(x0),1);
230:
231: if force
232:   fnow=feval(forc,tnow);
233: else
234:   fnow=zroforc;
235: end
236: x=zeros(n,nvar); x(1,:)=xnow'; fnext=fnow;
237:
238: % Main integration loop
239: for j=1:ns
240:   tnow=t0+(j-1)*h; tnext=tnow+h;
241:   if force
242:     fnext=feval(forc,tnext);
243:     di1=h12*(fnow-fnext);
244:     di2=h2*(feval(forc,tnow+b1)+ ...
245:             feval(forc,tnow+b2));
246:     z=mnv*[(di1+m*(h*vnow)); (di2-k*(h*xnow))];
247:     fnow=fnext;
248:   else
249:     z=mnv*[m*(h*vnow); -k*(h*xnow)];
250:   end
251:   vnext=vnow + z(1:nvar);
252:   xnext=xnow + z((nvar+1):2*nvar);
253:   jstep=jstep+1;
254:
255:   % Save results every incout steps
256:   if jstep == incout
257:     jstep=0; jrow=jrow+1; x(jrow,:)=xnext';
258:   end
259:
260:   % Update quantities for next step
261:   xnow=xnext; vnow=vnext; fnow=fnext;
262: end
263: if nargout==3
264:   tcp=etime(clock,tcp);
265: else
```

```
266:   tcp=[];
267: end
268:
269: %==========================================
270:
271: function [m,k]=cablemk(masses,lngths,gravty)
272: %
273: % [m,k]=cablemk(masses,lngths,gravty)
274: % ~~~~~~~~~~~~~~~~~~~~~~~~~~~~~~~~~~~
275: % Form the mass and stiffness matrices for
276: % the cable.
277: %
278: % masses     - vector of masses
279: % lngths     - vector of link lengths
280: % gravty     - gravity constant
281: % m,k        - mass and stiffness matrices
282: %
283: % User m functions called:  none.
284: %------------------------------------------
285:
286: m=diag(masses);
287: b=flipud(cumsum(flipud(masses(:))))* ...
288:   gravty./lngths;
289: n=length(masses); k=zeros(n,n); k(n,n)=b(n);
290: for i=1:n-1
291:   k(i,i)=b(i)+b(i+1); k(i,i+1)=-b(i+1);
292:   k(i+1,i)=k(i,i+1);
293: end
294:
295: %==========================================
296:
297: function plterror(xmr,t2,h2,x2,T2,H2,X2,...
298:          t4,h4,x4,T4,H4,X4,tr2,Tr2,tr4,Tr4)
299: % plterror(xmr,t2,h2,x2,T2,H2,X2,...
300: % ~~~~~~~~~~~~~~~~~~~~~~~~~~~~~~~~~~~
301: %          t4,h4,x4,T4,H4,X4,tr2,Tr2,tr4,Tr4)
302: % ~~~~~~~~~~~~~~~~~~~~~~~~~~~~~~
303: % Plots error measures showing how different
304: % integrators and time steps compare with
305: % the exact solution using modal response.
306: %
307: % User m functions called:  none
308: %------------------------------------------
309:
310: % Compare the maximum error in any component
```

```
311: % at each time with the largest deflection
312: % occurring during the complete time history
313: maxd=max(abs(xmr(:)));
314: er2=max(abs(x2-xmr)')/maxd;
315: Er2=max(abs(X2-xmr)')/maxd;
316: er4=max(abs(x4-xmr)')/maxd;
317: Er4=max(abs(X4-xmr)')/maxd;
318:
319: plot(t2,er2,'-',T2,Er2,'--');
320: title(['Solution Error For Implicit ',...
321:        '2nd Order Integrator']);
322: xlabel('time');
323: ylabel('solution error measure');
324: lg1=['h= ', num2str(h2),  ...
325:      ', relative cputime= ', num2str(tr2)];
326: lg2=['h= ', num2str(H2),  ...
327:      ', relative cputime= ', num2str(Tr2)];
328: legend(lg1,lg2,2); figure(gcf);
329: disp('Press [Enter] to continue'); pause
330: % print -deps deislne2
331:
332: plot(t4,er4,'-',T4,Er4,'--');
333: title(['Solution Error For Implicit ',...
334:        '4th Order Integrator']);
335: xlabel('time');
336: ylabel('solution error measure');
337: lg1=['h= ', num2str(h4),  ...
338:      ', relative cputime= ', num2str(tr4)];
339: lg2=['h= ', num2str(H4),  ...
340:      ', relative cputime= ', num2str(Tr4)];
341: legend(lg1,lg2,2); figure(gcf);
342: % print -deps deislne4
343: disp(' '), disp('All Done')
344:
345: %==============================================
346:
347: % function [t,u,mdvc,natfrq]=...
348: %               udfrevib(m,k,u0,v0,tmin,tmax,nt)
349: % See Appendix B
350:
```

Chapter 8

Integration of Nonlinear Initial Value Problems

8.1 General Concepts on Numerical Integration of Nonlinear Matrix Differential Equations

Methods for solving differential equations numerically are one of the most valuable analysis tools now available. Inexpensive computer power and user friendly software are stimulating wider use of digital simulation methods. At the same time, intelligent use of numerically integrated solutions requires appreciation of inherent limitations of the techniques employed. The present chapter discusses the widely used Runge-Kutta method and applies it to some specific examples.

When physical systems are described by mathematical models, it is common that various system parameters are only known approximately. For example, to predict the response of a building undergoing earthquake excitation, simplified formulations may be necessary to handle the elastic and frictional characteristics of the soil and the building. Our observation that simple models are used often to investigate behavior of complex systems does not necessarily amount to a rejection of such procedures. In fact, good engineering analysis depends critically on development of reliable models which can capture salient features of a process without employing unnecessary complexity. At the same time, analysts need to maintain proper caution regarding trustworthiness of answers produced with computer models. Nonlinear system response sometimes changes greatly when only small changes are made in the physical parameters. Scientists today realize that, in dealing with highly nonlinear phenomena such as weather prediction, it is simply impossible to make reliable long term forecasts [45] because of various unalterable factors. Among these are a) uncertainty about initial conditions, b) uncertainty about the adequacy of mathematical models describing relevant physical processes, c) uncertainty about error contributions arising from use of spatial and time discretizations in construction of approximate numerical solutions, and d) uncertainty about effects of arithmetic roundoff error. In light of the criticism and cautions being stated about the dangers of using numerical solutions, the thrust of the discussion is that idealized models must not be regarded as infallible, and no numerical solution should be accepted as credible without adequately investigating effects of parameter perturbation within uncertainty limits of the parameters. To illustrate how sensitive a system can be to initial conditions, we might consider a very simple model concerning motion of a pendulum of length ℓ given an initial velocity v_0 starting from a vertically downward position. If v_0 exceeds $2\sqrt{g\ell}$, the pendulum will reach a vertically upward position and will go over the top. If v_0 is less than $2\sqrt{g\ell}$, the vertically upward position is never reached. Instead, the pendulum oscillates about the bottom position. Consequently, initial velocities of $1.999\sqrt{g\ell}$ and $2.001\sqrt{g\ell}$ produce quite different system behavior with only a tiny change in initial velocity. Other examples illustrating the difficulties of computing the response of nonlinear systems are cited below. These examples are not chosen to discourage use of the powerful tools now available for numerical integration of differential equations. Instead, the intent is to encourage users of these methods to exercise proper caution so that confidence in the reliability of results is fully justified.

Many important physical processes are governed by differential equations. Typical cases include dynamics of rigid and flexible bodies, heat conduction, and electrical current flow. Solving a system of differential equations subject to known initial conditions allows us to predict the future behavior of the related physical system. Since very few important differential equations can be solved in closed form, approximations which are directly or indirectly founded on series expansion methods have been developed. The basic problem addressed is that of accurately computing $Y(t+h)$ when $Y(t)$ is known, along with a differential equation governing system behavior from time t to $(t+h)$. Recursive application of a satisfactory numerical approximation procedure, with possible adjustment of step-size to maintain accuracy and stability, allows approximate prediction of system response subsequent to the starting time.

Numerical methods for solving differential equations are important tools for analyzing engineering systems. Although valuable algorithms have been developed which facilitate construction of approximate solutions, all available methods are vulnerable to limitations inherent in the underlying approximation processes. The essence of the difficulty lies in the fact that, as long as a finite integration step-size is used, integration error occurs at each time step. These errors sometimes have an accumulative effect which grows exponentially and eventually destroys solution validity. To some extent, accuracy problems can be limited by regulating step-size to keep local error within a desired tolerance. Typically, decreasing an integration tolerance increases the time span over which a numerical solution is valid. However, high costs for supercomputer time to analyze large and complex systems sometimes preclude generation of long time histories which may be more expensive than is practically justifiable.

8.2 Runge-Kutta Methods and the ODE45 Integrator Provided in MATLAB

Formulation of one method to solve differential equations is discussed in this section. Suppose a function $y(x)$ satisfies a differential equation of the form $y'(x) = f(x, y)$, subject to $y(x_0) = y_0$, where f is a known differentiable function. We would like to compute an approximation of $y(x_0 + h)$ which agrees with a Taylor's series expansion up to a certain order of error. Hence,

$$y(x_0 + h) = \tilde{y}(x_0, h) + O(h^{n+1})$$

where $O(h^{n+1})$ denotes a quantity which decreases at least as fast as h^{n+1} for small h. Taylor's theorem allows us to write

$$\begin{aligned} y(x_0 + h) &= y(x_0) + y'(x_0)h + \frac{1}{2}y''(x_0)h^2 + O(h^3) \\ &= y_0 + f(x_0, y_0)h + \frac{1}{2}[f_x(x_0, y_0) + f_y(x_0, y_0)f_0]h^2 + O(h^3) \end{aligned}$$

where $f_0 = f(x_0, y_0)$. The last formula can be used to compute a second order approximation $\hat{y}(x_0+h)$, provided the partial derivatives f_x and f_y can be evaluated. However, this may be quite difficult since the function $f(x, y)$ may not even be known explicitly.

The idea leading to Runge-Kutta integration is to compute $y(x_0 + h)$ by making several evaluations of function f instead of having to differentiate that function. Let us seek an approximation in the form

$$\tilde{y}(x_0 + h) = y_0 + h[k_0 f_0 + k_1 f(x_0 + \alpha h, y_0 + \beta h f_0)].$$

We choose k_0, k_1, α, and β to make $\tilde{y}(x_0 + h)$ match the series expansion of $y(x)$ as well as possible. Since

$$f(x_0 + \alpha h, y_0 + \beta h f_0) = f_0 + [f_x(x_0, y_0)\alpha + f_y(x_0, y_0)f_0\beta]h + O(h^2),$$

we must have

$$\begin{aligned} \tilde{y}(x_0 + h) &= y_0 + h[(k_0 + k_1)f_0 + k_1\langle f_x(x_0, y_0)\alpha + f_y(x_0, y_0)\beta f_0\rangle]h + O(h^2) \\ &= y_0 + (k_0 + k_1)f_0 h + [f_x(x_0, y_0)\alpha k_1 + f_y(x_0, y_0)f_0\beta k_1]h^2 + O(h^3). \end{aligned}$$

The last relation shows that

$$y(x_0 + h) = \tilde{y}(x_0 + h) + O(h^3)$$

provided

$$k_0 + k_1 = 1 \ , \ \alpha k_1 = \frac{1}{2} \ , \ \beta k_1 = \frac{1}{2}.$$

This system of three equations in four unknowns has an infinite number of solutions; one of these is $k_0 = k_1 = \frac{1}{2}$, $\alpha = \beta = 1$. This implies that

$$y(x_0 + h) = y(x_0) + \frac{1}{2}[f_0 + f(x_0 + h, y_0 + hf_0)]h + O(h^3).$$

Neglecting the truncation error $O(h^3)$ gives a difference approximation known as Heun's method [61], which is classified as a second order Runge-Kutta method. Reducing the step-size by h reduces the truncation error by about a factor of $(\frac{1}{2})^3 = \frac{1}{8}$. Of course, the formula can be used recursively to compute approximations to $y(x_0 + h), y(x_0 + 2h), y(x_0 + 3h), \ldots$. In most instances, the solution accuracy decreases as the number of integration steps is increased and results eventually become unreliable. Decreasing h and taking more steps within a fixed time span helps, but this also has practical limits governed by computational time and arithmetic roundoff error.

The idea leading to Heun's method can be extended further to develop higher order formulas. One of the best known is the fourth order Runge-Kutta method described as follows

$$y(x_0 + h) = y(x_0) + h[k_1 + 2k_2 + 2k_3 + k_4]/6$$

where

$$k_1 = f(x_0, y_0) \ , \ k_2 = f(x_0 + \frac{h}{2}, y_0 + k_1\frac{h}{2}),$$

$$k_3 = f(x_0 + \frac{h}{2}, y_0 + k_2\frac{h}{2}) \ , \ k_4 = f(x_0 + h, y_0 + k_3 h).$$

The truncation error for this formula is order h^5; so, the error is reduced by about a factor of $\frac{1}{32}$ when the step-size is halved. The development of the fourth order Runge-Kutta method is algebraically quite complicated [43]. We note that accuracy of order four is achieved with four evaluations of f for each integration step. This situation does not extend to higher orders. For instance, an eighth order formula may require twelve evaluations per step. This price of more function evaluations may be worthwhile provided the resulting truncation error is small enough to permit much larger integration steps than could be achieved with formulas of lower order. MATLAB provides the function **ode45** which uses variable step-size and employs formulas of order four and five. (Note: In MATLAB 6.x the integrators can output results for an arbitrary time vector using, for instance, even time increments.)

8.3 Step-size Limits Necessary to Maintain Numerical Stability

It can be shown that, for many numerical integration methods, taking too large a step-size produces absurdly large results that increase exponentially with successive time steps. This phenomenon, known as numerical instability, can be illustrated with the simple differential equation

$$y'(t) = f(t, y) = \lambda y$$

which has the solution $y = ce^{\lambda t}$. If the real part of λ is positive, the solution becomes unbounded with increasing time. However, a pure imaginary λ produces a bounded oscillatory solution, whereas the solution decays exponentially for $\mathbf{real}(\lambda) < 0$. Applying Heun's method [43] gives

$$y(t+h) = y(t)\left[1 + (\lambda h) + \frac{(\lambda h)^2}{2}\right].$$

This shows that at each integration step the next value of y is obtained by multiplying the previous value by a factor

$$p = 1 + (\lambda h) + \frac{(\lambda h)^2}{2},$$

which agrees with the first three Taylor series terms of $e^{\lambda h}$. Clearly, the difference relation leads to

$$y_n = y_0 p^n.$$

As n increases, y_n will approach infinity unless $|p| \le 1$. This stability condition can be interpreted geometrically by regarding λh as a complex variable z and solving for all values of z such that

$$1 + z + \frac{z^2}{2} = \zeta e^{\imath\theta}\ ,\ |\zeta| \le 1\ ,\ 0 \le \theta \le 2\pi.$$

Taking $\zeta = 1$ identifies the boundary of the stability region, which is normally a closed curve lying in the left half of the complex plane. Of course, h is assumed to be positive and the real part of λ is nonpositive. Otherwise, even the exact solution would grow exponentially. For a given λ, the step-size h must be taken small enough to make $|\lambda h|$ lie within the stability zone. The larger $|\lambda|$ is, the smaller h must be to prevent numerical instability.

The idea illustrated by Heun's method can be easily extended to a Runge-Kutta method of arbitrary order. A Runge-Kutta method of order n reproduces the exact solution through terms of order n in the Taylor series expansion. The differential equation $y' = \lambda y$ implies

$$y(t+h) = y(t)e^{\lambda h}$$

and

$$e^{\lambda h} = \sum_{k=0}^{n} \frac{(\lambda h)^k}{k!} + O(h^{n+1}).$$

Consequently, points on the boundary of the stability region for a Runge-Kutta method of order n are found by solving the polynomial

$$1 - e^{\imath\theta} + \sum_{k=1}^{n} \frac{z^k}{k!} = 0$$

for a dense set of θ-values ranging from 0 to 2π. Using MATLAB's intrinsic function **roots** allows easy calculation of the polynomial roots which may be plotted to show the stability boundary. The following short program accomplishes the task. Program output for integrators of order four and six is shown in Figures 8.1 and 8.2. Note that the region for order 4 resembles a semicircle with radius close to 2.8. Using $|\lambda h| > 2.8$, with Runge-Kutta of order 4, would give results which rapidly become unstable. The figures also show that the stability region for Runge-Kutta of order 6 extends farther out on the negative real axis than Runge-Kutta of order 4 does. The root finding process also introduces some meaningless stability zones in the right half plane which should be ignored.

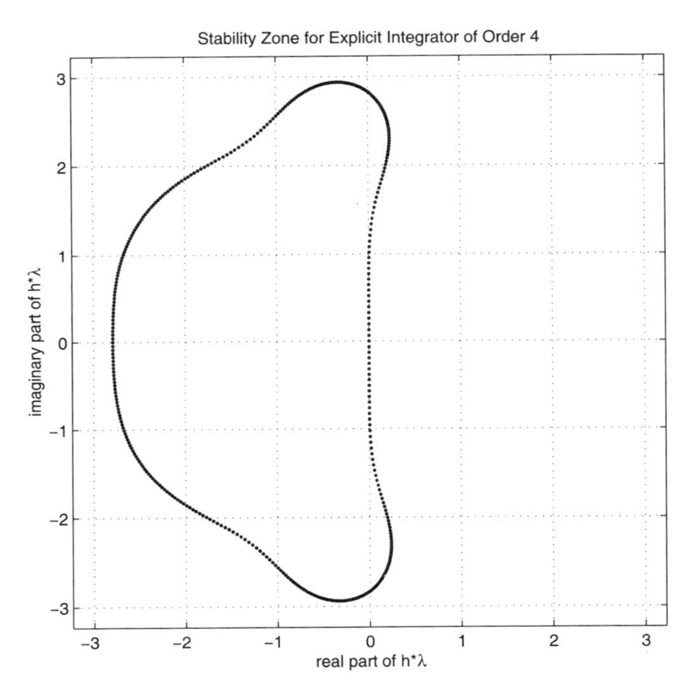

Figure 8.1: **Stability Zone for Explicit Integrator of Order 4**

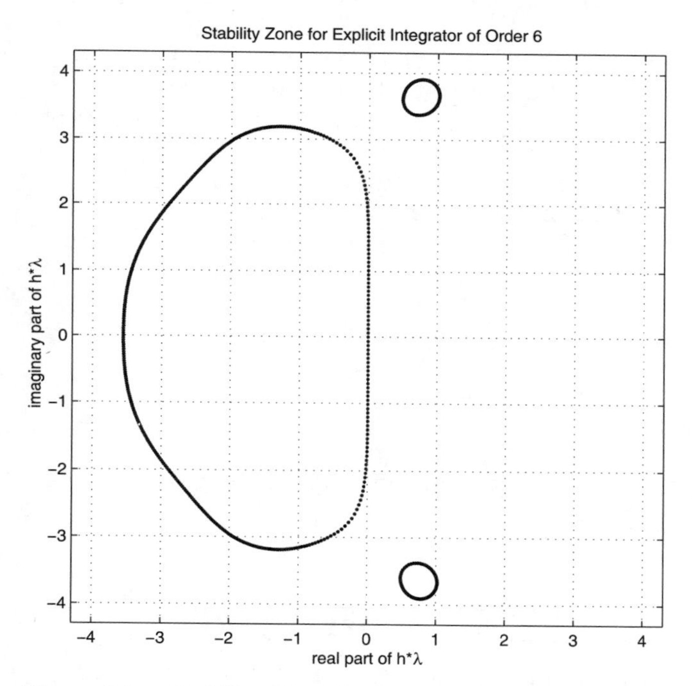

Figure 8.2: Stability Zone for Explicit Integrator of Order 6

MATLAB Example

Program rkdestab

```
1: % Example:  rkdestab
2: % ~~~~~~~~~~~~~~~~~~~
3: % This program plots the boundary of the region
4: % of the complex plane governing the maximum
5: % step size which may be used for stability of
6: % a Runge-Kutta integrator of arbitrary order.
7: %
8: % npts  - a value determining the number of
9: %         points computed on the stability
10: %         boundary of an explicit Runge-Kutta
11: %         integrator.
12: % xrang - controls the square window within
13: %         which the diagram is drawn.
14: %         [ -3, 3, -3, 3] is appropriate for
15: %         the fourth order integrator.
16: %
17: % User m functions required: none
18:
19: hold off; clf; close;
20: fprintf('\nSTABILITY REGION FOR AN ');
21: fprintf('EXPLICIT RUNGE-KUTTA');
22: fprintf('\n     INTEGRATOR OF ARBITRARY ');
23: fprintf('ORDER\n\n');
24: while 1
25: disp(' ')
26: nordr=input('Give the integrator order ? > ');
27: if isempty(nordr) | nordr==0, break; end
28: % fprintf('\nInput the number of points ');
29: % fprintf('used to define\n');
30: % npts=input('the boundary (100 is typical) ? > ');
31: npts=100;
32: r=zeros(npts,nordr); v=1./gamma(nordr+1:-1:2);
33: d=2*pi/(npts-1); i=sqrt(-1);
34:
35: % Generate polynomial roots to define the
36: % stability boundary
37: for j=1:npts
38:   % polynomial coefficients
39:   v(nordr+1)=1-exp(i*(j-1)*d);
40:   % complex roots
```

```
41:     t=roots(v); r(j,:)=t(:).';
42: end
43:
44: % Plot the boundary
45: rel=real(r(:)); img=imag(r(:));
46: w=1.1*max(abs([rel;img]));
47: zoom on; plot(rel,img,'.');
48: axis([-w,w,-w,w]); axis('square');
49: xlabel('real part of h*\lambda');
50: ylabel('imaginary part of h*\lambda');
51: ns=int2str(nordr);
52: st=['Stability Zone for Explicit ' ...
53:      'Integrator of Order ',ns];
54: title(st); grid on; figure(gcf);
55: % print -deps rkdestab
56: end
57:
58: disp(' '); disp('All Done');
```

8.4 Discussion of Procedures to Maintain Accuracy by Varying Integration Step-size

When we solve a differential equation numerically, our first inclination is to seek output at even increments of the independent variable. However, this is not the most natural form of output appropriate to maintain integration accuracy. Whenever solution components are changing rapidly, a small time step may be needed, whereas using a small time step might be quite inefficient at times where the solution remains smooth. Most modern ODE programs employ variable step-size algorithms which decrease the integration step-size whenever some local error tolerance is violated and conversely increase the step-size when the increase can be performed without loss of accuracy. If results at even time increments are needed, these can be determined by interpolation of the non-equidistant values. The differential equation integrators provide the capability to output results at an arbitrary vector of times over the integration interval.

Although the derivation of algorithms to regulate step-size is an important topic, development of these methods is not presented here. Several references [43, 46, 51, 61] discuss this topic with adequate detail. The primary objective in regulating step-size is to gain computational efficiency by taking as large a step-size as possible while maintaining accuracy and minimizing the number of function evaluations.

Practical problems involving a single first order differential equation are rarely encountered. More commonly, a system of second order equations occurs which is then transformed into a system involving twice as many first order equations. Several hundred, or even several thousand dependent variables may be involved. Evaluating the necessary time derivatives at a single time step may require computationally intensive tasks such as matrix inversion. Furthermore, performing this fundamental calculation several thousand times may be necessary in order to construct time responses over time intervals of practical interest. Integrating large systems of nonlinear differential equations is one of the most important and most resource intensive aspects of scientific computing.

Instead of deriving the algorithms used for step-size control in **ode45**, we will outline briefly the ideas employed to integrate $y'(t) = f(t, y)$ from t to $(t + h)$. It is helpful to think of y as a vector. For a given time step and y value, the program makes six evaluations of f. These values allow evaluation of two Runge-Kutta formulas, each having different truncation errors. These formulas permit estimation of the actual truncation error and proper step-size adjustment to control accuracy. If the estimated error is too large, the step-size is decreased until the error tolerance is satisfied or an error condition occurs because the necessary step-size has fallen below a set limit. If the estimated error is found to be smaller than necessary, the integration result is accepted and the step-size is increased for the next pass. Even though this type of process may not be extremely interesting to discuss, it is nevertheless an essential part of any well designed program for integrating differential equations numerically. Readers should become familiar with the error control features employed by ODE solvers. Printing and studying the code for **ode45** is worthwhile. Studying the convergence tolerance used in connection with function **odeset** is also instructive. It should be remembered that solutions generated with tools such as **ode45** are vulnerable to accumulated errors from roundoff and arithmetic truncation. Such errors usually render unreliable the results obtained sufficiently far from the starting time.

This chapter concludes with the analysis of several realistic nonlinear problems having certain properties of their exact solutions known. These known properties are compared with numerical results to assess error growth. The first problem involves an inverted pendulum for which the loading function produces a simple exact displacement function. Examples concerning top dynamics, a projectile trajectory, and a falling chain are presented.

8.5 Example on Forced Oscillations of an Inverted Pendulum

The inverted pendulum in Figure 8.3 involves a weightless rigid rod of length l which has a mass m attached to the end. Attached to the mass is a spring with stiffness constant k and an unstretched length of γl. The spring has length l when the pendulum is in the vertical position. Externally applied loads consist of a driving moment $M(t)$, the particle weight, and a viscous damping moment $cl^2\dot{\theta}$. The differential equation governing the motion of this system is

$$\ddot{\theta} = -(c/m)\dot{\theta} + (g/l)\sin(\theta) + M(t)/(ml^2) - (2k/m)\sin(\theta)(1 - \alpha/\lambda)$$

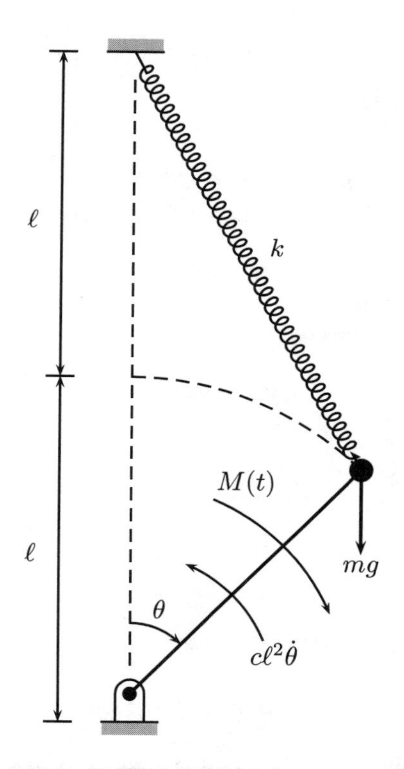

Figure 8.3: Forced Vibration of an Inverted Pendulum

where

$$\lambda = \sqrt{5 - 4\cos(\theta)}.$$

This system can be changed to a more convenient form by introducing dimensionless variables. We let $t = (\sqrt{l/g})\tau$ where τ is dimensionless time. Then

$$\ddot{\theta} = -\alpha\dot{\theta} + \sin(\theta) + P(\tau) - \beta\sin(\theta)(1 - \gamma/\lambda)$$

where

$$\begin{aligned}
\alpha &= (c/m)\sqrt{l/g} = \text{viscous damping factor}, \\
\beta &= 2(k/m)/(g/l), \\
\lambda &= \sqrt{5 - 4\cos(\theta)}, \\
\gamma &= (\text{unstretched spring length})/l, \\
P(\tau) &= M/(mgl) = \text{dimensionless driving moment}.
\end{aligned}$$

It is interesting to test how well a numerical method can reconstruct a known exact solution for a nonlinear function. Let us assume that the driving moment $M(\tau)$ produces a motion having the equation

$$\theta_e(\tau) = \theta_0 \sin(\omega\tau)$$

for arbitrary θ_0 and ω. Then

$$\dot{\theta}_e(\tau) = \omega\theta_0 \cos(\omega\tau)$$

and

$$\ddot{\theta}_e(\tau) = -\omega^2\theta_e.$$

Consequently, the necessary driving moment is

$$P(\tau) = -\omega^2\theta_e - \sin(\theta_e) + \gamma\omega\theta_0\cos(\omega\tau) + \beta\sin(\theta_e)\left[1 - \gamma/\sqrt{5 - 4\cos(\theta_e)}\right].$$

Applying this forcing function, along with the initial conditions

$$\theta(0) = 0\ ,\ \dot{\theta}(0) = \theta_0\omega$$

should return the solution $\theta = \theta_e(\tau)$. For a specific numerical example we choose $\theta_0 = \pi/8$, $\omega = 0.5$, and four different combinations of β, γ, and tol. The second order differential equation has the form $\ddot{\theta} = f(\tau, \theta, \dot{\theta})$. This is expressed as a first order matrix system by letting $y_1 = \theta$, $y_2 = \dot{\theta}$, which gives

$$\dot{y}_1 = y_2\ ,\ \dot{y}_2 = f(\tau, y_1, y_2).$$

A function describing the system for solution by **ode45** is provided at the end of this section. Parameters θ_0, ω_0, α, ζ, and β are passed as global variables.

We can examine how well the numerically integrated θ match θ_e by using the error measure

$$|\theta(\tau) - \theta_e(\tau)|.$$

Furthermore, the exact solution satisfies

$$\theta_e^2 + (\dot{\theta}_e/\omega)^2 = \theta_0^2.$$

Plotting $\dot{\theta}/(\theta_0\omega)$ on a horizontal axis and θ/θ_0 on a vertical axis should produce a unit circle. Violation of that condition signals loss of solution accuracy.

How certain physical parameters and numerical tolerances affect terms in this problem can be demonstrated by the following four data cases:

1. The spring is soft and initially unstretched. A liberal integration tolerance is used.

2. The spring is soft and initially unstretched. A stringent integration tolerance is used.

3. The spring is stiff and initially stretched. A liberal integration tolerance is used.

4. The spring is stiff and initially stretched. A stringent integration tolerance is used.

The curves in Figure 8.4 show the following facts:

1. When the spring is unstretched initially, the numerical solution goes unstable quickly.

2. Stretching the spring initially and increasing the spring constant improves numerical stability of the solution.

3. Decreasing the integration tolerance increases the time period over which the solution is valid.

An additional curve illustrating the numerical inaccuracy of results for Case 1 appears in Figure 8.5. A plot of $\theta(\tau)$ versus $\dot{\theta}(\tau)/\omega$ should produce a circle. However, solution points quickly depart from the desired locus.

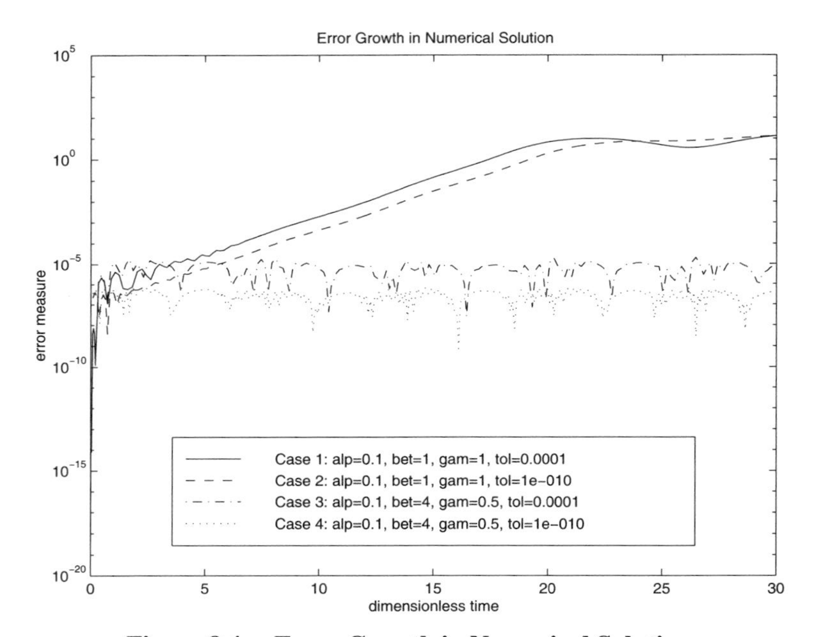

Figure 8.4: Error Growth in Numerical Solution

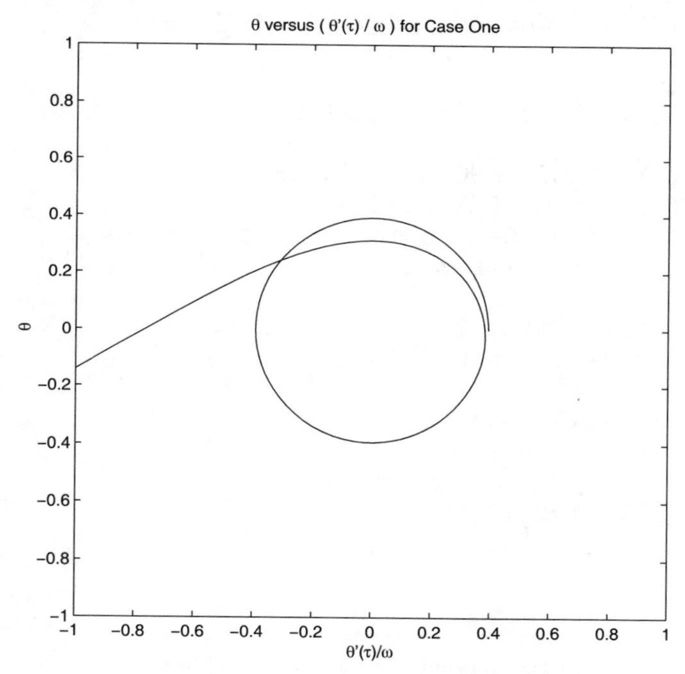

Figure 8.5: θ **versus** $(\theta'(\tau)/\omega)$ **for Case One**

MATLAB Example

Program prun

```
1: function prun
2: % Example: prun
3: % ~~~~~~~~~~~~~
4: % Dynamics of an inverted pendulum integrated
5: % by use of ode45.
6: %
7: % User m functions required: pinvert, mom
8:
9: global ncal
10: th0=pi/8; w=.5; tmax=30; ncal=0;
11:
12: fprintf('\nFORCED OSCILLATION OF AN ');
13: fprintf('INVERTED PENDULUM\n');
14: fprintf('\nNote: Generating four sets of\n');
15: fprintf('numerical results takes a while.\n');
16:
17: % loose spring with liberal tolerance
18: alp=0.1; bet=1.0; gam=1.0; tol=1.e-4;
19: a1=num2str(alp); b1=num2str(bet);
20: g1=num2str(gam); e1=num2str(tol);
21: options=odeset('RelTol',tol);
22: [t1,z1]= ...
23:   ode45(@pinvert,[0,tmax],[0;w*th0],...
24:         options,alp,bet,gam,th0,w);
25: n1=ncal; ncal=0;
26:
27: % loose spring with stringent tolerance
28: alp=0.1; bet=1.0; gam=1.0; tol=1.e-10;
29: a2=num2str(alp); b2=num2str(bet);
30: g2=num2str(gam); e2=num2str(tol);
31: options=odeset('RelTol',tol);
32: [t2,z2]= ...
33:   ode45(@pinvert,[0,tmax],[0;w*th0],...
34:         options,alp,bet,gam,th0,w);
35: n2=ncal; ncal=0;
36:
37: % tight spring with liberal tolerance
38: alp=0.1; bet=4.0; gam=0.5; tol=1.e-4;
39: a3=num2str(alp); b3=num2str(bet);
40: g3=num2str(gam); e3=num2str(tol);
```

```
41: options=odeset('RelTol',tol);
42: [t3,z3]= ...
43:   ode45(@pinvert,[0,tmax],[0;w*th0],...
44:          options,alp,bet,gam,th0,w);
45: n3=ncal; ncal=0;
46:
47: % tight spring with stringent tolerance
48: alp=0.1; bet=4.0; gam=0.5; tol=1.e-10;
49: a4=num2str(alp); b4=num2str(bet);
50: g4=num2str(gam); e4=num2str(tol);
51: options=odeset('RelTol',tol);
52: [t4,z4]= ...
53:   ode45(@pinvert,[0,tmax],[0;w*th0],...
54:          options,alp,bet,gam,th0,w);
55: n4=ncal; ncal=0; save pinvert.mat;
56:
57: % Plot results
58: clf; semilogy( ...
59:   t1,abs(z1(:,1)/th0-sin(w*t1)),'-r',...
60:   t2,abs(z2(:,1)/th0-sin(w*t2)),'--g',...
61:   t3,abs(z3(:,1)/th0-sin(w*t3)),'-.b',...
62:   t4,abs(z4(:,1)/th0-sin(w*t4)),':m');
63: title('Error Growth in Numerical Solution')
64: xlabel('dimensionless time');
65: ylabel('error measure');
66: c1=['Case 1: alp=',a1,', bet=',b1,', gam=', ...
67:     g1,', tol=',e1];
68: c2=['Case 2: alp=',a2,', bet=',b2,', gam=', ...
69:     g2,', tol=',e2];
70: c3=['Case 3: alp=',a3,', bet=',b3,', gam=', ...
71:     g3,', tol=',e3];
72: c4=['Case 4: alp=',a4,', bet=',b4,', gam=', ...
73:     g4,', tol=',e4];
74: legend(c1,c2,c3,c4,4); shg
75: dum=input('\nPress [Enter] to continue\n','s');
76: %print -deps pinvert
77:
78: % plot a phase diagram for case 1
79: clf; plot(z1(:,2)/w,z1(:,1));
80: axis('square'); axis([-1,1,-1,1]);
81: xlabel('\theta''(\tau)/\omega'); ylabel('\theta');
82: title(['\theta versus ( \theta''(\tau) / ' ...
83:        '\omega ) for Case One']); figure(gcf);
84: %print -deps crclplt
85: disp(' '); disp('All Done');
```

```
86:
87: %==============================================
88:
89: function zdot=pinvert(t,z,alp,bet,gam,th0,w)
90: %
91: % zdot=pinvert(t,z,alp,bet,gam,th0,w)
92: % ~~~~~~~~~~~~~~~~~~
93: % Equation of motion for the pendulum
94: %
95: % t    - time value
96: % z    - vector [theta ; thetadot]
97: % alp,bet,gam,th0,w
98: %      - physical parameters in the
99: %        differential equation
100: % zdot - time derivative of z
101: %
102: % User m functions called:  mom
103: %----------------------------------------------
104:
105: global ncal
106: ncal=ncal+1; th=z(1); thd=z(2);
107: c=cos(th); s=sin(th); lam=sqrt(5-4*c);
108: zdot=[thd; mom(t,alp,bet,gam,th0,w)+...
109:            s-alp*thd-bet*s*(1-gam/lam)];
110:
111: %==============================================
112:
113: function me=mom(t,alp,bet,gam,th0,w)
114: %
115: % me=mom(t,alp,bet,gam,th0,w)
116: % ~~~~~~~~~
117: % t  - time
118: % alp,bet,gam,th0,w
119: %    - physical parameters in the
120: %      differential equation
121: % me - driving moment needed to produce
122: %      exact solution
123: %
124: % User m functions called:  none.
125: %----------------------------------------------
126:
127: th=th0*sin(w*t);
128: thd=w*th0*cos(w*t); thdd=-th*w^2;
129: s=sin(th); c=cos(th); lam=sqrt(5-4*c);
130: me=thdd-s+alp*thd+bet*s*(1-gam/lam);
```

8.6 Dynamics of a Spinning Top

The dynamics of a symmetrical spinning top can be analyzed simply by computing the path followed by the gravity center in Cartesian coordinates. Consider a top spinning with its apex (or tip) constrained to remain at the origin. The gravity center lies at position r along the axis of symmetry and the only applied forces are the weight $-mg\hat{\boldsymbol{k}}$ through the gravity center and the support reaction at the tip of the top. The inertial properties involve a moment of inertia J_a about the symmetry axis and a transverse inertial moment J_t relative to an axis normal to the symmetry axis and passing through the apex of the top. The velocity of the gravity center and the angular velocity Ω are related by[1]

$$\boldsymbol{v} = \dot{\boldsymbol{r}} = \boldsymbol{\Omega} \times \boldsymbol{r}.$$

This implies that $\boldsymbol{\Omega}$ can be expressed in terms of radial and transverse components as

$$\boldsymbol{\Omega} = \ell^{-2} rv \times \boldsymbol{v} + \ell^{-1}\omega_a \boldsymbol{r}$$

where $\ell = |\boldsymbol{r}|$ and ω_a is the magnitude of the angular velocity component in the radial direction. The angular momentum with respect to the origin is therefore

$$\boldsymbol{H} = J_t \ell^{-2}\, \boldsymbol{r} \times \boldsymbol{v} + J_a \ell^{-1} \omega_a \boldsymbol{r}$$

and the potential plus kinetic energy is given by

$$K = mgz + \frac{J_t \ell^{-2}\, \boldsymbol{v} \cdot \boldsymbol{v} + J_a \omega_a^2}{2}$$

where z is the height of the gravity center above the origin.

The equations of motion can be found using the principle that the moment of all applied forces about the origin must equal the time rate of change of the corresponding angular momentum. Hence

$$\boldsymbol{M} = J_t \ell^{-2}\, \boldsymbol{r} \times \boldsymbol{a} + J_a \ell^{-1} \left[\omega_a \boldsymbol{v} + \dot{\omega}_a \boldsymbol{r}\right]$$

where $\boldsymbol{a} = \dot{\boldsymbol{v}} = \ddot{\boldsymbol{r}}$ is the total acceleration of the gravity center. The radial component of the last equation is obtainable by a dot product with $\boldsymbol{r}$ to give

$$\boldsymbol{r} \cdot \boldsymbol{M} = J_a \ell \dot{\omega}_a$$

where simplifications result because $\boldsymbol{r} \cdot (\boldsymbol{r} \times \boldsymbol{a}) = 0$ and $\boldsymbol{r} \cdot \boldsymbol{v} = 0$. The remaining components of $\boldsymbol{M}$ for the transverse direction result by taking $\boldsymbol{r} \times \boldsymbol{M}$ and noting that

$$\boldsymbol{r} \times (\boldsymbol{r} \times \boldsymbol{a}) = (\boldsymbol{r} \cdot \boldsymbol{a})\boldsymbol{r} - \ell^2 \boldsymbol{a} = -\ell^2 \boldsymbol{a}_t$$

[1] In this section the quantities v, r, Ω, H, M, and a all represent vector quantities.

where $\boldsymbol{a}_t$ is the vector component of total acceleration normal to the direction of $\boldsymbol{r}$. This leads to

$$\boldsymbol{r} \times \boldsymbol{M} = -J_t \boldsymbol{a}_t + J_a \ell^{-1} \omega_a \, \boldsymbol{r} \times \boldsymbol{v}.$$

Since the gravity center moves on a spherical surface of radius ℓ centered at the origin, the radial acceleration is given by

$$\boldsymbol{a}_r = -\boldsymbol{v} \cdot \boldsymbol{v} \, \ell^{-2} \boldsymbol{r}$$

and the total acceleration equation becomes

$$\boldsymbol{a} = -\frac{\boldsymbol{r} \times \boldsymbol{M}}{J_t} + \frac{J_a \ell^{-1} \omega_a}{J_t} \boldsymbol{r} \times \boldsymbol{v} - \boldsymbol{v} \cdot \boldsymbol{v} \, \ell^{-2} \boldsymbol{r}.$$

In the case studied here, only the body weight $-mg\hat{k}$ causes a moment about the origin so

$$\boldsymbol{M} = -mg \, \boldsymbol{r} \times \hat{\boldsymbol{k}} \; , \; \boldsymbol{r} \cdot \boldsymbol{M} = 0$$

and

$$\boldsymbol{r} \times \boldsymbol{M} = -mg \left[z\boldsymbol{r} - \ell^2 \hat{\boldsymbol{k}} \right].$$

The radial component of the moment equation simply gives $\dot{\omega}_a = 0$, so the axial component of angular velocity retains its initial value throughout the motion.

Integrating the differential equations

$$\dot{\boldsymbol{v}} = \boldsymbol{a} \; , \; \dot{\boldsymbol{r}} = \boldsymbol{v}$$

numerically subject to appropriate initial conditions produces a trajectory of the gravity center motion. The simple formulation presented here treats x, y, and z as if they were independent variables even though

$$x^2 + y^2 + z^2 = \ell^2 \; , \; x v_x + y v_y + z v_z = 0$$

are implied. The type of analysis traditionally used in advanced dynamics books [48] would employ Euler angles, thereby assuring exact satisfaction of $|\boldsymbol{r}| = \ell$. The accuracy of the solution method proposed here can be checked by finding a) whether the total energy of the system remains constant and b) whether the component of angular momentum in the z-direction remains constant. However, even when constraint conditions are satisfied exactly, reliability of numerical simulations of nonlinear systems over long time periods becomes questionable due to accumulated inaccuracies caused by arithmetic roundoff and the approximate nature of integration formulas.

The program **toprun** integrates the equations of motion and interprets the results. This program reads data to specify properties of a conical top along with the initial position and the angular velocity. Intrinsic function **ode45** is employed to integrate the motion equation defined in function **topde**. The path followed by the gravity center is plotted and error measures regarding conservation of energy and angular momentum are computed. Figures 8.6 and 8.7 show results for a top having properties given by the test case suggested in the interactive data input. A top which

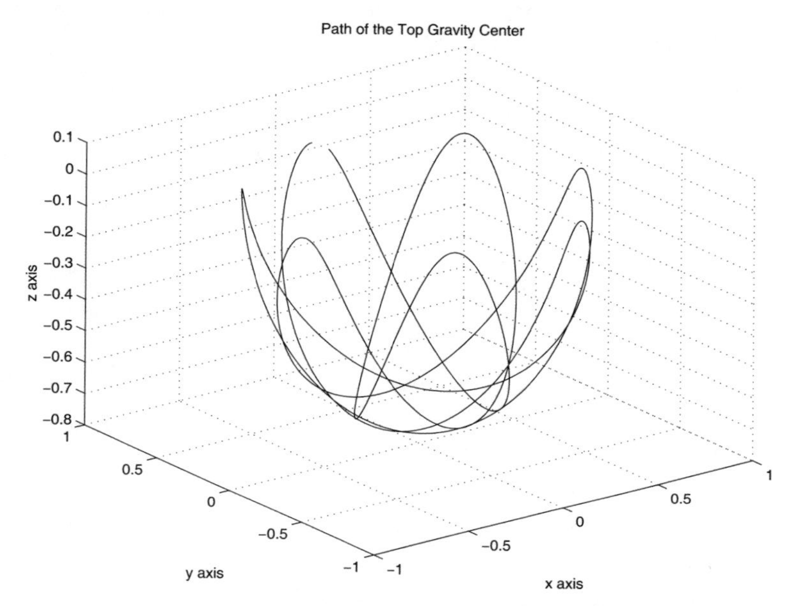

Figure 8.6: Path of the Top Gravity Center

has its symmetry axis initially horizontal along the y-axis is given an angular velocity of $[0, 10, 2]$. Integrating the equation of motion with an error tolerance of 10^{-8} leads to the response shown in the Figure 8.6. Error measures computed regarding the fluctuation in predicted values of total energy and angular momentum about the z-axis (Figure 8.7) fluctuate about one part in 100,000. It appears that the analysis employing Cartesian coordinates does produce good results.

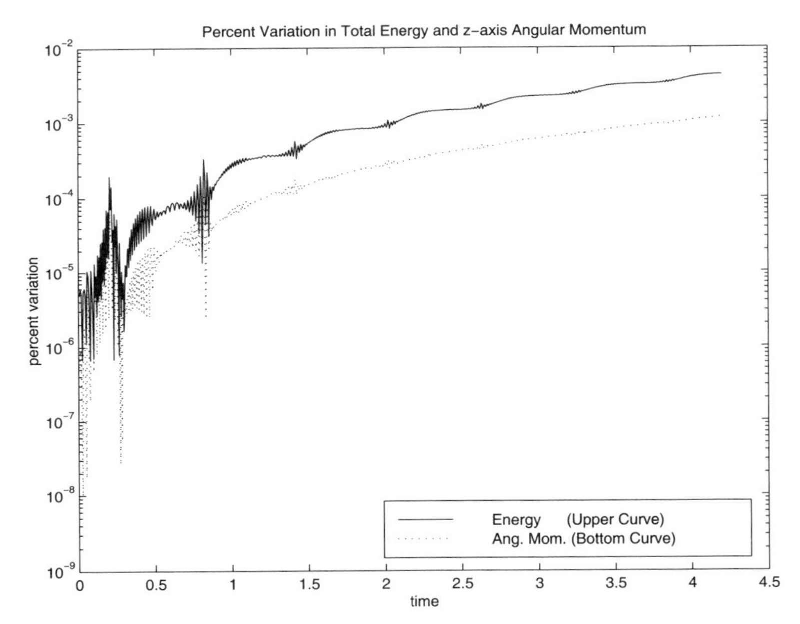

Figure 8.7: Variation in Total Energy and z-axis Angular Momentum

Program Output and Code

Program Toprun

```
1: function toprun
2: % Example: toprun
3: % ~~~~~~~~~~~~~~~~
4: %
5: % Example that analyzes the response of a
6: % spinning conical top.
7: %
8: % User m functions required:
9: %    topde, cubrange, inputv
10:
11: disp(' ');
12: disp(['*** Dynamics of a Homogeneous ', ...
13:       'Conical Top ***']); disp(' ');
14: disp(['Input the gravity constant and the ', ...
15:       'body weight (try 32.2,5)']);
16: [grav,wt]=inputv('? ');
17: mass=wt/grav; tmp=zeros(3,1);
18: disp(' ');
19: disp(['Input the height and base radius ', ...
20:       '(try 1,.5)']);
21: [ht,rb]=inputv('? '); len=.75*ht;
22: jtrans=3*mass/20*(rb*rb+4*ht*ht);
23: jaxial=3*mass*rb*rb/10;
24: disp(' ');
25: disp(['Input a vector along the initial ', ...
26:       'axis direction (try 0,1,0)']);
27: [tmp(1),tmp(2),tmp(3)]=inputv('? ');
28: e3=tmp(:)/norm(tmp); r0=len*e3;
29: disp(' ');
30: disp(['Input the initial angular velocity ', ...
31:       '(try 0,10,2)']);
32: [tmp(1),tmp(2),tmp(3)]=inputv('? '); omega0=tmp;
33: omegax=e3'*omega0(:); rdot0=cross(omega0,r0);
34: z0=[r0(:);rdot0(:)]; uz=[0;0;1];
35: c1=wt*len^2/jtrans; c2=omegax*jaxial/jtrans;
36: disp(' ');
37: disp(['Input tfinal,and the integration ', ...
38:       'tolerance (try 4.2, 1e-8)']);
39: [tfinl,tol]=inputv('? '); disp(' ');
40: fprintf( ...
```

```
41:    'Please wait for solution of equations.\n');
42:
43: % Integrate the equations of motion
44: odeoptn=odeset('RelTol',tol);
45: [tout,zout]=ode45(@topde,[0,tfinl],z0,...
46:                   odeoptn,uz,c1,c2);
47: t=tout; x=zout(:,1); y=zout(:,2); z=zout(:,3);
48: vx=zout(:,4); vy=zout(:,5); vz=zout(:,6);
49:
50: % Compute total energy and angular momentum
51: c3=jtrans/(len*len); taxial=jaxial/2*omegax^2;
52: r=zout(:,1:3)'; v=zout(:,4:6)';
53: etotal=(wt*r(3,:)+taxial+c3/2*sum(v.*v))';
54: h=(jaxial*omegax/len*r+c3*cross(r,v))';
55:
56: % Plot the path of the gravity center
57: clf; axis('equal');
58: axis(cubrange([x(:),y(:),z(:)])); plot3(x,y,z);
59: title('Path of the Top Gravity Center');
60: xlabel('x axis'); ylabel('y axis');
61: zlabel('z axis'); grid on; figure(gcf);
62: disp(' '); disp(...
63: 'Press [Enter] to plot error measures'), pause
64: % print -deps toppath
65: n=2:length(t);
66:
67: % Compute energy and angular momentum error
68: % quantities and plot results
69: et=etotal(1); enrger=abs(100*(etotal(n)-et)/et);
70: hzs=abs(h(1,3));
71: angmzer=abs(100*(h(n,3)-hzs)/hzs);
72: vec=[enrger(:);angmzer(:)];
73: minv=min(vec); maxv=max(vec);
74:
75: clf;
76: semilogy(t(n),enrger,'-r',t(n),angmzer,':m');
77: axis('normal'); xlabel('time');
78: ylabel('percent variation');
79: title(['Percent Variation in Total Energy ', ...
80:        'and z-axis Angular Momentum']);
81: legend(' Energy      (Upper Curve)', ...
82:        ' Ang. Mom. (Bottom Curve)',4);
83: figure(gcf), pause
84: % print -deps topvar
85:
```

```
86: disp(' '), disp('All Done')
87:
88: %==========================================
89:
90: function zdot=topde(t,z,uz,c1,c2)
91: %
92: % zdot=topde(t,z,uz,c1,c2)
93: % ~~~~~~~~~~~~~~~~~~~~~~~
94: %
95: % This function defines the equation of motion
96: % for a symmetrical top. The vector z equals
97: % [r(:);v(:)] which contains the Cartesian
98: % components of the gravity center radius and
99: % its velocity.
100: %
101: % t    - the time variable
102: % z    - the vector [x; y; z; vx; vy; vz]
103: % uz   - the vector [0;0;1]
104: % c1   - wt*len^2/jtrans
105: % c2   - omegax*jaxial/jtrans
106: %
107: % zdot - the time derivative of z
108: %
109: % User m functions called:  none
110: %------------------------------------------
111:
112: z=z(:); r=z(1:3); len=norm(r); ur=r/len;
113:
114: % Make certain the input velocity is
115: % perpendicular to r
116: v=z(4:6); v=v-(ur'*v)*ur;
117: vdot=-c1*(uz-ur*ur(3))+c2*cross(ur,v)- ...
118:      ((v'*v)/len)*ur;
119: zdot=[v;vdot];
120:
121: %==========================================
122:
123: % function varargout=inputv(prompt)
124: % See Appendix B
125:
126: % =========================================
127:
128: % function range=cubrange(xyz,ovrsiz)
129: % See Appendix B
```

8.7 Motion of a Projectile

The problem of aiming a projectile to strike a distant target involves integrating a system of differential equations governing the motion and adjusting the initial inclination angle to achieve the desired hit [101]. A reasonable model for the projectile motion assumes atmospheric drag proportional to the square of the velocity. Consequently, the equations of motion are

$$\dot{v}_x = -cvv_x \ , \ \dot{v}_y = -g - cvv_y \ , \ \dot{x} = v_x \ , \ \dot{y} = v_y$$

where g is the gravity constant and c is a ballistic coefficient depending on such physical properties as the projectile shape and air density.

The natural independent variable in the equations of motion is time. However, horizontal position x is a more desirable independent variable, since the target will be located at some distant point (x_f, y_f) relative to the initial position $(0, 0)$ where the projectile is launched. We can formulate the differential equations in terms of x by using the relationship

$$dx = v_x \, dt \text{ or } \frac{dt}{dx} = \frac{1}{v_x}.$$

Then

$$\frac{dy}{dx} = \frac{v_y}{v_x} \ , \ \frac{dv_y}{dt} = v_x \frac{dv_y}{dx} \ , \ \frac{dv_x}{dt} = v_x \frac{dv_x}{dx} \ ,$$

and the equations of motion become

$$\frac{dy}{dx} = \frac{v_y}{v_x} \ , \ \frac{dt}{dx} = \frac{1}{v_x} \ , \ \frac{dv_x}{dx} = -cv \ , \ \frac{dv_y}{dx} = \frac{-(g + cvv_y)}{v_x}.$$

Taking a vector z defined by

$$z = [v_x; \ v_y; \ y; \ t]$$

leads to a first order matrix differential equation

$$\frac{dz}{dx} = \frac{[-cvv_x; \ -(g + cvv_y); \ v_y; \ 1]}{v_x}$$

where

$$v = \sqrt{v_x^2 + v_y^2}.$$

The reader should note that an ill-posed problem can occur if the initial velocity of the projectile is not large enough so that the maximum desired value of x is reached before v_x is reduced to zero from atmospheric drag. Consequently, error checking is needed to handle such a circumstance. The functions **traject** and **projcteq** employ intrinsic function **ode45** to compute the projectile trajectory. Graphical results

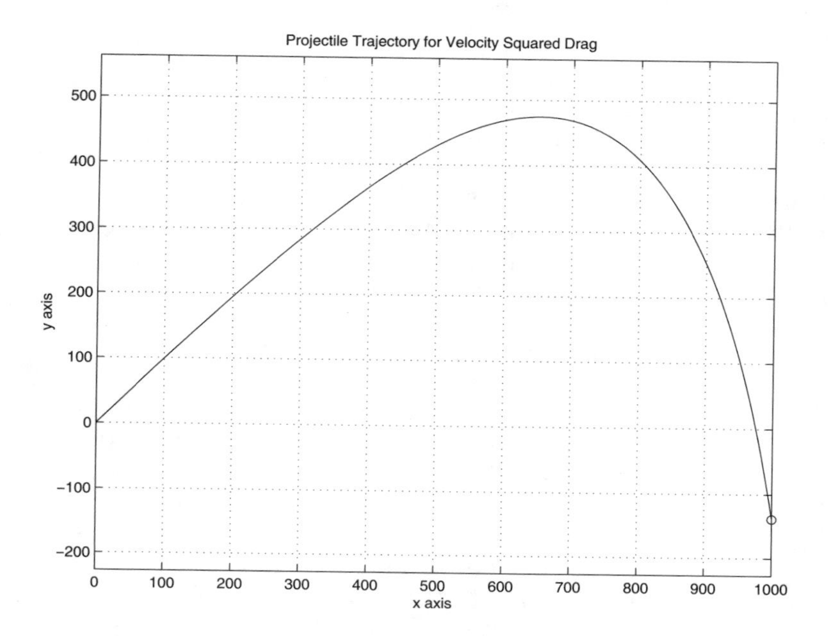

Figure 8.8: Projectile Trajectory for v^2 Drag Condition

produced by the default data case appear in Figure 8.8. The function **traject** will be employed again in Chapter 12 for an optimization problem where a search procedure is used to compute the initial inclination angle needed to hit a target at some specified distant position. In this section we simply provide the functions to integrate the equations of motion.

Program Output and Code

Function traject

```
1: function [y,x,t]=traject ...
2:          (angle,vinit,gravty,cdrag,xfinl,noplot)
3: % [y,x,t]=traject ...
4: %        (angle,vinit,gravty,cdrag,xfinl,noplot)
5: % ~~~~~~~~~~~~~~~~~~~~~~~~~~~~~~~~~~~~~~~~~~~~~~
6: %
7: % This function integrates the dynamical
8: % equations for a projectile subjected to
9: % gravity loading and atmospheric drag
10: % proportional to the square of the velocity.
```

```
11: %
12: % angle  - initial inclination of the
13: %          projectile in degrees
14: % vinit  - initial velocity of the projectile
15: %          (muzzle velocity)
16: % gravty - the gravitational constant
17: % cdrag  - drag coefficient specifying the drag
18: %          force per unit mass which equals
19: %          cdrag*velocity^2.
20: % xfinl  - the projectile is fired toward the
21: %          right from x=0.  xfinl is the
22: %          largest x value for which the
23: %          solution is computed. The initial
24: %          velocity must be large enough that
25: %          atmospheric damping does not reduce
26: %          the horizontal velocity to zero
27: %          before xfinl is reached.  Otherwise
28: %          an error termination will occur.
29: % noplot - plotting of the trajectory is
30: %          omitted when this parameter is
31: %          given an input value
32: %
33: % y,x,t  - the y, x and time vectors produced
34: %          by integrating the equations of
35: %          motion
36: %
37: % Global variables:
38: %
39: % grav,  - two constants replicating gravty and
40: % dragc    cdrag, for use in function projcteq
41: % vtol   - equal to vinit/1e6, used in projcteq
42: %          to check whether the horizontal
43: %          velocity has been reduced to zero
44: %
45: % User m functions called: projcteq
46:
47: global grav dragc vtol
48:
49: % Default data case generated when input is null
50: if nargin ==0
51:   angle=45; vinit=600; gravty=32.2;
52:   cdrag=0.002; xfinl=1000;
53: end;
54:
55: % Assign global variables and evaluate
```

```
56: % initial velocity
57: grav=gravty; dragc=cdrag; ang=pi/180*angle;
58: vtol=vinit/1e6;
59: z0=[vinit*cos(ang); vinit*sin(ang); 0; 0];
60:
61: % Integrate the equations of motion defined
62: % in function projcteq
63: deoptn=odeset('RelTol',1e-6);
64: [x,z]=ode45(@projcteq,[0,xfinl],z0,deoptn);
65:
66: y=z(:,3); t=z(:,4); n=length(x);
67: xf=x(n); yf=y(n);
68:
69: % Plot the trajectory curve
70: if nargin < 6
71:   plot(x,y,'-',xf,yf,'o');
72:   xlabel('x axis'); ylabel('y axis');
73:   title(['Projectile Trajectory for ', ...
74:          'Velocity Squared Drag']);
75:   axis('equal'); grid on; figure(gcf);
76:   % print -deps trajplot
77: end
78:
79: %==========================================
80:
81: function zp=projcteq(x,z)
82: %
83: % zp=projcteq(x,z)
84: % ~~~~~~~~~~~~~~~~
85: %
86: % This function defines the equation of motion
87: % for a projectile loaded by gravity and
88: % atmospheric drag proportional to the square
89: % of the velocity.
90: %
91: % x     -  the horizontal spatial variable
92: % z     -  a vector containing [vx; vy; y; t];
93: %
94: % zp    -  the derivative dz/dx which equals
95: %          [vx'(x); vy'(x); y'(x); t'(x)];
96: %
97: % Global variables:
98: %
99: % grav  -  the gravity constant
100: % dragc -  the drag coefficient divided by
```

```
101: %           gravity
102: % vtol    -  a global variable used to check
103: %            whether vx is zero
104: %
105: % User m functions called:  none
106: %----------------------------------------------
107:
108: global grav dragc vtol
109: vx=z(1); vy=z(2); v=sqrt(vx^2+vy^2);
110:
111: % Check to see whether drag reduced the
112: % horizontal velocity to zero before the
113: % xfinl was reached.
114: if abs(vx) < vtol
115:   disp(' ');
116:   disp('*************************************');
117:   disp('ERROR in function projcteq. The ');
118:   disp('  initial velocity of the projectile');
119:   disp('  was not large enough for xfinal to');
120:   disp('  be reached.');
121:   disp('EXECUTION IS TERMINATED.');
122:   disp('*************************************');
123:   disp(' '),error(' ');
124: end
125: zp=[-dragc*v; -(grav+dragc*v*vy)/vx; ...
126:      vy/vx; 1/vx];
```

8.8 Example on Dynamics of a Chain with Specified End Motion

The dynamics of flexible cables is often modeled using a chain of rigid links connected by frictionless joints. A chain having specified end motions illustrates the behavior of a system governed by nonlinear equations of motion and auxiliary algebraic constraints. In particular, we will study a gravity loaded cable fixed at both ends. The total cable length exceeds the distance between supports, so that the static deflection configuration resembles a catenary.

A simple derivation of the equations of motion employing principles of rigid body dynamics is given next. Readers not versed in principles of rigid body dynamics [48] may nevertheless understand the subsequent programs by analyzing the equations of motion which have a concise mathematical form. The numerical solutions vividly illustrate some numerical difficulties typically encountered in multibody dynamical studies. Such problems are both computationally intensive, as well as highly sensitive to accumulated effects of numerical error.

The mathematical model of interest is the two-dimensional motion of a cable (or chain) having n rigid links connected by frictionless joints. A typical link $\imath$ has its

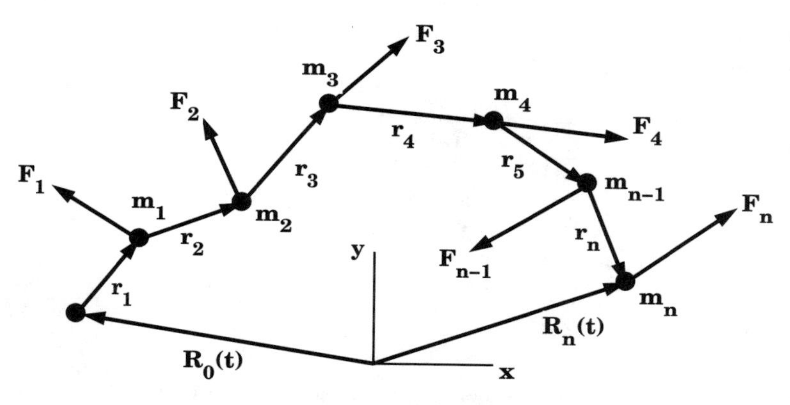

Figure 8.9: Chain with Specified End Motion

mass $m_\imath$ concentrated at one end. The geometry is depicted in Figure 8.9. The chain ends undergo specified motions $\boldsymbol{R}_0(t) = [X_0(t) \; ; \; Y_0(t)]$ for the first link and $\boldsymbol{R}_n(t) = [X_n(t) \; ; \; Y_n(t)]$ for the last link. The direction vector along link $\imath$ is described by $\boldsymbol{r}_\imath = [x_\imath \; ; \; y_\imath] = \ell_\imath[\cos(\theta_\imath) \; ; \; \sin(\theta_\imath)]$. We assume that each joint $\imath$ is subjected to a force $\boldsymbol{F}_\imath = [f_{x\imath} \; ; \; f_{y\imath}]$ where $0 \leq \imath \leq n$. Index values $\imath = 0$ and $\imath = n$ denote unknown constraint forces which must act at the outer ends of the first and last links to achieve the required end displacements. The forces applied at the interior joints are arbitrary. It is convenient to characterize the dynamics of each link in terms of its direction angle. Thus

$$\dot{\boldsymbol{r}}_\imath = \boldsymbol{r}_\imath'\dot{\theta}_\imath \;, \; \ddot{\boldsymbol{r}}_\imath = \boldsymbol{r}_\imath'\ddot{\theta}_\imath + \boldsymbol{r}_\imath''\dot{\theta}_\imath^2 = \boldsymbol{r}_\imath'\ddot{\theta}_\imath - \boldsymbol{r}_\imath\dot{\theta}_\imath^2$$

where primes and dots denote differentiation with respect to $\theta_\imath$ and t, respectively. Therefore

$$\dot{\boldsymbol{r}}_\imath = [-y_\imath \; ; \; x_\imath]\dot{\theta}_\imath \;, \; \ddot{\boldsymbol{r}}_\imath = [-\ddot{y}_\imath \; ; \; x_\imath]\ddot{\theta}_\imath - [x_\imath \; ; \; y_\imath]\dot{\theta}_\imath^2.$$

The global position vector of joint $\imath$ is

$$\boldsymbol{R}_\imath = \boldsymbol{R}_0 + \sum_{\jmath=1}^{\imath} \boldsymbol{r}_\jmath = \boldsymbol{R}_0 + \sum_{\jmath=1}^{n} < \imath - \jmath > \boldsymbol{r}_\jmath$$

where the symbol $< k >= 1$ for $k \geq 0$, and 0 for $k < 0$. Consequently, the velocity and acceleration of joint $\imath$ are

$$\dot{\boldsymbol{R}}_\imath = \dot{\boldsymbol{R}}_0 + \sum_{\jmath=1}^{n} < \imath - \jmath > \boldsymbol{r}_\jmath'\dot{\theta}_\jmath,$$

$$\ddot{\boldsymbol{R}}_\imath = \ddot{\boldsymbol{R}}_0 + \sum_{\jmath=1}^{n} < \imath - \jmath > \boldsymbol{r}_\jmath'\ddot{\theta}_\jmath - \sum_{\jmath=1}^{n} < \imath - \jmath > \boldsymbol{r}_\jmath\dot{\theta}_\jmath^2.$$

The ends of the chain each have specified motions; so not all of the inclination angles are independent. Consequently,

$$\sum_{\jmath=1}^{n} \boldsymbol{r}_\jmath = \boldsymbol{R}_n - \boldsymbol{R}_0,$$

$$\sum_{\jmath=1}^{n} \boldsymbol{r}'_\jmath \dot{\theta}_\jmath = \dot{\boldsymbol{R}}_n - \dot{\boldsymbol{R}}_0,$$

$$\sum_{\jmath=1}^{n} \boldsymbol{r}'_\jmath \ddot{\theta}_\jmath - \sum_{\jmath=1}^{n} \boldsymbol{r}_\jmath \dot{\theta}_\jmath^2 = \ddot{\boldsymbol{R}}_n - \ddot{\boldsymbol{R}}_0.$$

Combining the last constraint equation with equations of motion written for masses $m_1, \cdots, m_n$ yields a complete system of $(n+2)$ equations determining $\ddot{\theta}_1, \cdots, \ddot{\theta}_n$ and the components of $\boldsymbol{F}_n$. The fact that all masses are concentrated at frictionless joints shows that link $\imath$ is a two-force member carrying an internal load directed along $r_\imath$. Consequently, the D'Alembert principle [48] implies that the sum of all external and inertial loads from joints $\imath, \imath+1, \cdots, n$ must give a resultant passing through joint $\imath$ in the direction of $\boldsymbol{r}_\imath$. Since $\boldsymbol{r}'_\imath$ and $\boldsymbol{r}_\imath$ are perpendicular, requiring a vector to be in the direction of $r_\imath$ is equivalent to making it normal to $\boldsymbol{r}'_\imath$. Therefore

$$\boldsymbol{r}'_\imath \cdot \left[\sum_{\jmath=1}^{n} < \jmath - \imath > \left\{ \boldsymbol{F}_\jmath - m_\jmath \ddot{\boldsymbol{R}}_\jmath \right\} \right] = 0 \ , \ 1 \le \imath \le n.$$

The last n equations involve $\ddot{\theta}_\imath$ and two end force components f_{xn} and f_{yn}. Some algebraic rearrangement results in a matrix differential equation of concise form containing several auxiliary coefficients defined as follows:

$$b_\imath = \sum_{k=\imath}^{n} m_k \ , \ m_{\imath\jmath} = m_{\jmath\imath} = b_\imath \ , \ 1 \le \imath \le n \ , \ 1 \le \jmath \le \imath,$$

$$a_{\imath\jmath} = m_{\imath\jmath}(x_\imath x_\jmath + y_\imath y_\jmath) \ , \ 1 \le \imath \le n \ , \ 1 \le \jmath \le n$$
$$b_{\imath\jmath} = m_{\imath\jmath}(x_\imath y_\jmath - x_\jmath y_\imath) \ , \ 1 \le \imath \le n \ , \ 1 \le \jmath \le n$$

$$p_{x\imath} = \sum_{\jmath=\imath}^{n-1} f_{x\imath} \ , \ p_{y\imath} = \sum_{\jmath=\imath}^{n-1} f_{y\imath} \ , \ 1 \le \imath \le n.$$

For $\imath = n$, the last two sums mean $p_{xn} = p_{yn} = 0$. Furthermore, we denote the acceleration components of the chain ends as $\ddot{\boldsymbol{R}}_0 = [a_{xo} \ ; \ a_{y0}]$ and $\ddot{\boldsymbol{R}}_n = [a_{xn} \ ; \ a_{yn}]$. Using the various quantities just defined, the equations of motion become

$$\sum_{\jmath=1}^{n} a_{\imath\jmath} \ddot{\theta}_\jmath + y_\imath f_{xn} - x_\imath f_{yn} = \sum_{\jmath=1}^{n} b_{\imath\jmath} \dot{\theta}_\jmath^2 + x_\imath (p_{y\imath} - b_\imath a_{y0}) - y_\imath (p_{x\imath} - b_\imath a_{x0})$$
$$= e_\imath \ , \ 1 \le \imath \le n.$$

The remaining two components of the constraint equations completing the system are

$$\sum_{j=1}^{n} y_j \ddot{\theta}_j = -\sum_{j=1}^{n} x_j \dot{\theta}_j^2 - a_{xn} + a_{x0} = e_{n+1},$$
$$\sum_{j=1}^{n} x_j \ddot{\theta}_j = \sum_{j=1}^{n} y_j \dot{\theta}_j^2 + a_{yn} - a_{y0} \quad = e_{n+2}.$$

Consequently, we get the following symmetric matrix equation to solve for $\ddot{\theta}_1, \cdots, \ddot{\theta}_n$, f_{xn} and f_{yn}

$$\begin{bmatrix} A & X\,Y \\ X^T & 0\;\;0 \\ Y^T & 0\;\;0 \end{bmatrix} \begin{bmatrix} \ddot{\theta} \\ f_{xn} \\ -f_{yn} \end{bmatrix} = [\,E\,]$$

where X, Y, E and θ are column matrices, and the matrix $A = [a_{ij}]$ is symmetric. Because most numerical integrators for differential equations solve first order systems, it is convenient to employ the vector $\boldsymbol{Z} = [\theta\ ;\ \dot{\theta}]$ having $2n$ components. Then the differential equation $\dot{\boldsymbol{Z}} = \boldsymbol{H}(t, \boldsymbol{Z})$ is completely defined when $\ddot{\theta}$ has been computed for known $\boldsymbol{Z}$. The system is integrated numerically to give θ and $\dot{\theta}$ as functions of time. These quantities can then be used to compute the global Cartesian coordinates of the link configurations, thereby completely describing the time history of the chain.

The general equations of motion simplify somewhat when the chain ends are fixed and the external forces only involve gravity loads. Then $p_{xi} = 0$ and $p_{yi} = -g(b_i - b_n)$ which gives

$$\sum_{j=1}^{n} m_{ij}(x_i x_j + y_i y_j)\ddot{\theta}_j - x_i f_{yn} + y_i f_{xn} =$$

$$g(b_i - b_n) + \sum_{j=1}^{n} m_{ij}(x_i y_j - x_j y_i)\dot{\theta}_j^2 \;,\; 1 \le i \le n.$$

The last two equations to complete the set are:

$$\sum_{j=1}^{n} x_j \ddot{\theta}_j = \sum_{j=1}^{n} y_j \dot{\theta}_j^2 \;,\; \sum_{j=1}^{n} y_j \ddot{\theta}_j = -\sum_{j=1}^{n} x_j \dot{\theta}_j^2.$$

A program was written to simulate motion of a cable fixed at both ends and released from rest. The cable falls under the influence of gravity from an initially elevated position. Function **ode45** is used to perform the numerical integration. The program consists of three functions **cablenl**, **plotmotn**, and **equamo**. Function **cablenl** creates the data, calls **ode45** to perform the integration, and displays the output from the simulation. Function **plotmotn** plots the motion for specified time limits. Results can be shown using animation or plots superimposing successive positions of the cable. Most of the analysis in the program is performed in function **equamo** which forms the equations of motion which are passed to **ode45** for integration.

A configuration with eight identical links was specified. For simplicity, the total mass, total cable length, and gravity constant were all normalized to equal unity. The numerical integration error was controlled using a relative tolerance of 1e-6 and an absolute error tolerance of 1e-8. Results of the simulation appear below. Figure 8.10 shows cable positions during the early stages of motion when results of the numerical integration are reliable. However, the numerical solution eventually becomes worthless due to accumulated numerical inaccuracies yielding the motion predictions indicated in Figure 8.11. The nature of the error growth can be seen clearly in Figure 8.12 which plots the x-coordinate of the chain midpoint as a function of time. Since the chosen mass distribution and initial deflection is symmetrical about the middle, the subsequent motion will remain symmetrical unless the numerical solution becomes invalid. The x coordinate of the midpoint should remain at a constant value of $\sqrt{2}/4$, but it appears to abruptly go unstable near $t = 18$. More careful examination indicates that this numerical instability does not actually occur suddenly. Instead, it grows exponentially from the outset of the simulation. The error is caused by the accumulation of truncation errors intrinsic to the numerical integration process allowing errors at each step which are regulated within a small but finite tolerance. A global measure of symmetry loss of the y deflection pattern is plotted on a semilog scale in Figure 8.13. Note that the error curve has a nearly linear slope until the solution degenerates completely near $t = 18$. The reader can verify that choosing less stringent error tolerances produces solutions which become inaccurate sooner than $t = 18$. It should also be observed that this dynamical model exhibits another important characteristic of highly nonlinear systems, namely, extreme sensitivity to physical properties. Note that shortening the last link by only one part in ten thousand makes the system deflection quickly lose all appearance of symmetry by $t = 6$. Hence, two systems having nearly identical physical parameters and initial conditions may behave very differently a short time after motion is initiated. The conclusion implied is that analysts should thoroughly explore how parameter variations affect response predictions in nonlinear models.

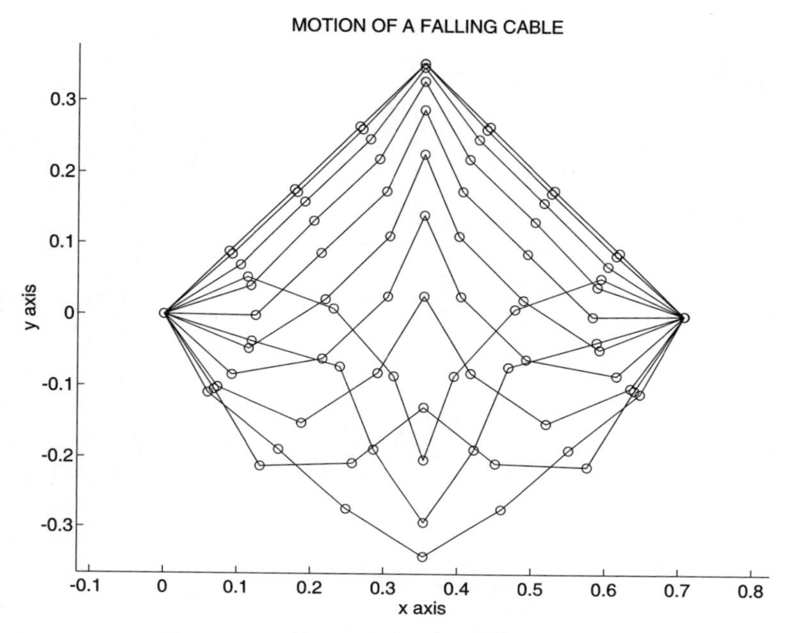

Figure 8.10: **Motion During Initial Phase**

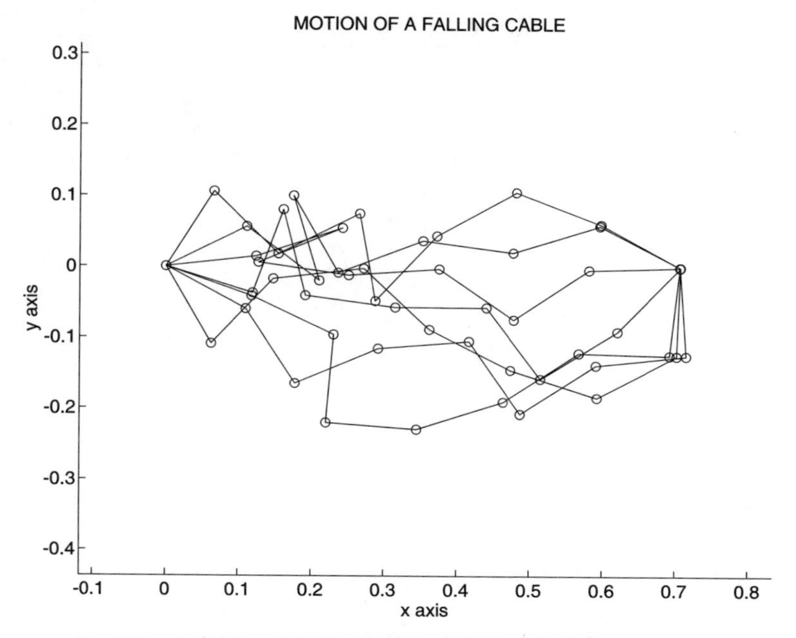

Figure 8.11: **Motion After Solution Degenerates**

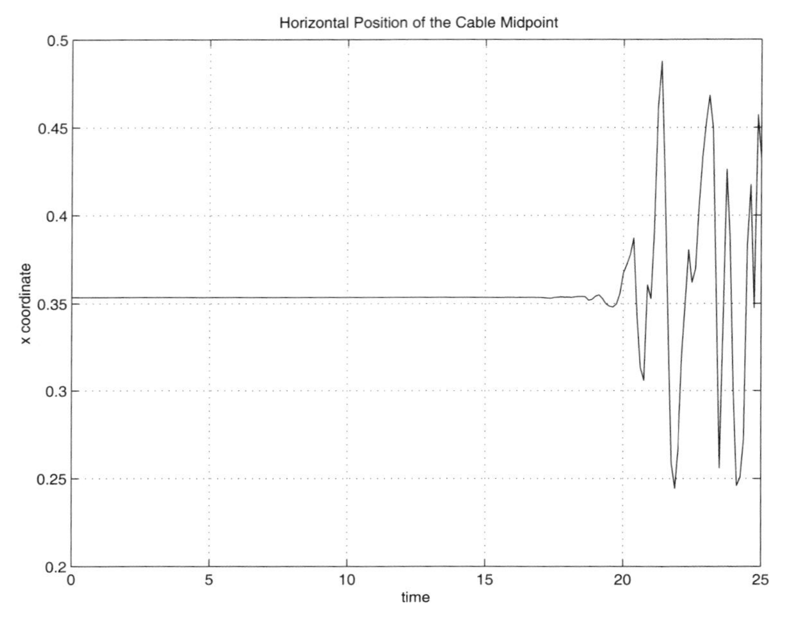

Figure 8.12: Horizontal Position of the Cable Midpoint

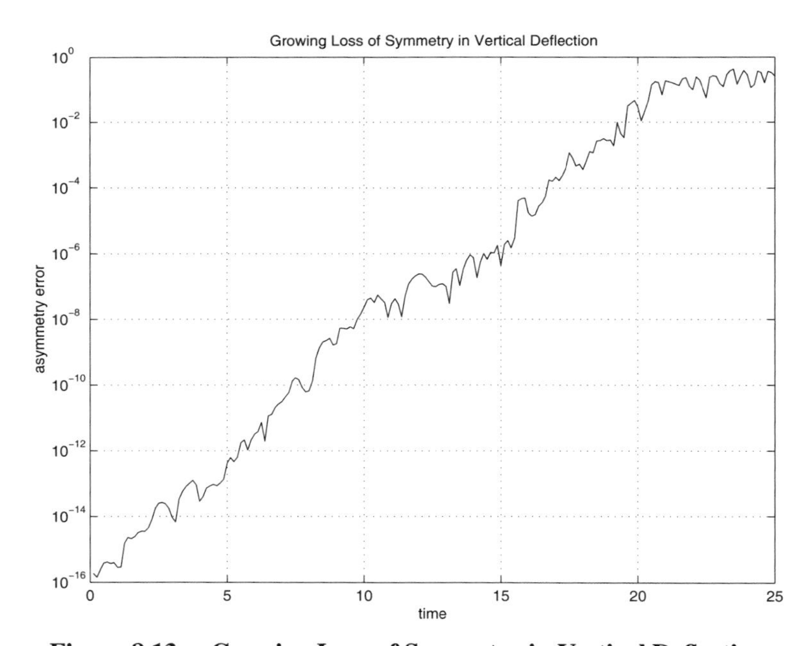

Figure 8.13: Growing Loss of Symmetry in Vertical Deflection

Example on Nonlinear Cable Motion

Program cablenl

```
1: function [t,x,y,theta,cptim]=cablenl
2: % [t,x,y,theta,cptim]=cablenl
3: % Example: cablenl
4: % ~~~~~~~~~~~~~~~~
5: % Numerical integration of the matrix
6: % differential equations for the nonlinear
7: % dynamics of a cable of rigid links with
8: % the outer ends of the cable fixed.
9: %
10: % t     - time vector for the solution
11: % x,y   - matrices with nodal coordinates
12: %         stored in the columns. The time
13: %         history of point j is in x(:,j)
14: %         and y(:,j)
15: % theta - matrix with inclination angles
16: %         stored in the columns
17: % cptim - number of seconds to integrate
18: %         the equations of motion
19: %
20: % User m functions required:
21: %         plotmotn, equamo
22:
23: clear all; close;
24:
25: % Make variables global for use by
26: % function equamo
27: global first_ n_ m_ len_ grav_ b_ mas_ py_
28:
29: fprintf('\nNONLINEAR DYNAMICS OF A ')
30: fprintf('FALLING CABLE\n')
31: fprintf(...
32: '\nNote: The calculations take awhile\n')
33:
34: % Set up data for a cable of n_ links,
35: % initially arranged in a triangular
36: % deflection configuration.
37:
38: % parameter controlling initialization of
39: % auxiliary parameters used in function
40: % equamo
```

```
41: first_=1;
42: % number of links in the cable
43: n_=8; n=n_; nh=n_/2;
44: % vector of lengths and gravity constant
45: len_=1/n*ones(n,1); grav_=1;
46: % vector of mass constants
47: m_=ones(1,n_)/n_;
48:
49: % initial position angles
50: th0=pi/4*[ones(nh,1);-ones(nh,1)];
51: td0=zeros(size(th0)); z0=[th0;td0];
52:
53: % time limits, integration tolerances,
54: % and the number of solution points
55: tmin=0; tmax=25; nt=201;
56: t=linspace(0,tmax,nt)';
57: tolrel=1e-6; tolabs=1e-8; len=len_;
58:
59: % Perform the numerical integration using a
60: % variable stepsize Runge-Kutta integrator
61: tic;
62: odetol=odeset('RelTol',tolrel,'AbsTol',tolabs);
63: [t,w]=ode45(@equamo,t,z0,odetol);
64: theta=w(:,1:n); cptim=toc;
65:
66: % Compute node point coordinates
67: Z=[zeros(nt,1),repmat(len',nt,1).*exp(i*theta)];
68: Z=cumsum(Z.').'; x=real(Z); y=imag(Z);
69:
70: % Plot the horizontal position of the midpoint
71: clf; plot(t,x(:,1+n_ /2));
72: ylabel('x coordinate'); xlabel('time')
73: title(['Horizontal Position of the ' ...
74:        'Cable Midpoint'])
75: grid on; figure(gcf);
76: % print -deps xmidl
77:
78: disp(' '), disp(...
79: 'Press [Enter] to see the error growth curve');
80: pause, close
81:
82: % Show error growth indicated by symmetry
83: % loss of the vertical deflection symmetry.
84: % An approximately linear trend on the semilog
85: % plot indicates exponential growth of the error.
```

```
86: unsymer=sqrt(sum((y-y(:,end:-1:1)).^2,2));
87: hold off; axis('normal'); clf;
88: semilogy(t,unsymer);
89: xlabel('time'); ylabel('asymmetry error');
90: title(['Growing Loss of Symmetry in ' ...
91:        'Vertical Deflection']);
92: grid on; figure(gcf);
93: % print -deps unsymerr
94:
95: disp(' '), disp(...
96: 'Press [Enter] to see the response animation');
97:
98: % Show animation of the cable response
99: disp(' ')
100: disp('The  motion can be animated or a trace')
101: disp('can be shown for successive positions')
102: disp(['between t = ',num2str(tmin),...
103:       ' and t = ',num2str(tmax)])
104:
105: % Plot the position for different times limits
106: titl='CABLE MOTION FOR T = ';
107: while 1
108:   disp(' '), disp(...
109:   ['Choose a plot option (1 <=> animate, ',...
110:    ' 2 <=> trace,'])
111:   opt=input('3 <=> stop)  > ? ');
112:   if opt==3, break, end
113:   disp(' '), disp(...
114:   'Give a time vector such as 0:.1:15')
115:   Tp=input('Time vector > ? ','s');
116:   if isempty(Tp), break, end
117:   tp=eval(Tp); tp=tp(:); T=[titl,Tp];
118:   xp=interp1q(t,x,tp); yp=interp1q(t,y,tp);
119:   if opt ==1, plotmotn(xp,yp,T)
120:   else, plotmotn(xp,yp,T,1), end
121: end
122: fprintf('\nAll Done\n')
123:
124: %===============================================
125:
126: function plotmotn(x,y,titl,isave)
127: %
128: % plotmotn(x,y,titl,isave)
129: % ~~~~~~~~~~~~~~~~~~~~~~~
130: % This function plots the cable time
```

```
131: % history described by coordinate values
132: % stored in the rows of matrices x and y.
133: %
134: % x,y   - matrices having successive rows
135: %         which describe position
136: %         configurations for the cable
137: % titl  - a title shown on the plots
138: % isave - parameter controlling the form
139: %         of output. When isave is not input,
140: %         only one position at a time is shown
141: %         in rapid succession to animate the
142: %         motion. If isave is given a value,
143: %         then successive are all shown at
144: %         once to illustrate a kinematic
145: %         trace of the motion history.
146: %
147: % User m functions called:  none
148: %----------------------------------------------
149: 
150: % Set a square window to contain all
151: % possible positions
152: [nt,n]=size(x);
153: if nargin==4, save =1; else, save=0; end
154: xmin=min(x(:)); xmax=max(x(:));
155: ymin=min(y(:)); ymax=max(y(:));
156: w=max(xmax-xmin,ymax-ymin)/2;
157: xmd=(xmin+xmax)/2; ymd=(ymin+ymax)/2;
158: hold off; clf; axis('normal'); axis('equal');
159: range=[xmd-w,xmd+w,ymd-w,ymd+w];
160: title(titl)
161: xlabel('x axis'); ylabel('y axis')
162: if save==0
163:   for j=1:nt
164:     xj=x(j,:); yj=y(j,:);
165:     plot(xj,yj,'-k',xj,yj,'ok');
166:     axis(range), axis off
167:     title(titl)
168:     figure(gcf), drawnow, pause(.1)
169:   end
170:   pause(2)
171: else
172:   hold off; close
173:   for j=1:nt
174:     xj=x(j,:); yj=y(j,:);
175:     plot(xj,yj,'-k',xj,yj,'ok');
```

```
176:     axis(range), axis off, hold on
177:   end
178:   title(titl)
179:   figure(gcf), drawnow, hold off, pause(2)
180: end
181:
182: % Save plot history for subsequent printing
183: % print -deps plotmotn
184:
185: %==========================================
186:
187: function zdot=equamo(t,z)
188: %
189: % zdot=equamo(t,z)
190: % ~~~~~~~~~~~~~~~~
191: % Equation of motion for a cable fixed at
192: % both ends and loaded by gravity forces only
193: %
194: % t          current time value
195: % z          column vector defined by
196: %            [thet(t);theta'(t)]
197: % zdot       column vector defined by
198: %            the concatenation
199: %            z'(t) = [theta'(t);theta''(t)]
200: %
201: % User m functions called:  none.
202: %------------------------------------------
203:
204: % Values accessed as global variables
205: global first_ n_ m_ len_ grav_ b_ mas_ py_
206:
207: % Initialize parameters first time
208: % function is called
209: if first_==1, first_=0;
210: % mass parameters
211:   m_=m_(:); b_=flipud(cumsum(flipud(m_)));
212:   mas_=b_(:,ones(n_,1));
213:   mas_=tril(mas_)+tril(mas_,-1)';
214: % load effects from gravity forces
215:   py_=-grav_*(b_-b_(n_));
216: end
217:
218: % Solve for zdot = [theta'(t); theta''(t)];
219: n=n_; len=len_;
220: th=z(1:n); td=z(n+1:2*n); td2=td.*td;
```

```
221: x=len.*cos(th); y=len.*sin(th);
222:
223: % Matrix of mass coefficients and
224: % constraint conditions
225: amat=[[mas_.*(x*x'+y*y'),x,y];
226:        [x,y;zeros(2,2)]'];
227:
228: % Right side vector involves applied forces
229: % and inertial terms
230: bmat=x*y'; bmat=mas_.*(bmat-bmat');
231:
232: % Solve for angular acceleration.
233: % Most computation occurs here.
234: soln=amat\[x.*py_+bmat*td2; y'*td2; -x'*td2];
235:
236: % Final result needed for use by the
237: % numerical integrator
238: zdot=[td; soln(1:n)];
```

8.9 Dynamics of an Elastic Chain

The preceding article analyzed a chain of rigid links requiring only one rotation angle per link. Next we study a similar model of an elastic chain involving several point masses connected by elastic springs which can only support tension. The equations of motion are easy to formulate in terms of the horizontal and vertical coordinates of each mass. The dimensionality needed to handle the elastic chain is twice that needed for a similar rigid link model. It is natural to utilize a three-dimensional model that easily simplifies for two dimensional motion.

Consider a chain having n mass particles

$$m_j, \ 1 \leq j \leq n$$

connected by $n+1$ springs having unstretched lengths

$$l_j, \ 1 \leq j \leq n+1.$$

The position of particle m_j is denoted by vector $\boldsymbol{r}_j$ with $\boldsymbol{r}_0(t)$ and $\boldsymbol{r}_{n+1}(t)$ signifying the outer end positions of the first and last springs, which are assumed to be known functions of time. Furthermore, concentrated forces $\boldsymbol{P}_j(t)$ are applied to the particles. The tensile force in spring number j is

$$\boldsymbol{T}_j = k_j\,(1 - l_j/L_j)\,(L_j > l_j)\,\boldsymbol{R}_j$$

where

$$\boldsymbol{R}_j = \boldsymbol{r}_{j+1} - \boldsymbol{r}_j, \ L_j = |\,\boldsymbol{R}_j|,$$

and k_j denotes a spring constant. Then the equations of motion are given by

$$\dot{\boldsymbol{r}}_j(t) = \boldsymbol{v}_j \ , \ \dot{\boldsymbol{v}}_j(t) = (\boldsymbol{P}_j + \boldsymbol{T}_j - \boldsymbol{T}_{j-1} - c_j \boldsymbol{v}_j) \, / \, m_j \ , \ 1 \leq j \leq n$$

where viscous drag forces defined by the particle velocities times damping coefficients c_j are included. These equations are easy to form using array operations. Furthermore, the two-dimensional case can be simplified further by using complex numbers to represent the particle positions.

A program was written to compute the response of a chain released from at rest in a horizontal position with the springs unstretched. The chain is subjected to gravity loading and the ends of the chain are rotated at constant speed around circular paths. The left and right ends rotate counterclockwise and clockwise respectively. A special case where the right end of the chain is free is provided by setting the last spring constant to zero. Another case where the chain ends do not move occurs when the radii of the end path motions are set to zero.

The following program called **sprnchan** computes the response of a chain with an arbitrary number of identical masses connected by identical springs. The radii and the rotation rate of the end motions, as well as the amount of viscous damping can be changed easily. Function **sprnchan** reads data from function **chaindata** and calls **ode45** to integrate the equations of motion which are formed with functions **spreqmof** and **endmo**. Using the output from **ode45**, function **plotmotn** provides visual descriptions of the response. The motion can be presented using either animation or by superimposing plots of successive positions of the chain in chosen time intervals. To run a different problem, the sample data function **chaindata** can be saved using a different name; and the variables n, tmax, nt, fixorfree, rend, omega, and cdamp can be changed appropriately. Furthermore, modifying the program to handle different variations of stiffness and mass, as well as different end conditions would be straightforward. Figures 8.14 and 8.15 show program results where 1) the left end of the chain was rotated and the right end was detached and 2) both ends of the chain were rotated simultaneously in opposite directions. The time response was computed for a maximum time value of 20, but the chosen time traces only show small subintervals chosen so that successive positions do not overlap excessively. Readers may find it interesting to observe the animation responses resulting from different data choices.

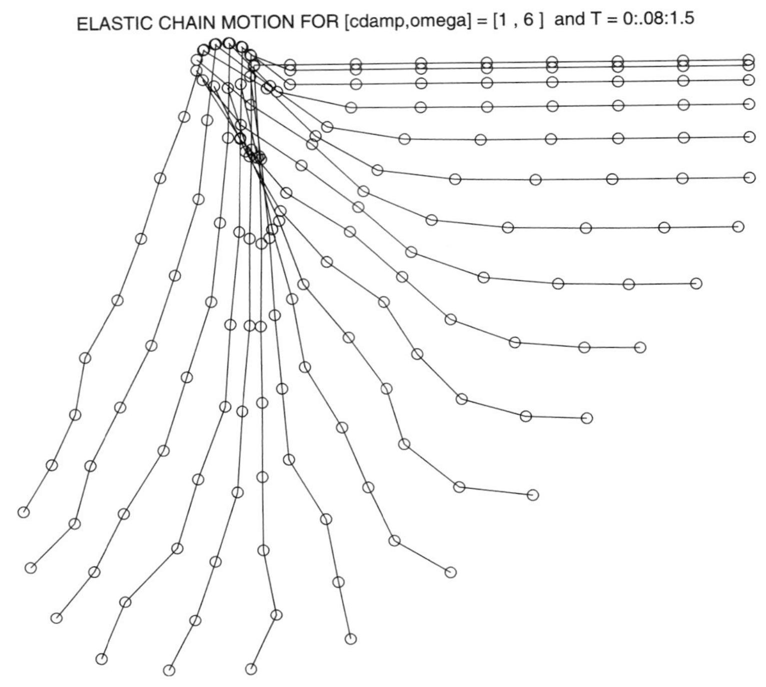

Figure 8.14: Chain with Left End Rotating and Right End Free

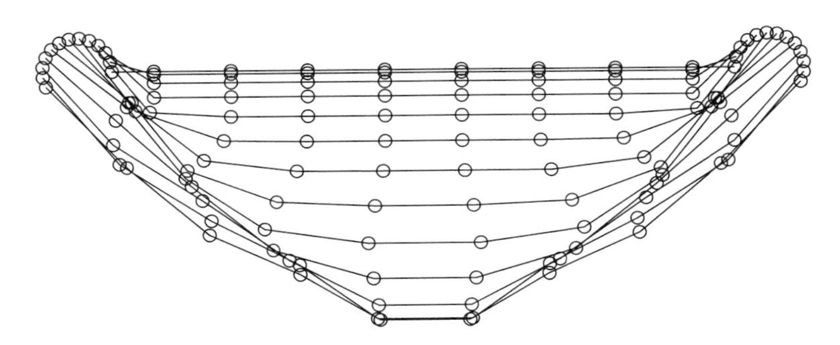

Figure 8.15: Chain with Both Ends Rotating

Program for Elastic Chain Dynamics

Program sprnchan

```
1: function [t,z,cptim]=sprnchan
2: %
3: % [t,z,cptim]=sprnchan
4: % ~~~~~~~~~~~~~~~~~~~~~
5: %       DYNAMIC SIMULATION OF AN ELASTIC CHAIN
6: % This program simulates the dynamics of an elastic
7: % chain modeled by a series of mass particles joined
8: % by elastic springs. The outer springs at each end
9: % are connected to foundations moving on circular
10: % paths at constant speed. The system is released from
11: % rest in a horizontal position. Forces on the system
12: % include gravity, linear viscous drag, and foundation
13: % motion. If the last spring in the chain is assigned
14: % zero stiffness, then the last particle is freed from
15: % the foundation and a swinging chain with the upper
16: % end  shaken is analyzed. The principal variables for
17: % the problem are listed below. Different data choices
18: % can be made by changing function chaindata.
19: %
20: % tlim    - vector of time values at which the
21: %           solution is computed
22: % m       - vector of mass values for the particles
23: % k       - vector of stiffness values for springs
24: %           connecting the particles. If the last
25: %           spring constant is set to zero, then the
26: %           right end constraint is removed
27: % L       - vector of unstretched spring lengths
28: % zend    - complex position coordinate of the outer
29: %           end of the last spring (assuming the outer
30: %           end of the first spring is held at z=0)
31: % zinit   - vector of complex initial displacement
32: %           values for each mass particle. Initial
33: %           velocity values are zero.
34: % fext    - vector of constant complex force components
35: %           applied to the individual masses
36: % c       - vector of damping coefficients specifying
37: %           drag on each particle linearly proportional
38: %           to the particle velocity
39: % tolrel  - relative error tolerance for function ode45
40: % tolabs  - absolute error tolerance for function ode45
```

```
41: % t        - vector of times returned by ode45
42: % z        - matrix of complex position and velocity
43: %            values returned by ode45. A typical row
44: %            z(j,:) gives the system position and
45: %            velocity for time t(j). The first half of
46: %            the row contains complex position values
47: %            and the last half contains velocity values
48: % omega    - frequency at which the ends of the chain
49: %            are shaken
50: % yend     - amplitude of the vertical motion of the
51: %            chain ends. If this is set to zero then
52: %            the chain ends do not move
53: % endmo    - the function defining the end motion of
54: %            the chain
55: % spreqmof - the function defining the equation of
56: %             motion to be integrated using ode45
57: %
58: % User m functions called:  chaindata, spreqmof,
59: %                           endmo, plotmotn
60: %----------------------------------------------
61:
62: global zend omega Rend
63:
64: fprintf('\nDYNAMICS OF A FALLING ELASTIC CHAIN\n\n')
65: disp('Give a file name to define the data. Try')
66: datname=input('chaindata as an example > ? ','s');
67: eval(['[n,tmax,nt,fixorfree,rend,omega,cdamp]=',...
68:         datname,';']);
69:
70: % The following data values are scaled in terms of
71: % the parameters returned by the data input function
72:
73: % Time vector for solution output
74: tmin=0; tlim=linspace(tmin,tmax,nt)';
75:
76: % Number of masses, gravity constant, mass vector
77: g=32.2; len0=1; mas=1/g; m=mas*ones(n,1);
78:
79: % Spring lengths and spring constants
80: L=len0*ones(n+1,1); ksp=5*mas*g*(n+1)/(2*len0);
81: k=ksp*ones(n+1,1);
82:
83: % If the far end of the chain is free, then the
84: % last spring constant is set equal to zero
85:  k(n+1)=fixorfree*k(n+1);
```

```
86:
87: % Viscous damping coefficients
88: c=cdamp*sqrt(mas*ksp)/40*ones(n,1);
89:
90: % Chain end position and initial position of
91: % each mass. Parameters concerning the end
92: % positions are passed as global variables.
93: % global zend omega Rend
94: zend=len0*(n+1); zinit=cumsum(L(1:n));
95: Rend=rend*zend;
96:
97: % Function name giving end position of the chain
98: re=@endmo;
99:
100: % Gravity forces and integration tolerance
101: fext=-i*g*m; tolrel=1e-6;  tolabs=1e-8;
102:
103: % Initial conditions for the ode45 integrator
104: n=length(m); r0=[zinit;zeros(n,1)];
105:
106: % Integrate equations of motion
107: options = odeset('reltol',tolrel,'abstol',tolabs);
108: fprintf('\nPlease Wait While the Equations\n')
109: fprintf('of Motion Are Being Integrated\n')
110: pause(1), tic;
111:
112: [t,r]=ode45(@spreqmof,tlim,r0,options,...
113:                          m,k,L,re,fext,c);
114:
115: cptim=toc; cpt=num2str(fix(10*cptim)/10);
116: fprintf(...
117: ['\nComputation time was ',cpt,' seconds\n'])
118:
119: % Extract displacement history and add
120: % end positions
121: R=endmo(t); z=[R(:,1),r(:,1:n)];
122: if k(n+1)~=0, z=[z,R(:,2)]; end
123: X=real(z); Y=imag(z);
124:
125: % Show animation or motion trace of the response.
126: % disp('Press [Enter] to continue'), pause
127: disp(' ')
128: disp('The  motion can be animated or a trace')
129: disp('can be shown for successive positions')
130: disp(['between t = ',num2str(tmin),...
```

```
131:          ' and t = ',num2str(tmax)])
132: titl=['ELASTIC CHAIN MOTION FOR ',...
133:        '[cdamp,omega] = [',num2str(cdamp),' , ',...
134: num2str(omega),' ]  and T = '];
135:
136: % Plot the position for different times limits
137: while 1
138:   disp(' '), disp(...
139:   ['Choose a plot option (1 <=> animate, ',...
140:    ' 2 <=> trace,'])
141:   opt=input('3 <=> stop)  > ? ');
142:   if opt==3, break, end
143:   disp(' '), disp(...
144:   'Give a time vector such as 0:.1:15')
145:   Tp=input('Time vector > ? ','s');
146:   if isempty(Tp), break, end
147:   tp=eval(Tp); tp=tp(:); T=[titl,Tp];
148:   xp=interp1q(t,X,tp); yp=interp1q(t,Y,tp);
149:   if opt ==1, plotmotn(xp,yp,T)
150:   else, plotmotn(xp,yp,T,1), end
151: end
152:
153: % Save plot history for subsequent printing
154: % print -deps plotmotn
155:
156: fprintf('\nAll Done\n')
157:
158: %==================================
159:
160: function [n,tmax,nt,fixorfree,rend,omega,...
161:                              cdamp]=chaindata
162: %
163: % [n,tmax,nt,fixorfree,rend,omega,...
164: %                            cdamp]=chaindata
165: % ~~~~~~~~~~~~~~~~~~~~~~~~~~~~~~~~~~~~~~~~~~~
166: % This example function creates data defining
167: % the chain. The function can be renamed and
168: % modified to handle different problems.
169:
170: n=8;            % Number or point masses
171: tmax=20;        % Maximum time for the solution
172: nt=401;         % Number of time values from 0 to tmax
173: fixorfree=0;    % Determines whether the right end
174:                 % position is controlled or free. Use
175:                 % zero for free or one for controlled.
```

```
176: rend=0.05;       % Amplitude factor for end motion. This
177:                  % can be zero if the ends are fixed.
178: omega=6;         % Frequency at which the ends are
179:                  % rotated.
180: cdamp=1;         % Coefficient regulating the amount of
181:                  % viscous damping. Reduce cdamp to give
182:                  % less damping.
183:
184: %==================================
185:
186: function rdot=spreqmof(t,r,m,k,L,re,fext,c)
187: %
188: % rdot=spreqmof(t,r,m,k,L,re,fext,c)
189: % ~~~~~~~~~~~~~~~~~~~~~~~~~~~~~~~~~~
190: % This function forms the two-dimensional equation
191: % of motion for a chain of spring-connected particles.
192: % The positions of the ends of the chain may be time
193: % dependent and are computed from a function named in
194: % the input parameter re. The applied external loading
195: % consists of constant loads on the particles and
196: % linear viscous damping proportional to the particle
197: % velocities. Data parameters for the problem are
198: % defined in a function file specified by the user.
199: % Function chaindata gives a typical example.
200: %
201: % t    - current value of time
202: % r    - vector containing complex displacements in
203: %        the top half and complex velocity components
204: %        in the bottom half
205: % m    - vector of particle masses
206: % k    - vector of spring constant values
207: % L    - vector of unstretched spring lengths
208: % re   - name of a function which returns the time
209: %        dependent complex position coordinate for
210: %        the ends of the chain
211: % fext - vector of constant force components applied
212: %        to the spring
213: % c    - vector of viscous damping coefficients for
214: %        the particles
215:
216: N=length(r); n=N/2; z=r(1:n); v=r(n+1:N);
217: R=feval(re,t);
218: zdif=diff([R(1);z;R(2)]); len=abs(zdif);
219: fsp=zdif./len.*((len-L).*(len-L>0)).*k; fdamp=-c.*v;
220: accel=(fext+fdamp+fsp(2:n+1)-fsp(1:n))./m;
```

```
221: rdot=[v;accel];
222:
223: %=====================================
224:
225: function rends=endmo(t)
226: %
227: % rends=endmo(t)
228: % ~~~~~~~~~~~~~~
229: % This function specifies the varying end positions.
230: % In this example the ends rotate at frequency omega
231: % around circles of radius Rend.
232: %
233: % User m functions called:  none
234: %---------------------------------------------
235:
236: global zend Rend omega
237:
238: s=Rend*exp(i*omega*t); rends=[s,zend-conj(s)];
239:
240: %=============================================
241:
242: % function plotmotn(x,y,titl,isave)
243: % See Appendix B
```

Chapter 9

Boundary Value Problems for Partial Differential Equations

9.1 Several Important Partial Differential Equations

Many physical phenomena are characterized by linear partial differential equations. Such equations are attractive to study because (a) principles of superposition apply in the sense that linear combinations of component solutions can often be used to build more general solutions and (b) finite difference or finite element approximations lead to systems of linear equations amenable to solution by matrix methods. The accompanying table lists several frequently encountered equations and some applications. We only show one- or two-dimensional forms, although some of these equations have relevant applications in three dimensions.

In most practical applications the differential equations must be solved within a finite region of space while simultaneously prescribing boundary conditions on the function and its derivatives. Furthermore, initial conditions may exist. In dealing with the initial value problem, we are trying to predict future system behavior when initial conditions, boundary conditions, and a governing physical process are known. Solutions to such problems are seldom obtainable in a closed finite form. Even when series solutions are developed, an infinite number of terms may be needed to provide generality. For example, the problem of transient heat conduction in a circular cylinder leads to an infinite series of Bessel functions employing characteristic values which can only be computed approximately. Hence, the notion of an "exact" solution expressed as an infinite series of transcendental functions is deceiving. At best, we can hope to produce results containing insignificantly small computation errors.

The present chapter applies eigenfunction series to solve nine problems. Examples involving the Laplace, wave, beam, and heat equations are given. Nonhomogeneous boundary conditions are dealt with in several instances. Animation is also provided whenever it is helpful to illustrate the nature of the solutions.

Equation	Equation Name	Applications
$u_{xx} + u_{yy} = \alpha u_t$	Heat	Transient heat conduction
$u_{xx} + u_{yy} = \alpha u_{tt}$	Wave	Transverse vibrations of membranes and other wave type phenomena
$u_{xx} + u_{yy} = 0$	Laplace	Steady-state heat conduction and electrostatics
$u_{xx} + u_{yy} = f(x, y)$	Poisson	Stress analysis of linearly elastic bodies
$u_{xx} + u_{yy} + \omega^2 u = 0$	Helmholtz	Steady-state harmonic vibration problems
$EIy_{xxxx} = -A\rho y_{tt} + f(x, t)$	Beam	Transverse flexural vibrations of elastic beams

9.2 Solving the Laplace Equation inside a Rectangular Region

Functions which satisfy Laplace's equation are encountered often in practice. Such functions are called harmonic; and the problem of determining a harmonic function subject to given boundary values is known as the Dirichlet problem [119]. In a few cases with simple geometries, the Dirichlet problem can be solved explicitly. One instance is a rectangular region with the boundary values of the function being expandable in a Fourier sine series. The following program employs the FFT to construct a solution for boundary values represented by piecewise linear interpolation. Surface and contour plots of the resulting field values are also presented.

The problem of interest satisfies the differential equation

$$\frac{\partial^2 u}{\partial x^2} + \frac{\partial^2 u}{\partial y^2} = 0 \quad , \quad 0 < x < a \quad , \quad 0 < y < b$$

with the boundary conditions of the form

$$\begin{aligned} u(x, 0) &= F(x) \quad , \quad 0 < x < a \, , \\ u(x, b) &= G(x) \quad , \quad 0 < x < a \, , \\ u(0, y) &= P(y) \quad , \quad 0 < y < b \, , \\ u(a, y) &= Q(y) \quad , \quad 0 < y < b \, . \end{aligned}$$

The series solution can be represented as

$$u(x, y) = \sum_{n=1}^{\infty} f_n a_n(x, y) + g_n a_n(x, b - y) + p_n b_n(x, y) + q_n b_n(a - x, y)$$

where

$$a_n(x,y) = \sin\left[\frac{n\pi x}{a}\right] \sinh\left[\frac{n\pi(b-y)}{a}\right] / \sinh\left[\frac{n\pi b}{a}\right],$$

$$b_n(x,y) = \sinh\left[\frac{n\pi(a-x)}{b}\right] \sin\left[\frac{n\pi y}{b}\right] / \sinh\left[\frac{n\pi a}{b}\right],$$

and the constants f_m, g_m, p_n, and q_n are coefficients in the Fourier sine expansions of the boundary value functions. This implies that

$$F(x) = \sum_{n=1}^{\infty} f_n \sin\left[\frac{n\pi x}{a}\right] \ , \ G(x) = \sum_{n=1}^{\infty} g_n \sin\left[\frac{n\pi x}{a}\right],$$

$$P(y) = \sum_{n=1}^{\infty} p_n \sin\left[\frac{n\pi y}{b}\right] \ , \ Q(y) = \sum_{n=1}^{\infty} q_n \sin\left[\frac{n\pi y}{b}\right].$$

The coefficients in the series can be computed by integration as

$$f_n = \frac{2}{a}\int_0^a F(x) \sin\left[\frac{n\pi x}{a}\right] dx \ , \ g_n = \frac{2}{a}\int_0^a G(x) \sin\left[\frac{n\pi x}{a}\right] dx,$$

$$p_n = \frac{2}{a}\int_0^b P(y) \sin\left[\frac{n\pi y}{b}\right] dy \ , \ q_n = \frac{2}{a}\int_0^b Q(y) \sin\left[\frac{n\pi y}{b}\right] dy,$$

or approximate coefficients can be obtained using the FFT. The latter approach is chosen here and the solution is evaluated for an arbitrary number of terms in the series.

The chosen problem solution has the disadvantage of employing eigenfunctions that vanish at the ends of the expansion intervals. Consequently, it is desirable to combine the series with an additional term allowing exact satisfaction of the corner conditions for cases where the boundary value functions for adjacent sides agree. This implies requirements such as $F(a) = Q(0)$ and three other similar conditions. It is evident that the function

$$u_p(x,y) = c_1 + c_2 x + c_3 y + c_4 xy$$

is harmonic and varies linearly along each side of the rectangle. Constants $c_1, \cdots, c_4$ can be computed to satisfy the corner values and the total solution is represented as u_p plus a series solution involving modified boundary conditions.

The following program **laplarec** solves the Dirichlet problem for the rectangle. Function values and gradient components are computed and plotted. Functions used in this program are described below. The example data set defined in the driver program was chosen to produce interesting surface and contour plots. Different boundary conditions can be handled by slight modifications of the input data. In this example 100 term series are used. Figure 9.1 through Figure 9.4 show function and gradient components, as well as a contour plot of function values. Readers may find it instructive to run the program and view these figures from different angles

laplarec	inputs data, calls computation modules, and plots results
datafunc	defines an example datacase
ulinbc	particular solution for linearly varying boundary conditions
recseris	sums the series for function and gradient values
sincof	generates coefficients in a Fourier sine series
lintrp	piecewise linear interpolation function allowing jump discontinuities

using the interactive figure rotating capability provided in MATLAB. Note that the figure showing the function gradient in the x direction used view([225,20]) to show clearly the jump discontinuity in this quantity.

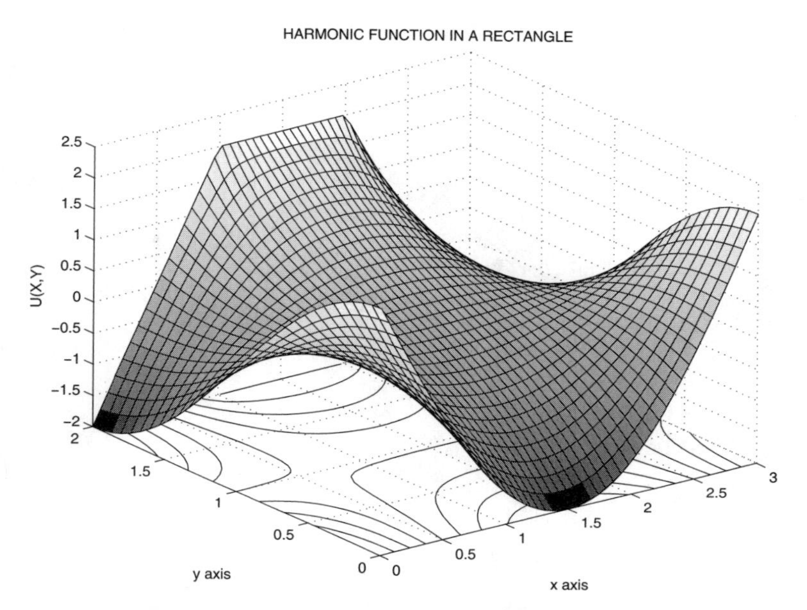

Figure 9.1: **Surface Plot of Function Values**

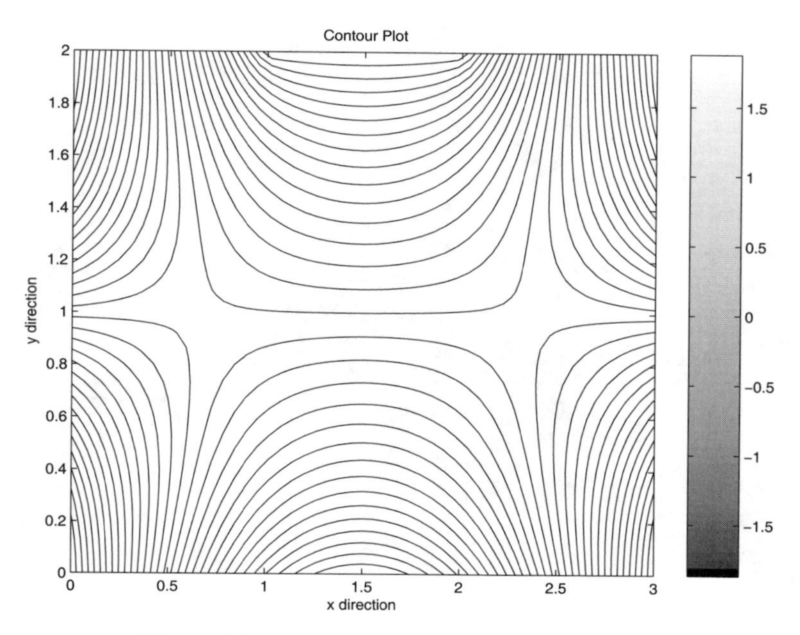

Figure 9.2: **Contour Plot of Function Values**

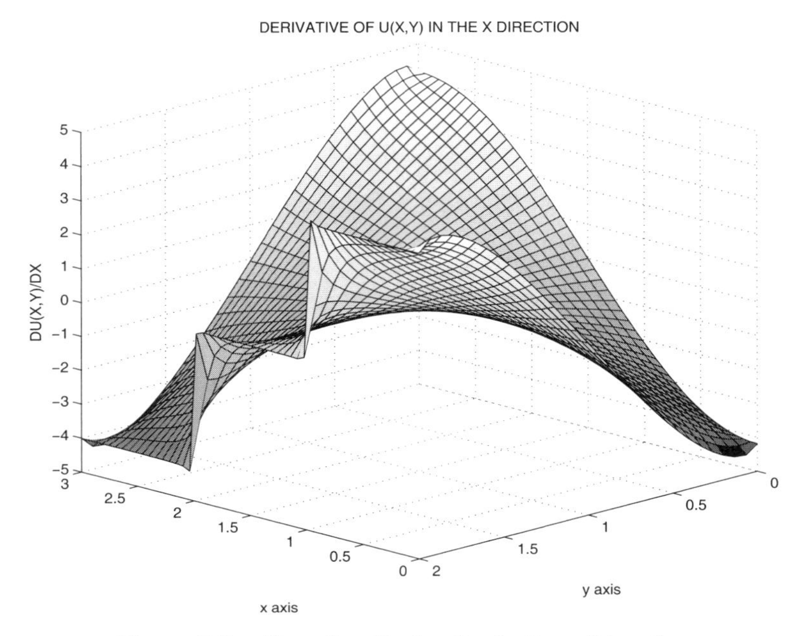

Figure 9.3: Function Derivative in the x Direction

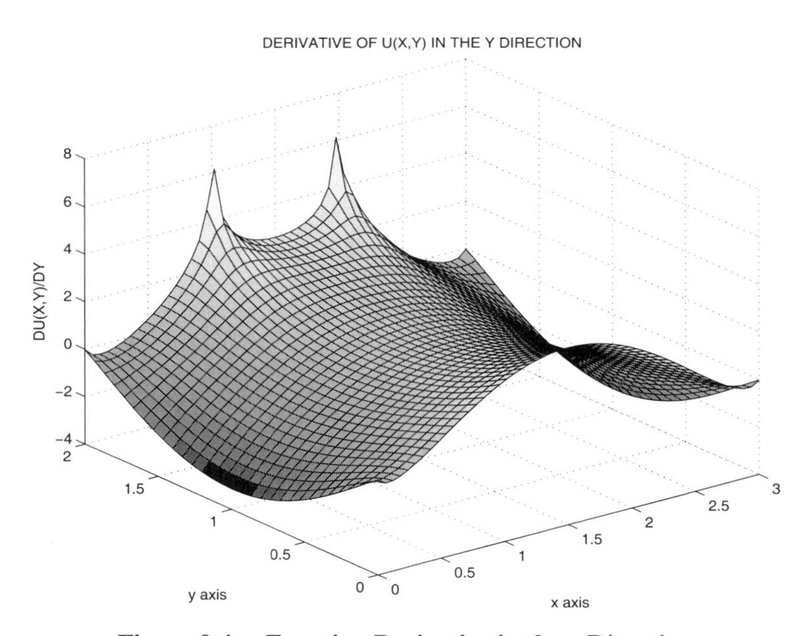

Figure 9.4: Function Derivative in the y Direction

MATLAB Example

Program laplarec

```
1: function [u,ux,uy,X,Y]=laplarec(...
2:                       ubot,utop,ulft,urht,a,b,nx,ny,N)
3: %
4: % [u,ux,uy,X,Y]=laplarec(...
5: %                  ubot,utop,ulft,urht,a,b,nx,ny,N)
6: % ~~~~~~~~~~~~~~~~~~~~~~~~~~~~~~~~~~~~~~~~~~~~~~~~~~~~
7: % This program evaluates a harmonic function and its
8: % first partial derivatives in a rectangular region.
9: % The method employs a Fourier series expansion.
10: % ubot   - defines the boundary values on the bottom
11: %          side. This can be an array in which
12: %          ubot(:,1) is x coordinates and ubot(:,2)
13: %          is function values. Values at intermediate
14: %          points are obtained by piecewise linear
15: %          interpolation. A character string giving
16: %          the name of a function can also be used.
17: %          Then the function is evualuated using 200
18: %          points along a side to convert ubot to an
19: %          array. Similar comments apply for utop,
20: %          ulft, and urht introduced below.
21: % utop   - boundary value definition on the top side
22: % ulft   - boundary value definition on the left side
23: % urht   - boundary value definition on the right side
24: % a,b    - rectangle dimensions in x and y directions
25: % nx,ny  - number of x and y values for which the
26: %          solution is evaluated
27: % N      - number of terms used in the Fourier series
28: % u      - function value for the solution
29: % ux,uy  - first partial derivatives of the solution
30: % X,Y    - coordinate point arrays where the solution
31: %          is evaluated
32: %
33: % User m functions used: datafunc ulinbc
34: %                        recseris ftsincof
35:
36: disp(' ')
37: disp('SOLVING THE LAPLACE EQUATION IN A RECTANGLE')
38: disp(' ')
39:
40: if nargin==0
```

```
41:    disp(...
42:       'Give the name of a function defining the data')
43:    datfun=input(...
44:       '(try datafunc as an example): > ? ','s');
45:    [ubot,utop,ulft,urht,a,b,nx,ny,N]=feval(datfun);
46: end
47:
48: % Create a grid to evaluate the solution
49: x=linspace(0,a,nx); y=linspace(0,b,ny);
50: [X,Y]=meshgrid(x,y); d=(a+b)/1e6;
51: xd=linspace(0,a,201)'; yd=linspace(0,b,201)';
52:
53: % Check whether boundary values are given using
54: % external functions. Convert these to arrays
55:
56: if isstr(ubot)
57:    ud=feval(ubot,xd); ubot=[xd,ud(:)];
58: end
59: if isstr(utop)
60:    ud=feval(utop,xd); utop=[xd,ud(:)];
61: end
62: if isstr(ulft)
63:    ud=feval(ulft,yd); ulft=[yd,ud(:)];
64: end
65: if isstr(urht)
66:    ud=feval(urht,yd); urht=[yd,ud(:)];
67: end
68:
69: % Determine function values at the corners
70: ub=interp1(ubot(:,1),ubot(:,2),[d,a-d]);
71: ut=interp1(utop(:,1),utop(:,2),[d,a-d]);
72: ul=interp1(ulft(:,1),ulft(:,2),[d,b-d]);
73: ur=interp1(urht(:,1),urht(:,2),[d,b-d]);
74: U=[ul(1)+ub(1),ub(2)+ur(1),ur(2)+ut(2),...
75:       ut(1)+ul(2)]/2;
76:
77: % Obtain a solution satisfying the corner
78: % values and varying linearly along the sides
79:
80: [v,vx,vy]=ulinbc(U,a,b,X,Y);
81:
82: % Reduce the corner values to zero to improve
83: % behavior of the Fourier series solution
84: % near the corners
85:
```

```
86: f=inline('u0+(u1-u0)/L*x','x','u0','u1','L');
87: ubot(:,2)=ubot(:,2)-f(ubot(:,1),U(1),U(2),a);
88: utop(:,2)=utop(:,2)-f(utop(:,1),U(4),U(3),a);
89: ulft(:,2)=ulft(:,2)-f(ulft(:,1),U(1),U(4),b);
90: urht(:,2)=urht(:,2)-f(urht(:,1),U(2),U(3),b);
91:
92: % Evaluate the series and combine results
93: % for the various component solutions
94:
95: [ub,ubx,uby]=recseris(ubot,a,b,1,x,y,N);
96: [ut,utx,uty]=recseris(utop,a,b,2,x,y,N);
97: [ul,ulx,uly]=recseris(ulft,a,b,3,x,y,N);
98: [ur,urx,ury]=recseris(urht,a,b,4,x,y,N);
99: u=v+ub+ut+ul+ur; ux=vx+ubx+utx+ulx+urx;
100: uy=vy+uby+uty+uly+ury; close
101:
102: % Show results graphically
103:
104: surfc(X,Y,u), xlabel('x axis'), ylabel('y axis')
105: zlabel('U(X,Y)')
106: title('HARMONIC FUNCTION IN A RECTANGLE')
107: shg, pause
108: % print -deps laprecsr
109:
110: contour(X,Y,u,30); title('Contour Plot');
111: xlabel('x direction'); ylabel('y direction');
112: colorbar, shg, pause
113: % print -deps laprecnt
114:
115: surf(X,Y,ux), xlabel('x axis'), ylabel('y axis')
116: zlabel('DU(X,Y)/DX')
117: title('DERIVATIVE OF U(X,Y) IN THE X DIRECTION')
118: shg, pause
119: % print -deps laprecdx
120:
121: surf(X,Y,uy), xlabel('x axis'), ylabel('y axis')
122: zlabel('DU(X,Y)/DY')
123: title('DERIVATIVE OF U(X,Y) IN THE Y DIRECTION')
124: % print -deps laprecdy
125: shg
126:
127: %==============================================
128:
129: function [ubot,utop,ulft,urht,a,b,...
130:                                   nx,ny,N]=datafunc
```

```
131: %
132: % [ubot,utop,ulft,urht,a,b,...
133: %          nx,ny,N]=datafunc
134: % ~~~~~~~~~~~~~~~~~~~~~~~~~~~~
135: % This is a sample data case which can be
136: % modified to apply to other examples
137: %
138: % ubot, utop - vectors of function values on the
139: %              bottom and top sides
140: % ulft, urht - vectors of function values on the
141: %              right and left sides
142: % a, b       - rectangle dimensions along the
143: %              x and y axis
144: % nx, ny     - number of grid values for the x
145: %              and y directions
146: % N          - number of terms used in the
147: %              Fourier series solution
148:
149: a=3; b=2; e=1e-5; N=100;
150: x=linspace(0,1,201)'; s=sin(pi*x);
151: c=cos(pi*x); ubot=[a*x,2-4*s];
152: utop=[a*x,interp1([0,1/3,2/3,1],...
153:       [-2,2,2,-2],x)];
154: ulft=[b*x,2*c]; urht=ulft; nx=51; ny=31;
155:
156: %=============================================
157:
158: function [u,ux,uy]=ulinbc(U,a,b,X,Y)
159: %
160: % [u,ux,uy]=ulinbc(U,a,b,X,Y)
161: % ~~~~~~~~~~~~~~~~~~~~~~~~~~~
162: % This function determines a harmonic function
163: % varying linearly along the sides of a rectangle
164: % with specified corner values
165: %
166: % U     - corner values of the harmonic function
167: %         [U(1),...U(4)] <=> corner coordinates
168: %         (0,0), (0,a), (a,b), (0,b)
169: % a,b   - rectangle dimensions in the x and y
170: %         directions
171: % X,Y   - array coordinates where the solution
172: %         is evaluated
173: % u     - function values evaluated for X,Y
174: % ux,uy - first derivative components evaluated
175: %         for the X,Y arrays
```

```
176:
177: c=[1,0,0,0;1,a,0,0;1,a,b,a*b;1,0,b,0;]\U(:);
178: u=c(1)+c(2)*X+c(3)*Y+c(4)*X.*Y;
179: ux=c(2)+c(4)*Y; uy=c(3)+c(4)*X;
180:
181: %==============================================
182:
183: function [u,ux,uy,X,Y]=recseris(udat,a,b,iside,x,y,N)
184: %
185: % [u,ux,uy,X,Y]=recseris(udat,a,b,iside,x,y,N)
186: % ~~~~~~~~~~~~~~~~~~~~~~~~~~~~~~~~~~~~~~~~~~~~~
187: % This function computes a function harmonic in
188: % a rectangle with general function values given
189: % on one side and zero function values on the
190: % other three sides.
191: % udat     - a data array to determine the function
192: %            values by piecewise linear interpolation
193: %            along the side having nonzero values.
194: %            udat(:,1) contains either x or y values
195: %            along a side, and udat(:,2) contains
196: %            corresponding function values
197: % a,b      - side lengths for the x and y directions
198: % iside    - an index indicating the side for which
199: %            function values are given.
200: %            [1,2,3,4]<=>[bottom,top,left,right]
201: % x,y        data vectors defining a grid
202: %            [X,Y]=meshgrid(x,y) on which the function
203: %            and its first partial derivatives are
204: %            computed
205: % N        - number of series terms used (up to 500)
206: % u,ux,uy  - arrays of values of the harmonic function
207: %            and its first partial derivatives
208: % X,Y        arrays of coordinate values for which
209: %            function values were computed.
210:
211: x=x(:)'; y=y(:); ny=length(y); N=min(N,500);
212: if iside<3, period=2*a; else, period=2*b; end
213: c=ftsincof(udat,period); n=1:N; c=c(n);
214: if iside<3      % top or bottom sides
215:    npa=pi/a*n; c=c./(1-exp(-2*b*npa));
216:    sx=sin(npa(:)*x); cx=cos(npa(:)*x);
217:    if iside==1 % bottom side
218:       dy=exp(-y*npa); ey=exp(-(2*b-y)*npa);
219:       u=repmat(c,ny,1).*(dy-ey)*sx;
220:       c=repmat(c.*npa,ny,1);
```

```
221:       ux=c.*(dy-ey)*cx; uy=-c.*(dy+ey)*sx;
222:    else          % top side
223:       dy=exp((y-b)*npa); ey=exp(-(y+b)*npa);
224:       u=repmat(c,ny,1).*(dy-ey)*sx;
225:       c=repmat(c.*npa,ny,1);
226:       ux=c.*(dy-ey)*cx; uy=c.*(dy+ey)*sx;
227:    end
228: else             % left or right sides
229:    npb=pi/b*n; c=c./(1-exp(-2*a*npb));
230:    sy=sin(y*npb); cy=cos(y*npb);
231:    if iside==3 % left side
232:       dx=exp(-npb(:)*x);
233:       ex=exp(-npb(:)*(2*a-x));
234:       u=repmat(c,ny,1).*sy*(dx-ex);
235:       c=repmat(c.*npb,ny,1);
236:       ux=c.*sy*(-dx-ex); uy=c.*cy*(dx-ex);
237:    else          % right side
238:       dx=exp(-npb(:)*(a-x));
239:       ex=exp(-npb(:)*(a+x));
240:       u=repmat(c,ny,1).*sy*(dx-ex);
241:       c=repmat(c.*npb,ny,1);
242:       ux=c.*sy*(dx+ex); uy=c.*cy*(dx-ex);
243:    end
244: end
245: [X,Y]=meshgrid(x,y);
246: 
247: %============================================
248: 
249: function c=ftsincof(y,period)
250: %
251: % c=ftsincof(y,period)
252: % ~~~~~~~~~~~~~~~~~~~~
253: % This function computes 500 Fourier sine
254: % coefficients for a piecewise linear
255: % function defined by a data array
256: % y      - an array defining the function
257: %          over half a period as
258: %          Y(x)=interp1(y(:,1),y(:,2),x)
259: % period - the period of the function
260: %
261: xft=linspace(0,period/2,513);
262: uft=interp1(y(:,1),y(:,2)/512,xft);
263: c=fft([uft,-uft(512:-1:2)]);
264: c=-imag(c(2:501));
```

9.3 The Vibrating String

Transverse motion of a tightly stretched string illustrates one of the simplest occurrences of one-dimensional wave propagation. The transverse deflection satisfies the wave equation

$$a^2 \frac{\partial^2 u}{\partial X^2} = \frac{\partial^2 u}{\partial T^2}$$

where $u(X, T)$ satisfies initial conditions

$$u(X, 0) = F(X) \; , \; \frac{\partial u(X, 0)}{\partial T} = G(X)$$

with boundary conditions

$$u(0, T) = 0 \; , \; u(\ell, T) = 0$$

where ℓ is the string length. If we introduce the dimensionless variables $x = X/\ell$ and $t = T/(\ell/a)$ the differential equation becomes

$$u_{xx} = u_{tt}$$

where subscripts denote partial differentiation. The boundary conditions become

$$u(0, t) = u(1, t) = 0$$

and the initial conditions become

$$u(x, 0) = f(x) \; , \; u_t(x, 0) = g(x).$$

Let us consider the case where the string is released from rest initially so $g(x) = 0$. The solution can be found in series form as

$$u(x, t) = \sum_{n=1}^{\infty} a_n p_n(x) \cos(\omega_n t)$$

where ω_n are natural frequencies and satisfaction of the differential equation of motion requires

$$p_n''(x) + \omega_n^2 p_n(x) = 0$$

so

$$p_n = A_n \sin(\omega_n x) + B_n \cos(\omega_n x).$$

The boundary condition $p_n(0) = B_n = 0$ and $p_n(1) = A_n \sin(\omega_n)$ requires $A_n \neq 0$ and $\omega_n = n\pi$, where n is an integer. This leads to a solution in the form

$$u(x, t) = \sum_{n=1}^{\infty} a_n \sin(n\pi x) \cos(n\pi t).$$

The remaining condition on initial conditions requires

$$\sum_{n=1}^{\infty} a_n \sin(n\pi x) = f(x) \ , \ 0 < x < 1.$$

Therefore, the coefficients a_n are obtainable from an odd-valued Fourier series expansion of $f(x)$ vanishing at $x = 0$ and $x = 2$. We see that $f(-x) = -f(x)$ and $f(x+2) = f(x)$, and the coefficients are obtainable by integration as

$$a_n = 2 \int_0^1 f(x) \sin(n\pi x) \, dx.$$

However, an easier way to compute the coefficients is to use the FFT. A solution will be given for an arbitrary piecewise linear initial condition.

Before implementing the Fourier series solution, let us digress briefly to examine the case of an infinite string governed by

$$a^2 u_{XX} = u_{TT} \ , \ -\infty < X < \infty$$

and initial conditions

$$u(X,0) = F(X) \ , \ u_T(X,0) = G(X).$$

The reader can verify directly that the solution of this problem is given by

$$u(X,T) = \frac{1}{2}\left[F(X - aT) + F(X + aT)\right] + \frac{1}{2a}\int_{X-aT}^{X+aT} G(x) \, dx.$$

When the string is released from rest, $G(X)$ is zero and the solution reduces to

$$\frac{F(X - aT) + F(X + aT)}{2}$$

which shows that the initial deflection splits into two parts with one half translating to the left at speed a and the other half moving to the right at speed a. This solution can also be adapted to solve the problem for a string of length ℓ fixed at each end. The condition $u(0,T) = 0$ implies

$$F(-aT) = -F(aT)$$

which shows that $F(X)$ must be odd valued. Similarly, $u(\ell, T) = 0$ requires

$$F(\ell - aT) + F(\ell + aT) = 0.$$

Combining this condition with $F(X) = -F(X)$ shows that

$$F(X + 2\ell) = F(X)$$

so, $F(X)$ must have a period of 2ℓ. In the string of length ℓ, $F(X)$ is only known for $0 < X < \ell$, and we must take

$$F(X) = -F(2\ell - X)\ ,\ \ell < X < 2\ell.$$

Furthermore the solution has the form

$$u(X,T) = \frac{F(x_p) + F(x_m)}{2}$$

where $x_p = X + aT$ and $x_m = X - aT$. The quantity x_p will always be positive and x_m can be both positive and negative. The necessary sign change and periodicity can be achieved by evaluating $F(X)$ as

$$\mathbf{sign}(X).{*}F(\mathbf{rem}(\mathbf{abs}(X)), 2 * \ell)$$

where **rem** is the intrinsic remainder function used in the exact solution implemented in function **strngwav** presented earlier in section 2.7.

A computer program employing the Fourier series solution was written for an initial deflection that is piecewise linear. The series solution allows the user to select varying numbers of terms in the series to examine how well the initial deflection configuration is represented by a truncated sine series. A function animating the time response shows clearly how the initial deflection splits in two parts translating in opposite directions. In the Fourier solution, dimensionless variables are employed to make the string length and the wave speed both equal one. Consequently, the time required for the motion to exactly return to the starting position equals two, representing how long it takes for a disturbance to propagate from one end to the other and back. When the motion is observed for $0 < x < 1$, it is evident that waves reflected from a wall are inverted. The program employs the following functions.

stringft	function to input initial deflection data
sincof	uses **fft** to generate coefficients in a sine series
initdefl	defines the initial deflection by piecewise linear interpolation
strvib	evaluates the series solution for general x and t
smotion	animates the string motion
inputv	facilitates interactive data input
lintrp	performs interpolation to evaluate a piecewise linear function

Results are shown below for a string which was deflected initially in a square wave. The example was chosen to illustrate the approximation produced when a small number of Fourier coefficients, in this case 30, is used. Ripples are clearly evident in the surface plot of $u(x,t)$ in Figure 9.5. The deflection configuration of

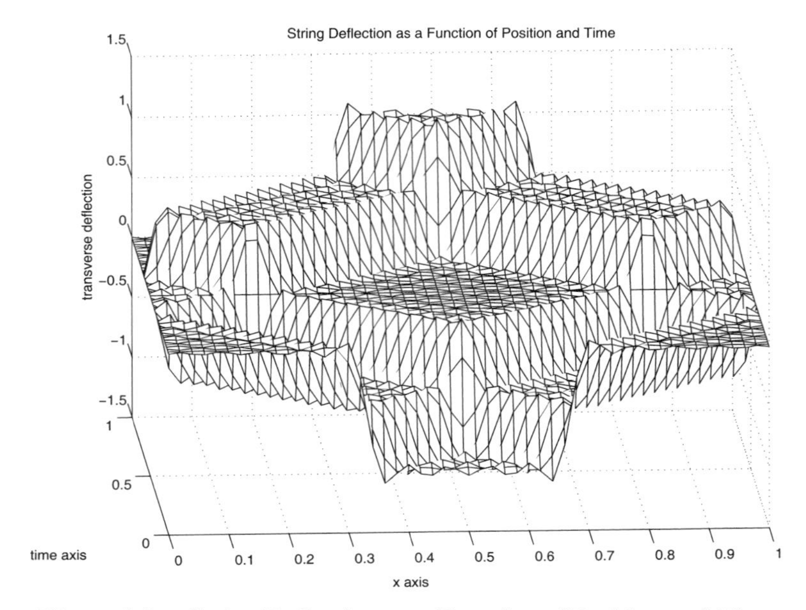

Figure 9.5: String Deflection as a Function of Position and Time

the string at $t = 1$ when the initial deflection form has passed through half a period of motion appears in Figure 9.6. One other example given in Figure 9.7 shows the deflection surface produced using 100 series terms and a triangular initial deflection pattern. The surface describes $u(x, t)$ through one period of motion.

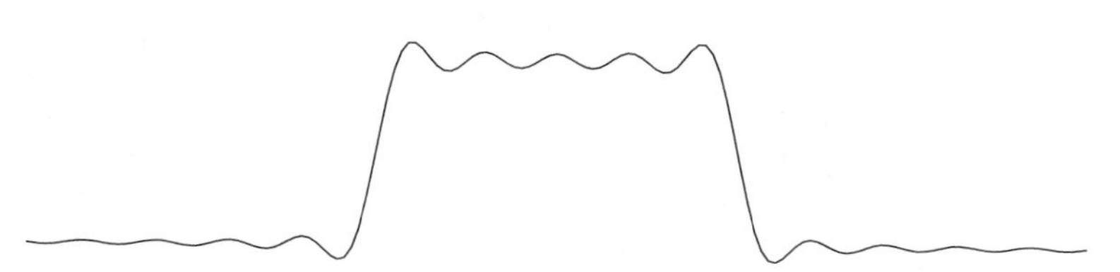

Figure 9.6: Wave Propagation in a String

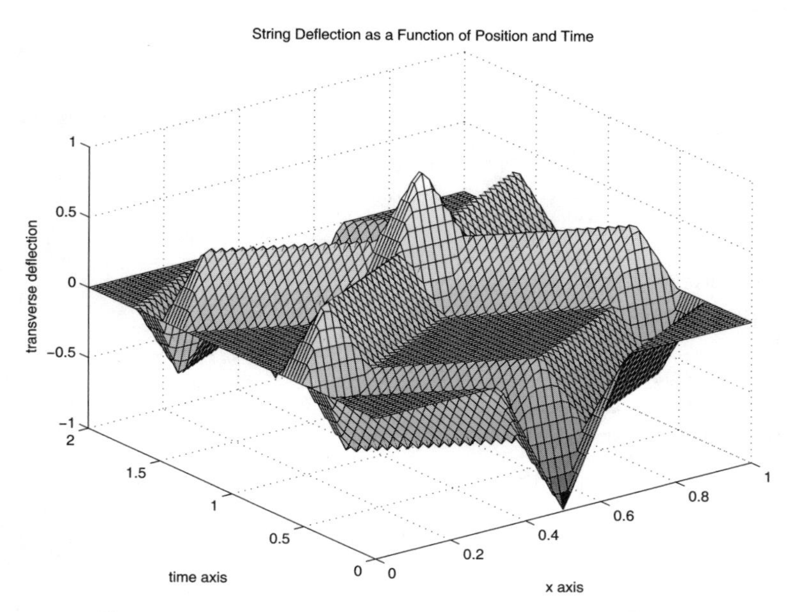

Figure 9.7: Surface for Triangular Initial Deflection

Program Output and Code

Output from Example stringft

```
>> stringft;

  FOURIER SERIES SOLUTION FOR WAVES
IN A STRING WITH LINEARLY INTERPOLATED
  INITIAL DEFLECTION AND FIXED ENDS

Enter the number of interior data points (the fixed
end point coordinates are added automatically)
? 4

The string stretches between fixed endpoints at
x=zero and x=one.

Enter 4 sets of x,y to specify interior
initial deflections (one pair per line)

 ? .33,0
 ? .33,-1
 ? .67,-1
 ? .67,0

Give the number of series terms
and the maximum value of t
(give 0,0 to stop)
 ? 30,1

Press [Enter] to
see the animation

Give the number of series terms
and the maximum value of t
(give 0,0 to stop)
 ? 0,0
>>
```

String Vibration Program

```
1: function [x,t,y]=stringft(Xdat,Ydat)
2: %
```

```
 3: % Example: [x,t,y]=stringft(Xdat,Ydat)
 4: % ~~~~~~~~~~~~~~~~
 5: % This program analyzes wave motion in a string
 6: % having arbitrary piecewise linear initial
 7: % deflection. A Fourier series expansion is used
 8: % to construct the solution
 9: %
10: % Xdat,Ydat -data vectors defining the initial
11: %            deflections at interior points. The
12: %            deflections at x=0 and x=1 are set
13: %            to xero automatically. For example,
14: %            try Xdat=[.2,.3,.7,.8],
15: %                Ydat=[0,-1,-1,0]
16: %
17: % x,t,y     - arrays containing the position, time
18: %            and deflection values
19: %
20: % User m functions required:
21: %    sincof, initdefl, strvib, smotion, inputv,
22: %    lintrp
23: 
24: global xdat ydat
25: 
26: disp(' '), disp( ...
27:   '  FOURIER SERIES SOLUTION FOR WAVES')
28: disp(....
29:   'IN A STRING WITH LINEARLY INTERPOLATED')
30: disp(...'
31:   '  INITIAL DEFLECTION AND FIXED ENDS')
32: if nargin==0
33:   disp(' ')
34:   disp(['Enter the number of interior ',...
35:         'data points (the fixed'])
36:   disp(['end point coordinates are ',...
37:         'added automatically)'])
38:   n=input('? '); if isempty(n), break, end
39:   xdat=zeros(n+2,1); ydat=xdat; xdat(n+2)=1;
40:   disp(' ')
41:   disp(['The string stretches between ',...
42:         'fixed endpoints at'])
43:   disp('x=zero and x=one. '),disp(' ')
44:   disp(['Enter ',num2str(n),...
45:        ' sets of x,y to specify interior'])
46:   disp(['initial deflections ',...
47:       '(one pair per line)']), disp(' ')
```

```
48:   for j=2:n+1,[xdat(j),ydat(j)]=inputv; end;
49: else
50:   xdat=[0;Xdat(:);1]; ydat=[0;Ydat(:);0];
51: end
52:
53: a=sincof(@initdefl,1,1024); % sine coefficients
54: nx=51; x=linspace(0,1,nx);
55: xx=linspace(0,1,151);
56:
57: while 1
58:   disp(' ')
59:   disp('Give the number of series terms')
60:   disp('and the maximum value of t')
61:   disp('(give 0,0 to stop)')
62:   [ntrms,tmax]=inputv;
63:   if isnan(ntrms)| norm([ntrms,tmax])==0
64:   break, end
65:   nt=ceil(nx*tmax); t=linspace(0,tmax,nt);
66:   y=strvib(a,t,x,1,ntrms); % time history
67:   yy=strvib(a,t,xx,1,ntrms);
68:   [xo,to]=meshgrid(x,t);
69:   hold off; surf(xo,to,y);
70:   grid on; colormap([1 1 1]);
71:   %colormap([127/255 1 212/255]);
72:   xlabel('x axis'); ylabel('time axis');
73:   zlabel('transverse deflection');
74:   title(['String Deflection as a Function ', ...
75:     'of Position and Time']);
76:   disp(' '), disp('Press [Enter] to')
77:   disp('see the animation'), shg, pause
78:   % print -deps strdefl
79:   smotion(xx,yy,'Wave Propagation in a String');
80:   disp(''); pause(1);
81: end
82: % print -deps strwave
83:
84: %==============================================
85:
86: function y=initdefl(x)
87: %
88: % y=initdefl(x)
89: % ~~~~~~~~~~~~~
90: % This function defines the linearly
91: % interpolated initial deflection
92: % configuration.
```

```
93: %
94: % x - a vector of points at which the initial
95: %     deflection is to be computed
96: %
97: % y - transverse initial deflection value for
98: %     argument x
99: %
100: % xdat, ydat - global data vectors used for
101: %              linear interpolation
102: %
103: % User m functions required:  lintrp
104: %----------------------------------------------
105:
106: global xdat ydat
107: y=lintrp(xdat,ydat,x);
108:
109: %==============================================
110:
111: function y=strvib(a,t,x,hp,n)
112: %
113: % y=strvib(a,t,x,hp,n)
114: % ~~~~~~~~~~~~~~~~~~~~~
115: % Sum the Fourier series for the string motion.
116: %
117: % a   - Fourier coefficients of initial
118: %       deflection
119: % t,x - vectors of time and position values
120: % hp  - the half period for the series
121: %       expansion
122: % n   - the number of series terms used
123: %
124: % y   - matrix with y(i,j) equal to the
125: %       deflection at position x(i) and
126: %       time t(j)
127: %
128: % User m functions required: none
129: %----------------------------------------------
130:
131: w=pi/hp*(1:n); a=a(1:n); a=a(:)';
132: x=x(:); t=t(:)';
133: y=((a(ones(length(x),1),:).* ...
134:   sin(x*w))*cos(w(:)*t))';
135:
136: %==============================================
137:
```

```
138: function smotion(x,y,titl)
139: %
140: % smotion(x,y,titl)
141: % ~~~~~~~~~~~~~~~~~
142: % This function animates the string motion.
143: %
144: % x    - a vector of position values along the
145: %        string
146: % y    - a matrix of transverse deflection
147: %        values where successive rows give
148: %        deflections at successive times
149: % titl - a title shown on the plot (optional)
150: %
151: % User m functions required: none
152: %----------------------------------------------
153:
154: if nargin < 3, titl=' '; end
155: xmin=min(x); xmax=max(x);
156: ymin=min(y(:)); ymax=max(y(:));
157: [nt,nx]=size(y); clf reset;
158: for j=1:nt
159:   plot(x,y(j,:),'k');
160:   axis([xmin,xmax,2*ymin,2*ymax]);
161:   axis('off'); title(titl);
162:   drawnow; figure(gcf); pause(.1)
163: end
164:
165: %=============================================
166:
167: function a=sincof(func,hafper,nft)
168: %
169: % a=sincof(func,hafper,nft)
170: % ~~~~~~~~~~~~~~~~~~~~~~~~~
171: % This function calculates the sine
172: % coefficients.
173: %
174: % func   - the name of a function  defined over
175: %          a half period
176: % hafper - the length of the half period of the
177: %          function
178: % nft    - the number of function values used
179: %          in the Fourier series
180: %
181: % a      - the vector of Fourier sine series
182: %          coefficients
```

```
183: %
184: % User m functions required:  none
185: %----------------------------------------
186:
187: n2=nft/2; x=hafper/n2*(0:n2);
188: y=feval(func,x); y=y(:);
189: a=fft([y;-y(n2:-1:2)]); a=-imag(a(2:n2))/n2;
190:
191: %========================================
192:
193: % function y=lintrp(xd,yd,x)
194: % See Appendix B
195:
196: %========================================
197:
198: % function varargout=inputv(prompt)
199: % See Appendix B
```

9.4 Force Moving on an Elastic String

The behavior of a semi-infinite string acted on by a moving transverse force illustrates an interesting aspect of wave propagation. Consider a taut string initially at rest and un-deflected when a force moving at constant speed is applied. This simple example shows how a wave front moves ahead of the force when the velocity of wave propagation in the string exceeds the speed of the force, but the force acts at the front of the disturbance when the force moves faster than the wave speed of the string. The governing differential equations, initial conditions, and boundary conditions are:

$$a^2 u_{xx}(x,t) = u_{tt}(x,t) + \frac{F_0}{\rho}\delta(x - vt) \ , \ t > 0 \ , \ 0 < x < \infty,$$

$$u(0,t) = 0 \ , \ u(\infty,t) = 0,$$

$$u(x,0) = 0 \ , \ u_t(x,0) = 0 \ , \ 0 < x < \infty.$$

In these equations a is the speed of wave propagation in the string and v is the speed at which a concentrated downward force F_0 moves toward the right along the string, ρ is the mass per unit length of the string, and δ is the Dirac delta function. This problem can be solved using the Fourier sine transform pair defined by

$$U(p,t) = \int_0^\infty u(x,t)\sin(px)dx \ , \ u(x,t) = \frac{2}{\pi}\int_0^\infty U(p,t)\sin(px)dp.$$

Transforming the differential equation and initial conditions, and making use of the boundary conditions gives

$$-p^2a^2U(p,t) = U_{tt}(p,t) + \frac{F_0}{\rho}\sin(pvt) \ , \ U(p,0) = 0 \ , \ U_t(p,0) = 0.$$

It follows that

$$U(p,t) = \frac{F_0}{a\rho(a^2 - v^2)}\left[\frac{v\sin(apt) - a\sin(vpt)}{p^2}\right],$$

provided $v \neq a$. Applying the inverse transformation then gives the desired displacement response as

$$u(x,t) = \frac{F_0}{2a\rho(a^2 - v^2)}\left[\ (v-a)x - v\,|x - at| + a\,|x - vt|\ \right].$$

9.4.1 Computer Analysis

The following MATLAB program analyzes the response predicted by the last equation. A surface plot shows $u(x,t)$. Positions of the force at successive times are also marked by a heavy dark line superimposed on the surface. Then an animation shows the string deflection and the point of action of the force throughout the chosen time interval. As the force moves along the string, no deflection occurs ahead of the force if the speed of the force exceeds the speed of wave propagation for the string. Otherwise, a disturbance propagates ahead of the force at the wave speed of the string. Graphical results from the program are shown first. Then the computer code is listed.

Let us first consider what happens when the force moves slower than the wave speed. Taking so $v = 1.0, \ \ a = 1.2$ gives the following results in Figure 9.8. Since the point of application of the load is denoted by an arrow, it is clear from the last figure that the disturbance moves ahead of the load when the load moves slower than the wave speed for the string. Next consider what happens when the force moves faster than the wave speed for the string. For example taking $v = 1$, $a = 0.80$ gives significantly different output. In this instance, no disturbance occurs at a point until the load passes the point. This case is illustrated in Figure 9.9. The reader may find it instructive to run the program for different combinations of force speed and wave speed. The program does not account for the case where v exactly equals a, but these values can be taken close enough together to see what the limiting case will give. We simply increase a to 1.00001 times a.

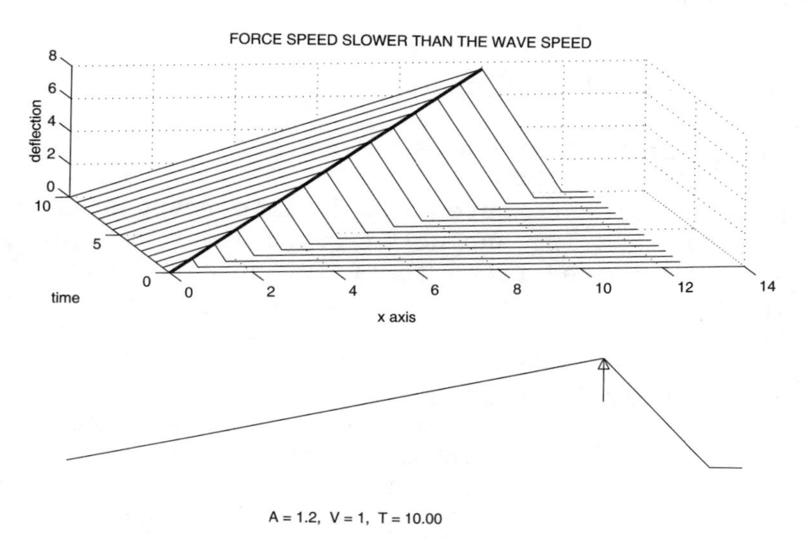

Figure 9.8: **Force Moving Slower than the Wave Speed**

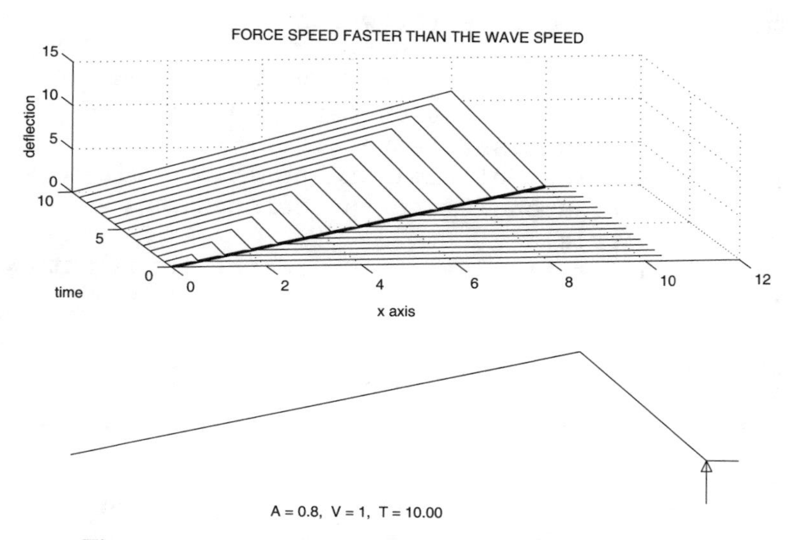

Figure 9.9: **Force Moving Faster than the Wave Speed**

Program forcmove

```
1: function [u,X,T,uf,t]=forcmove(a,v,tmax,nt)
2: %
3: % [u,X,T,uf,t]=forcmove(a,v,tmax,nt)
4: % ~~~~~~~~~~~~~~~~~~~~~~~~~~~~~~~~~~~
5: % This function computes the dynamic response
6: % of a taut string subjected to an upward
7: % directed concentrated force moving along the
8: % string at constant speed. The string is
9: % fixed at x=0 and x=+infinity. The system
10: % is initially at rest when the force starts
11: % moving toward the right from the left end. If
12: % the force speed exceeds the wave propagation
13: % speed, then no disturbance occurs ahead of
14: % the force. If the force speed is slower
15: % than the wave propagation speed, then the
16: % deflection propagates ahead of the force at
17: % the wave propagation speed.
18: %
19: % v      - speed of the moving load
20: % a      - speed of wave propagation in the
21: %          string
22: % tmax   - maximum time for which the
23: %          solution is computed
24: % u      - matrix of deflection values where
25: %          time and position vary row-wise and
26: %          column-wise, respectively
27: % T,X    - matrices of time and position values
28: %          corresponding to the deflection
29: %          matrix U
30: % uf     - deflection values where the force acts
31: % t      - vector of times (same as columns of T)
32: %
33: % User m functions used: ustring
34:
35: if nargin==0, a=.8; v=1; tmax=10; nt=15; end
36:
37: if a>v
38:   titl='FORCE SPEED SLOWER THAN THE WAVE SPEED';
39: elseif a<v
40:   titl='FORCE SPEED FASTER THAN THE WAVE SPEED';
41: else
```

```
42:    titl='FORCE SPEED EQUAL TO THE WAVE SPEED';
43:    a=v*1.00001;
44: end
45:
46: % Obtain solution values and plot results
47: [u,X,T,uf,t]=ustring(a,v,tmax,nt);
48: if a>v, xf=X(:,2); uf=u(:,2); xw=X(:,3);
49: else, xf=X(:,3); uf=u(:,3); end
50: close, subplot(211)
51: waterfall(X,T,-u), xlabel('x axis')
52: ylabel('time'), zlabel('deflection')
53:
54: title(titl), grid on, hold on
55: % plot3(xf,t,-uf,'.k',xf,t,-uf,'k')
56: plot3(xf,t,-uf,'k','linewidth',2);
57: colormap([0 0 0]), view([-10,30]), shg
58: umin=min(u(:)); umax=max(u(:)); xmax=X(1,4);
59: range=[0,xmax,2*umin,2*umax]; hold on
60: Titl=['A = ',num2str(a),',  V = ',num2str(v),...
61:       ',  T = %4.2f']; subplot(212) , axis off
62:
63: % Use a dense set of points for animation
64: nt=80; [uu,XX,TT,uuf,tt]=ustring(a,v,tmax,nt);
65: umax=max(abs(uu(:))); uu=uu/umax; uuf=uuf/umax;
66: XX=XX/xmax; range=[0,1,-1,1]; h=.4;
67: arx=h*[0,.02,-.02,0,0]; ary=h*[0,.25,.25,0,1];
68: for j=1:nt
69:    uj=uu(j,:); xj=XX(j,:);
70:    xfj=v/xmax*tt(j); ufj=uuf(j);
71:    plot(xj,-uj,'k',xfj+arx,-ufj-ary,'-k')
72:    axis off, time=(sprintf(Titl,tt(j)));
73:    text(.3,-.5,time), axis(range), drawnow
74:    pause(.05), figure(gcf), if j<nt, cla, end
75: end
76: % print -deps forcmove
77: hold off; subplot
78:
79: %==============================================
80:
81: function [u,X,T,uf,t]=ustring(a,v,tmax,nt)
82: %
83: % [u,X,T,uf,t]=ustring(a,v,tmax,nt)
84: % ~~~~~~~~~~~~~~~~~~~~~~~~~~~~~~~~~~
85: % This function computes the deflection u(x,t)
86: % of a semi-infinite string subjected to a
```

```
87: % moving force. The equation for the normalized
88: % deflection is
89: % u(x,t)=1/a/(a^2-v^2)*((v-a-v*abs(x-a*t)...
90: %                          +a*abs(x-v*t));
91: % a    - speed of wave propagation in the string
92: % v    - speed of the force moving to the right
93: % tmax - maximum time for computing the solution
94: % nt   - number of time values
95: % uu   - array of displacement values normalized
96: %        by dividing by a factor equal to the force
97: %        magnitude over twice the density per unit
98: %        length. Position varies column-wise and
99: %        time varies row-wise in the array.
100: % X,T  - position and time arrays for the solution
101: % uf   - deflection vector under the force
102: % t    - time vector for the solution (same as the
103: %        columns of T)
104: %
105: t=linspace(0,tmax,nt)'; xmax=1.05*tmax*max(a,v);
106: u=zeros(nt,4); nx=4; X=zeros(nt,nx); X(:,nx)=xmax;
107: c=1/a/(a^2-v^2); xw=a*t; xf=v*t; T=repmat(t,1,4);
108: uw=c*xw*(v-a+abs(v-a)); uf=c*xf*(v-a-abs(v-a));
109: if a>v, X(:,2)=xf; X(:,3)=xw; u(:,2)=uf;
110: else, X(:,2)=xw; X(:,3)=xf; u(:,2)=uw; end
```

9.5 Waves in Rectangular or Circular Membranes

Wave propagation in two dimensions is illustrated well by the transverse vibration of an elastic membrane. Membrane dynamics is discussed here for general boundary shapes. Then specific solutions are given for rectangular and circular membranes subjected to a harmonically varying surface force. In the next chapter, natural mode vibrations of an elliptical membrane are also discussed. We consider a membrane occupying an area S of the x, y plane bounded by a curve L where the deflection is zero. The differential equation, boundary conditions, and initial conditions governing the transverse deflection $U(x, y, t)$ are

$$\nabla^2 U = c^{-2} U_{tt} - P(x, y, t) \ , \ (x, y) \in \ S,$$

$$U(x, y, 0) = U_0(x, y) \ , \ U_t(x, y, 0) = V_0(x, y) \ , \ (x, y) \in S,$$

$$U(x, y, t) = 0 \ , \ (x, y) \in L.$$

The parameter c is the speed of wave propagation in the membrane and P is the applied normal load per unit area divided by the membrane tension per unit length.

When $P = 0$, the motion is resolvable into a series of normal mode vibrations [22] of the form $u_n(x, y) \sin(\Omega_n t + \epsilon_n)$ satisfying

$$\nabla^2 u_n(x, y) = -\Lambda_n^2 \, u_n(x, y), \quad (x, y) \in S \ , \ u_n(x, y) = 0 \ , \ (x, y) \in L$$

where $\Lambda_n = \Omega_n / c$ is a positive real frequency parameter, and u_n satisfies

$$\iint u_n(x, y) \, u_m(x, y) \, dx dy = C_n \delta_{nm} \ , \ C_n = \iint u_n(x, y)^2 dx dy$$

where δ_{nm} is the Kronecker delta symbol. If the initial displacement and initial velocity are representable by a series of the modal functions, then the homogeneous solution satisfying general initial conditions is

$$U(x, y, t) = \sum_{n=1}^{\infty} u_n(x, y) \left[A_n \cos(\Omega_n t) + B_n \sin(\Omega_n t) / \Omega_n \right]$$

where

$$A_n = \iint U_0(x, y) \, u_n(x, y) \, dx dy / C_n \ , \ B_n = \iint V_0(x, y) \, u_n(x, y) \, dx dy / C_n.$$

The nonhomogeneous case will be treated where the applied normal force on the membrane varies harmonically as

$$P(x, y, t) = p(x, y) \cos(\Omega \, t)$$

and Ω does not match any natural frequency of the membrane. We assume that the membrane is initially at rest with zero deflection and $p(x, y)$ is expandable as

$$p(x, y) = \sum_{n=1}^{\infty} P_n u_n(x, y) \, dx dy \ , \ P_n = \iint p(x, y) \, u_n(x, y) \, dx dy / C_n.$$

Then the forced response solution satisfying zero initial conditions is found to be

$$U(x, y, t) = \sum_{n=1}^{\infty} \frac{P_n}{\Lambda^2 - \Lambda_n^2} \, u_n(x, y) \left[\cos(\Omega \, t) - \cos(\Omega_n t) \right].$$

This equation shows clearly that when the frequency of the forcing function is close to any one of the natural frequencies, then large deflection amplitudes can occur.

Next we turn to specific solutions for rectangular and circular membranes. Consider the normal mode functions for a rectangular region defined by $0 \leq x \leq a$, $0 \leq y \leq b$. It can be shown that the modal functions are

$$u_{nm}(x, y) = \sin(n\pi x / a) \sin(m\pi y / b) \ , \ \Omega_{nm} = c\pi \sqrt{(n/a)^2 + (m/b)^2}$$

and $C_n = ab/4$. In the simple case where the applied surface force is a concentrated load applied at (x_0, y_0), then

$$p(x, y) = p_0 \delta(x - x_0) \, \delta(y - y_0)$$

where δ is the Dirac delta function. The series solution for a forced response solution is found to be

$$U(x,y,t) = c^2 \sum_{n=1}^{\infty} \sum_{m=1}^{\infty} \frac{P_{nm}}{\Omega^2 - \Omega_{nm}^2} \sin(\frac{n\pi x}{a}) \sin(\frac{m\pi y}{b}) \left[\cos(\Omega\, t) - \cos(\Omega_{nm} t)\right]$$

with

$$P_{nm} = \frac{4p_0}{ab} \sin(n\pi x_0/a) \sin(m\pi y_0/b).$$

A similar kind of solution is obtainable as a series of Bessel functions when the membrane is circular. Transforming the wave equation to polar coordinates (r, θ) gives

$$U_{rr} + r^{-1}U_r + r^{-2}U_{\theta\theta} = c^{-2}U_{tt} - P(r,\theta,t)\ ,\ 0 \le r \le a\ ,\ -\pi \le \theta \le \pi\ ,\ t > 0.$$

To reduce the algebraic complexity of the series solution developed below, it is helpful to introduce dimensionless variables $\rho = r/a$ and $\tau = ct/a$. Then the boundary value problem involving a harmonic forcing function becomes

$$U_{\rho\rho} + \rho^{-1}U_\rho + \rho^{-2}U_{\theta\theta} = U_{\tau\tau} - p(\rho,\theta)\sin(\omega\,\tau)\ ,\ 0 \le \rho \le 1\ ,\ -\pi \le \theta \le \pi\ ,\ \tau > 0,$$

$$U(\rho,\theta,0) = 0\ ,\ U_\tau(\rho,\theta,0) = 0,$$

where $\omega = \Omega\, a/c$. The modal functions for this problem are

$$u_{nm}(\rho,\theta) = J_n(\lambda_{nm}\rho)\cos(n\theta + \epsilon_n)$$

involving the integer order Bessel functions, with λ_{nm} being the mth positive root of $J_n(\rho)$. These modal functions satisfy the orthogonality conditions discussed above and we employ the series expansion

$$p(\rho,\theta) = \sum_{n=0}^{\infty} \sum_{m=1}^{\infty} J_n(\lambda_{nm}\rho)\ \mathbf{real}(A_{nm}e^{in\theta})$$

where

$$A_{nm} = \frac{2}{\pi(1+\delta_{n0})\, J_{n+1}^2(\lambda_{nm})} \int_{-\pi}^{\pi} \int_0^1 p(\rho,\theta)\rho\, J_n(\lambda_{nm}\rho) e^{-in\theta} d\rho\, d\theta.$$

Then the forced response solution becomes

$$U(\rho,\theta,\tau) = \sum_{n=0}^{\infty} \sum_{m=1}^{\infty} \frac{J_n(\lambda_{nm}\rho)}{\omega^2 - \lambda_{nm}^2}\ \text{real}(A_{nm}e^{in\theta}) \left[\cos(\omega\tau) - \cos(\lambda_{nm}\tau)\right].$$

In the special case where a concentrated force acts at $\rho = \rho_0$, $\theta = 0$, so that

$$p(\rho,\theta) = p_0\delta(\rho - \rho_0)\delta(\theta),$$

then evaluating the double integral gives

$$A_{nm} = p_0\rho_0 J_n(\lambda nm\rho_0)$$

and $\mathbf{real}(A_{nm}e^{-in\theta})$ simplifies to $A_{nm}\cos(n\theta)$.

9.5.1 Computer Formulation

Program **membwave** was written to depict wave propagation in a rectangular or circular membrane. Input data specifies information on membrane dimensions, forcing function frequency, force position coordinates, wave speed, and maximum time for solution generation. The primary computation tasks involve summing the double series defining the solutions. In the case of the circular membrane, the Bessel function roots determining the natural frequencies must also be computed. The various program modules are listed in the following table.

membwave	reads data, calls other computational modules, and outputs time response
memrecwv	sums the series for dynamic response of a rectangular membrane
memcirwv	calls **besjroot** to obtain the natural frequencies and sums the series for the circular membrane response
besjroot	computes a table of Bessel function roots
membanim	animates the dynamic response of the membrane

9.5.2 Input Data for Program membwave

Listed below are data cases showing animations of both rectangular and circular membranes. Waves propagate outward in a circular pattern from the point of application of the oscillating concentrated load. The membrane response becomes more complex as waves reflect from all parts of the boundary. In order to fully appreciate the propagating wave phenomenon, readers should run the program for several combinations of forcing function frequency and maximum time. The two surface plots below show deflected positions before waves have reached the entire boundary, so some parts of the membrane surface still remain undisturbed.

```
>> membwave;

WAVE MOTION IN A RECTANGULAR OR CIRCULAR
  MEMBRANE HAVING AN OSCILLATING LOAD

Select the geometry type:
Enter 1 for a rectangle, 2 for a circle > ? 1

Specify the rectangle dimensions:
Give values for a,b > ? 2,1

Give coordinates (x0,y0) where the
```

```
force act. Enter x0,y0 > ? 1.5,.5

Enter the wave speed > ? 1

The first forty-two natural frequencies are:
    3.5124    4.4429    5.6636    6.4766    7.0248    7.0248
    7.8540    8.4590    8.8858    9.5548    9.9346    9.9346
   10.0580   10.5372   11.3272   11.3272   11.4356   12.2683
   12.6642   12.6642   12.9531   12.9531   13.3286   13.4209
   14.0496   14.0496   14.4820   14.4820   14.8189   15.4705
   15.7080   15.7080   15.7863   16.0190   16.0190   16.3996
   16.6978   16.9180   16.9180   16.9908   17.5620   17.5620

Input the frequency of the forcing function ? 17.5

Input the maximum solution evaluation time.
 > ? 5

Press return for animation
or enter 0 to stop > ?

Press return for animation
or enter 0 to stop > ? 0

All done

>> membwave;

WAVE MOTION IN A RECTANGULAR OR CIRCULAR
  MEMBRANE HAVING AN OSCILLATING LOAD

Select the geoemtry type:
Enter 1 for a rectangle, 2 for a circle > ? 2

The circle radius equals one. Give the radial
distance r0 from the circle center to the
force > ? .5

Enter the wave speed > ? 1

The first forty-two natural frequencies are:
    2.4048    3.8317    5.1356    5.5201    6.3801    7.0156
    7.5883    8.4173    8.6537    8.7715    9.7611    9.9362
   10.1735   11.0647   11.0864   11.6199   11.7916   12.2251
   12.3385   13.0152   13.3237   13.3543   13.5893   14.3726
```

```
    14.4755    14.7960    14.8213    14.9309    15.5898    15.7002
    16.0378    16.2234    16.4707    16.6983    17.0037    17.2412
    17.6159    17.8014    17.9599    18.0711    18.2876    18.4335

Input the frequency of the forcing function ? 17.5

Input the maximum solution evaluation time.
 > ? 5

Press return for animation
or enter 0 to stop > ?

Press return for animation
or enter 0 to stop > ? 0

All done
>>
```

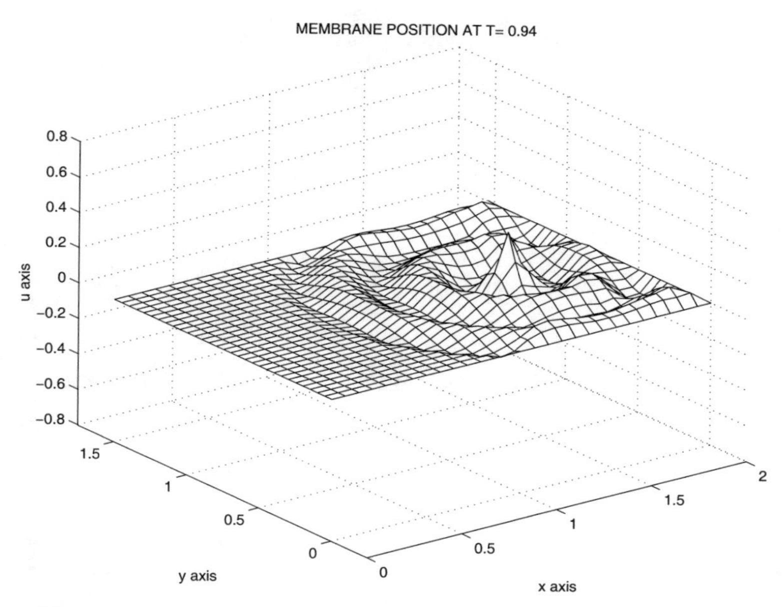

Figure 9.10: **Wave Propagation in a Rectangular Membrane**

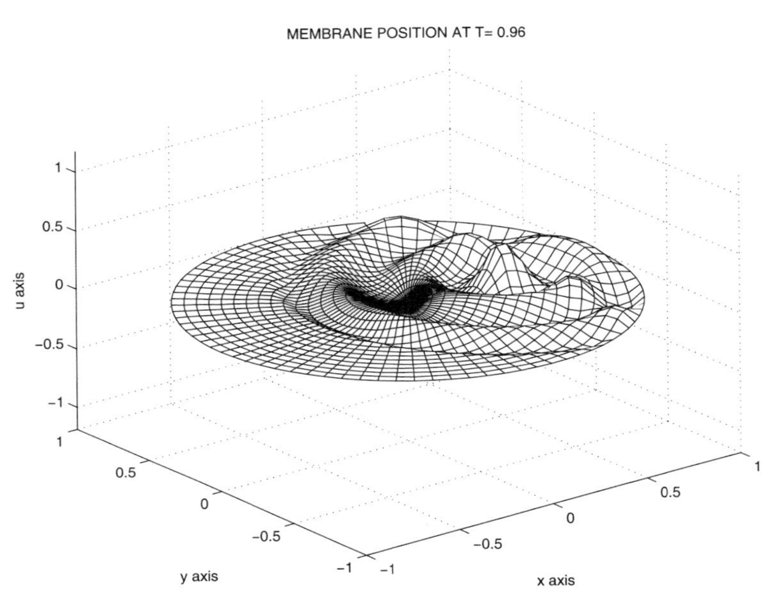

Figure 9.11: Wave Propagation in a Circular Membrane

Program membwave

```
1: function [u,x,y,t]= membwave(type,dims,alp,w,tmax)
2: %
3: % [u,x,y,t]=membwave(type,dims,alp,w,tmax)
4: % ~~~~~~~~~~~~~~~~~~~~~~~~~~~~~~~~~~~~~~~~~~~~~~~~~
5: % This program illustrates waves propagating in
6: % a membrane of rectangular or circular shape
7: % with an oscillatory concentrated load acting at
8: % an arbitrary interior point. The membrane has
9: % fixed edges and is initially undeflected and
10: % at rest. The response u(x,y,t) is computed and
11: % animated plots depicting the motion are shown.
12: %
13: % type    -  1 for rectangle, 2 for circle
14: % dims    -  vector giving problem dimensions. For
15: %            type=1, dims=[a,b,x0,y0] where a and
16: %            b are rectangle dimensions along the
17: %            x and y axes. Also the oscillating
18: %            force acts at (x0,y0). For type=2,
19: %            a circular membrane of unit radius is
20: %            analyzed with the concentrated force
21: %            acting at (r0,0) where r0=dims(1);
22: % alp     -  wave propagation velocity in the
23: %            membrane
24: % w       -  frequency of the applied force. This
25: %            can be zero if the force is constant.
26: % x0,y0   -  coordinates of the point where
27: %            the force acts
28: % x,y,t   -  vectors of position and time values
29: %            for evaluation of the solution
30: % u       -  an array of size [length(x),...
31: %                               length(y),length(t)]
32: %            in which u(i,j,k) contains the
33: %            normalized displacement at
34: %            y(i),x(j),t(k). The displacement is
35: %            normalized by dividing by
36: %            max(abs(u(:)))
37:
38: disp(' ')
39: disp('WAVE MOTION IN A RECTANGULAR OR CIRCULAR')
40: disp('  MEMBRANE HAVING AN OSCILLATING LOAD')
41:
```

```
42: if nargin > 0 % Data passed through the call list
43:   % must specify: type, dims, alp, w, tmax
44:   % Typical values are: a=2; b=1; alp=1;
45:   % w=18.4; x0=1; y0=0.5; tmax=5;
46:   if type==1
47:     a=dims(1); b=dims(2); x0=dims(3); y0=dims(4);
48:     [u,x,y,t]=memrecwv(a,b,alp,w,x0,y0,tmax);
49:   else
50:     r0=dims(1);
51:   end
52: else % Interactive data input
53:
54:   disp(' '), disp('Select the geometry type:')
55:   type=input(['Enter 1 for a rectangle, ',...
56:               '2 for a circle > ? ']);
57:   if type ==1
58:     disp(' ')
59:     disp('Specify the rectangle dimensions:')
60:     s=input('Give values for a,b > ? ','s');
61:     s=eval(['[',s,']']); a=s(1); b=s(2);
62:     disp(' ')
63:     disp('Give coordinates (x0,y0) where the')
64:     s=input('force acts. Enter x0,y0 > ? ','s');
65:     s=eval(['[',s,']']); x0=s(1); y0=s(2);
66:     disp(' '), alp=input('Enter the wave speed > ? ');
67:
68:     N=40; M=40; pan=pi/a*(1:N)'; pbm=pi/b*(1:M);
69:     W=alp*sqrt(repmat(pan.^2,1,M)+repmat(pbm.^2,N,1));
70:     wsort=sort(W(:)); wsort=reshape(wsort(1:42),6,7)';
71:     disp(' ')
72:     disp(['The first forty-two natural ',...
73:           'frequencies are:'])
74:     disp(wsort)
75:     w=input(...
76:     'Input the frequency of the forcing function ? ');
77:
78:   else
79:     disp(' '), disp(...
80:     'The circle radius equals one. Give the radial')
81:     disp(...
82:     'distance r0 from the circle center to the')
83:     r0=input('force > ? ');
84:
85:     disp(' '), alp=input('Enter the wave speed > ? ');
86:
```

```
87:      % First 42 Bessel function roots
88:      wsort=alp*[...
89:        2.4048  3.8317  5.1356  5.5201  6.3801  7.0156
90:        7.5883  8.4173  8.6537  8.7715  9.7611  9.9362
91:       10.1735 11.0647 11.0864 11.6199 11.7916 12.2251
92:       12.3385 13.0152 13.3237 13.3543 13.5893 14.3726
93:       14.4755 14.7960 14.8213 14.9309 15.5898 15.7002
94:       16.0378 16.2234 16.4707 16.6983 17.0037 17.2412
95:       17.6159 17.8014 17.9599 18.0711 18.2876 18.4335];
96:
97:      disp(' '), disp(['The first forty-two ',...
98:                      'natural frequencies are:'])
99:      disp(wsort)
100:     w=input(...
101:     'Input the frequency of the forcing function ? ');
102:   end
103:    disp(' ')
104:    disp('Input the maximum solution evaluation time.')
105:    tmax=input(' > ? ');
106: end
107:
108: if type==1
109:   [u,x,y,t]=memrecwv(a,b,alp,w,x0,y0,tmax);
110: else
111:   th=linspace(0,2*pi,81); r=linspace(0,1,20);
112:   [u,x,y,t]=memcirwv(r,th,r0,alp,w,tmax);
113: end
114:
115: % Animate the solution
116: membanim(u,x,y,t);
117:
118: %=================================================
119:
120: function [u,x,y,t]= memrecwv(a,b,alp,w,x0,y0,tmax)
121: %
122: % [u,x,y,t]=memrecwv(a,b,alp,w,x0,y0,tmax)
123: % ~~~~~~~~~~~~~~~~~~~~~~~~~~~~~~~~~~~~~~~~~
124: % This function illustrates wave motion in a
125: % rectangular membrane subjected to a concentrated
126: % oscillatory force applied at an arbitrary
127: % interior point. The membrane has fixed edges
128: % and is initially at rest in an undeflected
129: % position. The resulting response u(x,y,t)is
130: % computed and a plot of the motion is shown.
131: % a,b     -  side dimensions of the rectangle
```

```
132: % alp      -  wave propagation velocity in the
133: %             membrane
134: % w        -  frequency of the applied force. This
135: %             can be zero if the force is constant.
136: % x0,y0    -  coordinates of the point where
137: %             the force acts
138: % x,y,t    -  vectors of position and time values
139: %             for evaluation of the solution
140: % u        -  an array of size [length(y),...
141: %             length(x),length(t)] in which u(i,j,k)
142: %             contains the normalized displacement
143: %             corresponding to y(i), x(j), t(k). The
144: %             displacement is normalized by dividing
145: %             by max(abs(u(:))).
146: %
147: % The solution is a double Fourier series of form
148: %
149: % u(x,y,t)=Sum(A(n,m,x,y,t), n=1..N, m=1..M)
150: % where
151: % A(n,m,x,y,t)=sin(n*pi*x0/a)*sin(n*pi*x/a)*...
152: %               sin(m*pi*y0/b)*sin(m*pi*y/b)*...
153: %               (cos(w*t)-cos(W(n,m)*t))/...
154: %               ( w^2-W(n,m)^2)
155: % and the membrane natural frequencies are
156: % W(n,m)=pi*alp*sqrt((n/a)^2+(m/b)^2)
157: 
158: if nargin==0
159:   a=2; b=1; alp=1; tmax=3; w=13; x0=1.5; y0=0.5;
160: end
161: if a<b
162:    nx=31; ny=round(b/a*21); ny=ny+rem(ny+1,2);
163: else
164:    ny=31; nx=round(a/b*21); nx=nx+rem(nx+1,2);
165: end
166: x=linspace(0,a,nx); y=linspace(0,b,ny);
167: 
168: N=40; M=40; pan=pi/a*(1:N)'; pbm=pi/b*(1:M);
169: W=alp*sqrt(repmat(pan.^2,1,M)+repmat(pbm.^2,N,1));
170: wsort=sort(W(:)); wsort=reshape(wsort(1:30),5,6)';
171: Nt=ceil(40*tmax*alp/min(a,b));
172: t=tmax/(Nt-1)*(0:Nt-1);
173: 
174: % Evaluate fixed terms in the series solution
175: mat=sin(x0*pan)*sin(y0*pbm)./(w^2-W.^2);
176: sxn=sin(x(:)*pan'); smy=sin(pbm'*y(:)');
```

```
177:
178: u=zeros(ny,nx,Nt);
179: for j=1:Nt
180:   A=mat.*(cos(w*t(j))-cos(W*t(j)));
181:   uj=sxn*(A*smy); u(:,:,j)=uj';
182: end
183:
184: %==============================================
185:
186: function [u,x,y,t,r,th]=memcirwv(r,th,r0,alp,w,tmax)
187: %
188: % [u,x,y,t,r,th]=memcirwv(r,th,r0,alp,w,tmax)
189: % ~~~~~~~~~~~~~~~~~~~~~~~~~~~~~~~~~~~~~~~~~~~~
190: % This function computes the wave response in a
191: % circular membrane having an oscillating force
192: % applied at a point on the radius along the
193: % positive x axis.
194: %
195: % r,th - vectors of radius and polar angle values
196: % r0   - radial position of the concentrated force
197: % w    - frequency of the applied force
198: % tmax - maximum time for computing the solution
199: %
200: % User m function used: besjroot
201:
202: if nargin==0
203:   r0=.4; w=15.5; th=linspace(0,2*pi,81);
204:   r=linspace(0,1,21); alp=1;
205: end
206:
207: Nt=ceil(20*alp*tmax);  t=tmax/(Nt-1)*(0:Nt-1);
208:
209: % Compute the Bessel function roots needed in
210: % the series solution. This takes a while.
211: lam=besjroot(0:20,20,1e-3);
212:
213: % Compute the series coefficients
214: [nj,nk]=size(lam); r=r(:)'; nr=length(r);
215: th=th(:); nth=length(th); nt=length(t);
216: N=repmat((0:nj-1)',1,nk); Nvec=N(:)';
217: c=besselj(N,lam*r0)./(besselj(...
218:    N+1,lam).^2.*(lam.^2-w^2));
219: c(1,:)=c(1,:)/2; c=c(:)';
220:
221: % Sum the series of Bessel functions
```

```
222: lamvec=lam(:)'; wlam=w./lamvec;
223: c=cos(th*Nvec).*repmat(c,nth,1);
224: rmat=besselj(repmat(Nvec',1,nr),lamvec'*r);
225: u=zeros(nth,nr,nt);
226: for k=1:nt
227:   tvec=-cos(w*t(k))+cos(lamvec*t(k));
228:   u(:,:,k)=c.*repmat(tvec,nth,1)*rmat;
229: end
230: u=2/pi*u; x=cos(th)*r; y=sin(th)*r;
231:
232: %===============================================
233:
234: function rts=besjroot(norder,nrts,tol)
235: %
236: % rts=besjroot(norder,nrts,tol)
237: % ~~~~~~~~~~~~~~~~~~~~~~~~~~~~~~
238: % This function computes an array of positive roots
239: % of the integer order Bessel functions besselj of
240: % the first kind for various orders. A chosen number
241: % of roots is computed for each order
242: % norder - a vector of function orders for which
243: %          roots are to be computed. Taking 3:5
244: %          for norder would use orders 3,4 and 5.
245: % nrts   - the number of positive roots computed for
246: %          each order. Roots at x=0 are ignored.
247: % rts    - an array of roots having length(norder)
248: %          rows and nrts columns. The element in
249: %          column k and row i is the k'th root of
250: %          the function besselj(norder(i),x).
251: % tol    - error tolerance for root computation.
252:
253: if nargin<3, tol=1e-5; end
254: jn=inline('besselj(n,x)','x','n');
255: N=length(norder); rts=ones(N,nrts)*nan;
256: opt=optimset('TolFun',tol,'TolX',tol);
257: for k=1:N
258:    n=norder(k); xmax=1.25*pi*(nrts-1/4+n/2);
259:    xsrch=.1:pi/4:xmax; fb=besselj(n,xsrch);
260:    nf=length(fb); K=find(fb(1:nf-1).*fb(2:nf)<=0);
261:    if length(K)<nrts
262:       disp('Search error in function besjroot')
263:       rts=nan; return
264:    else
265:       K=K(1:nrts);
266:       for i=1:nrts
```

```
267:          interval=xsrch(K(i):K(i)+1);
268:          rts(k,i)=fzero(jn,interval,opt,n);
269:       end
270:    end
271: end
272:
273: %===================================================
274:
275: function membanim(u,x,y,t)
276: %
277: % function membanim(u,x,y,t)
278: % ~~~~~~~~~~~~~~~~~~~~~~~~~~
279: % This function animates the motion of a
280: % vibrating membrane
281: %
282: % u    array in which component u(i,j,k) is the
283: %      displacement for y(i),x(j),t(k)
284: % x,y  arrays of x and y coordinates
285: % t    vector of time values
286:
287: % Compute the plot range
288: if nargin==0;
289:   [u,x,y,t]=memrecwv(2,1,1,15.5,1.5,.5,5);
290: end
291: xmin=min(x(:)); xmax=max(x(:));
292: ymin=min(y(:)); ymax=max(y(:));
293: xmid=(xmin+xmax)/2; ymid=(ymin+ymax)/2;
294: d=max(xmax-xmin,ymax-ymin)/2; Nt=length(t);
295: range=[xmid-d,xmid+d,ymid-d,ymid+d,...
296:        3*min(u(:)),3*max(u(:))];
297:
298: while 1 % Show the animation repeatedly
299:   disp(' '), disp('Press return for animation')
300:   dumy=input('or enter 0 to stop > ? ','s');
301:   if ~isempty(dumy)
302:     disp(' '), disp('All done'), break
303:   end
304:
305:   % Plot positions for successive times
306:   for j=1:Nt
307:     surf(x,y,u(:,:,j)), axis(range)
308:     xlabel('x axis'), ylabel('y axis')
309:     zlabel('u axis'), titl=sprintf(...
310:     'MEMBRANE POSITION AT T=%5.2f',t(j));
311:     title(titl), colormap([1 1 1])
```

```
312:        colormap([127/255 1 212/255])
313:        % axis off
314:        drawnow, shg, pause(.1)
315:     end
316: end
```

9.6 Wave Propagation in a Beam with an Impact Moment Applied to One End

Analyzing the dynamic response caused when a time dependent moment acts on the end of an Euler beam involves a boundary value problem for a fourth order linear partial differential equation. In the following example we consider a beam of uniform cross section which is pin-ended (hinged at the ends) and is initially at rest. Suddenly, a harmonically varying moment $M_0 \cos(\Omega_0 T)$ is applied to the right end as shown in Figure 9.12. Determination of the resulting displacement and bending moment in the beam is desired. Let U be the transverse displacement, X the longitudinal distance

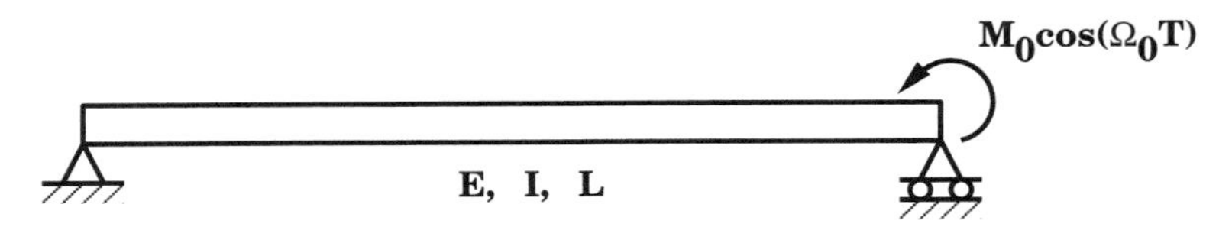

Figure 9.12: Beam Geometry and Loading

from the right end, and T the time. The differential equation, boundary conditions, and initial conditions characterizing the problem are

$$EI\frac{\partial^4 U}{\partial X^4} = -A\rho\frac{\partial^2 U}{\partial T^2}\ ,\ 0 < X < L\ ,\ T > 0,$$

$$U(0,T) = 0\ ,\ \frac{\partial^2 U}{\partial X^2}(0,T) = 0\ ,\ U(L,T) = 0\ ,\ \frac{\partial^2 U}{\partial X^2}(L,T) = M_0\cos(\Omega_0 T)/(EI),$$

$$U(0,T) = 0\ ,\ \frac{\partial U}{\partial T}(0,T) = 0,$$

where L is the beam length, EI is the product of the elastic modulus and the moment of inertia, and $A\rho$ is the product of the cross section area and the mass density.

This problem can be represented more conveniently by introducing dimensionless variables

$$x = \frac{X}{L}\ ,\ t = \sqrt{\frac{EI}{A\rho}\frac{T}{L^2}}\ ,\ u = \frac{EI}{M_0 L^2}U\ ,\ \omega = \sqrt{\frac{A\rho}{EI}}L^2\Omega_0\ ,\ m = \frac{\partial^2 u}{\partial x^2}.$$

The new boundary value problem is then

$$\frac{\partial^4 u}{\partial x^4} = -\frac{\partial^2 u}{\partial t^2} \,,\ 0 < x < 1 \,,\ t > 0,$$

$$u(0,t) = 0 \,,\ \frac{\partial^2 u}{\partial x^2}(0,t) = 0 \,,\ u(1,t) = 0 \,,\ \frac{\partial^2 u}{\partial x^2}(1,t) = \cos(\omega t),$$
$$u(x,0) = 0 \,,\ \frac{\partial u}{\partial t}(x,0) = 0 \,,\ 0 < x < 1.$$

The problem can be solved by combining a particular solution w which satisfies the differential equation and nonhomogeneous boundary conditions with a homogeneous solution in series form which satisfies the differential equation and homogeneous boundary conditions. Thus we have $u = w + v$. The particular solution can be found in the form

$$w = f(x)\cos(\omega t)$$

where $f(x)$ satisfies

$$f''''(x) = \omega^2 f(x)$$

and

$$f(0) = f''(0) = f(1) = 0 \,,\ f''(1) = 1.$$

This ordinary differential equation is solvable as

$$f(x) = \sum_{k=1}^{4} c_k e^{s_k x}$$

where

$$s_k = \sqrt{\omega}\, e^{\pi\imath(k-1)/2}$$

and $\imath = \sqrt{-1}$. The boundary conditions require

$$\sum_{k=1}^{4} c_k = 0 \,,\ \sum_{k=1}^{4} s_k^2 c_k = 1 \,,\ \sum_{k=1}^{4} c_k\, e^{s_k} = 0 \,,\ \sum_{k=1}^{4} c_k s_k^2 e^{s_k} = 0.$$

Solving these simultaneous equations determines the particular solution. The initial displacement for the particular solution can be expanded in a Fourier series as an odd valued function of period 2. Hence we can write

$$w(0,t) = f(x) = \sum_{k=-\infty}^{\infty} c_k\, e^{\imath\pi k x} = \sum_{k=1}^{\infty} a_k \sin(k\pi x) \,,\ \frac{\partial w}{\partial t}(0,t) = 0$$

involving complex Fourier coefficients, c_k, and $a_k = -2\,\mathbf{imag}(c_k)$. The homogeneous solution is representable as

$$v(x,t) = -\sum_{k=1}^{\infty} a_k \cos(\pi^2 k^2 t)\sin(k\pi x)$$

so that $w + v$ combine to satisfy the desired initial conditions of zero displacement and velocity.

Of course, perfect satisfaction of the initial conditions cannot be achieved without taking an infinite number of terms in the Fourier series. However, the series converges very rapidly because the coefficients are of order n^{-3}. When a hundred or more terms are used, an approximate solution produces results which satisfy the differential equation and boundary conditions, and which insignificantly violate the initial displacement condition. It is important to remember the nature of this error when examining the bending moment results presented below. Effects of high frequency components are very evident in the moment. Despite the oscillatory character of the moments, these results are exact for the initial displacement conditions produced by the truncated series. These displacements agree closely with the exact solution.

A program was written to evaluate the series solution to compute displacements and moments as functions of position and time. Plots and surfaces showing these quantities are presented along with timewise animations of the displacement and moment across the span. The computation involves the following steps:

1. Evaluate $f(x)$;
2. Expand $f(x)$ using the FFT to get coefficients for the homogeneous series solution;
3. Combine the particular and homogeneous solution by summing the series for any number of terms desired by the user;
4. Plot u and m for selected times;
5. Plot surfaces showing $u(x,t)$ and $m(x,t)$;
6. Show animated plots of u and m.

The principal parts of the program are shown in the table below.

bemimpac	reads data and creates graphical output
beamresp	converts material property data to dimensionless form and calls **ndbemrsp**
ndbemrsp	construct the solution using Fourier series
sumser	sums the series for displacement and moment
animate	animates the time history of displacement and moment

The numerical results show the response for a beam subjected to a moment close to the first natural frequency of the beam. It can be shown that, in the dimensionless problem, the system of equations defining the particular solution becomes singular

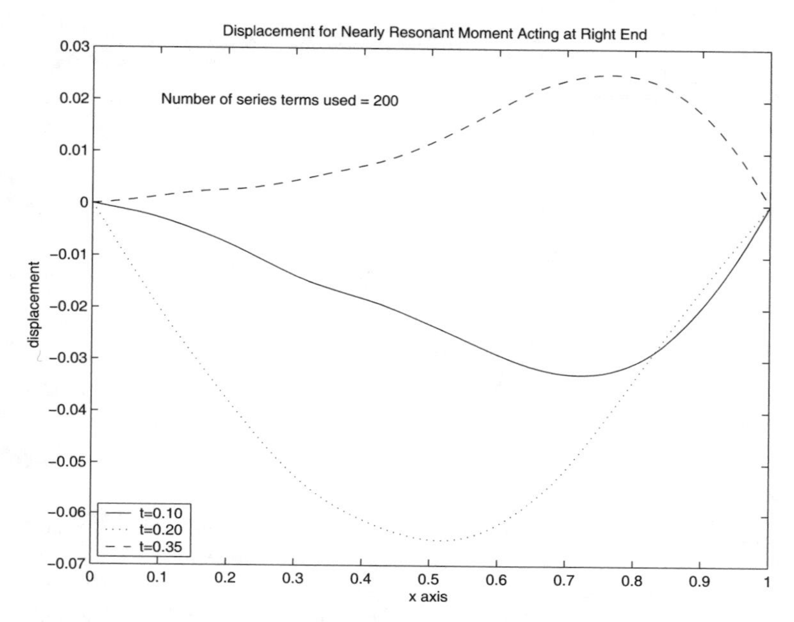

Figure 9.13: **Displacement Due to Impact Moment at Right End**

when ω assumes values of the form $k^2\pi^2$ for integer k. In that instance the series solution provided here will fail. However, values of ω near to resonance can be used to show how the displacements and moments quickly become large. In our example we let EI, $A\rho$, l, and M_0 all equal unity, and $\omega = 0.95\pi^2$. Figures 9.13 and 9.14 show displacement and bending moment patterns shortly after motion is initiated. The surfaces in Figures 9.15 and 9.16 also show how the displacement and moment grow quickly with increasing time. The reader may find it interesting to run the program for various choices of ω and observe how dramatically the chosen forcing frequency affects results.

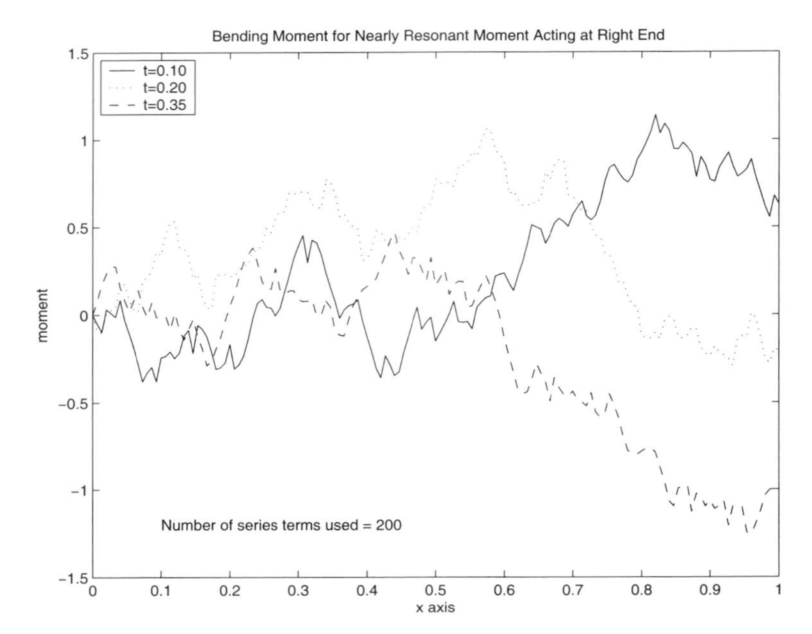

Figure 9.14: Bending Moment in the Beam

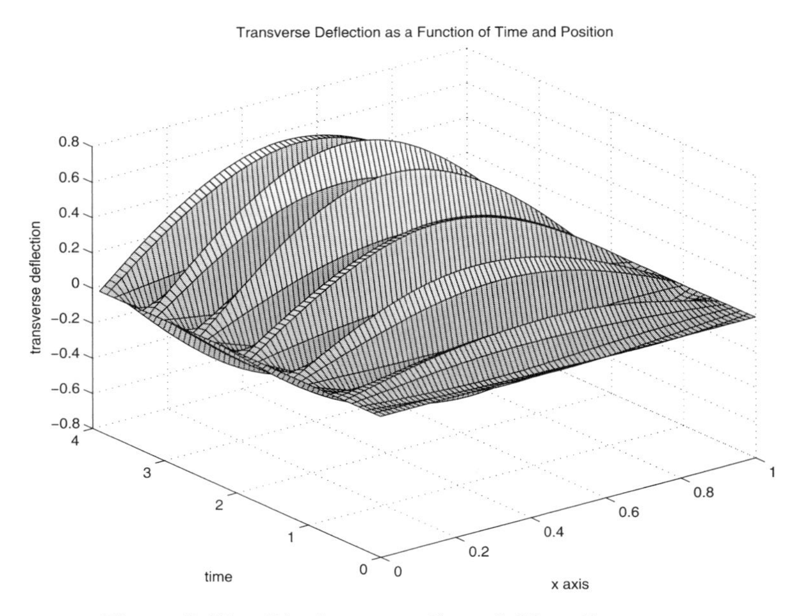

Figure 9.15: Displacement Growth Near Resonance

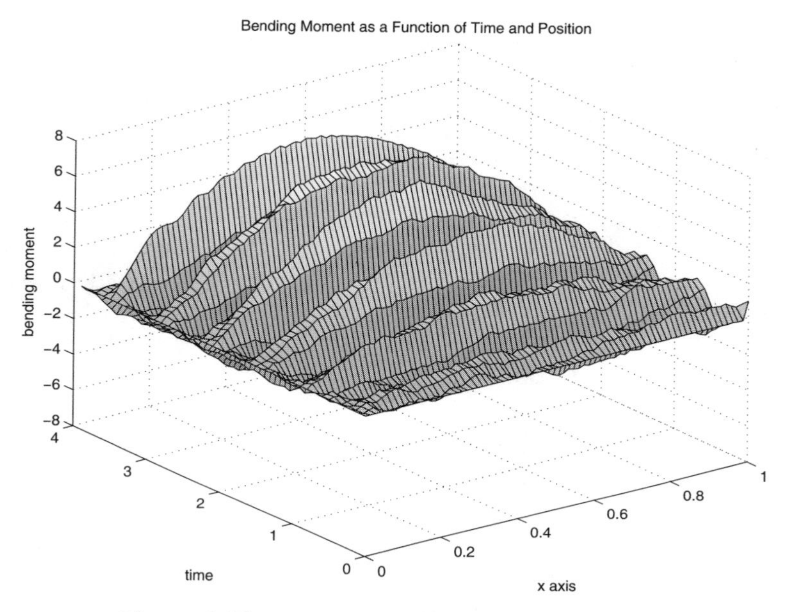

Figure 9.16: **Moment Growth Near Resonance**

MATLAB Example

Program bemimpac

```
 1: function bemimpac
 2: % Example: bemimpac
 3: % ~~~~~~~~~~~~~~~~~
 4: % This program analyzes an impact dynamics
 5: % problem for an elastic Euler beam of
 6: % constant cross section which is simply
 7: % supported at each end. The beam is initially
 8: % at rest when a harmonically varying moment
 9: % m0*cos(w0*t) is applied to the right end.
10: % The resulting transverse displacement and
11: % bending moment are computed.  The
12: % displacement and moment are plotted as
13: % functions of x for the three time values.
14: % Animated plots of the entire displacement
15: % and moment history are also given.
16: %
17: % User m functions required:
```

```
18: %     beamresp, beamanim, sumser, ndbemrsp
19:
20: fprintf('\nDYNAMICS OF A BEAM WITH AN ');
21: fprintf('OSCILLATING END MOMENT\n');
22: ei=1; arho=1; len=1; m0=1; w0=.90*pi^2;
23: tmin=0; tmax=5; nt=101;
24: xmin=0; xmax=len; nx=151; ntrms=200;
25: [t,x,displ,mom]=beamresp(ei,arho,len,m0,w0,...
26:                 tmin,tmax,nt,xmin,xmax,nx,ntrms);
27: disp(' ')
28: disp('Press [Enter] to see the deflection')
29: disp('for three positions'), pause
30:
31: np=[3 5 8]; clf; pltsave=0;
32: dip=displ(np,:); mop=mom(np,:);
33: plot(x,dip(1,:),'-k',x,dip(2,:),':b',...
34:      x,dip(3,:),'--r');
35: xlabel('x axis'); ylabel('displacement');
36: hh=gca;
37: r(1:2)=get(hh,'XLim'); r(3:4)=get(hh,'YLim');
38: xp=r(1)+(r(2)-r(1))/10;
39: dp=r(4)-(r(4)-r(3))/10;
40: tstr=['Displacement for Nearly Resonant' ...
41:       ' Moment Acting at Right End'];
42: title(tstr);
43: text(xp,dp,['Number of series terms ' ...
44:             'used = ',int2str(ntrms)]);
45: legend('t=0.10','t=0.20','t=0.35',3)
46: disp(' ')
47: disp('Press [Enter] to the bending moment')
48: disp('for three positions')
49: shg; pause
50: if pltsave, print -deps 3positns, end
51:
52: clf;
53: plot(x,mop(1,:),'-k',x,mop(2,:),':b',...
54:      x,mop(3,:),'--r');
55: h=gca;
56: r(1:2)=get(h,'XLim'); r(3:4)=get(h,'YLim');
57: mp=r(3)+(r(4)-r(3))/10;
58: xlabel('x axis'); ylabel('moment');
59: tstr=['Bending Moment for Nearly Resonant' ...
60:       ' Moment Acting at Right End'];
61: title(tstr);
62: text(xp,mp,['Number of series terms ' ...
```

```
63:                'used = ',int2str(ntrms)]);
64: legend('t=0.10','t=0.20','t=0.35',2),
65: disp(' '), disp(...
66: 'Press [Enter] to see the deflections surface')
67: shg, pause
68: if pltsave, print -deps 3moments, end
69:
70: inct=2; incx=2;
71: ht=0.75; it=1:inct:.8*nt; ix=1:incx:nx;
72: tt=t(it); xx=x(ix);
73: dd=displ(it,ix); mm=mom(it,ix);
74: a=surf(xx,tt,dd);
75: tstr=['Transverse Deflection as a ' ...
76:       'Function of Time and Position'];
77: title(tstr);
78: xlabel('x axis'); ylabel('time');
79: zlabel('transverse deflection');
80: disp(' '), disp(['Press [Enter] to ',...
81: 'see the bending moment surface'])
82: shg, pause
83: if pltsave, print -deps bdeflsrf, end
84:
85: a=surf(xx,tt,mm);
86: title(['Bending Moment as a Function ' ...
87:        'of Time and Position'])
88: xlabel('x axis'); ylabel('time');
89: zlabel('bending moment'); disp(' ')
90: disp('Press [Enter] to see animation of');
91: disp('the beam deflection'), shg, pause
92: if pltsave, print -deps bmomsrf, end
93: beamanim(x,displ,.1,'Transverse Deflection', ...
94:          'x axis','deflection'), disp(' ')
95: disp('Press [Enter] to see animation');
96: disp('of the bending moment'); pause
97: beamanim(x,mom,.1,'Bending Moment History', ...
98:          'x axis','moment');
99: fprintf('\nAll Done\n'); close;
100:
101: %=========================================
102:
103: function [t,x,displ,mom]= ...
104:       beamresp(ei,arho,len,m0,w0,tmin,tmax, ...
105:                nt,xmin,xmax,nx,ntrms)
106: %
107: % [t,x,displ,mom]=beamresp(ei,arho,len,m0, ...
```

```
108: %            w0,tmin,tmax,nt,xmin,xmax,nx,ntrms)
109: % ~~~~~~~~~~~~~~~~~~~~~~~~~~~~~~~~~~~~~~~~~~~~~~~~~~~~
110: % This function evaluates the time dependent
111: % displacement and moment in a constant
112: % cross section, simply supported beam which
113: % is initially at rest when a harmonically
114: % varying moment is suddenly applied at the
115: % right end.  The resulting time histories of
116: % displacement and moment are computed.
117: %
118: % ei        - modulus of elasticity times
119: %             moment of inertia
120: % arho      - mass per unit length of the
121: %             beam
122: % len       - beam length
123: % m0,w0     - amplitude and frequency of the
124: %             harmonically varying right end
125: %             moment
126: % tmin,tmax - minimum and maximum times for
127: %             the solution
128: % nt        - number of evenly spaced
129: %             solution times
130: % xmin,xmax - minimum and maximum position
131: %             coordinates for the solution.
132: %             These values should lie between
133: %             zero and len (x=0 and x=len at
134: %             the left and right ends).
135: % nx        - number of evenly spaced solution
136: %             positions
137: % ntrms     - number of terms used in the
138: %             Fourier sine series
139: % t         - vector of nt equally spaced time
140: %             values varying from tmin to tmax
141: % x         - vector of nx equally spaced
142: %             position values varying from
143: %             xmin to xmax
144: % displ     - matrix of transverse
145: %             displacements with time varying
146: %             from row to row, and position
147: %             varying from column to column
148: % mom       - matrix of bending moments with
149: %             time varying from row to row,
150: %             and position varying from column
151: %             to column
152: %
```

```
153: % User m functions called:  ndbemrsp
154: %----------------------------------------------
155:
156: tcof=sqrt(arho/ei)*len^2; dcof=m0*len^2/ei;
157: tmin=tmin/tcof; tmax=tmax/tcof; w=w0*tcof;
158: xmin=xmin/len; xmax=xmax/len;
159: [t,x,displ,mom]=...
160: ndbemrsp(w,tmin,tmax,nt,xmin,xmax,nx,ntrms);
161: t=t*tcof; x=x*len;
162: displ=displ*dcof; mom=mom*m0;
163:
164: %==============================================
165:
166: function beamanim(x,u,tpause,titl,xlabl,ylabl)
167: %
168: % beamanim(x,u,tpause,titl,xlabl,ylabl,save)
169: % ~~~~~~~~~~~~~~~~~~~~~~~~~~~~~~~~~~~~~~~~~~~~
170: % This function draws an animated plot of data
171: % values stored in array u.  The different
172: % columns of u correspond to position values
173: % in vector x.  The successive rows of u
174: % correspond to different times. Parameter
175: % tpause controls the speed of animation.
176: %
177: %  u      - matrix of values to animate plots
178: %           of u versus x
179: %  x      - spatial positions for different
180: %           columns of u
181: %  tpause - clock seconds between output of
182: %           frames. The default is .1 secs
183: %           when tpause is left out. When
184: %           tpause=0, a new frame appears
185: %           when the user presses any key.
186: %  titl   - graph title
187: %  xlabl  - label for horizontal axis
188: %  ylabl  - label for vertical axis
189: %
190: % User m functions called:  none
191: %----------------------------------------------
192:
193: if nargin<6, ylabl=''; end;
194: if nargin<5, xlabl=''; end
195: if nargin<4, titl=''; end;
196: if nargin<3, tpause=.1; end;
197:
```

```
198: [ntime,nxpts]=size(u);
199: umin=min(u(:)); umax=max(u(:));
200: udif=umax-umin; uavg=.5*(umin+umax);
201: xmin=min(x); xmax=max(x);
202: xdif=xmax-xmin; xavg=.5*(xmin+xmax);
203: xwmin=xavg-.55*xdif; xwmax=xavg+.55*xdif;
204: uwmin=uavg-.55*udif; uwmax=uavg+.55*udif; clf;
205: axis([xwmin,xwmax,uwmin,uwmax]); title(titl);
206: xlabel(xlabl); ylabel(ylabl); hold on;
207:
208: for j=1:ntime
209:   ut=u(j,:);
210:   plot(x,ut,'-'); axis('off'); figure(gcf);
211:   if tpause==0
212:     pause;
213:   else
214:     pause(tpause);
215:   end
216:   if j==ntime, break, else, cla; end
217: end
218: % print -deps cntltrac
219: hold off; clf;
220:
221: %==============================================
222:
223: function [u,t,x] = sumser(a,b,c,funt,funx, ...
224:                    tmin,tmax,nt,xmin,xmax,nx)
225: %
226: % [u,t,x] = sumser(a,b,c,funt,funx,tmin, ...
227: %                  tmax,nt,xmin,xmax,nx)
228: % ~~~~~~~~~~~~~~~~~~~~~~~~~~~~~~~~~~~~~~~~~~~
229: % This function evaluates a function U(t,x)
230: % which is defined by a finite series. The
231: % series is evaluated for t and x values taken
232: % on a rectangular grid network. The matrix u
233: % has elements specified by the following
234: % series summation:
235: %
236: % u(i,j)  =   sum(  a(k)*funt(t(i)*b(k))*...
237: %            k=1:nsum
238: %                          funx(c(k)*x(j))
239: %
240: % where nsum is the length of each of the
241: % vectors a, b, and c.
242: %
```

```
243: % a,b,c            - vectors of coefficients in
244: %                    the series
245: % funt,funx        - handles of functions accepting
246: %                    matrix argument.  funt is
247: %                    evaluated for an argument of
248: %                    the form funt(t*b) where t is
249: %                    a column and b is a row. funx
250: %                    is evaluated for an argument
251: %                    of the form funx(c*x) where
252: %                    c is a column and x is a row.
253: % tmin,tmax,nt - produces vector t with nt
254: %                    evenly spaced values between
255: %                    tmin and tmax
256: % xmin,xmax,nx - produces vector x with nx
257: %                    evenly spaced values between
258: %                    xmin and xmax
259: % u              - the nt by nx matrix
260: %                    containing values of the
261: %                    series evaluated at t(i),x(j),
262: %                    for i=1:nt and j=1:nx
263: % t,x            - column vectors containing t
264: %                    and x values. These output
265: %                    values are optional.
266: %
267: % User m functions called:  none.
268: %----------------------------------------------
269:
270: tt=(tmin:(tmax-tmin)/(nt-1):tmax)';
271: xx=(xmin:(xmax-xmin)/(nx-1):xmax); a=a(:).';
272: u=a(ones(nt,1),:).*feval(funt,tt*b(:).')*...
273:   feval(funx,c(:)*xx);
274: if nargout>1, t=tt; x=xx'; end
275:
276: %==============================================
277:
278: function [t,x,displ,mom]= ...
279:      ndbemrsp(w,tmin,tmax,nt,xmin,xmax,nx,ntrms)
280: %
281: % [t,x,displ,mom]=ndbemrsp(w,tmin,tmax,nt,...
282: %                            xmin,xmax,nx,ntrms)
283: % ~~~~~~~~~~~~~~~~~~~~~~~~~~~~~~~~~~~~~~~~~~~~
284: % This function evaluates the nondimensional
285: % displacement and moment in a constant
286: % cross section, simply supported beam which
287: % is initially at rest when a harmonically
```

```
288: % varying moment of frequency w is suddenly
289: % applied at the right end. The resulting
290: % time history is computed.
291: %
292: % w            - frequency of the harmonically
293: %                varying end moment
294: % tmin,tmax    - minimum and maximum
295: %                dimensionless times
296: % nt           - number of evenly spaced
297: %                solution times
298: % xmin,xmax    - minimum and maximum
299: %                dimensionless position
300: %                coordinates. These values
301: %                should lie between zero and
302: %                one (x=0 and x=1 give the
303: %                left and right ends).
304: % nx           - number of evenly spaced
305: %                solution positions
306: % ntrms        - number of terms used in the
307: %                Fourier sine series
308: % t            - vector of nt equally spaced
309: %                time values varying from
310: %                tmin to tmax
311: % x            - vector of nx equally spaced
312: %                position values varying
313: %                from xmin to xmax
314: % displ        - matrix of dimensionless
315: %                displacements with time
316: %                varying from row to row,
317: %                and position varying from
318: %                column to column
319: % mom          - matrix of dimensionless
320: %                bending moments with time
321: %                varying from row to row, and
322: %                position varying from column
323: %                to column
324: %
325: % User m functions called:  sumser
326: %----------------------------------------------
327:
328: if nargin < 8, w=0; end; nft=512; nh=nft/2;
329: xft=1/nh*(0:nh)';
330: x=xmin+(xmax-xmin)/(nx-1)*(0:nx-1)';
331: t=tmin+(tmax-tmin)/(nt-1)*(0:nt-1)';
332: cwt=cos(w*t);
```

```
333:
334: % Get particular solution for nonhomogeneous
335: % end condition
336: if w ==0 % Case for a constant end moment
337:   cp=[1 0 0 0; 0 0 2 0; 1 1 1 1; 0 0 2 6]\ ...
338:      [0;0;0;1];
339:   yp=[ones(size(x)), x, x.^2, x.^3]*cp; yp=yp';
340:   mp=[zeros(nx,2), 2*ones(nx,1), 6*x]*cp;
341:   mp=mp';
342:   ypft=[ones(size(xft)), xft, xft.^2, xft.^3]*cp;
343:
344: % Case where end moment oscillates
345: % with frequency w
346: else
347:   s=sqrt(w)*[1, i, -1, -i]; es=exp(s);
348:   cp=[ones(1,4); s.^2; es; es.*s.^2]\ ...
349:      [0; 0; 0; 1];
350:   yp=real(exp(x*s)*cp); yp=yp';
351:   mp=real(exp(x*s)*(cp.*s(:).^2)); mp=mp';
352:   ypft=real(exp(xft*s)*cp);
353: end
354:
355: % Fourier coefficients for
356: % particular solution
357: yft=-fft([ypft;-ypft(nh:-1:2)])/nft;
358:
359: % Sine series coefficients for
360: % homogeneous solution
361: acof=-2*imag(yft(2:ntrms+1));
362: ccof=pi*(1:ntrms)'; bcof=ccof.^2;
363:
364: % Sum series to evaluate Fourier
365: % series part of solution. Then combine
366: % with the particular solution.
367: displ=sumser(acof,bcof,ccof,@cos,@sin,...
368:                 tmin,tmax,nt,xmin,xmax,nx);
369: displ=displ+cwt*yp; acof=acof.*bcof;
370: mom=sumser(acof,bcof,ccof,'cos','sin',...
371:                 tmin,tmax,nt,xmin,xmax,nx);
372: mom=-mom+cwt*mp;
```

9.7 Forced Vibration of a Pile Embedded in an Elastic Medium

Structures are often supported by piles embedded in soil foundations. The response of these systems, when the foundation is shaken in the manner occurring in an earthquake, has considerable practical interest. Let us examine a simple model approximating a single pile connected to an overlying structure. The pile is treated as a beam of uniform cross section buried in an elastic medium. An attached mass at the top causes inertial resistance to translation and rotation. The beam, shown in Figure 9.19 in a deflected position, has length ℓ with $x = 0$ denoting the lower end and $x = \ell$ denoting the top. Rotating the member 90° from the vertical is done to agree with the coordinate referencing traditionally used in beam analysis. We are interested in the steady-state response when the foundation displacement is $y_o \cos(\omega t)$. For convenience we use a complex valued forcing function and get the final results by taking the real part of the complex valued solution. The transverse bending response is to be computed when the surrounding elastic medium has an oscillatory motion of the form

$$y_f = y_o e^{i\omega t}.$$

The differential equation governing transverse oscillations of the beam is

$$EI\frac{\partial^4 y(x,t)}{\partial x^4} = -A\rho\frac{\partial^2 y(x,t)}{\partial t^2} + k\left(y_o e^{i\omega t} - y\right)$$

where EI is the product of the elastic modulus and the inertial moment of the beam, $A\rho$ is the product of the cross section area and the mass per unit volume, and k describes the foundation stiffness in terms of force per unit length per unit of transverse deflection. The shear V and moment M in the beam are related to the deflection $y(x,t)$ by

$$V = EI\frac{\partial^3 y(x,t)}{\partial x^3}\ ,\ M = EI\frac{\partial^2 y(x,t)}{\partial x^2}.$$

In the current analysis we consider forced response of frequency ω described in the form

$$y(x,t) = f(x)e^{i\omega t}$$

so that

$$V = EIf'''(x)e^{i\omega t}\ ,\ M = EIf''(x)e^{i\omega t}.$$

The boundary conditions at $x = 0$ require vanishing moment and shear:

$$f''(0) = 0\ ,\ f'''(0) = 0.$$

The boundary conditions at $x = \ell$ are more involved because inertial resistance of the end mass must be handled. We assume that the gravity center of the end mass is located along the axis of the beam at a distance h above the top end. Furthermore,

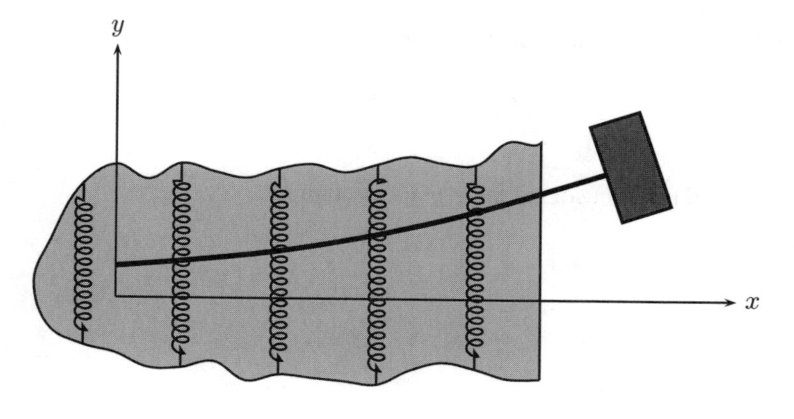

Figure 9.17: Forced Vibration of a Pile in an Elastic Medium

the attached body has a mass m_o and inertial moment $\jmath_o$ about its gravity center. The angular acceleration $\ddot{\theta}$ and the transverse acceleration a_m are expressible as

$$\ddot{\theta} = \frac{\partial^3 y(\ell, t)}{\partial x \partial t^2} = -\omega^2 f'(\ell) e^{i\omega t}$$

and

$$a_m = \frac{\partial^2 y(\ell, t)}{\partial t^2} + h\ddot{\theta} = -\omega^2 e^{i\omega t} \left[f(\ell) + h f'(\ell) \right].$$

Writing equations of motion for the end mass gives

$$m_o a_m = V(\ell, t) \text{ and } \jmath_o \ddot{\theta}_m = -hV(\ell, t) - M(\ell, t).$$

Representing these conditions in terms of $f(x)$ yields

$$-\omega^2 m_o [f(\ell) + hf'(\ell)] = EIf'''(\ell) \text{ and } \omega^2 \jmath_o f'(\ell) = EI[f''(\ell) + hf'''(\ell)].$$

Furthermore, the factor $e^{i\omega t}$ cancels out of the differential equation

$$EIf''''(x) = (A\rho\omega^2 - k) f(x) + y_o k.$$

The general solution of this fourth order linear differential equation is expressed as

$$f(x) = \frac{y_o k}{k - A\rho\omega^2} \left[1 + \sum_{\jmath=1}^{4} c_\jmath e^{s_\jmath x} \right]$$

where $s_\jmath$ are complex roots given by

$$s_\jmath = \left(\frac{A\rho\omega^2 - k}{EI} \right)^{1/4} e^{i(\jmath - 1)\pi/2}, \ \jmath = 1, 2, 3, 4.$$

The conditions of zero moment and shear at $x = 0$ lead to

$$\sum_{j=1}^{4} s_j^2 c_j = 0 \, , \; \sum_{j=0}^{4} s_j^3 c_j = 0.$$

The shear and moment conditions at $x = \ell$ require

$$\sum_{j=1}^{4} s_j^3 e^{s_j \ell} c_j = -m_o \omega^2 \left[1 + \sum_{j=1}^{4} (1 + h s_j) e^{s_j \ell} c_j \right]$$

and

$$\sum_{j=1}^{4} (s_j^2 + h s_j^3) e^{s_j \ell} c_j = j_o \omega^2 \sum_{j=1}^{4} s_j e^{s_j \ell} c_j.$$

The system of four simultaneous equations can be solved for $c_1, \ldots, c_4$. Then the forced response solution corresponding to a foundation motion

$$\textbf{real} \left(y_o e^{i\omega t} \right) = y_o \cos(\omega t)$$

is given by

$$y(x, t) = \textbf{real} \left(f(x) e^{i\omega t} \right)$$

where $f(x)$ is complex valued.

The function **pilevibs** evaluates the displacement, moment, and shear for $0 \leq x \leq \ell$, $0 \leq t \leq 2\pi/\omega$. Surface plots of these quantities are shown in Figures 9.18 through 9.20. Figure 9.21 is a single frame from an animation depicting how the pile and the attached mass move.

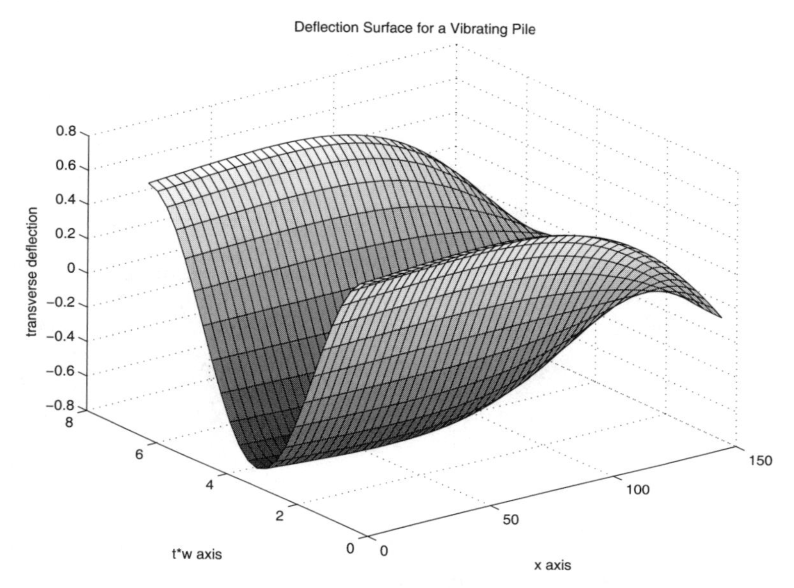

Figure 9.18: Deflection Surface for a Vibrating Pile

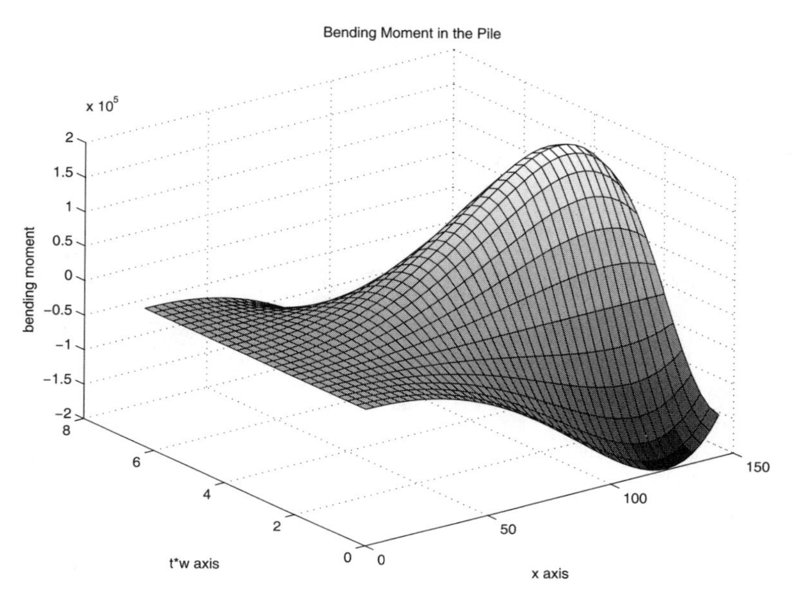

Figure 9.19: Bending Moment in a Vibrating Pile

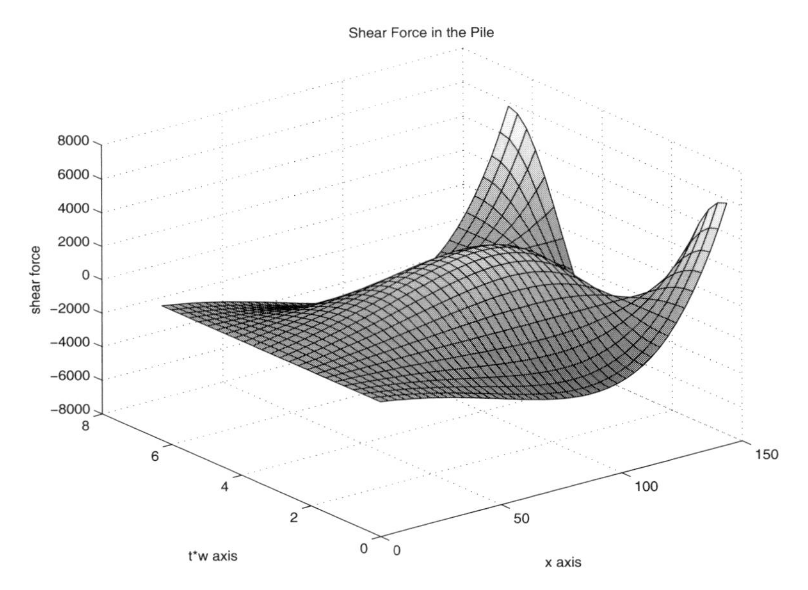

Figure 9.20: Shear Force in a Vibrating Pile

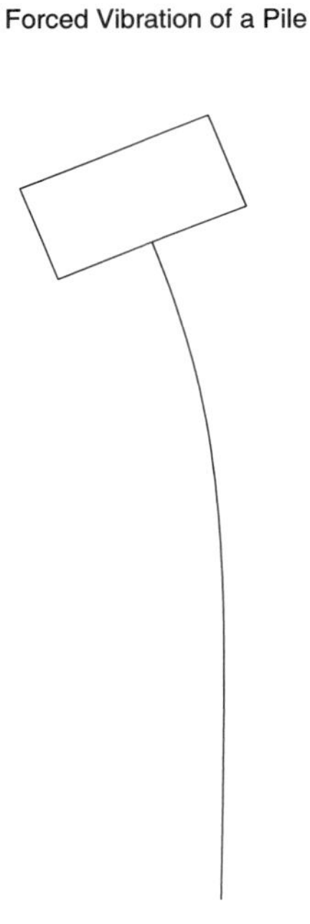

Figure 9.21: Frame from Pile Animation

Program Output and Code

Program pilevibs

```
1: function pilevibs
2: % Example: pilevibs
3: % ~~~~~~~~~~~~~~~~~
4: % The routine is used to solve an example
5: % problem using function pvibs. The example
6: % involves a steel pile 144 inches long which
7: % has a square cross section of 4 inch depth.
8: % The pile is immersed in soil having an elastic
9: % modulus of 200 psi. The attached mass weighs
10: % 736 lb. The foundation is shaken at an
11: % amplitude of 0.5 inch with a frequency of
12: % 20 cycles per second.
13: %
14: % User m functions required: pvibs
15:
16: clear;
17: L=144; d=4; a=d^2; I=d^4/12; e=30e6; ei=e*I;
18: g=32.2*12; Density_steel=0.284;
19: rho=Density_steel/g;
20: Cap_w=36; Cap_h=18; Cap_t=4;
21: m0=Cap_w*Cap_h*Cap_t*rho;
22: j0=m0/12*(Cap_h^2+Cap_w^2);
23: h=Cap_h/2; arho=a*rho;
24: e_soil=200; k=e_soil*d; y0=0.5; w=40*pi;
25: nx=42; nt=25;
26:
27: [t,x,y,m,v]= ...
28:   pvibs(y0,ei,arho,L,k,w,h,m0,j0,nx,nt);
29:
30: %=============================================
31:
32: function [t,x,y,m,v]= ...
33:           pvibs(y0,ei,arho,L,k,w,h,m0,j0,nx,nt)
34: %
35: % [t,x,y,m,v]=pvibs ...
36: %             (y0,ei,arho,L,k,w,h,m0,j0,nx,nt)
37: % ~~~~~~~~~~~~~~~~~~~~~~~~~~~~~~~~~~~~~~~~~~~~
38: %
39: % This function computes the forced harmonic
40: % response of a pile buried in an oscillating
```

```
41: % elastic medium. The lower end of the pile is
42: % free from shear and moment. The top of the
43: % pile carries an attached body having general
44: % mass and inertial properties. The elastic
45: % foundation is given a horizontal oscillation
46: % of the form
47: %
48: %   yf=real(y0*exp(i*w*t))
49: %
50: % The resulting transverse forced response of
51: % the pile is expressed as
52: %
53: %   y(x,t)=real(f(x)*exp(i*w*t))
54: %
55: % where f(x) is a complex valued function. The
56: % bending moment and shear force in the pile
57: % are also computed.
58: %
59: % y0   - amplitude of the foundation oscillation
60: % ei   - product of moment of inertia and
61: %        elastic modulus for the pile
62: % arho - mass per unit length of the pile
63: % L    - pile length
64: % k    - the elastic resistance constant for the
65: %        foundation described as force per unit
66: %        length per unit of transverse
67: %        deflection
68: % w    - the circular frequency of the
69: %        foundation oscillation which vibrates
70: %        like real(y0*exp(i*w*t))
71: % h    - the vertical distance above the pile
72: %        upper end to the gravity center of the
73: %        attached body
74: % m0   - the mass of the attached body
75: % j0   - the mass moment of inertia of the
76: %        attached body with respect to its
77: %        gravity center
78: % nx   - the number of equidistant values along
79: %        the pile at which the solution is
80: %        computed
81: % nt   - the number of values of t values at
82: %        which the solution is computed such
83: %        that 0 <= w*t <= 2*pi
84: %
85: % t    - a vector of time values such that the
```

```
86: %          pile moves through a full period of
87: %          motion. This means 0 <= t <= 2*pi/w
88: % x      - a vector of x values with 0 <= x <= L
89: % y      - the transverse deflection y(x,t) for
90: %          the pile with t varying from row to
91: %          row, and x varying from column to
92: %          column
93: % m,v    - matrices giving values bending moment
94: %          and shear force
95: %
96: % User m functions called: none
97: %----------------------------------------------
98:
99: % Default data for a steel pile 144 inches long
100: if nargin==0
101:   y0=0.5; ei=64e7; arho=0.0118; L=144; k=800;
102:   w=125.6637; h=9; m0=1.9051; j0=257.1876;
103:   nx=42; nt=25;
104: end
105:
106: w2=w^2; x=linspace(0,L,nx)';
107: t=linspace(0,2*pi/w,nt);
108:
109: % Evaluate characteristic roots and complex
110: % exponentials
111: s=((arho*w2-k)/ei)^(1/4)*[1,i,-1,-i];
112: s2=s.^2; s3=s2.*s;
113: c0=y0*k/(k-w2*arho); esl=exp(s*L);
114: esx=exp(x*s); eiwt=exp(i*w*t);
115:
116: % Solve for coefficients to satisfy the
117: % boundary conditions
118: c=[s2; s3; esl.*(h*s3+s2-j0*w2/ei*s); ...
119:     esl.*(s3+m0*w2/ei*(1+h*s))]\ ...
120:     [0;0;0;-c0*m0*w2/ei];
121:
122: % Compute the deflection, moment and shear
123: y=real((c0+esx*c)*eiwt)';
124: ype=real(s.*esl*c*eiwt)';
125: m=real(ei*s2(ones(nx,1),:).*esx*c*eiwt)';
126: v=real(ei*s3(ones(nx,1),:).*esx*c*eiwt)';
127: t=t'; x=x'; hold off; clf;
128:
129: % Make surface plots showing the deflection,
130: % moment, and shear over a complete period of
```

```
131: % the motion
132: surf(x,t*w,y);
133: xlabel('x axis'); ylabel('t*w axis');
134: zlabel('transverse deflection');
135: title('Deflection Surface for a Vibrating Pile');
136: grid on; figure(gcf)
137: % print -deps pilesurf
138: disp('Press [Enter] to continue'), pause
139:
140: surf(x,t*w,m);
141: xlabel('x axis'); ylabel('t*w axis');
142: zlabel('bending moment');
143: title('Bending Moment in the Pile')
144: grid on; figure(gcf)
145: % print -deps pilemom;
146: disp('Press [Enter] to continue'), pause
147:
148: surf(x,t*w,v);
149: xlabel('x axis'); ylabel('t*w axis');
150: zlabel('shear force');
151: title('Shear Force in the Pile');
152: grid on; figure(gcf)
153: % print -deps pilesher
154: disp('Press [Enter] to see animation'), pause
155:
156: % Draw an animation depicting the pile response
157: % to the oscillation of the foundation
158: fu=.10/max(y(:)); p=[-0.70, 0.70, -.1, 1.3];
159: u=fu*y; upe=fu*L*ype; d=.15;
160: xm=[0,0,1,1,0,0]*d;
161: ym=[0,-1,-1,1,1,0]*d; zm=xm+i*ym;
162: close;
163: for jj=1:4
164:   for j=1:nt
165:     z=exp(i*atan(upe(j)))*zm;
166:     xx=real(z); yy=imag(z);
167:     ut=[u(j,:),u(j,nx)+yy]; xt=[x/L,1+xx];
168:     plot(ut,xt,'-'); axis(p); axis('square');
169:     title('Forced Vibration of a Pile');
170:     axis('off'); drawnow; figure(gcf);
171:   end
172: end
173: % print -deps pileanim
174: fprintf('\nAll Done\n');
```

9.8 Transient Heat Conduction in a One-Dimensional Slab

Let us analyze the temperature history in a slab which has the left side insulated while the right side temperature varies sinusoidally according to $U_0 \sin(\Omega T)$. The initial temperature in the slab is specified to be zero. The pertinent boundary value problem is

$$\alpha \frac{\partial^2 U}{\partial X^2}(X,T) = \frac{\partial U}{\partial T}(X,T)\ ,\ 0 < X < \ell\ ,\ T > 0,$$

$$\frac{\partial U}{\partial X}(0,T) = 0\ ,\ U(\ell,T) = U_0 \sin(\Omega T),$$

$$U(X,0) = 0\ ,\ 0 < X < \ell$$

where U, X, T, and α are, respectively, the temperature, position, time, and thermal diffusivity.

The problem can be converted to dimensionless form by letting

$$u = \frac{U}{U_0}\ ,\ x = \frac{X}{\ell}\ ,\ t = \frac{\alpha T}{\ell^2}\ ,\ \omega = \frac{\Omega \ell^2}{\alpha}.$$

Then we get

$$\frac{\partial^2 u}{\partial x^2} = \frac{\partial u}{\partial x}\ ,\ 0 < x < 1\ ,\ t > 0,$$

$$\frac{\partial u}{\partial x}(0,t) = 0\ ,\ u(1,t) = \textbf{imag}\left(e^{i\omega t}\right)\ ,\ u(x,0) = 0.$$

The solution consists of two parts as $u = w + v$, where w is a particular solution satisfying the differential equation and nonhomogeneous boundary conditions, and v is a solution satisfying homogeneous boundary conditions and specified to impose the desired zero initial temperature when combined with w. The appropriate form for the particular solution is

$$w = \textbf{imag}\left[f(x)e^{i\omega t}\right].$$

Making w satisfy the heat equation requires

$$f''(x) = i w f(x).$$

Consequently

$$f(x) = c_1 \sin(\phi x) + c_2 \cos(\phi x)$$

where $\phi = \sqrt{-\imath\omega}$. The conditions of zero gradient at $x = 0$ and unit function value at $x = 1$ determine c_1 and c_2. We get the particular solution as

$$w = \textbf{imag}\left[\frac{\cos(\phi x)}{\cos(\phi)} e^{i\omega t}\right].$$

This forced response solution evaluated at $t = 0$ yields

$$w(x,0) = \textbf{imag}\left[\frac{\cos(\phi x)}{\cos(\phi)}\right].$$

The general solution of the heat equation satisfying zero gradient at $x = 0$ and zero function value at $x = 1$ is found to be

$$v(x,t) = \sum_{n=1}^{\infty} a_n \cos(\lambda_n x) e^{-\lambda_n^2 t}$$

where $\lambda_n = \pi(2n-1)/2$. To make the initial temperature equal zero in the combined solution, the coefficients a_n are chosen to satisfy

$$\sum_{n=1}^{\infty} a_n \cos(\lambda_n x) = -\,\textbf{imag}\left[\frac{\cos(\phi x)}{\cos(\phi)}\right].$$

The orthogonality of the functions $\cos(\lambda_n x)$ implies

$$a_n = -2\int_0^1 \textbf{imag}\left[\frac{\cos(\phi x)}{\cos(\phi)}\right]\cos(\lambda_n x)dx$$

which can be integrated to give

$$a_n = -\,\textbf{imag}\left[\frac{(\sin(\lambda_n + \phi)/(\lambda_n + \phi) + \sin(\lambda_n - \phi)/(\lambda_n - \phi))}{\cos(\phi)}\right].$$

This completely determines the solution. Taking any finite number of terms in the series produces an approximate solution exactly satisfying the differential equation and boundary conditions. Exact satisfaction of the zero initial condition would theoretically require an infinite number of series terms. However, the terms in the series decrease like $O(1/n^3)$ and using a 250-term series produces initial temperature values not exceeding 10^{-6}. Thus, the finite series is satisfactory for practical purposes.

The above equations were evaluated in a function called **heat**. Function **slabheat** was also written to plot numerical results. The code and resulting Figures 9.23 and 9.24 appear below. This example illustrates nicely how well MATLAB handles complex arithmetic and complex valued functions.

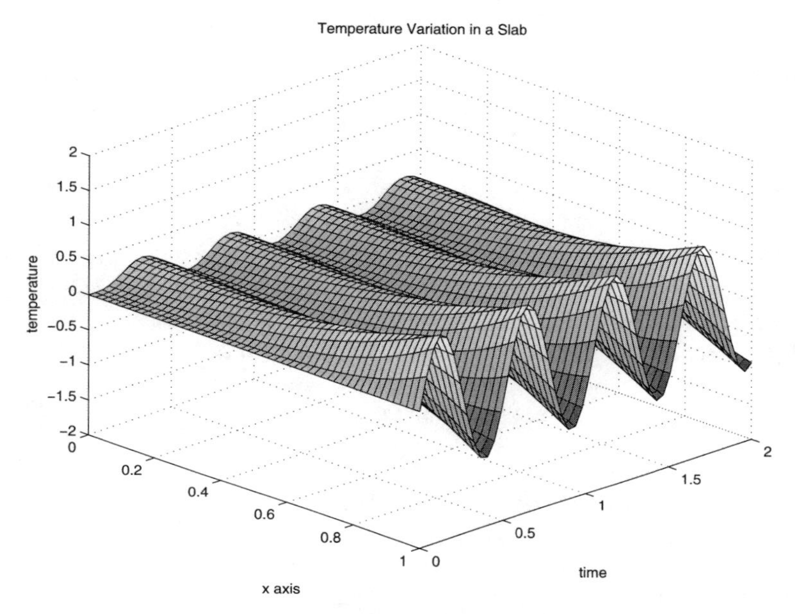

Figure 9.22: **Temperature Variation in a Slab**

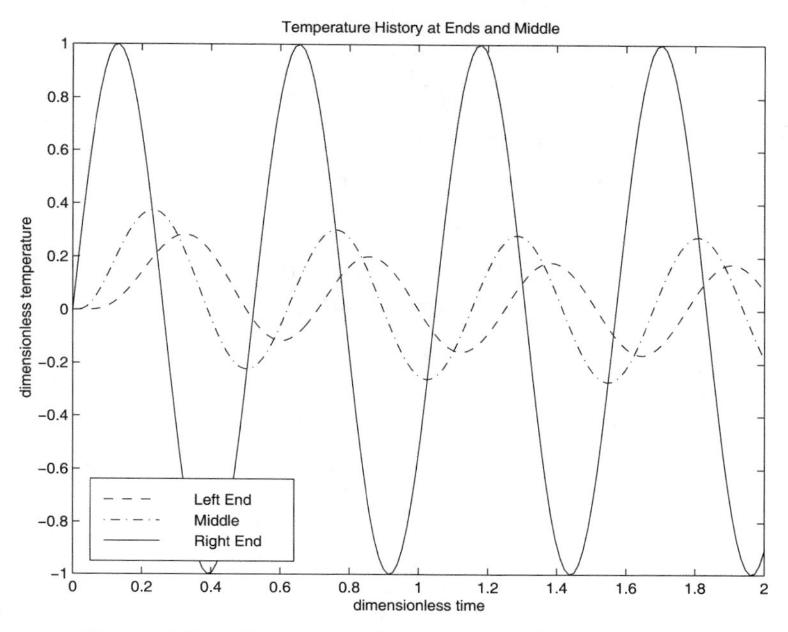

Figure 9.23: **Temperature History at Ends and Middle**

Heat Conduction Program

Program slabheat

```
1: function slabheat
2: % Example: slabheat
3: % ~~~~~~~~~~~~~~~~~~
4: % This program computes the temperature
5: % variation in a one-dimensional slab with
6: % the left end insulated and the right end
7: % given a temperature variation sin(w*t).
8: %
9: % User m functions required:  heat
10:
11: [u1,t1,x1]=heat(12,0,2,50,0,1,51,250);
12: surf(x1,t1,u1); axis([0 1 0 2 -2 2]);
13: title('Temperature Variation in a Slab');
14: xlabel('x axis'); ylabel('time');
15: zlabel('temperature'); view([45,30])
16: colormap('default'), shg
17: disp(' '), disp('Press [Enter] to continue')
18: pause
19: % print -deps tempsurf
20:
21: [u2,t2,x2]=heat(12,0,2,150,0,1,3,250);
22: plot(t2,u2(:,1),'--',t2,u2(:,2),':', ...
23:      t2,u2(:,3),'-');
24: title(['Temperature History at Ends' ...
25:        ' and Middle']);
26: xlabel('dimensionless time');
27: ylabel('dimensionless temperature');
28: text1='Left End'; text2='Middle';
29: text3='Right End';
30: legend(text1,text2,text3,3); shg
31: % print -deps templot
32: disp(' '), disp('All Done');
33:
34: %==============================================
35:
36: function [u,t,x]= ...
37:          heat(w,tmin,tmax,nt,xmin,xmax,nx,nsum)
38: %
39: %[u,t,x]=heat(w,tmin,tmax,nt,xmin,xmax,nx,nsum)
40: %~~~~~~~~~~~~~~~~~~~~~~~~~~~~~~~~~~~~~~~~~~~~~~
```

```
41: % This function evaluates transient heat
42: % conduction in a slab which has the left end
43: % (x=0) insulated and has the right end (x=1)
44: % subjected to a temperature variation
45: % sin(w*t). The initial temperature of the slab
46: % is zero.
47: %
48: % w            - frequency of the right side
49: %                temperature variation
50: % tmin,tmax    - time limits for solution
51: % nt           - number of uniformly spaced
52: %                time values used
53: % xmin,xmax    - position limits for solution.
54: %                Values should lie between zero
55: %                and one.
56: % nx           - number of equidistant x values
57: % nsum         - number of terms used in the
58: %                series solution
59: % u            - matrix of temperature values.
60: %                Time varies from row to row.
61: %                x varies from column to column.
62: % t,x          - vectors of time and x values
63: %
64: % User m functions called:  none.
65: %----------------------------------------------
66:
67: t=tmin+(tmax-tmin)/(nt-1)*(0:nt-1);
68: x=xmin+(xmax-xmin)/(nx-1)*(0:nx-1)';
69: W=sqrt(-i*w); ln=pi*((1:nsum)-1/2);
70: v1=ln+W; v2=ln-W;
71: a=-imag((sin(v1)./v1+sin(v2)./v2)/cos(W));
72: u=imag(cos(W*x)*exp(i*w*t)/cos(W))+ ...
73:   (a(ones(nx,1),:).*cos(x*ln))* ...
74:   exp(-ln(:).^2*t);
75: u=u'; t=t(:);
```

9.9 Transient Heat Conduction in a Circular Cylinder with Spatially Varying Boundary Temperature

9.9.1 Problem Formulation

Transient heat conduction in a circular cylinder can be analyzed using an infinite series of Bessel functions. Consider a cylinder having an initial temperature distribution $u_0(r,\theta)$ when the boundary is suddenly given a temperature variation $f(\theta)$ depending on the polar angle but independent of time. The problem is conveniently formulated in polar coordinates using dimensionless radius and time variables. The differential equation, boundary conditions, and initial conditions are as follows:

$$u_{rr} + \frac{1}{r}u_r + \frac{1}{r^2}u_{\theta\theta} = u_t \ , \ 0 \le r \le 1 \ , \ t > 0,$$

$$u(1,\theta,t) = f(\theta) = \sum_{n=-\infty}^{\infty} f_n e^{in\vartheta} \ , \ 0 \le \theta \le 2\pi,$$

$$u(r,\theta,0) = u_0(r,\theta) \ , \ 0 \le r \le 1 \ , \ 0 \le \theta \le 2\pi.$$

With the boundary condition expressed as a complex Fourier series, the steady-state solution satisfying the differential equation and the boundary conditions is

$$v(r,\theta) = -f_0 + 2 \ \mathbf{real}\left(\sum_{n=0}^{\infty} f_n z^n\right) \ \text{where } z = re^{i\theta}.$$

The total solution is the steady-state solution combined with a transient solution $w(r,\theta,t)$ chosen to satisfy the initial condition and boundary conditions expressed as

$$w(r,\theta,0) = u_0(r,\theta) - v(r,\theta) \ , \ w(1,\theta,t) = 0.$$

The transient solution is a Fourier-Bessel series involving double subscripted coefficients depending on the functions $v(r,\theta)$ and $u_0(r,\theta)$. It is found that

$$w(r,\theta,t) = \sum_{n=0}^{\infty}\sum_{k=1}^{\infty} J_n(\lambda_{nk}r)\left[A_{nk}\cos(n\theta) + B_{nk}\sin(n\theta)\right]\exp(-\lambda_{nk}^2 t)$$

where, for $n > 0$ and $k \ge 1$, we have

$$A_{nk} + iB_{nk} = C_{nk} = \frac{2}{\pi J_{n+1}^2(\lambda_{nk})}\int_0^{2\pi}\int_0^1 w(r,\theta,0) r J_n(\lambda_{nk}r)\exp(in\theta)\,dr\,d\theta$$

with λ_{nk} denoting the k'^{th} positive root of $J_n(r)$. The last formula almost applies for $n = 0$ except that $A_{0k} = C_{0k}/2$ and $B_{0k} = 0$. The coefficients for $n = 0$ pertain to the radially symmetric case independent of the polar angle. Evaluating this series solution involves several steps which are: 1) Expanding the boundary condition in a complex Fourier series to obtain the steady-state solution; 2) Determining the zeros of the integer order Bessel functions $J_n(r)$; 3) Computing the series coefficients by numerical integration; and 4) Summing the series solution for various (r,θ) values with enough terms being used in the series to assure adequate satisfaction of the initial conditions and boundary conditions.

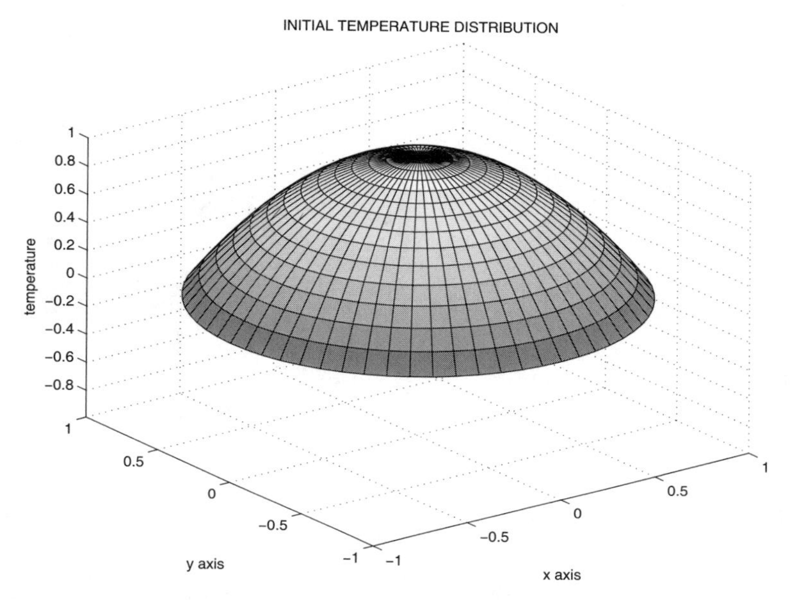

Figure 9.24: **Initial Temperature**

9.9.2 Computer Formulation

A computer program was written to analyze the time dependent temperature field. The program specifies general initial temperature and boundary temperature. The series solution is evaluated on a polar coordinate grid and an animation of the temperature variation from initial to steady state is shown. The program modules include: 1) **heatcyln** which calls the computational modules and plots results; 2) **besjtabl** returns Bessel function roots used in the series solution; 3) **tempinit** specifies the initial temperature field; 4) **tempstdy** computes the steady state solution; 5) **tempdif** computes the difference in the initial and the final temperature fields; 6) **foubesco** evaluates coefficients in the Fourier-Bessel series; and (7) **tempsum** sums the Fourier-Bessel series for a vector of time values. Figures 9.25 through 9.28 show the initial, final, and two intermediate temperature states. The program animates the temperature history so the transition from initial to steady-state can be visualized.

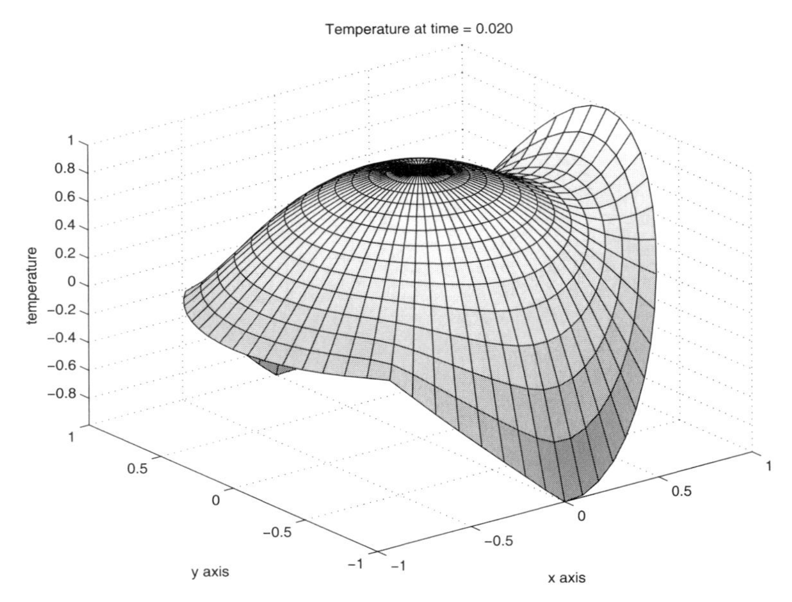

Figure 9.25: **Temperature at t=0.02**

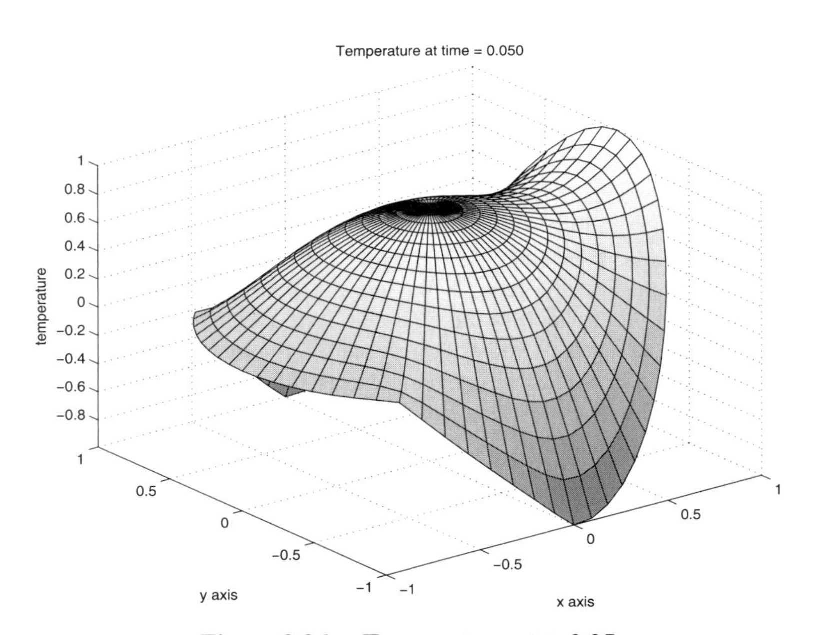

Figure 9.26: **Temperature at t=0.05**

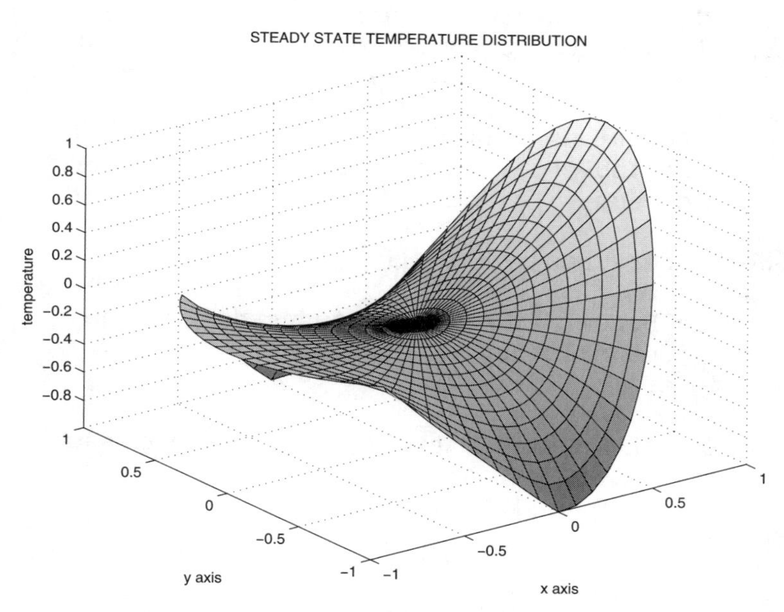

Figure 9.27: **Steady State Temperature**

Program heatcyln

```
1: function heatcyln
2: %
3: % heatcyln
4: % ~~~~~~~~
5: % This program analyzes the time varying temperature
6: % history in a circular cylinder which initially has
7: % a radially symmetric temperature varying para-
8: % bolically. Then a spatially varying but constant
9: % boundary temperature distribution is imposed. The
10: % total solution is composed of a harmonic steady
11: % state solution plus a transient component given by
12: % a Fourier-Bessel series.
13: % User functions called:
14: %     besjtabl, tempinit, tempstdy, foubesco,
15: %     tempsum,  tempdif,  gcquad
16:
17: global ubdry besjrt
18:
19: % Obtain Bessel function roots needed in the
20: % transient solution
21: besjrt=besjtabl(0:20,20);
22:
23: % Define the steady state temperature imposed
24: % on the outer boundary for t>0
25: th=linspace(0,pi,100)';
26: ud=cos(2*th).*(th<=pi/2)+...
27: (-3+4/pi*th).*(th>pi/2&th<3*pi/4);
28: ud=[ud;ud(end-1:-1:1)];
29: ubdry=[linspace(0,360,199)',ud];
30: theta=linspace(0,2*pi,65);
31: r=linspace(0,1,15);
32:
33: % Compute and plot the initial and final
34: % temperature fields
35: [uinit,z]=tempinit(theta,r);
36: [usteady,z]=tempstdy(theta,r);
37: umin=min([usteady(:);uinit(:)]);
38: umax=max([usteady(:);uinit(:)]);
39: range=[-1,1,-1,1,umin,umax];
40: x=real(z); y=imag(z);
41:
```

```
42: surf(x,y,uinit), colormap('default')
43: title('INITIAL TEMPERATURE DISTRIBUTION')
44: xlabel('x axis'), ylabel('y axis')
45: zlabel('temperature'), axis(range), disp(' ')
46: disp('Press [Enter] to see the steady')
47: disp('state temperature distribution')
48: shg, pause, disp(' ')
49: % print -deps tempinit
50:
51: surf(x,y,usteady)
52: title('STEADY STATE TEMPERATURE DISTRIBUTION')
53: xlabel('x axis'), ylabel('y axis')
54: zlabel('temperature'), axis(range), shg
55: % print -deps tempstdy
56:
57: % Compute coefficients used in the Fourier-
58: % Bessel series for the transient solution
59: [c,lam,cptim]=foubesco(@tempdif,20,20,40,128);
60:
61: % Set a time interval sufficient to nearly
62: % reach steady state
63: tmax=.4; nt=81; t=linspace(0,tmax,nt);
64:
65: % Evaluate the transient solution
66: [u,tsum]=tempsum(c,theta,r,t,lam);
67: u(:,:,1)=uinit-usteady;
68:
69: % Plot time history for the total solution
70: while 1
71:   disp('Press [Enter] to see the animation')
72:   disp('or enter 0 to stop'), v=input('> ? ');
73:   if isempty(v), v=1; end
74:   if v~=1, break, end
75:   for j=1:nt
76:     utotal=usteady+u(:,:,j);
77:     surf(x,y,utotal)
78:     titl=sprintf(['Temperature at time =',...
79:        '%6.3f'],t(j)); title(titl)
80:     xlabel('x axis'), ylabel('y axis')
81:     zlabel('temperature'), axis(range);
82:     drawnow; shg, pause(.3)
83:   end
84: end
85:
86: %=========================================
```

```
87:
88: function [u,z]=tempstdy(theta,r)
89: %
90: % [u,z]=tempstdy(theta,r)
    % ~~~~~~~~~~~~~~~~~~~~~~~
91: %
92: % Steady state temperature distribution in a
93: % circular cylinder of unit radius with
94: % piecewise linear boundary values
95: % described in global array ubdry.
96: global ubdry
97:
98: thft=2*pi/(1024)*(0:1023); n=100;
99: ufft=interp1(pi/180*ubdry(:,1),...
100:              ubdry(:,2)/1024,thft);
101: c=fft(ufft); z=exp(i*theta(:))*r(:)';
102: u=-real(c(1))+2*real(...
103:    polyval(c(n:-1:1),z));
104:
105: %==============================================
106:
107: function [u,z]=tempinit(theta,r)
108: %
109: % [u,z]=tempinit(theta,r)
     % ~~~~~~~~~~~~~~~~~~~~~~~
110: %
111: % Initial temperature varying parabolically
112: % with the radius
113: theta=theta(:); r=r(:)'; z=exp(i*theta)*r;
114: u=ones(length(theta),1)*(1-r.^2);
115:
116: %==============================================
117:
118: function [u,z]=tempdif(theta,r)
119: %
120: % [u,z]=tempdif(theta,r)
     % ~~~~~~~~~~~~~~~~~~~~~~
121: %
122: % Difference between the steady state temp-
123: % erature and the initial temperature
124: u1=tempstdy(theta,r); [u2,z]=tempinit(theta,r);
125: u=u2-u1;
126:
127: %==============================================
128:
129: function [c,lam,cptim]=foubesco(...
130:                         f,nord,nrts,nrquad,nft)
131: %
```

```
132: % [c,lam,cptim]=foubesco(f,nord,nrts,nrquad,nft)
133: % ~~~~~~~~~~~~~~~~~~~~~~~~~~~~~~~~~~~~~~~~~~~~~~~
134: % Fourier-Bessel coefficients computed using the
135: % FFT
136: global besjrt
137: if nargin<5, nft=128; end
138: if nargin<4, nrquad=50; end
139: if nargin<3, nrts=10; end
140: if nargin<2, nord=10; end
141: if nargin==0, f='fbes'; end
142: tic; lam=besjrt(1:nord,1:nrts);
143: c=zeros(nord,nrts);
144: [dummy,r,w]=gcquad([],0,1,nrquad,1);
145: r=r(:)'; w=w(:)'; th=2*pi/nft*(0:nft-1)';
146: fmat=fft(feval(f,th,r));
147: fmat=fmat(1:nord,:).*repmat(r.*w,nord,1);
148: for n=1:nord
149:   for k=1:nrts
150:     lnk=lam(n,k);
151:     v=sum(fmat(n,:).*besselj(n-1,lnk*r));
152:     c(n,k)=4*v/nft/besselj(n,lnk).^2;
153:   end
154: end
155: c(1,:)=c(1,:)/2; cptim=toc;
156:
157: %==============================================
158:
159: function [u,tcpu]=tempsum(c,th,r,t,lam)
160: %
161: % [u,tsum]=tempsum(c,th,r,t,lam)
162: %
163: % This function sums a Fourier-Bessel series
164: % for transient temperature history in a circular
165: % cylinder with given initial conditions and
166: % zero temperature at the boundary. The series
167: % has the form
168: % u(theta,r,t)=sum({n=0:nord-1),k=1:nrts},...
169: % besselj(n,lam(n+1,k)*r)*real(...
170: % c(n+1,k)*exp(i*(n+1)*theta))*...
171: % exp(-lam(n+1,k)^2*t), where
172: % besselj(n-1,lam(n,k))=0 and
173: % [nord,nrts]=size(c)
174: %
175: % c    - the series coefficients for the initial
176: %        temperature distribution obtained using
```

```
177: %          function foubesco
178: % th   - vector or theta values between
179: %          zero and 2*pi
180: % r    - vector of radius values between
181: %          zero and one
182: % lam  - matrix of bessel function roots.
183: %          If this argument is omitted, then
184: %          function besjroot is called to
185: %          compute the roots
186: % u    - a three-dimensional array of function
187: %          values where u(i,j,k) contains the
188: %          temperature for theta(i), r(j), t(k)
189: % tcpu - computation time in seconds
190:
191: tic; [nord,nrts]=size(c);
192: if nargin<5, lam=besjroot(0:nord-1,nrts); end
193: th=th(:); nth=length(th); r=r(:)'; nr=length(r);
194: nt=length(t); N=repmat((0:nord-1)',1,nrts);
195: N=N(:)'; c=c(:).'; lam=lam(:); lam2=-(lam.^2)';
196: u=zeros(nth,nr,nt); thmat=exp(i*th*N);
197: besmat=besselj(repmat(N',1,nr),lam*r);
198: for I=1:nt
199:   C=c.*exp(lam2*t(I));
200:   u(:,:,I)=real(thmat.*repmat(C,nth,1))*besmat;
201: end
202: tcpu=toc;
203:
204: %==============================================
205:
206: function r=besjtabl(nordr,nrts)
207: %
208: % r=besjtable(nordr,nrts)
209: % ~~~~~~~~~~~~~~~~~~~~~~~
210: % This function returns a table for roots of
211: % besselj(n,x)=0 accurate to about five digits.
212: % r(k,:) - contains the first 20 positive roots of
213: %            besselj(k-1,x)=0; for k=1:21
214: % nordr  - a vector of function orders lying
215: %            between 0 and 20
216: % nrts   - the highest root order not to exceed
217: %            the twentieth positive root
218:
219: if nargin==0, nordr=0:20; nrts=20; end
220: if max(nordr)>20 | nrts>20, r=nan; return; end
221: r=[2.4048 21.6415 40.7729 33.7758 53.7383 73.2731
```

222:	3.8317	22.9452	42.0679	35.3323	55.1847	74.6738
223:	5.1356	24.2339	43.3551	36.8629	56.6196	76.0673
224:	6.3801	25.5094	44.6349	38.3705	58.0436	77.4536
225:	7.5883	26.7733	45.9076	39.8577	59.4575	78.8337
226:	8.7715	28.0267	47.1740	41.3263	60.8617	80.2071
227:	9.9362	29.2706	48.4345	42.7784	62.2572	81.5752
228:	11.0864	30.5060	24.3525	44.2154	63.6441	55.7655
229:	12.2251	31.7334	25.9037	45.6384	65.0231	57.3275
230:	13.3543	32.9537	27.4206	47.0487	66.3943	58.8730
231:	14.4755	34.1672	28.9084	48.4475	67.7586	60.4033
232:	15.5898	35.3747	30.3710	49.8346	69.1159	61.9193
233:	16.6983	36.5764	31.8117	51.2120	70.4668	63.4221
234:	17.8014	37.7729	33.2330	52.5798	71.8113	64.9128
235:	18.9000	14.9309	34.6371	53.9382	46.3412	66.3913
236:	19.9944	16.4707	36.0257	55.2892	47.9015	67.8594
237:	21.0852	17.9599	37.4001	56.6319	49.4422	69.3172
238:	22.1725	19.4094	38.7618	57.9672	50.9651	70.7653
239:	23.2568	20.8269	40.1118	59.2953	52.4716	72.2044
240:	24.3383	22.2178	41.4511	60.6170	53.9631	73.6347
241:	25.4171	23.5861	42.7804	61.9323	55.4405	75.0567
242:	5.5201	24.9350	44.1006	36.9171	56.9052	76.4710
243:	7.0156	26.2668	45.4122	38.4748	58.3579	77.8779
244:	8.4173	27.5839	46.7158	40.0085	59.7991	79.2776
245:	9.7611	28.8874	48.0122	41.5208	61.2302	80.6706
246:	11.0647	30.1790	49.3012	43.0138	62.6513	82.0570
247:	12.3385	31.4600	50.5836	44.4893	64.0629	83.4373
248:	13.5893	32.7310	51.8600	45.9489	65.4659	84.8116
249:	14.8213	33.9932	27.4935	47.3941	66.8607	58.9070
250:	16.0378	35.2471	29.0469	48.8259	68.2474	60.4695
251:	17.2412	36.4934	30.5692	50.2453	69.6268	62.0162
252:	18.4335	37.7327	32.0649	51.6533	70.9988	63.5484
253:	19.6160	38.9654	33.5372	53.0504	72.3637	65.0671
254:	20.7899	40.1921	34.9887	54.4378	73.7235	66.5730
255:	21.9563	41.4131	36.4220	55.8157	75.0763	68.0665
256:	23.1158	18.0711	37.8387	57.1850	49.4826	69.5496
257:	24.2692	19.6159	39.2405	58.5458	51.0436	71.0219
258:	25.4170	21.1170	40.6286	59.8990	52.5861	72.4843
259:	26.5598	22.5828	42.0041	61.2448	54.1117	73.9369
260:	27.6979	24.0190	43.3684	62.5840	55.6217	75.3814
261:	28.8317	25.4303	44.7220	63.9158	57.1174	76.8170
262:	29.9616	26.8202	46.0655	65.2418	58.5996	78.2440
263:	8.6537	28.1912	47.4003	40.0584	60.0694	79.6643
264:	10.1735	29.5456	48.7265	41.6171	61.5277	81.0769
265:	11.6199	30.8854	50.0446	43.1535	62.9751	82.4825
266:	13.0152	32.2119	51.3552	44.6698	64.4123	83.8815

```
267:     14.3726 33.5265 52.6589 46.1679 65.8399 85.2738
268:     15.7002 34.8300 53.9559 47.6493 67.2577 86.6603
269:     17.0037 36.1237 55.2466 49.1157 68.6681 88.0408
270:     18.2876 37.4081 30.6346 50.5681 70.0699 62.0485
271:     19.5546 38.6843 32.1897 52.0077 71.4639 63.6114
272:     20.8070 39.9526 33.7166 53.4352 72.8506 65.1593
273:     22.0470 41.2135 35.2187 54.8517 74.2302 66.6933
274:     23.2758 42.4678 36.6990 56.2576 75.6032 68.2142
275:     24.4949 43.7155 38.1598 57.6538 76.9699 69.7230
276:     25.7051 44.9577 39.6032 59.0409 78.3305 71.2205
277:     26.9074 21.2117 41.0308 60.4194 52.6241 72.7065
278:     28.1024 22.7601 42.4439 61.7893 54.1856 74.1827
279:     29.2909 24.2702 43.8439 63.1524 55.7297 75.6493
280:     30.4733 25.7482 45.2315 64.5084 57.2577 77.1067
281:     31.6501 27.1990 46.6081 65.8564 58.7709 78.5555
282:     32.8218 28.6266 47.9743 67.1982 60.2703 79.9960
283:     33.9887 30.0337 49.3308 68.5339 61.7567 81.4291
284:     11.7916 31.4228 50.6782 43.1998 63.2313 82.8535
285:     13.3237 32.7958 52.0172 44.7593 64.6947 84.2714
286:     14.7960 34.1543 53.3483 46.2980 66.1476 85.6825
287:     16.2234 35.4999 54.6719 47.8178 67.5905 87.0870
288:     17.6159 36.8336 55.9885 49.3204 69.0240 88.4846
289:     18.9801 38.1563 57.2984 50.8072 70.4486 89.8772
290:     20.3208 39.4692 58.6020 52.2794 71.8648 91.2635];
291: r=reshape(r(:),21,20); r=r(1+nordr,1:nrts);
292:
293: %============================================
294:
295: % function [val,bp,wf]=gcquad(func,xlow,...
296: %                    xhigh,nquad,mparts,varargin)
297: % See Appendix B
```

9.10 Torsional Stresses in a Beam of Rectangular Cross Section

Elastic beams of uniform cross section are commonly used structural members. Evaluation of the stresses caused when beams undergo torsional moments depends on finding a particular type of complex valued function. This function is analytic inside the beam cross section and has its imaginary part known on the boundary [72]. The shear stresses τ_{XZ} and τ_{YZ} are obtained from the stress function $f(z)$ of the complex variable $z = x + iy$ according to

$$\frac{\tau_{ZX} - i\tau_{ZY}}{\mu\alpha} = f'(z) - i\bar{z}$$

where μ is the shear modulus and α is the twist per unit length. In the case for a simply connected cross section, such as a rectangle or a semicircle, the necessary boundary condition is

$$\mathbf{imag}[f(z)] = \frac{1}{2}|z|^2$$

at all boundary points. It can also be shown that the torsional moment causes the beam cross section to warp. The warped shape is given by the real part of $f(z)$.

The geometry we will analyze is rectangular. As long as the ratio of side length remains fairly close to unity, $f(z)$ can be well approximated by

$$f(z) = i\sum_{j=1}^{n} c_j \left(\frac{z}{s}\right)^{2j-2}$$

where $c_1, \ldots, c_n$ are real coefficients computed to satisfy the boundary conditions in the least square sense. The parameter s is used for scaling to prevent occurrence of large numbers when n becomes large. We take a rectangle with sides parallel to the coordinate axes and assume side lengths of $2a$ and $2b$ for the horizontal and vertical directions, respectively. The scaling parameter will be chosen as the larger of a and b. The boundary conditions state that for any point z_i on the boundary we should have

$$\sum_{j=1}^{n} c_j \,\mathbf{real}\left[(\frac{z_i}{s})^{2j-2}\right] = \frac{1}{2}|z_i|^2.$$

Once the series coefficients are found, then shear stresses are computed as

$$\frac{\tau_{XZ} - i\tau_{YZ}}{\mu\alpha} = -i\bar{z} + 2is^{-1}\sum_{j=2}^{n}(j-1)c_j\left(\frac{z}{s}\right)^{2j-3}$$

A program was written to compute stresses in a rectangular beam and to show graphically the cross section warping and the dimensionless stress values. The program is short and the necessary calculations are almost self explanatory. It is worthwhile to observe, however, the ease with which MATLAB handles complex functions. Note how intrinsic function **linspace** is used to generate boundary data and **meshgrid** is used to generate a grid of complex values (see lines 50, 51, 72, 73, and 74 of function **recstrs**). The sample problem employs a rectangle of dimension 2 units by 4 units. The maximum stress occurs at the middle of the longest side. Figures 9.28 through 9.31 plot the results of this analysis.

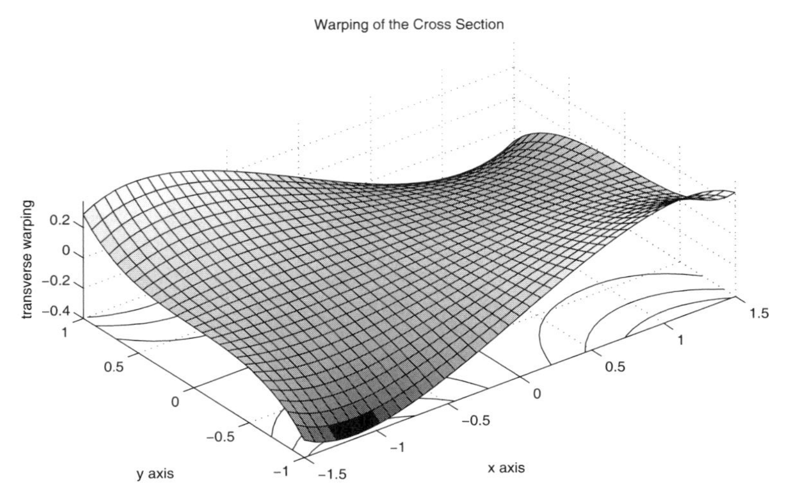

Figure 9.28: Warping of the Cross Section

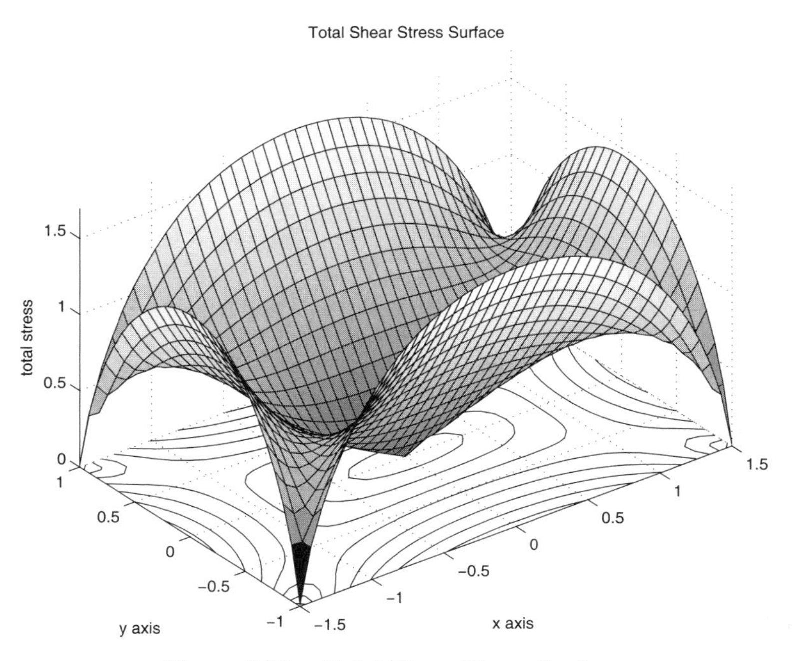

Figure 9.29: Total Shear Stress Surface

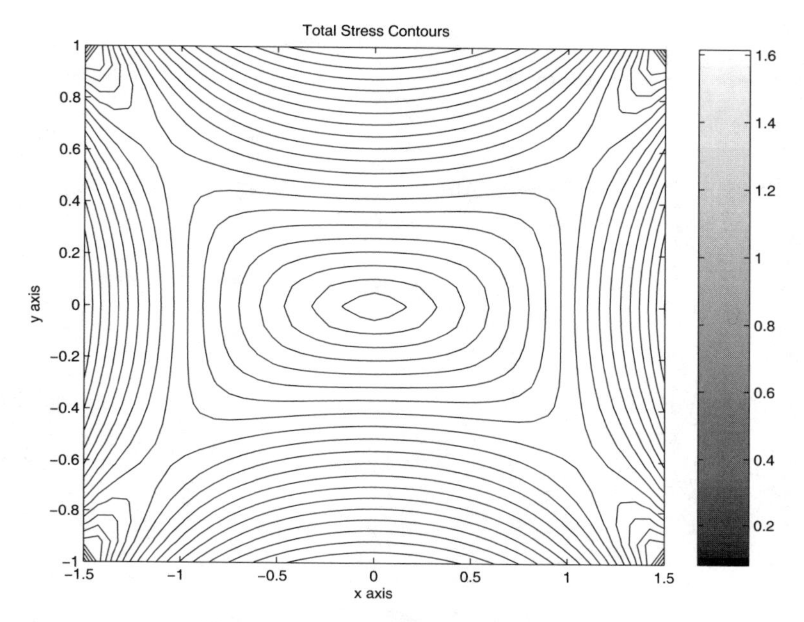

Figure 9.30: Total Stress Contours

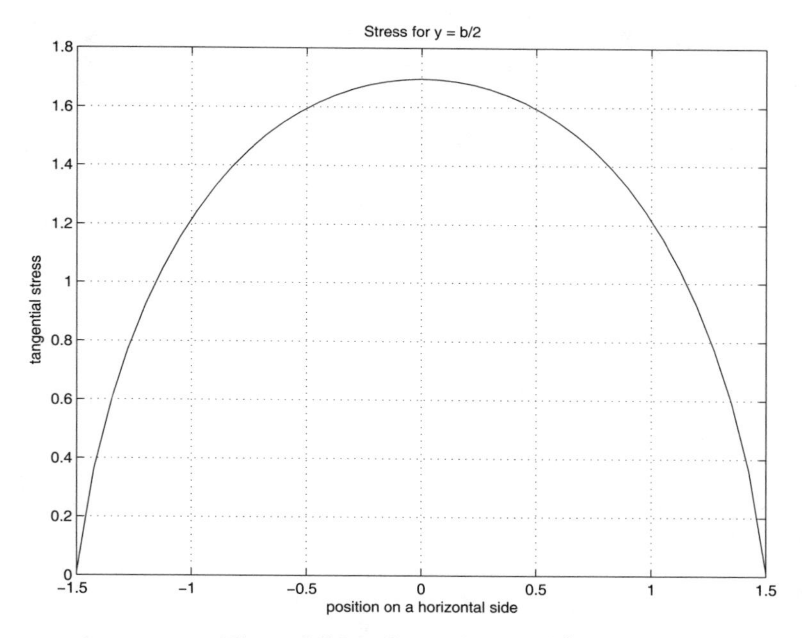

Figure 9.31: Stress for $y = b/2$

MATLAB Example

Output from Torsion Example

```
>> rector;

===  TORSIONAL STRESS CALCULATION IN A RECTANGULAR  ===
===      BEAM USING LEAST SQUARE APPROXIMATION      ===

Input the lengths of the horizontal and the vertical sides
(make the long side horizontal)
> ? 3,2

Input the number of terms used in the stress function
(30 terms is usually enough)
> ? 30

Press [Enter] to plot
the warping surface

Press [[Enter]] to plot the
total stress surface

Press [Enter] to plot the
stress contours

Press [Enter] to plot the maximum
stress on a rectangle side

The Maximum Shear Stress is 1.6951
at x = 0 and y = 1

All Done
>>
```

Program rector

```
1: function rector
2: % Example:  rector
3: % ~~~~~~~~~~~~~~~~~
4: % This program uses point matching to obtain an
5: % approximate solution for torsional stresses
6: % in a Saint Venant beam having a rectangular
7: % cross section. The complex stress function is
```

```
 8: % analytic inside the rectangle and has its
 9: % real part equal to abs(z*z)/2 on the
10: % boundary. The problem is solved approximately
11: % using a polynomial stress function which fits
12: % the boundary condition in the least square
13: % sense. Surfaces and contour curves describing
14: % the stress and deformation pattern in the
15: % beam cross section are drawn.
16: %
17: % User m functions required: recstrs
18:
19: clear;
20: fprintf('\n===  TORSIONAL STRESS CALCULATION');
21: fprintf(' IN A RECTANGULAR  ===');
22: fprintf('\n===      BEAM USING LEAST SQUARE ');
23: fprintf('APPROXIMATION      ===\n');
24: fprintf('\nInput the lengths of the ');
25: fprintf('horizontal and the vertical sides\n');
26: fprintf('(make the long side horizontal)\n');
27: u=input('> ? ','s'); u=eval(['[',u,']']);
28: a=u(1)/2; b=u(2)/2;
29:
30: % The boundary conditions are approximated in
31: % terms of the number of least square points
32: % used along the sides
33: nsegb=100; nsega=ceil(a/b*nsegb);
34: nsega=fix(nsega/2); nsegb=fix(nsegb/2);
35: fprintf('\nInput the number of terms ');
36: fprintf('used in the stress function');
37: fprintf('\n(30 terms is usually enough)\n');
38: ntrms=input('> ? ');
39:
40: % Define a grid for evaluation of stresses.
41: % Include the middle of each side.
42: nx=41; ny=fix(b/a*nx); ny=ny+1-rem(ny,2);
43:
44: [c,phi,stres,z] = ...
45:   recstrs(a,nsega,b,nsegb,ntrms,nx,ny);
46: [smax,k]=max(abs(stres(:))); zmax=z(:);
47: zmax=zmax(k); xmax=abs(real(zmax));
48: ymax=abs(imag(zmax));
49: disp(' '), disp(['The Maximum Shear ',...
50:                  'Stress is ',num2str(smax)]);
51: disp(['at x = ',num2str(xmax),' and y = ',...
52:                  num2str(ymax)]);
```

```
53: disp(' '); disp('All Done');
54:
55: %=============================================
56:
57: function [c,phi,stres,z]=...
58:   recstrs(a,nsega,b,nsegb,ntrms,nxout,nyout)
59: %
60: % [c,phi,stres,z]=...
61: %   recstrs(a,nsega,b,nsegb,ntrms,nxout,nyout)
62: % ~~~~~~~~~~~~~~~~~~~~~~~~~~~~~~~~~~~~~~~~~~~~
63: % This function uses least square fitting to
64: % obtain an approximate solution for torsional
65: % stresses in a Saint Venant beam having a
66: % rectangular cross section. The complex stress
67: % function is analytic inside the rectangle
68: % and has its real part equal to abs(z*z)/2 on
69: % the boundary. The problem is solved
70: % approximately using a polynomial stress
71: % function which fits the boundary condition
72: % in the least square sense. The beam is 2*a
73: % wide parallel to the x axis and 2*b deep
74: % parallel to the y axis. The shear stresses
75: % in the beam are given by the stress formula:
76: %
77: % (tauzx-i*tauzy)/(mu*alpha) = -i*conj(z)+f'(z)
78: %
79: % where
80: %
81: %   f(z)=i*sum( c(j)*z^(2*j-2), j=1:ntrms )
82: %
83: % and c(j) are real.
84: %
85: % a,b     - half the side lengths of the
86: %           horizontal and vertical sides
87: % nsega,  - numbers of subintervals used to
88: % nsegb     form the least square equations
89: % ntrms   - number of terms used in the
90: %           polynomial stress function
91: % nxout,  - number of grid points used to
92: % nyout     evaluate output
93: % c       - coefficients defining the stress
94: %           function
95: % phi     - values of the membrane function
96: % stres   - array of complex stress values
97: % z       - complex point array at which
```

```
98: %                stresses are found
99: %
100: % User m functions called:  none
101: %----------------------------------------------
102:
103: % Generate vector zbdry of boundary points
104: % for point matching.
105: zbdry=[a+i*b/nsega*(0:nsega-1)';
106:        i*b+a/nsegb*(nsegb:-1:0)'];
107:
108: % Determine a scaling parameter used to
109: % prevent occurrence of large numbers when
110: % high powers of z are used
111: s=max(a,b);
112:
113: % Form the least square equations to impose
114: % the boundary conditions.
115: neq=length(zbdry); amat=ones(neq,ntrms);
116: ztmp=(zbdry/s).^2; bvec=.5*abs(zbdry).^2;
117: for j=2:ntrms
118:   amat(:,j)=amat(:,j-1).*ztmp;
119: end
120:
121: % Solve the least square equations.
122: amat=real(amat); c=pinv(amat)*bvec;
123:
124: % Generate grid points to evaluate
125: % the solution.
126: xsid=linspace(-a,a,nxout);
127: ysid=linspace(-b,b,nyout);
128: [xg,yg]=meshgrid(xsid,ysid);
129: z=xg+i*yg; zz=(z/s).^2;
130:
131: % Evaluate the warping function
132: phi=-imag(polyval(flipud(c),zz));
133:
134: % Evaluate stresses and plot results
135: cc=(2*(1:ntrms)-2)'.*c;
136: stres=-i*conj(z)+i* ...
137:       polyval(flipud(cc),zz)./(z+eps*(z==0));
138: am=num2str(-a);ap=num2str(a);
139: bm=num2str(-b);bp=num2str(b);
140:
141: % Plot results
142: disp(' '), disp('Press [Enter] to plot')
```

```
143: disp('the warping surface'), pause
144: [pa,k]=max(abs(phi(:)));
145: Phi=a/4*sign(phi(k))/phi(k)*phi;
146: close, colormap('default')
147: surfc(xg,yg,Phi)
148: title('Warping of the Cross Section')
149: xlabel('x axis'), ylabel('y axis')
150: zlabel('transverse warping'); axis('equal')
151: shg, disp(' ')
152: disp('Press [[Enter]] to plot the')
153: disp('total stress surface'), pause
154: % print -deps warpsurf
155: 
156: surfc(xg,yg,abs(stres));
157: title('Total Shear Stress Surface')
158: xlabel('x axis'); ylabel('y axis')
159: zlabel('total stress'), axis('equal'), shg
160: disp(' '), disp('Press [Enter] to plot the')
161: disp('stress contours'), pause
162: % print -deps rectorst
163: 
164: contour(xg,yg,abs(stres),20); colorbar
165: title('Total Stress Contours');
166: xlabel('x axis'); ylabel('y axis')
167: shg, disp(' ')
168: disp('Press [Enter] to plot the maximum')
169: disp('stress on a rectangle side'), pause
170: % print -deps torcontu
171: 
172: plot(xsid,abs(stres(1,:)),'k');
173: grid; ylabel('tangential stress');
174: xlabel('position on a horizontal side');
175: title('Stress for y = b/2'); shg
176: % print -deps torstsid
```

Chapter 10

Eigenvalue Problems and Applications

10.1 Introduction

Eigenvalue problems occur often in mechanics, especially linear system dynamics, and elastic stability. Usually nontrivial solutions are sought for homogeneous systems of differential equations. For a few simple systems like the elastic string, or a rectangular membrane, the eigenvalues and eigenfunctions can be determined exactly. More often, some discretization methods such as finite difference or finite element methods are employed to reduce the system to a linear algebraic form which is numerically solvable. Several eigenvalue problems analyzed in earlier chapters reduced easily to algebraic form where the function **eig** could immediately produce the desired results. The present chapter deals with several instances where reduction to eigenvalue problems is more involved. We will also make some comparisons of exact, finite difference, and finite element analyses. Among the physical systems studied are Euler beams and columns, two-dimensional trusses, and elliptical membranes.

10.2 Approximation Accuracy in a Simple Eigenvalue Problem

One of the simplest but useful eigenvalue problems concerns determining nontrivial solutions of

$$y''(x) + \lambda^2 y(x) = 0, \ y(0) = y(1) = 0.$$

The eigenvalues and eigenfunctions are

$$y_n = \sin(n\pi x), \ 0 \le x \le 1, \ \text{where } \lambda_n = n\pi, \ n = 1, 2, 3, \ldots$$

It is instructive to examine the answers obtained for this problem using finite differences and spline approximations. We introduce a set of node points defined by

$$x_j = j\Delta, \ j = 0, 1, 2, \ldots, N+1, \ \Delta = 1/(N+1).$$

Then a finite difference description for the differential equation and boundary conditions is

$$y_{j-1} - 2y_j + y_{j+1} + \omega^2 y_j = 0, \;\; 1 \leq j \leq N, \;\;\; y_0 = y_{N+1} = 0, \, \omega = \Delta\lambda.$$

Solving the linear difference equation gives

$$\lambda_n^d = 2(N+1)\sin\left(\frac{\pi n}{2(N+1)}\right), \;\; n = 1, \ldots, N,$$

$$y_j^d = \sin\left(\frac{\pi j n}{N+1}\right), \;\; n = 1, \ldots, N, \;\;\; j = 0, \ldots, N+1$$

where the superscript d indicates a finite difference result. The ratio of the approximate eigenvalues to the exact eigenvalues is

$$\lambda_n^d \,/\, \lambda_n = \sin\left(\frac{\pi n}{2(N+1)}\right) \,/\, \left(\frac{\pi n}{2(N+1)}\right).$$

So, for large enough M, we get $\lambda_1^d \,/\, \lambda_1 = 1$ and $\lambda_N^d \,/\, \lambda_N = \frac{2}{\pi} \approx 0.63$. The smallest eigenvalue is quite accurate, but the largest eigenvalue is too low by about thirty-seven percent. This implies that the finite difference method is not very good for computing high order eigenvalues. For instance, to get $\lambda_{100}^d \,/\, \lambda_{100} = 0.999$ requires a rather high value of $N = 2027$.

An alternate approach to the finite difference method is to use a series representation

$$y(x) = \sum_{k=1}^{N} f_k(x)\, c_k$$

where the $f_k(x)$ vanish at the end points. We then seek a least-squares approximate solution imposing

$$\sum_{k=1}^{N} f_k''(\xi_j) c_k + \lambda^2 \sum_{k=1}^{N} f_k(\xi_j)\, c_k = 0$$

for a set of collocation points ξ_j, $j = 1 \ldots M$ with M taken much larger than N. With the matrix form of the last equation denoted as $B\,C + \lambda^2 A\,C = 0$, we make the error orthogonal to the columns of matrix A and get the resulting eigenvalue problem

$$(A \backslash B)\, C + \lambda^2\, C = 0$$

employing the generalized inverse of A. A short program **eigverr** written to compare the accuracy of the finite difference and the spline algorithms produced Figure 10.1. The program is also listed. The spline approximation method gives quite accurate results, particularly if no more than half of the computed eigenvalues are used.

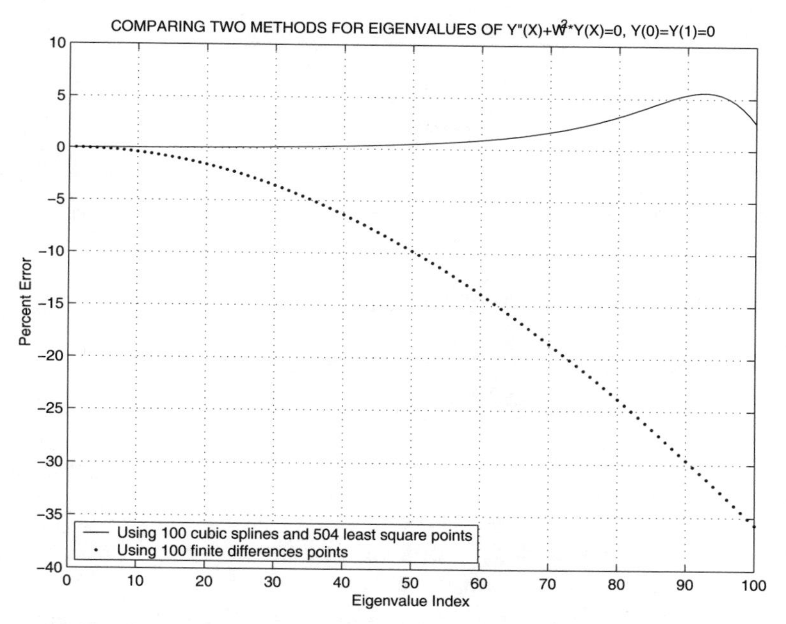

Figure 10.1: Comparing an eigenvalue computation using the least squares method and a second order finite differences method

Program eigverr

```
1: function eigverr(nfd,nspl,kseg)
2: % eigverr(nfd,nspl,kseg)
3: % This function compares two methods of computing
4: % eigenvalues corresponding to
5: %
6: % y"(x)+w^2*y(x)=0, y(0)=y(1)=0.
7: %
8: % Results are obtained using 1) finite differences
9: % and 2) cubic splines.
10: %
11: % nfd -  number of interior points used for the
12: %        finite difference equations
13: % nspl - number of interior points used for the
14: %        spline functions.
15: % kseg - the number of interior spline points is
16: %        kseg*(nspl+1)+nspl
17:
18: if nargin==0, nfd=100; nspl=100; kseg=4; end
19: [ws,es]=spleig(nspl,kseg); [wd,ed]=findieig(nfd);
20: str=['COMPARING TWO METHODS FOR EIGENVALUES ',...
21:      'OF Y"(X)+W^2*Y(X)=0, Y(0)=Y(1)=0'];
22: plot(1:nspl,es,'k-',1:nfd,ed,'k.')
23: title(str), xlabel('Eigenvalue Index')
24: ylabel('Percent Error'), Nfd=num2str(nfd);
25: Ns=num2str(nspl); M=num2str(nspl+(nspl+1)*kseg);
26: legend(['Using ',Ns,' cubic splines and ',...
27:         M,' least square points'],...
28:   ['Using ',Nfd,' finite differences points'],3)
29: grid on, shg
30: % print -deps eigverr
31:
32: %=========================================
33:
34: function [w,pcterr]=findieig(n)
35: % [w,pcterr]=findieig(n)
36: % This function determines eigenvalues of
37: % y''(x)+w^2*y(x)=0, y(0)=y(1)=0
38: % The solution uses an n point finite
39: % difference approximation
40: if nargin==0, n=100; end
41: a=2*eye(n,n)-diag(ones(n-1,1),1)...
```

```
42:    -diag(ones(n-1,1),-1);
43: w=(n+1)*sqrt(sort(eig(a))); we=pi*(1:n)';
44: pcterr=100*(w-we)./we;
45:
46: %==========================================
47:
48: function [w,pcterr]=spleig(n,nseg)
49: % [w,pcterr]=spleig(n,nseg)
50: % This function determines eigenvalues of
51: % y''(x)+w^2*y(x)=0, y(0)=y(1)=0
52: % The solution uses n spline basis functions
53: % and nseg*(n+1)+n least square points
54:
55: if nargin==0, n=100; nseg=1; end
56: nls=(n+1)*nseg+n; xls=(1:nls)'/(nls+1);
57: a=zeros(nls,n); b=a;
58: for k=1:n
59:    a(:,k)=splnf(k,n,1,xls,2);
60:    b(:,k)=splnf(k,n,1,xls);
61: end
62: w=sqrt(sort(eig(-b\a))); we=pi*(1:n)';
63: pcterr=100*(w-we)./we;
64:
65: %==========================================
66:
67: function y=splnf(n,N,len,x,ideriv)
68: % y=splnf(n,N,len,x,ideriv)
69: % This function computes the spline basis
70: % functions and derivatives
71: xd=len/(N+1)*(0:N+1)'; yd=zeros(N+2,1);
72: yd(n+1)=1;
73: if nargin<5, y=spline(xd,yd,x);
74: elseif ideriv==1, y=splined(xd,yd,x);
75: else, y=splined(xd,yd,x,2); end
76:
77: %==========================================
78:
79: % function val=splined(xd,yd,x,if2)
80: % See Appendix B
```

10.3 Stress Transformation and Principal Coordinates

The state of stress at a point in a three-dimensional continuum is described in terms of a symmetric 3 x 3 matrix $t = [t(\imath, \jmath)]$ where $t(\imath, \jmath)$ denotes the stress component in the direction of the $x_\imath$ axis on the plane with it normal in the direction of the $x_\jmath$ axis [9]. Suppose we introduce a rotation of axes defined by matrix b such that row $b(\imath, :)$ represents the components of a unit vector along the new $\tilde{x}_\imath$ axis measured relative to the initial reference state. It can be shown that the stress matrix $\tilde{t}$ corresponding to the new axis system can be computed by the transformation

$$\tilde{t} = btb^T.$$

Sometimes it is desirable to locate a set of reference axes such that $\tilde{t}$ is diagonal, in which case the diagonal components of $\tilde{t}$ represent the extremal values of normal stress. This means that seeking maximum or minimum normal stress on a plane leads to the same condition as requiring zero shear stress on the plane. The eigenfunction operation

```
[eigvecs,eigvals]=\beig(t);
```

applied to a symmetric matrix t produces an orthonormal set of eigenvectors stored in the columns of `eigvecs`, and a diagonal matrix `eigvals` having the eigenvalues on the diagonal. These matrices satisfy

$$\texttt{eigvecs}^T\ \texttt{t}\ \texttt{eigvecs} = \texttt{eigvals}.$$

Consequently, the rotation matrix b needed to transform to principal axes is simply the transpose of the matrix of orthonormalized eigenvectors. In other words, the eigenvectors of the stress tensor give the unit normals to the planes on which the normal stresses are extremal and the shear stresses are zero. The function **prnstres** performs the principal axis transformation.

10.3.1 Principal Stress Program

Function prnstres

```
1: function [pstres,pvecs]=prnstres(stress)
2: % [pstres,pvecs]=prnstres(stress)
3: % ~~~~~~~~~~~~~~~~~~~~~~~~~~~~~~~~~~~~
4: %
5: % This function computes principal stresses
6: % and principal stress directions for a three-
```

```
7: % dimensional stress state.
8: %
9: % stress - a vector defining the stress
10: %          components in the order
11: %          [sxx,syy,szz,sxy,sxz,syz]
12: %
13: % pstres - the principal stresses arranged in
14: %          ascending order
15: % pvecs  - the transformation matrix defining
16: %          the orientation of the principal
17: %          axis system.  The rows of this
18: %          matrix define the surface normals to
19: %          the planes on which the extremal
20: %          normal stresses act
21: %
22: % User m functions called:  none
23:
24: s=stress(:)';
25: s=([s([1 4 5]); s([4 2 6]); s([5 6 3])]);
26: [pvecs,pstres]=eig(s);
27: [pstres,k]=sort(diag(pstres));
28: pvecs=pvecs(:,k)';
29: if det(pvecs)<0, pvecs(3,:)=-pvecs(3,:); end
```

10.3.2 Principal Axes of the Inertia Tensor

A rigid body dynamics application quite similar to principal stress analysis occurs in the kinetic energy computation for a rigid body rotating with angular velocity $\omega = [\omega_x; \omega_y; \omega_z]$ about the reference origin [48]. The kinetic energy, K, of the body can be obtained using the formula

$$K = \frac{1}{2}\omega^T J \omega$$

with the inertia tensor J computed as

$$J = \iiint_V \rho \left[I \boldsymbol{r}^T \boldsymbol{r} - \boldsymbol{r}\boldsymbol{r}^T \right] dV,$$

where ρ is the mass per unit volume, I is the identity matrix, and $\boldsymbol{r}$ is the Cartesian radius vector. The inertia tensor is characterized by a symmetric matrix expressed in component form as

$$J = \iiint_V \begin{bmatrix} y^2+z^2 & -xy & -xz \\ -xy & x^2+z^2 & -yz \\ -xz & -yz & x^2+y^2 \end{bmatrix} dxdydz.$$

Under the rotation transformation

$$\tilde{r} = br \quad \text{with} \quad b^T b = I,$$

we can see that the inertia tensor transforms as

$$\tilde{J} = bJb^T$$

which is identical to the transformation law for the stress component matrix discussed earlier. Consequently, the inertia tensor will also possess principal axes which make the off-diagonal components zero. The kinetic energy is expressed more simply as

$$K = \frac{1}{2}\left(\omega_1^2 J_{11} + \omega_2^2 J_{22} + \omega_3^2 J_{33}\right)$$

where the components of ω and J must be referred to the principal axes. The function **prnstres** can also be used to locate principal axes of the inertia tensor since the same transformations apply. As an example of principal axis computation, consider the inertia tensor for a cube of side length A and mass M which has a corner at $(0,0,0)$ and edges along the coordinate axes. The inertia tensor is found to be

$$J = \begin{bmatrix} 2/3 & -1/4 & -1/4 \\ -1/4 & 2/3 & -1/4 \\ -1/4 & -1/4 & 2/3 \end{bmatrix} MA^2.$$

The computation

```
[pvl,pvc]=prnstres([2/3,2/3,2/3,-1/4,-1/4,-1/4]);
```

produces the results

$$\texttt{pvl} = \begin{bmatrix} 0.1667 \\ 0.9167 \\ 0.9167 \end{bmatrix}, \quad \texttt{pvc} = \begin{bmatrix} -0.5574 & -0.5574 & -0.5574 \\ -0.1543 & 0.7715 & -0.6172 \\ 0.8018 & -0.2673 & -0.5345 \end{bmatrix}.$$

This shows that the smallest possible inertial component equals $1/6 (\approx 0.1667)$ about the diagonal line through the origin while the maximal inertial moments of $11/12 (\approx 0.9167)$ occur about the axes normal to the diagonal.

10.4 Vibration of Truss Structures

Trusses are a familiar type of structure used in diverse applications such as bridges, roof supports, and power transmission towers. These structures can be envisioned as a series of nodal points among which various axially loaded members are connected. These members are assumed to act like linearly elastic springs supporting tension or compression. Typically, displacement constraints apply at one or more points to prevent movement of the truss from its supports. The natural frequencies and mode shapes of two-dimensional trusses are computed when the member properties are known and the loads of interest arise from inertial forces occurring during vibration. A similar analysis pertaining to statically loaded trusses has been published recently [102].

Consider an axially loaded member of constant cross section connected between nodes $\imath$ and $\jmath$ which have displacement components $(u_\imath, v_\imath)$ and $(u_\jmath, v_\jmath)$ as indicated in Figure 10.2. The member length is given by

$$\ell = \sqrt{(x_\jmath - x_\imath)^2 + (y_\jmath - y_\imath)^2},$$

and the member inclination is quantified by the trigonometric functions

$$c = \cos\theta = \frac{x_\jmath - x_\imath}{\ell} \quad \text{and} \quad s = \sin\theta = \frac{y_\jmath - y_\imath}{\ell}.$$

The axial extension for small deflections is

$$\Delta = (u_\jmath - u_\imath)c + (v_\jmath - v_\imath)s.$$

The axial force needed to extend a member having length ℓ, elastic modulus E, and cross section area A is given by

$$P_{\imath\jmath} = \frac{AE}{\ell}\Delta = \frac{AE}{\ell}[-c,\ -s,\ c,\ s]\,u_{\imath\jmath}$$

where

$$u_{\imath\jmath} = [u_\imath;\ v_\imath;\ u_\jmath;\ v_\jmath]$$

is a column matrix describing the nodal displacements of the member ends. The corresponding end forces are represented by

$$F_{\imath\jmath} = [F_{\imath x};\ F_{\imath y};\ F_{\jmath x};\ F_{\jmath y}] = P_{\imath\jmath}\,[-c,\ -s,\ c,\ s],$$

so that the end forces and end displacements are related by the matrix equation

$$F_{\imath\jmath} = K_{\imath\jmath}U_{\imath\jmath},$$

where the element stiffness matrix is

$$K_{\imath\jmath} = \frac{AE}{\ell}[-c;\ -s;\ c;\ s]\,[-c,\ -s,\ c,\ s].$$

In regard to mass effects in a member, we will assume that any transverse motion is negligible and half of the mass of each member can be lumped at each end. Hence the mass placed at each end would be $A\rho\ell/2$ where ρ is the mass per unit volume.

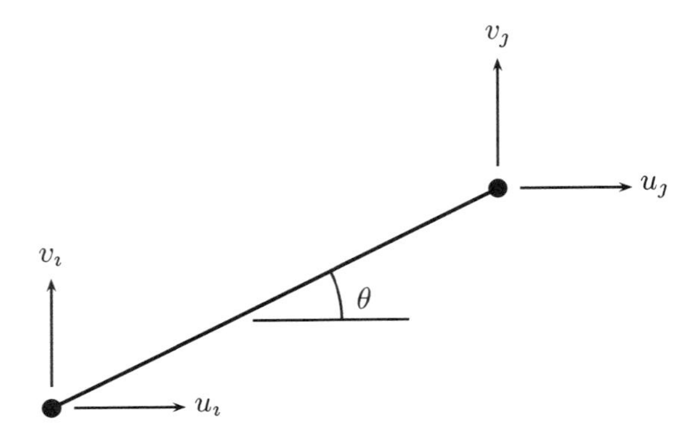

Figure 10.2: Typical Truss Element

The deflection of a truss with n nodal points can be represented using a generalized displacement vector and a generalized nodal force vector:

$$U = [u_1;\ v_1;\ u_2;\ v_2;\ \ldots;\ u_n;\ v_n]\ ,\ F = [F_{1x};\ F_{1y};\ F_{2x};\ F_{2y};\ \ldots;\ F_{nx};\ F_{ny}]\,.$$

When the contributions of all members in the network are assembled together, a global matrix relation results in the form

$$F = KU$$

where K is called the global stiffness matrix. Before we formulate procedures for assembling the global stiffness matrix, dynamical aspects of the problem will be discussed.

In the current application, the applied nodal forces are attributable to the acceleration of masses located at the nodes and to support reactions at points where displacement constraints occur. The mass concentrated at each node will equal half the sum of the masses of all members connected to the node. According to D'Alembert's principle [48] a particle having mass m and acceleration $\ddot{u}$ is statically equivalent to a force $-m\ddot{u}$. So, the equation of motion for the truss, without accounting for support reactions, is

$$KU = -M\ddot{U}$$

where M is a global mass matrix given by

$$M = \mathbf{diag}\left([m_1;\ m_1;\ m_2;\ m_2;\ \ldots;\ m_n;\ m_n]\right)$$

with m_i denoting the mass concentrated at the i'th node. The equation of motion $M\ddot{U} + KU = 0$ will also be subjected to constraint equations arising when some points are fixed or have roller supports. This type of support implies a matrix equation of the form $CU = 0$.

Natural frequency analysis investigates states-of-motion where each node of the structure simultaneously moves with simple harmonic motion of the same frequency. This means solutions are sought of the form

$$U = X\cos(\omega t)$$

where ω denotes a natural frequency and X is a modal vector describing the deflection pattern for the corresponding frequency. The assumed mode of motion implies $\ddot{U} = -\lambda U$ where $\lambda = \omega^2$. We are led to an eigenvalue problem of the form

$$KX = \lambda MX$$

with a side constraint $CX = 0$ needed to satisfy support conditions.

MATLAB provides the intrinsic functions **eig** and **null** which deal with the solution to this problem effectively. Using function **null** we can write

$$X = QY$$

where Q has columns that are an orthonormal basis for the null space of matrix C. Expressing the eigenvalue equation in terms of Y and multiplying both sides by Q^T gives

$$K_o Y = \lambda M_o Y$$

where

$$K_o = Q^T K Q \text{ and } M_o = Q^T M Q.$$

It can be shown from physical considerations that, in general, K and M are symmetric matrices such that K has real non-negative eigenvalues and M has real positive eigenvalues. This implies that M_o can be factored as

$$M_o = N^T N$$

where N is an upper triangular matrix. Then the eigenvalue problem can be rewritten as

$$K_1 Z = \lambda Z \; , \; Y = NZ \; , \; K_1 = \left(N^T\right)^{-1} K_o N^{-1}.$$

Because matrix K_1 will be real and symmetric, the intrinsic function **eig** generates orthonormal eigenvectors. The function **eigsym** used by program **trusvibs** produces a set of eigenvectors in the columns of X which satisfy generalized orthogonality conditions of the form

$$X^T M X = I \text{ and } X^T K X = \Lambda,$$

where Λ is a diagonal matrix containing the squares of the natural frequencies arranged in ascending order. The calculations performed in function **eigsym** illustrate the excellent matrix manipulative features that MATLAB embodies.

Before we discuss a physical example, the problem of assembling the global stiffness matrix will be addressed. It is helpful to think of all nodal displacements as if they were known and then compute the nodal forces by adding the stiffness contributions of all elements. Although the total force at each node results only from the forces in members touching the node, it is better to accumulate force contributions on an element-by-element basis instead of working node by node. For example, a member connecting node $\imath$ and node $\jmath$ will involve displacement components at row positions $2\imath - 1, 2\imath, 2\jmath - 1$, and $2\jmath$ in the global displacement vector and force components at similar positions in the generalized force matrix. Because principles of superposition apply, the stiffness contributions of individual members can be added, one member at a time, into the global stiffness matrix. This process is implemented in function **assemble** which also forms the mass matrix. First, selected points constrained to have zero displacement components are specified. Next the global stiffness and mass matrices are formed. This is followed by an eigenvalue analysis which yields the natural frequencies and the modal vectors. Finally the motion associated with each vibration mode is described by superimposing on the coordinates of each nodal point a multiple of the corresponding modal vector varying sinusoidally with time. Redrawing the structure produces an appearance of animated motion.

The complete program has several functions which should be studied individually for complete understanding of the methods developed. These functions and their purposes are summarized in the following table.

trusvibs	reads data and guides interactive input to animate the various vibration modes
crossdat	function typifying the nodal and element data to define a problem
assemble	assembles the global stiffness and mass data matrices
elmstf	forms the stiffness matrix and calculates the volume of an individual member
eigc	forms the constraint equations implied when selected displacement components are set to zero
eigsym	solves the constrained eigenvalue problem pertaining to the global stiffness and mass matrices
trifacsm	factors a positive definite matrix into upper and lower global triangular parts
drawtrus	draws the truss in deflected positions
cubrange	a utility routine to determine a window for drawing the truss without scale distortion

The data in function **crossdat** contains the information for node points, element data, and constraint conditions needed to define a problem. Once the data values are read, mode shapes and frequencies are computed and the user is allowed to observe the animation of modes ordered from the lowest to the highest frequency. The number of modes produced equals twice the number of nodal points minus the number of constraint conditions. The plot in Figure 10.3 shows mode eleven for the sample problem. This mode has no special significance aside from the interesting deflection pattern produced. The reader may find it instructive to run the program and select several modes by using input such as `3:5` or a single mode by specifying a single mode number.

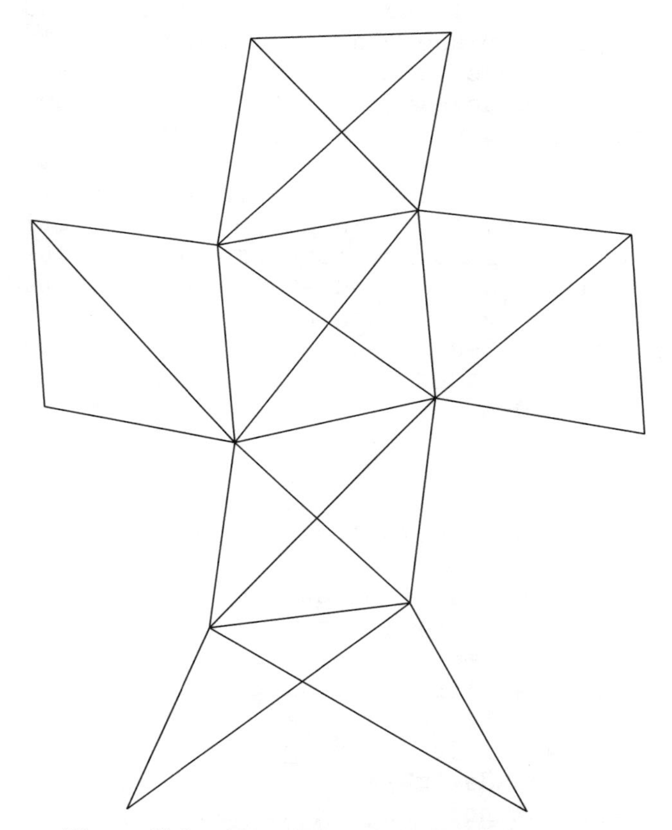

Figure 10.3: Truss Vibration Mode Number 11

10.4.1 Truss Vibration Program

Program trusvibs

```
1: function trusvibs
2: % Example: trusvibs
3: % ~~~~~~~~~~~~~~~~~~
4: %
5: % This program analyzes natural vibration modes
6: % for a general plane pin-connected truss. The
7: % direct stiffness method is employed in
8: % conjunction with eigenvalue calculation to
9: % evaluate the natural frequencies and mode
10: % shapes. The truss is defined in terms of a
11: % set of nodal coordinates and truss members
12: % connected to different nodal points. Global
13: % stiffness and mass matrices are formed. Then
14: % the frequencies and mode shapes are computed
15: % with provision for imposing zero deflection
16: % at selected nodes. The user is then allowed
17: % to observe animated motion of the various
18: % vibration modes.
19: %
20: % User m functions called:
21: %         eigsym, crossdat, drawtrus, eigc,
22: %         assemble, elmstf, cubrange
23:
24: global x y inode jnode elast area rho idux iduy
25: kf=1; idux=[]; iduy=[]; disp(' ')
26: disp(['Modal Vibrations for a Pin ', ...
27:       'Connected Truss']); disp(' ');
28:
29: % A sample data file defining a problem is
30: % given in crossdat.m
31: disp(['Give the name of a function which ', ...
32:       'creates your input data']);
33: disp(['Do not include .m in the name ', ...
34:       '(use crossdat as an example)']);
35: filename=input('>? ','s');
36: eval(filename); disp(' ');
37:
38: % Assemble the global stiffness and
39: % mass matrices
40: [stiff,masmat]= ...
```

```
41:   assemble(x,y,inode,jnode,area,elast,rho);
42:
43: % Compute natural frequencies and modal vectors
44: % accounting for the fixed nodes
45: ifixed=[2*idux(:)-1; 2*iduy(:)];
46: [modvcs,eigval]=eigc(stiff,masmat,ifixed);
47: natfreqs=sqrt(eigval);
48:
49: % Set parameters used in modal animation
50: nsteps=31; s=sin(linspace(0,6.5*pi,nsteps));
51: x=x(:); y=y(:); np=2*length(x);
52: bigxy=max(abs([x;y])); scafac=.05*bigxy;
53: highmod=size(modvcs,2); hm=num2str(highmod);
54:
55: % Show animated plots of the vibration modes
56: while 1
57:   disp('Give the mode numbers to be animated?');
58:   disp(['Do not exceed a total of ',hm, ...
59:         ' modes.']); disp('Input 0 to stop');
60:   if kf==1, disp(['Try 1:',hm]); kf=kf+1; end
61:   str=input('>? ','s');
62:   nmode=eval(['[',str,']']);
63:   nmode=nmode(find(nmode<=highmod));
64:   if sum(nmode)==0; break; end
65:   % Animate the various vibration modes
66:   hold off; clf; ovrsiz=1.1;
67:   w=cubrange([x(:),y(:)],ovrsiz);
68:   axis(w); axis('square'); axis('off'); hold on;
69:   for kk=1:length(nmode)  % Loop over each mode
70:     kkn=nmode(kk);
71:     titl=['Truss Vibration Mode Number ', ...
72:           num2str(kkn)];
73:     dd=modvcs(:,kkn); mdd=max(abs(dd));
74:     dx=dd(1:2:np); dy=dd(2:2:np);
75:     clf; pause(1);
76:     % Loop through several cycles of motion
77:     for jj=1:nsteps
78:       sf=scafac*s(jj)/mdd;
79:       xd=x+sf*dx; yd=y+sf*dy; clf;
80:       axis(w); axis('square'); axis('off');
81:       drawtrus(xd,yd,inode,jnode); title(titl);
82:       drawnow; figure(gcf);
83:     end
84:   end
85: end
```

```
86: disp(' ');
87:
88: %==============================================
89:
90: function crossdat
91: % [inode,jnode,elast,area,rho]=crossdat
92: % This function creates data for the truss
93: % vibration program. It can serve as a model
94: % for other configurations by changing the
95: % function name and data quantities
96: % Data set: crossdat
97: % ~~~~~~~~~~~~~~~~~~
98: %
99: % Data specifying a cross-shaped truss.
100: %
101: %----------------------------------------------
102:
103: global x y inode jnode elast area rho idux iduy
104:
105: % Nodal point data are defined by:
106: %   x - a vector of x coordinates
107: %   y - a vector of y coordinates
108: x=10*[.5 2.5 1 2 0 1 2 3 0 1 2 3 1 2];
109: y=10*[ 0   0 1 1 2 2 2 2 3 3 3 3 4 4];
110:
111: % Element data are defined by:
112: %   inode - index vector defining the I-nodes
113: %   jnode - index vector defining the J-nodes
114: %   elast - vector of elastic modulus values
115: %   area  - vector of cross section area values
116: %   rho   - vector of mass per unit volume
117: %           values
118: inode=[1 1 2 2 3 3 4 3 4 5 6 7 5 6 6 6 7 7 7 ...
119:        8 9 10 11 10 11 10 11 13];
120: jnode=[3 4 3 4 4 6 6 7 7 6 7 8 9 9 10 11 10 ...
121:        11 12 12 10 11 12 13 13 14 14 14];
122: elast=3e7*ones(1,28);
123: area=ones(1,28); rho=ones(1,28);
124:
125: % Any points constrained against displacement
126: % are defined by:
127: %   idux - indices of nodes having zero
128: %          x-displacement
129: %   iduy - indices of nodes having zero
130: %          y-displacement
```

```
131: idux=[1 2]; iduy=[1 2];
132:
133: %==============================================
134:
135: function drawtrus(x,y,i,j)
136: %
137: % drawtrus(x,y,i,j)
138: % ~~~~~~~~~~~~~~~~~
139: %
140: % This function draws a truss defined by nodal
141: % coordinates defined in x,y and member indices
142: % defined in i,j.
143: %
144: % User m functions called: none
145: %----------------------------------------------
146:
147: hold on;
148: for k=1:length(i)
149:   plot([x(i(k)),x(j(k))],[y(i(k)),y(j(k))]);
150: end
151:
152: %==============================================
153:
154: function [vecs,eigvals]=eigc(k,m,idzero)
155: %
156: % [vecs,eigvals]=eigc(k,m,idzero)
157: % ~~~~~~~~~~~~~~~~~~~~~~~~~~~~~~~
158: % This function computes eigenvalues and
159: % eigenvectors for the problem
160: %             k*x=eigval*m*x
161: % with some components of x constrained to
162: % equal zero. The imposed constraint is
163: %             x(idzero(j))=0
164: % for each component identified by the index
165: % matrix idzero.
166: %
167: % k       - a real symmetric stiffness matrix
168: % m       - a positive definite symmetric mass
169: %           matrix
170: % idzero  - the vector of indices identifying
171: %           components to be made zero
172: %
173: % vecs    - eigenvectors for the constrained
174: %           problem. If matrix k has dimension
175: %           n by n and the length of idzero is
```

```
176: %              m (with m<n), then vecs will be a
177: %              set on n-m vectors in n space
178: % eigvals -  eigenvalues for the constrained
179: %              problem. These are all real.
180: %
181: % User m functions called:  eigsym
182: %----------------------------------------------
183: 
184: n=size(k,1); j=1:n; j(idzero)=[];
185: c=eye(n,n); c(j,:)=[];
186: [vecs,eigvals]=eigsym((k+k')/2, (m+m')/2, c);
187: 
188: %==============================================
189: 
190: function [evecs,eigvals]=eigsym(k,m,c)
191: %
192: % [evecs,eigvals]=eigsym(k,m,c)
193: % ~~~~~~~~~~~~~~~~~~~~~~~~~~~~~~
194: % This function solves the constrained
195: % eigenvalue problem
196: %    k*x=(lambda)*m*x, with c*x=0.
197: % Matrix k must be real symmetric and matrix
198: % m must be symmetric and positive definite;
199: % otherwise, computed results will be wrong.
200: %
201: % k       - a real symmetric matrix
202: % m       - a real symmetric positive
203: %           definite matrix
204: % c       - a matrix defining the constraint
205: %           condition c*x=0. This matrix is
206: %           omitted if no constraint exists.
207: %
208: % evecs   - matrix of eigenvectors orthogonal
209: %           with respect to k and m. The
210: %           following relations apply:
211: %           evecs'*m*evecs=identity_matrix
212: %           evecs'*k*evecs=diag(eigvals).
213: % eigvals - a vector of the eigenvalues
214: %           sorted in increasing order
215: %
216: % User m functions called: none
217: %----------------------------------------------
218: 
219: if nargin==3
220:   q=null(c); m=q'*m*q; k=q'*k*q;
```

```
221: end
222: u=chol(m); k=u'\k/u; k=(k+k')/2;
223: [evecs,eigvals]=eig(k);
224: [eigvals,j]=sort(diag(eigvals));
225: evecs=evecs(:,j); evecs=u\evecs;
226: if nargin==3, evecs=q*evecs; end
227:
228: %==========================================
229:
230: function [stif,masmat]= ...
231:   assemble(x,y,id,jd,a,e,rho)
232: %
233: % [stif,masmat]=assemble(x,y,id,jd,a,e,rho)
234: % ~~~~~~~~~~~~~~~~~~~~~~~~~~~~~~~~~~~~~~~~~~
235: %
236: % This function assembles the global
237: % stiffness matrix and mass matrix for a
238: % plane truss structure. The mass density of
239: % each element equals unity.
240: %
241: % x,y    - nodal coordinate vectors
242: % id,jd  - nodal indices of members
243: % a,e    - areas and elastic moduli of members
244: % rho    - mass per unit volume of members
245: %
246: % stif   - global stiffness matrix
247: % masmat - global mass matrix
248: %
249: % User m functions called: elmstf
250: %------------------------------------------
251:
252: numnod=length(x); numelm=length(a);
253: id=id(:); jd=jd(:);
254: stif=zeros(2*numnod); masmat=stif;
255: ij=[2*id-1,2*id,2*jd-1,2*jd];
256: for k=1:numelm, kk=ij(k,:);
257:   [stfk,volmk]= ...
258:     elmstf(x,y,a(k),e(k),id(k),jd(k));
259:   stif(kk,kk)=stif(kk,kk)+stfk;
260:   masmat(kk,kk)=masmat(kk,kk)+ ...
261:                 rho(k)*volmk/2*eye(4,4);
262: end
263:
264: %==========================================
265:
```

```
266: function [k,vol]=elmstf(x,y,a,e,i,j)
267: %
268: % [k,vol]=elmstf(x,y,a,e,i,j)
269: % ~~~~~~~~~~~~~~~~~~~~~~~~~~~~~
270: %
271: % This function forms the stiffness matrix for
272: % a truss element. The member volume is also
273: % obtained.
274: %
275: % User m functions called: none
276: %----------------------------------------------
277:
278: xx=x(j)-x(i); yy=y(j)-y(i);
279: L=norm([xx,yy]); vol=a*L;
280: c=xx/L; s=yy/L; k=a*e/L*[-c;-s;c;s]*[-c,-s,c,s];
281:
282: %==============================================
283:
284: % function range=cubrange(xyz,ovrsiz)
285: % See Appendix B
```

10.5 Buckling of Axially Loaded Columns

Computing the buckling load and deflection curve for a slender axially loaded column leads to an interesting type of eigenvalue problem. Let us analyze a column of length L subjected to a critical value of axial load P just large enough to hold the column in a deflected configuration. Reducing the load below the critical value will allow the column to straighten out, whereas increasing the load above the buckling value will result in a structural failure. To prevent sudden collapse of structures using axially loaded members, designers must be able to calculate buckling loads corresponding to various end constraints. We will present an analysis allowing the flexural rigidity EI to vary along the length. Four common types of end conditions of interest are shown in Figure 10.4. For each of these systems we will assume that the coordinate origin is at the left end of the column[1] with $y(0) = 0$. Cases I and II involve statically determinate columns. Cases III and IV are different because unknown end reactions occur in the boundary conditions.

All four problems lead to a homogeneous linear differential equation subjected to homogeneous boundary conditions. All of these cases possess a trivial solution where $y(x)$ vanishes identically. However, the solutions of practical interest involve a nonzero deflection configuration which is only possible when P equals the buckling load. Finite difference methods can be used to accurately approximate

[1] Although columns are usually positioned vertically, we show them as horizontal for convenience.

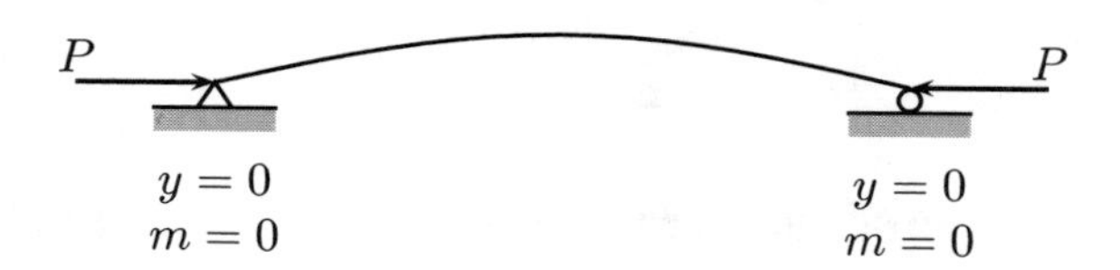

I) Pinned-Pinned

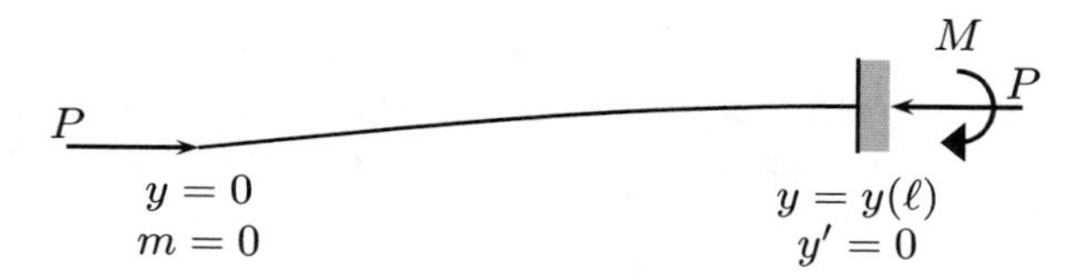

II) Free-Fixed

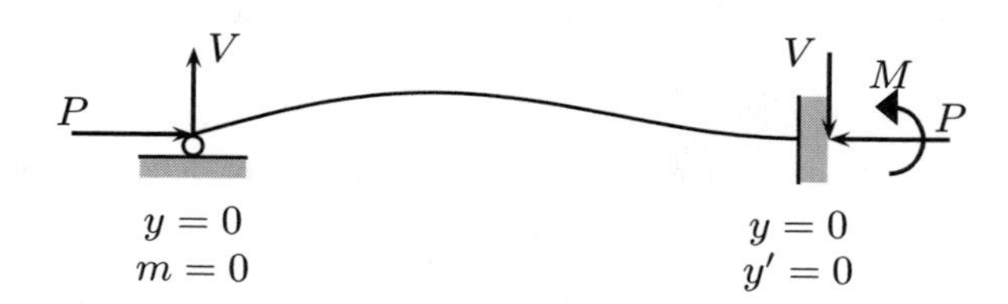

III) Pinned-Fixed

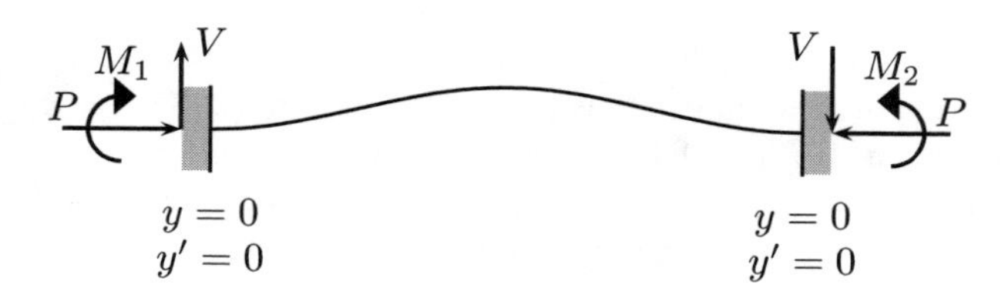

IV) Fixed-Fixed

Figure 10.4: **Buckling Configurations**

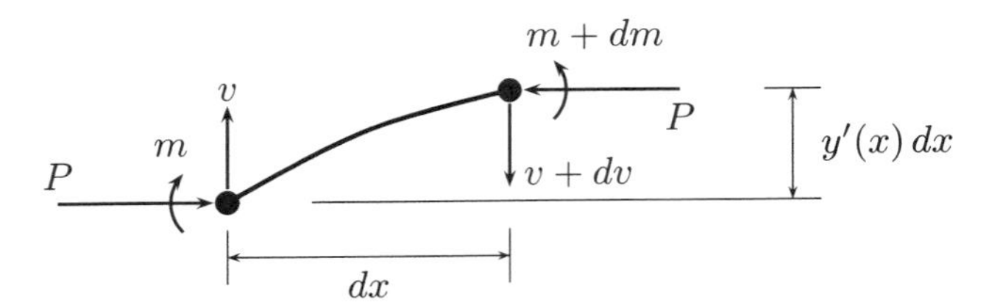

Figure 10.5: Beam Element Subjected to Axial Load

the differential equation and boundary conditions. In this manner we obtain a linear algebraic eigenvalue problem subjected to side constraints characterized by an underdetermined system of linear simultaneous equations.

Consider a beam element relating the bending moment m, the transverse shear v, the axial load P, and the transverse deflection y as shown in Figure 10.5. Equilibrium considerations imply

$$v'(x) = 0 \ , \ m'(x) + Py'(x) = v.$$

Since no transverse external loading acts on the column between the end supports, the shear v is constant. Differentiating the moment equation gives

$$m''(x) + Py''(x) = 0.$$

Furthermore, flexural deformation theory of slender elastic beams implies

$$EIy''(x) = m(x),$$

which leads to the following homogeneous differential equation governing the bending moment

$$EIm''(x) + Pm(x) = 0.$$

We need to find values of P allowing nontrivial solutions of this differential equation subject to the required homogeneous boundary conditions. The four types of end conditions shown in Figure 10.4 impose both deflection and moment conditions at the ends. Cases I and II can be formulated completely in terms of displacements because moment conditions evidently imply

$$EIy''(x) = m = -Py.$$

To handle cases III and IV, we need to relate the displacement and slope conditions at the ends to the bending moment. Let us denote the function $1/(EI)$ as $k(x)$ so that

$$y''(x) = k(x)m(x).$$

Integration gives

$$y'(x) = y'(0) + \int_0^x k(\xi)m(\xi)\, d\xi$$

and

$$y(x) = y(0) + y'(0)x + \int_0^x (x - \xi)k(\xi)m(\xi)\,d\xi.$$

The boundary conditions for the pinned-fixed case require that

$$\text{a)}\quad y(0) = 0, \quad \text{b)}\quad y'(L) = 0, \quad \text{c)}\quad y(L) = 0.$$

Condition *b*) requires

$$y'(0) = -\int_0^L k(\xi)m(\xi)\,d\xi,$$

whereas *a*) and *c*) combined lead to

$$y(L) = y(0) - L\int_0^L km\,d\xi + \int_0^L (L - \xi)km\,d\xi.$$

Consequently for Cases III and IV the governing equation is

$$EIm''(x) + Pm(x) = 0.$$

The boundary conditions for Case III are

$$m(0) = 0 \text{ and } \int_0^L xk(x)m(x)\,dx = 0.$$

The boundary conditions for Case IV are handled similarly. Since we must have $y'(0) = y'(L) = 0$ and $y(0) = y(L) = 0$, the conditions are

$$\int_0^L k(x)m(x)\,dx = 0 \text{ and } \int_0^L xk(x)m(x)\,dx = 0.$$

The results for each case require a nontrivial solution of a homogeneous differential equation satisfying homogeneous boundary conditions as summarized in the table below.

Each of these boundary value problems can be transformed to linear algebraic form by choosing a set of evenly spaced grid points across the span and approximating $y''(x)$ by finite differences. It follows from Taylor's series that

$$y''(x) = \frac{y(x-h) - 2y(x) + y(x+h)}{h^2} + O(h^2).$$

For sufficiently small h, we neglect the truncation error and write

$$y_j'' = \frac{y_{j-1} - 2y_j + y_{j+1}}{h^2}$$

where y_j is the approximation to y at $x = x_j = jh$ for $1 \le j \le n$, where the stepsize $h = L/(n+1)$. Thus we have

$$\frac{(EI)_j[y_{j-1} - 2y_j + y_{j+1}]}{h^2} + Py_j = 0$$

	Case	Differential Equation	Boundary Conditions
I:	pinned-pinned	$EIy''(x) + Py(x) = 0$	$y(0) = 0$ $y(L) = 0$
II:	free-fixed	$EIy''(x) + Py(x) = 0$	$y(0) = 0$ $y'(L) = 0$
III:	pinned-fixed	$EIm''(x) + Pm(x) = 0$	$m(0) = 0$ $\int_0^L k(x)m(x)\,dx = 0$
IV:	fixed-fixed	$EIm''(x) + Pm(x) = 0$	$\int_0^L k(x)m(x)\,dx = 0$ $\int_0^L xk(x)m(x)\,dx = 0$

Buckling Problem Summary

for Cases I or II, and

$$\frac{(EI)_j[m_{j-1} - 2m_j + m_{j+1}]}{h^2} + Pm_j = 0$$

for Cases III or IV. At the left end, either y or m is zero in all cases. Case I also has $y(L) = y_{n+1} = 0$. Case II requires $y'(L) = 0$. This is approximated in finite difference form as

$$y_{n+1} = \frac{4y_n - y_{n-1}}{3}$$

which implies for Case II that

$$y_n'' = \frac{2(y_{n-1} - y_n)}{3h^2}.$$

Cases III and IV are slightly more involved than I and II . The condition that

$$\int_0^L \frac{mx}{EI}\,dx = 0$$

can be formulated using the trapezoidal rule to give

$$b_1 \; * \; [m_1, \ldots, m_n, m_{n+1}]^T = 0,$$

where the asterisk indicates matrix multiplication involving a row matrix b_1 defined by

$$b_1 = [1, 1, \ldots, 1, 1/2] \; .* \; [x_1, x_2, \ldots, x_n, L] \; ./ \; [EI_1, \ldots, EI_n, EI_{n+1}].$$

Similarly, the condition

$$\int_0^L \frac{m}{EI}\, dx = 0$$

leads to

$$b_2 \, * \, [m_1, \ldots, m_n]^T + \frac{1}{2}\left[\frac{m_0}{EI_0} + \frac{m_{n+1}}{EI_{n+1}}\right] = 0$$

with

$$b_2 = \left[\frac{1}{EI_1}, \ldots, \frac{1}{EI_n}\right].$$

The first of these equations involving b_1 allows m_{n+1} to be eliminated in Case III, whereas the two equations involving b_1 and b_2 allow elimination of m_0 and m_{n+1} (the moments at $x = 0$ and $x = L$) for Case IV. Hence, in all cases, we are led to an eigenvalue problem typified as

$$EI_j(-m_{j-1} + 2m_j - m_{j+1}) = \lambda m_j$$

with $\lambda = h^2 P$, and we understand that the equations for $j = 1$ and $j = n$ may require modification to account for pertinent boundary conditions. We are led to solve

$$Am = \lambda m$$

where the desired buckling loads are associated with the smallest positive eigenvalue of matrix A. Cases I and II lead directly to the deflection curve forms. However, Cases III and IV require that the deflection curve be computed from the trapezoidal rule as

$$y'(x) = y'(0) + \int_0^x \frac{m}{EI}\, dx$$

and

$$y(x) = y(0) + y'(0) + x\int_0^x \frac{m}{EI}\, dx - \int_0^x \frac{mx}{EI}\, dx.$$

The deflection curves can be normalized to make $y_{\max}$ equal unity. This completes the formulation needed in the buckling analysis for all four cases studied. These solutions have been implemented in the program described later in this section. An example, which is solvable exactly, will be discussed next to demonstrate that the finite difference formulation actually produces good results.

10.5.1 Example for a Linearly Tapered Circular Cross Section

Consider a column with circular cross section tapered linearly from diameter h_1 at $x = 0$ to diameter h_2 at $x = L$. The moment of inertia is given by

$$I = \frac{\pi d^4}{64},$$

which leads to

$$EI = E_o I_o \left(1 + \frac{sx}{L}\right)^4,$$

where

$$s = \frac{h_2 - h_1}{h_1} \quad , \quad I_o = \frac{\pi h_1^4}{64}$$

and E_o is the elastic modulus which is assumed to have a constant value. The differential equation governing the moment in all cases (and for y in Case I or II) is

$$\left(1 + \frac{sx}{L}\right)^4 m''(x) + \frac{P}{E_o I_o} m(x) = 0.$$

This equation can be reduced to a simpler form by making a change of variables. Let us replace x and $m(x)$ by t and $g(t)$ defined by

$$t = \left(1 + \frac{sx}{L}\right)^{-1} \quad , \quad g(t) = t\, m(x).$$

The differential equation for $g(t)$ is found to be

$$g''(t) + \lambda^2 g(t) = 0 \text{ where } \lambda = \frac{L}{|s|}\sqrt{\frac{P}{E_o I_o}}.$$

Therefore,

$$m(x) = \left(1 + \frac{sx}{L}\right)\left[c_1 \sin\left(\frac{\lambda}{1 + \frac{sx}{L}}\right) + c_2 \cos\left(\frac{\lambda}{1 + \frac{sx}{L}}\right)\right]$$

where c_1 and c_2 are arbitrary constants found by imposing the boundary conditions. We will determine these constants for Cases I, II, and III. Case IV can be solved similarly and is left as an exercise for the reader.

To deal with Cases I, II, and III it is convenient to begin with a solution that vanishes at $x = 0$. A function satisfying this requirement has the form

$$m(x) = \left(1 + \frac{sx}{L}\right) \sin\left(\frac{\lambda}{1 + \frac{sx}{L}} - \lambda\right).$$

This equation can also represent the deflection curve for Cases I and II or the moment curve for Case III. Imposition of the remaining boundary conditions leads to an eigenvalue equation which is used to determine λ and the buckling load P. The deflection curve for Case I is taken as

$$y(x) = \left(1 + \frac{sx}{L}\right) \sin\left(\frac{\lambda}{1 + \frac{sx}{L}} - \lambda\right)$$

and the requirement that $y(L) = 0$ yields

$$\frac{\lambda s}{1 + s} = \left(\frac{s}{1 + s}\right)\left(\frac{L}{s}\sqrt{\frac{P}{E_o I_o}}\right) = \pi.$$

This means that the buckling load is

$$P = \frac{\pi^2 E_o I_o}{L^2}(1 + s)^2 \text{ where } s = \frac{h_2 - h_1}{h_1}.$$

Therefore the buckling load for the tapered column ($s \neq 0$) is simply obtained by multiplying the buckling load for the constant cross section column ($s = 0$) by a factor

$$(1+s)^2 = \left(\frac{h_2}{h_1}\right)^2.$$

This is also true for Cases III and IV, but is not true for Case II. Let us derive the characteristic equation for Case III. The constraint condition for the pinned-fixed case requires

$$\int_0^L \frac{x\, m(x)}{EI}\, dx = 0.$$

So we need

$$\int_0^L x\left(1+\frac{sx}{L}\right)^{-3} \sin\left(\frac{\lambda}{1+\frac{sx}{L}} - \lambda\right) dx = 0.$$

This equation can be integrated using the substitution $(1 + sx/L)^{-1} = t$. This leads to a characteristic equation of the form

$$\theta = \tan\theta\ ,\ \theta = \frac{\lambda s}{1+s} = \frac{L}{1+s}\sqrt{\frac{P}{E_o I_o}}.$$

The smallest positive root of this equation is $\theta = 4.4934$, which yields

$$P = \frac{20.1906 E_o I_o}{L^2}\,(1+s)^2 \ \text{ for Case III.}$$

Further analysis produces

$$P = \frac{4\pi^2 E_o I_o}{L^2}\,(1+s)^2 \ \text{ for Case IV.}$$

The characteristic equation for Case II can be obtained by starting with the Case I deflection equation and imposing the condition $y'(L) = 0$. This leads to

$$s\sin\theta + \theta\cos\theta = 0\ ,\ \theta = \frac{L}{1+s}\sqrt{\frac{P}{E_o I_o}}.$$

When $s = 0$, the smallest positive root of this equation is $\theta = \pi/2$. Therefore, the buckling load (when $s = 0$) is

$$P = \frac{\pi^2 E_o I_o}{4L^2}$$

for Case II, and the dependence on s found in the other cases does not hold for the free-fixed problem.

10.5.2 Numerical Results

The function **colbuc**, which uses the above relationships, was written to analyze variable depth columns using any of the four types of end conditions discussed. The program allows a piecewise linear variation of EI. The program employs the function **lintrp** for interpolation and the function **trapsum** to perform trapezoidal rule integration. Comparisons were made with results presented by Beer and Johnston [9] and a comprehensive handbook on stability [19]. We will present some examples to show how well the program works. It is known that a column of length L and constant cross section stiffness $E_o I_o$ has buckling loads of

$$\frac{\pi^2 E_o I_o}{L^2}, \quad \frac{\pi^2 E_o I_o}{(2L)^2}, \quad \frac{\pi^2 E_o I_o}{(0.6992L)^2}, \quad \frac{\pi^2 E_o I_o}{(0.5L)^2}$$

for the pinned-pinned, the free-fixed, the pinned-fixed, and the fixed-fixed end conditions respectively. These cases were verified using the program **colbuc**. Let us illustrate the capability of the program to approximately handle a discontinuous cross section change. We analyze a column twenty inches long consisting of a ten inch section pinned at the outer end and joined to a ten inch long section which is considered rigid and fixed at the outer end. We use $E_o I_o = 1$ for the flexible section and $E_o I_o = 10000$ for the rigid section. This configuration should behave much like a pinned-fixed column of length 100 with a buckling load of $(\pi/6.992)^2 = 0.2019$.

Using 100 segments (`nseq=100`) the program yields a value of 0.1976, which agrees within 2.2% of the expected value. A graph of the computed deflection configuration is shown in Figure 10.6. The code necessary to solve this problem is:

```
ei=[1 0; 1 10; 10000 10; 10000 20];
nseg=100; endc=3; len=20;
[p,y,x]=colbuc(len,ei,nseg,endc)
```

For a second example we consider a ten inch long column of circular cross section which is tapered from a one inch diameter at one end to a two inch diameter at the other end. We employ a fixed-fixed end condition and use $E_o = 1$. The theoretical results for this configuration indicate a buckling load of $\pi^3/400 = 0.07752$.

Using 100 segments the program produces a value of 0.07728, which agrees within 0.3% of the exact result. The code to generate this result utilizes function **eilt**:

```
ei=eilt(1,2,10,101,1);
[p,y,x]=colbuc(10,ei,100,4);
```

The examples presented illustrate the effectiveness of using finite difference methods in conjunction with the intrinsic eigenvalue solver in MATLAB to compute buck-

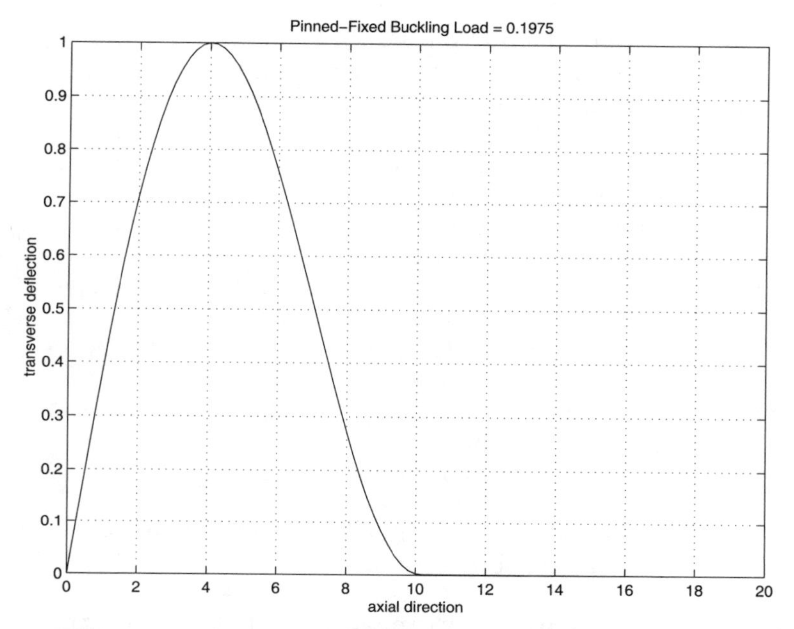

Figure 10.6: Analysis of Discontinuous Pinned-Fixed Column

ling loads. Furthermore, the provision for piecewise linear EI variation provided in the program is adequate to handle various column shapes.

Program Output and Code

Function colbuc

```
1: function [p,y,x]=colbuc(len,ei,nseg,endc)
2: % [p,y,x]=colbuc(len,ei,nseg,endc)
3: % ~~~~~~~~~~~~~~~~~~~~~~~~~~~~~~~~~~
4: %
5: % This function determines the Euler buckling
6: % load for a slender column of variable cross
7: % section which can have any one of four
8: % constraint conditions at the column ends.
9: %
10: % len  - the column length
11: % ei   - the product of Young's modulus and the
12: %        cross section moment of inertia. This
13: %        quantity is defined as a piecewise
```

```
14: %        linear function specified at one or
15: %        more points along the length.  ei(:,1)
16: %        contains ei values at points
17: %        corresponding to x values given in
18: %        ei(:,2). Values at intermediate points
19: %        are computed by linear interpolation
20: %        using function lintrp which allows
21: %        jump discontinuities in ei.
22: % nseg - the number of segments into which the
23: %        column is divided to perform finite
24: %        difference calculations.The stepsize h
25: %        equals len/nseg.
26: % endc - a parameter specifying the type of end
27: %        condition chosen.
28: %          endc=1, both ends pinned
29: %          endc=2, x=0 free, x=len fixed
30: %          endc=3, x=0 pinned, x=len fixed
31: %          endc=4, both ends fixed
32: %
33: % p    - the Euler buckling load of the column
34: % x,y  - vectors describing the shape of the
35: %        column in the buckled mode. x varies
36: %        between 0 and len. y is normalized to
37: %        have a maximum value of one.
38: %
39: % User m functions called:  lintrp, trapsum
40:
41: if nargin==0;
42:   ei=[1 0; 1 10; 1000 10; 1000 20];
43:   nseg=100; endc=3; len=20;
44: end
45:
46: % If the column has constant cross section,
47: % then ei can be given as a single number.
48: % Also, use at least 20 segments to assure
49: % that computed results will be reasonable.
50: if size(ei,1) < 2
51:   ei=[ei(1,1),0; ei(1,1),len];
52: end
53: nseg=max(nseg,30);
54:
55: if endc==1
56: % pinned-pinned case (y=0 at x=0 and x=len)
57:   str='Pinned-Pinned Buckling Load = ';
58:   h=len/nseg; n=nseg-1; x=linspace(h,len-h,n);
```

```
59:   eiv=lintrp(ei(:,2),ei(:,1),x);
60:   a=-diag(ones(n-1,1),1);
61:   a=a+a'+diag(2*ones(n,1));
62:   [yvecs,pvals]=eig(diag(eiv/h^2)*a);
63:   pvals=diag(pvals);
64:   % Discard any spurious nonpositive eigenvalues
65:   j=find(pvals<=0);
66:   if length(j)>0, pvals(j)=[]; yvecs(:,j)=[]; end
67:   [p,k]=min(pvals); y=[0;yvecs(:,k);0];
68:   [ym,j]=max(abs(y)); y=y/y(j); x=[0;x(:);len];
69: elseif endc==2
70: % free-fixed case (y=0 at x=0 and y'=0 at x=len)
71:   str='Free-Fixed Buckling Load = ';
72:   h=len/nseg; n=nseg-1; x=linspace(h,len-h,n);
73:   eiv=lintrp(ei(:,2),ei(:,1),x);
74:   a=-diag(ones(n-1,1),1);
75:   a=a+a'+diag(2*ones(n,1));
76:   % Zero slope at x=len implies
77:   % y(n+1)=4/3*y(n)-1/3*y(n-1). This
78:   % leads to y''(n)=(y(n-1)-y(n))*2/(3*h^2).
79:   a(n,[n-1,n])=[-2/3,2/3];
80:   [yvecs,pvals]=eig(diag(eiv/h^2)*a);
81:   pvals=diag(pvals);
82:   % Discard any spurious nonpositive eigenvalues
83:   j=find(pvals<=0);
84:   if length(j)>0, pvals(j)=[]; yvecs(:,j)=[]; end
85:   [p,k]=min(pvals); y=yvecs(:,k);
86:   y=[0;y;4*y(n)/3-y(n-1)/3]; [ym,j]=max(abs(y));
87:   y=y/y(j); x=[0;x(:);len];
88: elseif endc==3
89: % pinned-fixed case
90: % (y=0 at x=0 and x=len, y'=0 at x=len)
91:   str='Pinned-Fixed Buckling Load = ';
92:   h=len/nseg; n=nseg; x=linspace(h,len,n);
93:   eiv=lintrp(ei(:,2),ei(:,1),x);
94:   a=-diag(ones(n-1,1),1);
95:   a=a+a'+diag(2*ones(n,1));
96:   % Use a five point backward difference
97:   % approximation for the second derivative
98:   % at x=len.
99:   v=-[35/12,-26/3,19/2,-14/3,11/12];
100:  a(n,n:-1:n-4)=v; a=diag(eiv/h^2)*a;
101:  % Form the equation requiring zero deflection
102:  %   at x=len.
103:  b=x(:)'.*[ones(1,n-1),1/2]./eiv(:)';
```

```
104:   % Impose the homogeneous boundary condition
105:   q=null(b); [z,pvals]=eig(q'*a*q);
106:   pvals=diag(pvals);
107:   % Discard any spurious nonpositive eigenvalues
108:   k=find(pvals<=0);
109:   if length(k)>0, pvals(k)=[]; z(:,k)=[]; end;
110:   vecs=q*z; [p,k]=min(pvals); mom=[0;vecs(:,k)];
111:   % Compute the slope and deflection from
112:   %   moment values.
113:   yp=trapsum(0,len,mom./[1;eiv(:)]);
114:   yp=yp-yp(n+1);  y=trapsum(0,len,yp);
115:   [ym,j]=max(abs(y)); y=y/y(j); x=[0;x(:)];
116: else
117: % fixed-fixed case
118: % (y and y' both zero at each end)
119:   str='Fixed-Fixed Buckling Load = ';
120:   h=len/nseg; n=nseg+1; x=linspace(0,len,n);
121:   eiv=lintrp(ei(:,2),ei(:,1),x);
122:   a=-diag(ones(n-1,1),1);
123:   a=a+a'+diag(2*ones(n,1));
124:   % Use five point forward and backward
125:   % difference approximations for the second
126:   % derivatives at each end.
127:   v=-[35/12,-26/3,19/2,-14/3,11/12];
128:   a(1,1:5)=v; a(n,n:-1:n-4)=v;
129:   a=diag(eiv/h^2)*a;
130:   % Write homogeneous equations to make the
131:   % slope and deflection vanish at x=len.
132:   b=[1/2,ones(1,n-2),1/2]./eiv(:)';
133:   b=[b;x(:)'.*b];
134:   % Impose the homogeneous boundary conditions
135:   q=null(b); [z,pvals]=eig(q'*a*q);
136:   pvals=diag(pvals);
137:   % Discard any spurious nonpositive eigenvalues
138:   k=find(pvals<=0);
139:   if length(k>0), pvals(k)=[]; z(:,k)=[]; end;
140:   vecs=q*z; [p,k]=min(pvals); mom=vecs(:,k);
141:   % Compute the moment and slope from moment
142:   % values.
143:   yp=trapsum(0,len,mom./eiv(:));
144:   y=trapsum(0,len,yp);
145:   [ym,j]=max(abs(y)); y=y/y(j);
146: end
147:
148: close;
```

```
149: plot(x,y); grid on;
150: xlabel('axial direction');
151: ylabel('transverse deflection');
152: title([str,num2str(p)]); figure(gcf);
153: print -deps buck
154:
155: %===============================================
156:
157: function v=trapsum(a,b,y,n)
158: %
159: % v=trapsum(a,b,y,n)
160: % ~~~~~~~~~~~~~~~~~~
161: %
162: % This function evaluates:
163: %
164: %   integral(a=>x, y(x)*dx) for a<=x<=b
165: %
166: % by the trapezoidal rule (which assumes linear
167: % function variation between succesive function
168: % values).
169: %
170: % a,b - limits of integration
171: % y   - integrand which can be a vector valued
172: %       function returning a matrix such that
173: %       function values vary from row to row.
174: %       It can also be input as a matrix with
175: %       the row size being the number of
176: %       function values and the column size
177: %       being the number of components in the
178: %       vector function.
179: % n   - the number of function values used to
180: %       perform the integration.  When y is a
181: %       matrix then n is computed as the number
182: %       of rows in matrix y.
183: %
184: % v   - integral value
185: %
186: % User m functions called:  none
187: %-----------------------------------------------
188:
189: if isstr(y)
190:   % y is an externally defined function
191:   x=linspace(a,b,n)'; h=x(2)-x(1);
192:   Y=feval(y,x); % Function values must vary in
193:                 % row order rather than column
```

```
194:                     % order or computed results
195:                     % will be wrong.
196:   m=size(Y,2);
197: else
198:   % y is column vector or a matrix
199:   Y=y; [n,m]=size(Y); h=(b-a)/(n-1);
200: end
201: v=[zeros(1,m); ...
202:   h/2*cumsum(Y(1:n-1,:)+Y(2:n,:))];
203:
204: %==============================================
205:
206: function ei=eilt(h1,h2,L,n,E)
207: %
208: % ei=eilt(h1,h2,L,n,E)
209: % ~~~~~~~~~~~~~~~~~~~~
210: %
211: % This function computes the moment of inertia
212: % along a linearly tapered circular cross
213: % section and then uses that value to produce
214: % the product EI.
215: %
216: % h1,h2 - column diameters at each end
217: % L     - column length
218: % n     - number of points at which ei is
219: %         computed
220: % E     - Young's modulus
221: %
222: % ei    - vector of EI values along column
223: %
224: % User m functions called:  none
225: %----------------------------------------------
226:
227: if nargin<5, E=1; end;
228: x=linspace(0,L,n)';
229: ei=E*pi/64*(h1+(h2-h1)/L*x).^4;
230: ei=[ei(:),x(:)];
231:
232: %==============================================
233:
234: % function y=lintrp(xd,yd,x)
235: % See Appendix B
```

10.6 Accuracy Comparison for Euler Beam Natural Frequencies by Finite Element and Finite Difference Methods

Next we consider three different methods of natural frequency computation for a cantilever beam. Comparisons are made among results from: a) the solution of the frequency equation for the true continuum model; b) the approximation of the equations of motion using finite differences to replace the spatial derivatives; and c) the use of finite element methods yielding a piecewise cubic spatial interpolation of the displacement field. The first method is less appealing as a general tool than the last two methods because the frequency equation is difficult to obtain for geometries of variable cross section. Frequencies found using finite difference and finite element methods are compared with results from the exact model; and it is observed that the finite element method produces results that are superior to those from finite differences for comparable degrees of freedom. In addition, the natural frequencies and mode shapes given by finite elements are used to compute and animate the system response produced when a beam, initially at rest, is suddenly subjected to two concentrated loads.

10.6.1 Mathematical Formulation

The differential equation governing transverse vibrations of an elastic beam of constant depth is [69]

$$EI\frac{\partial^4 Y}{\partial X^4} = -\rho\frac{\partial^2 Y}{\partial T^2} + W(X,T) \quad 0 \le X \le \ell, \quad T \ge 0$$

where

$Y(X,T)$	–	transverse displacement,
X	–	horizontal position along the beam length,
T	–	time,
EI	–	product of moment of inertia and Young's modulus,
ρ	–	mass per unit length of the beam,
$W(X,T)$	–	external applied force per unit length.

In the present study, we consider the cantilever beam shown in Figure 10.7, having end conditions which are

$$Y(0,T) = 0\,, \quad \frac{\partial Y(0,T)}{\partial X} = 0\,, \quad EI\frac{\partial^2 Y(\ell,T)}{\partial X^2} = M_E(T)\,, \text{and } EI\frac{\partial^3 Y(\ell,T)}{\partial X^3} = V_E(T).$$

This problem can be expressed more concisely using dimensionless variables

$$x = \frac{X}{\ell}\,, \quad y = \frac{Y}{\ell} \text{ and } t = \sqrt{\frac{EI}{\rho}}\left(\frac{T}{\ell^2}\right).$$

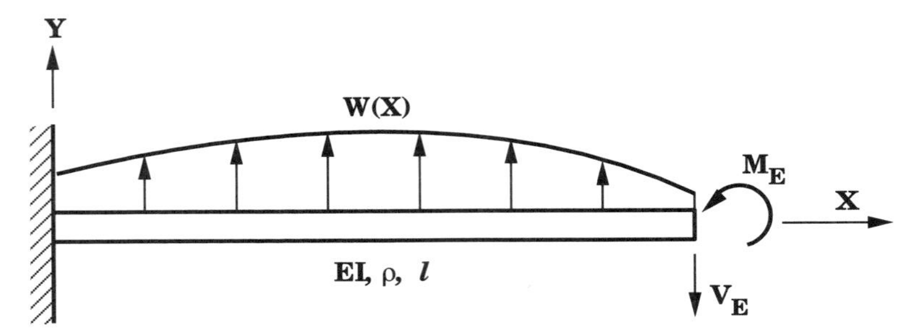

Figure 10.7: Cantilever Beam Subjected to Impact Loading

Then the differential equation becomes

$$\frac{\partial^4 y}{\partial x^4} = -\frac{\partial^2 y}{\partial t^2} + w(x, t),$$

and the boundary conditions reduce to

$$y(0, t) = 0\,, \quad \frac{\partial y}{\partial x}(0, t) = 0\,, \quad \frac{\partial^2 y}{\partial x^2}(1, t) = m_e(t) \text{ and } \frac{\partial^3 y}{\partial x^3}(1, t) = v_e(t)$$

where

$$w = (W\ell^3)/(EI)\,, \quad m_e = (M_E\ell)/(EI) \text{ and } v_e = (V_E\ell^2)/(EI).$$

The natural frequencies of the system are obtained by computing homogeneous solutions of the form $y(x, t) = f(x)\sin(\omega t)$ which exist when $w = m_e = v_e = 0$. This implies

$$\frac{d^4 f}{dx^4} = \lambda^4 f \text{ where } \lambda = \sqrt{\omega},$$

subject to

$$f(0) = 0\,, \; f'(0) = 0\,, \; f''(1) = 0\,, \; f'''(1) = 0.$$

The solution satisfying this fourth order differential equation with homogeneous boundary conditions has the form

$$f = [\cos(\lambda x) - \cosh(\lambda x)][\sin(\lambda) + \sinh(\lambda)] - [\sin(\lambda x) - \sinh(\lambda x)][\cos(\lambda) + \cosh(\lambda)],$$

where λ satisfies the frequency equation

$$p(\lambda) = \cos(\lambda) + 1/\cosh(\lambda) = 0.$$

Although the roots cannot be obtained explicitly, asymptotic approximations exist for large n:

$$\lambda_n = (2k - 1)\pi/2.$$

These estimates can be used as the starting points for finding approximate roots of the frequency equation using Newton's method:

$$\lambda_{NEW} = \lambda_{OLD} - p(\lambda_{OLD})/p'(\lambda_{OLD}).$$

The exact solution will be used to compare related results produced by finite difference and finite element methods. First we consider finite differences. The following difference formulas have a quadratic truncation error derivable from Taylor's series [1]:

$$\begin{aligned}
y'(x) &= [-y(x-h) + y(x+h)]/(2h), \\
y''(x) &= [y(x-h) - 2y(x) + y(x+h)]/h^2, \\
y'''(x) &= [-y(x-2h) + 2y(x-h) - 2y(x+h) + y(x+2h)]/(2h^3), \\
y''''(x) &= [y(x-2h) - 4y(x-h) + 6y(x) - 4y(x+h) + y(x+2h)]/h^4.
\end{aligned}$$

The step-size is $h = 1/n$ so that $x_\jmath = \jmath h$, $0 \le \jmath \le n$, where x_0 is at the left end and x_n is at the right end of the beam. It is desirable to include additional fictitious points x_{-1}, x_{n+1} and x_{n+2}. Then the left end conditions imply

$$y_0 = y_1 \text{ and } y_{-1} = y_1,$$

and the right end conditions imply

$$y_{n+1} = -y_{n-1} + 2y_n \text{ and } y_{n+2} = y_{n-2} - 4y_{n-1} + 4y_n.$$

Using these relations, the algebraic eigenvalue problem derived from the difference approximation is

$$\begin{aligned}
7y_1 - 4y_2 + y_3 &= \tilde{\lambda} y_1, \\
-4y_1 + 6y_2 - 4y_3 + y_4 &= \tilde{\lambda} f y_2, \\
y_{\jmath-2} - 4y_{\jmath-1} + 6y_\jmath - 4y_{\jmath+1} + y_{\jmath+2} &= \tilde{\lambda} y_\jmath \ , \ 2 < \jmath < (n-1), \\
y_{n-3} - 4y_{n-2} + 5y_{n-1} - 2y_n &= \tilde{\lambda} y_{n-1}, \\
2y_{n-2} - 4y_{n-1} + 2y_n &= \tilde{\lambda} y_n,
\end{aligned}$$

where $\tilde{\lambda} = h^4 \lambda$.

The finite element method leads to a similar problem involving global mass and stiffness matrices [54]. When we consider a single beam element of mass m and length ℓ, the elemental mass and stiffness matrices found using a cubically varying displacement approximation are

$$M_e = \frac{m}{420} \begin{bmatrix} 156 & 22\ell & 54 & -13\ell \\ 22\ell & 4\ell^2 & 13\ell & -3\ell^2 \\ 54 & 13\ell & 156 & -22\ell \\ -13\ell & -3\ell^2 & -22\ell & 4\ell^2 \end{bmatrix}, \quad K_e = \frac{EI}{\ell^3} \begin{bmatrix} 6 & 3\ell & -6 & 3\ell \\ 3\ell & 2\ell^2 & -3\ell & \ell^2 \\ -6 & -3\ell & 6 & -3\ell \\ 3\ell & \ell^2 & -3\ell & 2\ell^2 \end{bmatrix},$$

and the elemental equation of motion has the form

$$M_e Y_e'' + K_e Y_e = F_e$$

where

$$Y_e = [Y_1, Y_1^{'}, Y_2, Y_2^{'}]^T \text{ and } F_e = [F_1, M_1, F_2, M_2]^T$$

are generalized elemental displacement and force vectors. The global equation of motion is obtained as an assembly of element matrices and has the form

$$MY'' + KY = F.$$

A system with N elements involves $N + 1$ nodal points. For the cantilever beam studied here both Y_0 and $Y_0^{'}$ are zero. So removing these two variables leaves a system of $n = 2N$ unknowns. The solution of this equation in the case of a non-resonant harmonic forcing function will be discussed further. The matrix analog of the simple harmonic equation is

$$M\ddot{Y} + KY = F_1 \cos(\omega t) + F_2, \sin(\omega t)$$

with initial conditions

$$Y(0) = Y_0 \text{ and } \dot{Y}(0) = V_0.$$

The solution of this differential equation is the sum of a particular solution and a homogeneous solution:

$$Y = Y_P + Y_H,$$

where

$$Y_H = Y_1 \cos(\omega t) + Y_2 \sin(\omega t)$$

with

$$Y_j = (K - \omega^2 M)^{-1} F_j \qquad j = 1, 2.$$

This assumes that $K - \omega^2 M$ is nonsingular. The homogeneous equation satisfies the initial conditions

$$Y_H(0) = Y_0 - Y_1 \ , \ \dot{Y}_H(0) = V_0 - \omega Y_2.$$

The homogeneous solution components have the form

$$Y_{jH} = U_j \cos(\omega_j t + \phi_j)$$

where ω_j and U_j are natural frequencies and modal vectors satisfying the eigenvalue equation

$$KU_j = \omega_j^2 MU_j.$$

Consequently, the homogeneous solution completing the modal response is

$$Y_H(t) = \sum_{j=1}^{n} U_j[\cos(\omega_j t)c_j + \sin(\omega_j t)d_j/\omega_j]$$

where c_j and d_j are computed to satisfy the initial conditions which require

$$C = U^{-1}(Y_0 - Y_1) \text{ and } D = U^{-1}(V_0 - \omega Y_2).$$

The next section presents the MATLAB program. Natural frequencies from finite difference and finite element matrices are compared and modal vectors from the finite element method are used to analyze a time response problem.

10.6.2 Discussion of the Code

A program was written to compare exact frequencies from the original continuous beam model with approximations produced using finite differences and finite elements. The finite element results were also employed to calculate a time response by modal superposition for any structure that has general mass and stiffness matrices, and is subjected to loads which are constant or harmonically varying.

The code below is fairly long because various MATLAB capabilities are applied to three different solution methods. The following function summary involves nine functions, several of which were used earlier in the text.

cbfreq	driver to input data, call computation modules, and print results
cbfrqnwm	function to compute exact natural frequencies by Newton's method for root calculation
cbfrqfdm	forms equations of motion using finite differences and calls **eig** to compute natural frequencies
cbfrqfem	uses the finite element method to form the equation of motion and calls **eig** to compute natural frequencies and modal vectors
frud	function which solves the structural dynamics equation by methods developed in Chapter 7
examplmo	evaluates the response caused when a downward load at the middle and an upward load at the free end are applied
animate	plots successive positions of the beam to animate the motion
plotsave	plots the beam frequencies for the three methods. Also plots percent errors showing how accurate finite element and finite difference methods are
inputv	reads a sequence of numbers

Table 10.2: Functions Used in the Beam Code

Several characteristics of the functions assembled for this program are worth examining in detail. The next table contains remarks relevant to the code.

Routine	Line	Operation
Output		Natural frequencies are printed along with error percentages. The output shown here has been extracted from the actual output to show only the highest and lowest frequencies.
		continued on next page

continued from previous page

Routine	Line	Operation
cbfrqnwm	99	Asymptotic estimates are used to start a Newton method iteration.
	102-108	Root corrections are carried out for all roots until the correction to any root is sufficiently small.
cbfrqfdm	135-136	The equations of motion are formed without corrections for end conditions.
	138-145	End conditions are applied.
	149*150	**eig** computes the frequencies.
cbfrqfem	182-186	Form elemental mass matrix.
	189-192	Form elemental stiffness matrix.
	198-201	Global equations of motion are formed using an element by element loop.
	205	Boundary conditions are applied requiring zero displacement and slope at the left end, and zero moment and shear at the right end.
	208-214	Frequencies and modal vectors are computed. Note that modal vector computation is made optional since this takes longer than only computing frequencies.
frud		Compute time response by modal superposition. Theoretical details pertaining to this function appear in Chapter 7.
examplmo	292-296	The time step and maximum time for response calculation is selected.
	300-301	Function **frud** is used to compute displacement and rotation response. Only displacement is saved.
	304-307	Free end displacement is plotted.
	314-319	A surface showing displacement as a function of position and time is shown.
	324-326	Function **animate** is called.
animate	364-369	Window limits are determined.
	373-381	Each position is plotted. Then it is erased before proceeding to the next position.
plotsave		Plot and save graphs showing the frequencies and error percentages.

Table 10.3: Description of Code in Example

10.6.3 Numerical Results

The dimensionless frequency estimates from the finite difference and the finite element methods were compared for various numbers of degrees-of-freedom. Typical

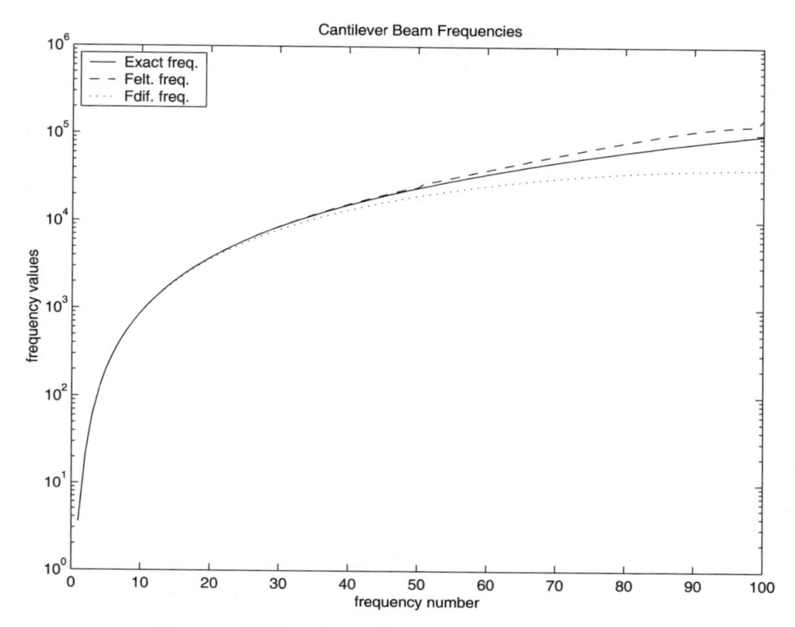

Figure 10.8: Cantilever Beam Frequencies

program output for $n = 100$ is shown at the end of this section. The frequency results and error percentages are shown in Figures 10.8 and 10.9. It is evident that the finite difference frequencies are consistently low and the finite element results are consistently high. The finite difference estimates degrade smoothly with increasing order. The finite element frequencies are surprisingly accurate for ω_k when $k < n/2$. At $k = n/2$ and $k = n$, the finite element error jumps sharply. This peculiar error jump halfway through the spectrum has also been observed in [54]. The most important and useful result seen from Figure 10.9 is that in order to obtain a particular number of frequencies, say N, which are accurate within 3.5%, it is necessary to employ more than $2N$ elements and keep only half of the predicted values.

The final result presented is the time response of a beam which is initially at rest when a concentrated downward load of five units is applied at the middle and a one unit upward load is applied at the free end. The time history was computed using function **frud**. Figure 10.10 shows the time history of the free end. Figure 10.11 is a surface plot illustrating how the deflection pattern changes with time. Finally, Figure 10.12 shows successive deflection positions produced by function **animate**. The output was obtained by suppressing the graph clearing option for successive configurations.

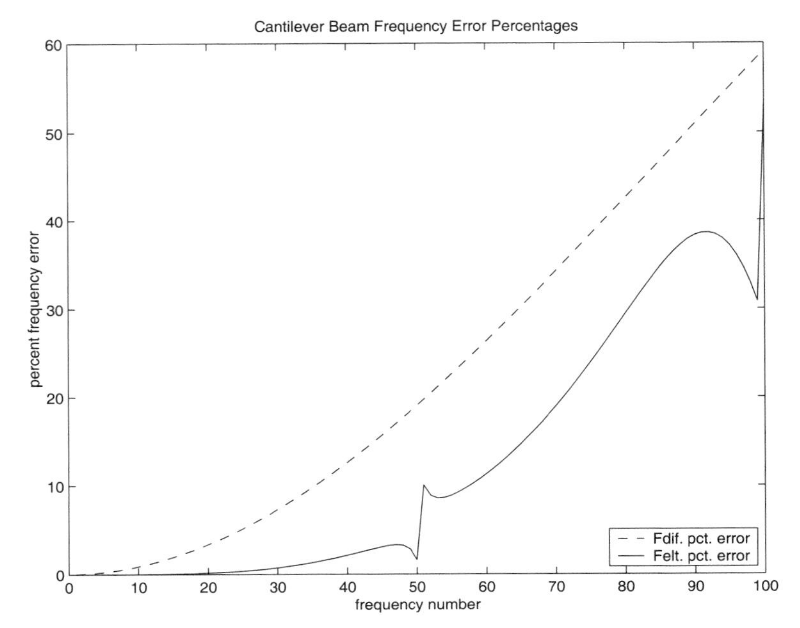

Figure 10.9: Cantilever Beam Frequency Error Percentages

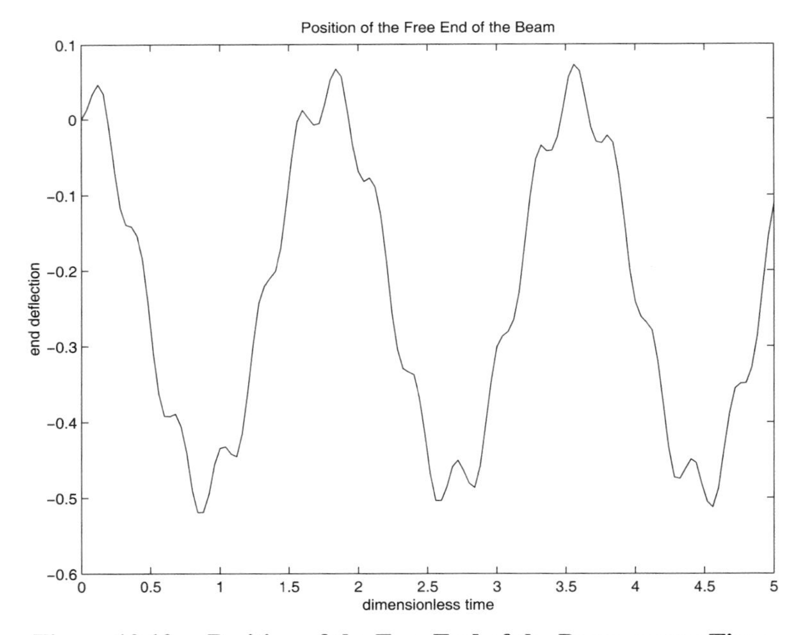

Figure 10.10: Position of the Free End of the Beam versus Time

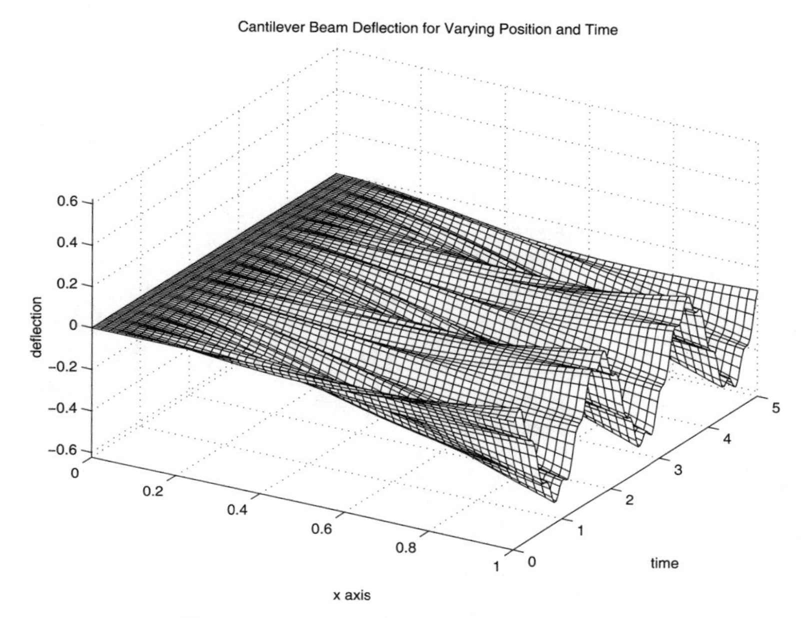

Figure 10.11: **Beam Deflection History**

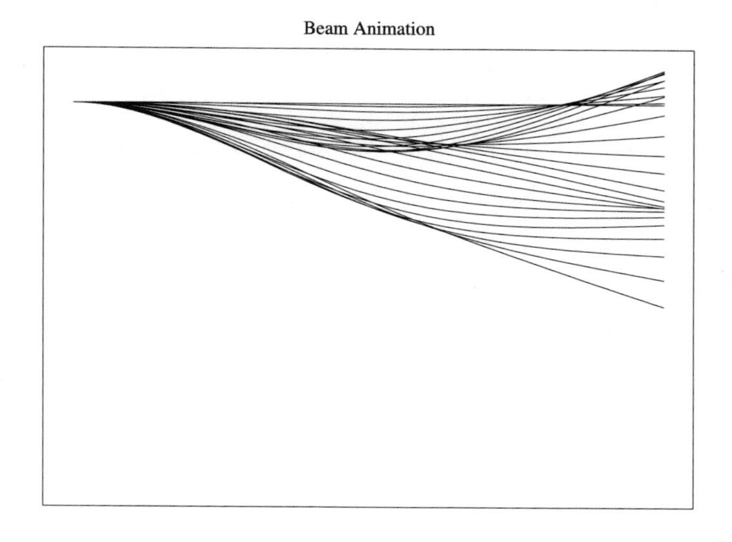

Figure 10.12: **Beam Animation**

MATLAB Example

Output from Example

```
>> cbfreq

CANTILEVER BEAM FREQUENCIES BY FINITE DIFFERENCE AND
          FINITE ELEMENT APPROXIMATION

Give the number of frequencies to be computed
(use an even number greater than 2)
? > 100

 freq.   exact.          fdif.      fd. pct.     felt.      fe. pct.
number   freq.           freq.       error       freq.       error

   1   3.51602e+00   3.51572e+00   -0.008   3.51602e+00   0.000
   2   2.20345e+01   2.20250e+01   -0.043   2.20345e+01   0.000
   3   6.16972e+01   6.16414e+01   -0.090   6.16972e+01   0.000
   4   1.20902e+02   1.20714e+02   -0.155   1.20902e+02   0.000
   5   1.99860e+02   1.99386e+02   -0.237   1.99860e+02   0.000
   6   2.98556e+02   2.97558e+02   -0.334   2.98558e+02   0.001
   7   4.16991e+02   4.15123e+02   -0.448   4.16999e+02   0.002
   8   5.55165e+02   5.51957e+02   -0.578   5.55184e+02   0.003
   9   7.13079e+02   7.07918e+02   -0.724   7.13119e+02   0.006
  10   8.90732e+02   8.82842e+02   -0.886   8.90809e+02   0.009
  11   1.08812e+03   1.07655e+03   -1.064   1.08826e+03   0.013
  12   1.30526e+03   1.28884e+03   -1.257   1.30550e+03   0.019
  13   1.54213e+03   1.51950e+03   -1.467   1.54252e+03   0.026
  14   1.79874e+03   1.76830e+03   -1.692   1.79937e+03   0.035
  15   2.07508e+03   2.03497e+03   -1.933   2.07605e+03   0.047
  16   2.37117e+03   2.31926e+03   -2.189   2.37261e+03   0.061
  17   2.68700e+03   2.62088e+03   -2.461   2.68908e+03   0.077
  18   3.02257e+03   2.93951e+03   -2.748   3.02551e+03   0.098
  19   3.37787e+03   3.27486e+03   -3.050   3.38197e+03   0.121
  20   3.75292e+03   3.62657e+03   -3.367   3.75851e+03   0.149

       ======  INTERMEDIATE LINES OF OUTPUT DELETED  ======

  90   7.90580e+04   3.88340e+04  -50.879   1.09328e+05  38.288
  91   8.08345e+04   3.90347e+04  -51.710   1.11989e+05  38.541
  92   8.26308e+04   3.92169e+04  -52.540   1.14512e+05  38.582
  93   8.44468e+04   3.93804e+04  -53.367   1.16860e+05  38.384
  94   8.62825e+04   3.95250e+04  -54.191   1.18999e+05  37.917
  95   8.81380e+04   3.96507e+04  -55.013   1.20889e+05  37.159
  96   9.00133e+04   3.97572e+04  -55.832   1.22496e+05  36.086
  97   9.19082e+04   3.98445e+04  -56.648   1.23786e+05  34.684
  98   9.38229e+04   3.99125e+04  -57.460   1.24730e+05  32.941
  99   9.57574e+04   3.99611e+04  -58.268   1.25305e+05  30.857
 100   9.77116e+04   3.99903e+04  -59.073   1.49694e+05  53.200

Evaluate the time response from two
concentrated loads.  One downward at the
middle and one upward at the free end.

input the time step and the maximum time
(0.04 and 5.0) are typical. Use 0,0 to stop
```

```
? .04,5

Evaluate the time response resulting from a
concentrated downward load at the middle and
an upward end load.

input the time step and the maximum time
(0.04 and 5.0) are typical. Use 0,0 to stop

? 0,0
```

Program cbfrq

```
1: function cbfreq
2: % Example:  cbfreq
3: % ~~~~~~~~~~~~~~~~
4: % This program computes approximate natural
5: % frequencies of a uniform depth cantilever
6: % beam using finite difference and finite
7: % element methods. Error results are presented
8: % which demonstrate that the finite element
9: % method is much more accurate than the finite
10: % difference method when the same matrix orders
11: % are used in computation of the eigenvalues.
12: %
13: % User m functions required:
14: %   cbfrqnwm, cbfrqfdm, cbfrqfem, frud,
15: %   examplmo, beamanim, plotsave, inputv
16:
17: clear, fprintf('\n\n')
18: fprintf('CANTILEVER BEAM FREQUENCIES BY ')
19: fprintf('FINITE DIFFERENCE AND')
20: fprintf(...
21: '\n          FINITE ELEMENT APPROXIMATION\n')
22:
23: fprintf('\nGive the number of frequencies ')
24: fprintf('to be computed')
25: fprintf('\n(use an even number greater ')
26: fprintf('than 2)\n'), n=input('? > ');
27: if rem(n,2) ~= 0, n=n+1; end
28:
29: % Exact frequencies from solution of
30: % the frequency equation
31: wex = cbfrqnwm(n,1e-12);
32:
```

```
33: % Frequencies for the finite
34: % difference solution
35: wfd = cbfrqfdm(n);
36:
37: % Frequencies, modal vectors, mass matrix,
38: % and stiffness matrix from the finite
39: % element solution.
40: nelts=n/2; [wfe,mv,mm,kk] = cbfrqfem(nelts);
41: pefdm=(wfd-wex)./(.01*wex);
42: pefem=(wfe-wex)./(.01*wex);
43:
44: nlines=17; nloop=round(n/nlines);
45: v=[(1:n)',wex,wfd,pefdm,wfe,pefem];
46: disp(' '), lo=1;
47: t1=[' freq.     exact.        fdif.' ...
48:     '         fd. pct.'];
49: t1=[t1,'    felt.       fe. pct.'];
50: t2=['number     freq.         freq.' ...
51:     '          error  '];
52: t2=[t2,'    freq.        error  '];
53: while lo < n
54:   disp(t1),disp(t2)
55:   hi=min(lo+nlines-1,n);
56:   for j=lo:hi
57:     s1=sprintf('\n %4.0f %13.5e %13.5e', ...
58:                v(j,1),v(j,2),v(j,3));
59:     s2=sprintf(' %9.3f %13.5e %9.3f', ...
60:                v(j,4),v(j,5),v(j,6));
61:     fprintf([s1,s2])
62:   end
63:   fprintf('\n\nPress [Enter] to continue\n\n');
64:   pause;
65:   lo=lo+nlines;
66: end
67: plotsave(wex,wfd,pefdm,wfe,pefem)
68: nfe=length(wfe); nmidl=nfe/2;
69: if rem(nmidl,2)==0, nmidl=nmidl+1; end
70: x0=zeros(nfe,1); v0=x0; w=0;
71: f1=zeros(nfe,1); f2=f1; f1(nfe-1)=1;
72: f1(nmidl)=-5;
73: xsav=examplmo(mm,kk,f1,f2,x0,v0,wfe,mv);
74: close; fprintf('All Done\n')
75:
76: %=============================================
77:
```

```
78: function z=cbfrqnwm(n,tol)
79: %
80: % z=cbfrqnwm(n,tol)
81: % ~~~~~~~~~~~~~~~~~
82: % Cantilever beam frequencies by Newton's
83: % method.  Zeros of
84: %        f(z) = cos(z) + 1/cosh(z)
85: % are computed.
86: %
87: % n   - Number of frequencies required
88: % tol - Error tolerance for terminating
89: %       the iteration
90: % z   - Dimensionless frequencies are the
91: %       squares of the roots of f(z)=0
92: %
93: % User m functions called:  none
94: %----------------------------------------------
95:
96: if nargin ==1, tol=1.e-5; end
97:
98: % Base initial estimates on the asymptotic
99: % form of the frequency equation
100: zbegin=((1:n)-.5)'*pi; zbegin(1)=1.875; big=10;
101:
102: % Start Newton iteration
103: while big > tol
104:   t=exp(-zbegin); tt=t.*t;
105:   f=cos(zbegin)+2*t./(1+tt);
106:   fp=-sin(zbegin)-2*t.*(1-tt)./(1+tt).^2;
107:   delz=-f./fp;
108:   z=zbegin+delz; big=max(abs(delz)); zbegin=z;
109: end
110: z=z.*z;
111:
112: %==============================================
113:
114: function [wfindif,mat]=cbfrqfdm(n)
115: %
116: % [wfindif,mat]=cbfrqfdm(n)
117: % ~~~~~~~~~~~~~~~~~~~~~~~~~
118: % This function computes approximate cantilever
119: % beam frequencies by the finite difference
120: % method. The truncation error for the
121: % differential equation and boundary
122: % conditions are of order h^2.
```

```
123: %
124: % n       - Number of frequencies to be
125: %           computed
126: % wfindif - Approximate frequencies in
127: %           dimensionless form
128: % mat     - Matrix having eigenvalues which
129: %           are the square roots of the
130: %           frequencies
131: %
132: % User m functions called:  none
133: %----------------------------------------------
134:
135: % Form the primary part of the frequency matrix
136: mat=3*diag(ones(n,1))-4*diag(ones(n-1,1),1)+...
137:     diag(ones(n-2,1),2); mat=(mat+mat');
138:
139: % Impose left end boundary conditions
140: % y(0)=0 and y'(0)=0
141: mat(1,[1:3])=[7,-4,1]; mat(2,[1:4])=[-4,6,-4,1];
142:
143: % Impose right end boundary conditions
144: % y''(1)=0 and y'''(1)=0
145: mat(n-1,[n-3:n])=[1,-4,5,-2];
146: mat(n,[n-2:n])=[2,-4,2];
147:
148: % Compute approximate frequencies and
149: % sort these values
150: w=eig(mat); w=sort(w); h=1/n;
151: wfindif=sqrt(w)/(h*h);
152:
153: %==============================================
154:
155: function [wfem,modvecs,mm,kk]= ...
156:                      cbfrqfem(nelts,mas,len,ei)
157: %
158: % [wfem,modvecs,mm,kk]=
159: %                    cbfrqfem(nelts,mas,len,ei)
160: % ~~~~~~~~~~~~~~~~~~~~~~~~~~~~~~~~~~~~~~~~~~~~~
161: % Determination of natural frequencies of a
162: % uniform depth cantilever beam by the Finite
163: % Element Method.
164: %
165: %  nelts   - number of elements in the beam
166: %  mas     - total beam mass
167: %  len     - total beam length
```

```
168: %  ei       - elastic modulus times moment
169: %             of inertia
170: %  wfem     - dimensionless circular frequencies
171: %  modvecs  - modal vector matrix
172: %  mm,kk    - reduced mass and stiffness
173: %             matrices
174: %
175: % User m functions called:  none
176: %----------------------------------------------
177:
178: if nargin==1, mas=1; len=1; ei=1; end
179: n=nelts; le=len/n; me=mas/n;
180: c1=6/le^2; c2=3/le; c3=2*ei/le;
181:
182: % element mass matrix
183: masselt=me/420* ...
184:         [  156,   22*le,     54,  -13*le
185:           22*le,  4*le^2,  13*le, -3*le^2
186:             54,   13*le,    156,  -22*le
187:          -13*le, -3*le^2, -22*le,  4*le^2];
188:
189: % element stiffness matrix
190: stifelt=c3*[ c1,  c2,  -c1,  c2
191:              c2,   2,  -c2,   1
192:             -c1, -c2,   c1, -c2
193:              c2,   1,  -c2,   2];
194:
195: ndof=2*(n+1); jj=0:3;
196: mm=zeros(ndof);  kk=zeros(ndof);
197:
198: % Assemble equations
199: for i=1:n
200:   j=2*i-1+jj; mm(j,j)=mm(j,j)+masselt;
201:   kk(j,j)=kk(j,j)+stifelt;
202: end
203:
204: % Remove degrees of freedom for zero
205: % deflection and zero slope at the left end.
206: mm=mm(3:ndof,3:ndof); kk=kk(3:ndof,3:ndof);
207:
208: % Compute frequencies
209: if nargout ==1
210:   wfem=sqrt(sort(real(eig(mm\kk))));
211: else
212:   [modvecs,wfem]=eig(mm\kk);
```

```
213:   [wfem,id]=sort(diag(wfem));
214:   wfem=sqrt(wfem); modvecs=modvecs(:,id);
215: end
216:
217: %==============================================
218:
219: function [t,x]= ...
220:         frud(m,k,f1,f2,w,x0,v0,wn,modvc,h,tmax)
221: %
222: % [t,x]=frud(m,k,f1,f2,w,x0,v0,wn,modvc,h,tmax)
223: % ~~~~~~~~~~~~~~~~~~~~~~~~~~~~~~~~~~~~~~~~~~~~~~
224: % This function employs modal superposition
225: % to solve
226: %
227: %    m*x'' + k*x = f1*cos(w*t) + f2*sin(w*t)
228: %
229: % m,k     - mass and stiffness matrices
230: % f1,f2   - amplitude vectors for the forcing
231: %           function
232: % w       - forcing frequency not matching any
233: %           natural frequency component in wn
234: % wn      - vector of natural frequency values
235: % x0,v0   - initial displacement and velocity
236: %           vectors
237: % modvc   - matrix with modal vectors as its
238: %           columns
239: % h,tmax  - time step and maximum time for
240: %           evaluation of the solution
241: % t       - column of times at which the
242: %           solution is computed
243: % x       - solution matrix in which row j
244: %           is the solution vector at
245: %           time t(j)
246: %
247: % User m functions called:  none
248: %----------------------------------------------
249:
250: t=0:h:tmax; nt=length(t); nx=length(x0);
251: wn=wn(:); wnt=wn*t;
252:
253: % Evaluate the particular solution.
254: x12=(k-(w*w)*m)\[f1,f2];
255: x1=x12(:,1); x2=x12(:,2);
256: xp=x1*cos(w*t)+x2*sin(w*t);
257:
```

```
258: % Evaluate the homogeneous solution.
259: cof=modvc\[x0-x1,v0-w*x2];
260: c1=cof(:,1)'; c2=(cof(:,2)./wn)';
261: xh=(modvc.*c1(ones(1,nx),:))*cos(wnt)+...
262:    (modvc.*c2(ones(1,nx),:))*sin(wnt);
263:
264: % Combine the particular and
265: % homogeneous solutions.
266: t=t(:); x=(xp+xh)';
267:
268: %==============================================
269:
270: function x=examplmo(mm,kk,f1,f2,x0,v0,wfe,mv)
271: %
272: % x=examplmo(mm,kk,f1,f2,x0,v0,wfe,mv)
273: % ~~~~~~~~~~~~~~~~~~~~~~~~~~~~~~~~~~~~~~
274: % Evaluate the response caused when a downward
275: % load at the middle and an upward load at the
276: % free end is applied.
277: %
278: % mm, kk - mass and stiffness matrices
279: % f1, f2 - forcing function magnitudes
280: % x0, v0 - initial position and velocity
281: % wfe    - forcing function frequency
282: % mv     - matrix of modal vectors
283: %
284: % User m functions called:  frud, beamanim, inputv
285: %----------------------------------------------
286:
287: w=0; n=length(x0); t0=0; x=[];
288: s1=['\nEvaluate the time response from two',...
289:   '\nconcentrated loads. One downward at the',...
290:   '\nmiddle and one upward at the free end.'];
291: while 1
292:   fprintf(s1), fprintf('\n\n')
293:   fprintf('Input the time step and ')
294:   fprintf('the maximum time ')
295:   fprintf('\n(0.04 and 5.0) are typical.')
296:   fprintf(' Use 0,0 to stop\n')
297:   [h,tmax]=inputv;
298:   if norm([h,tmax])==0 | isnan(h), return, end
299:   disp(' ')
300:
301:   [t,x]= ...
302:      frud(mm,kk,f1,f2,w,x0,v0,wfe,mv,h,tmax);
```

```
303:    x=x(:,1:2:n-1); x=[zeros(length(t),1),x];
304:    [nt,nc]=size(x); hdist=linspace(0,1,nc);
305: 
306:    clf, plot(t,x(:,nc),'k-')
307:    title('Position of the Free End of the Beam')
308:    xlabel('dimensionless time')
309:    ylabel('end deflection'), figure(gcf)
310:    disp('Press [Enter] for a surface plot of')
311:    disp('transverse deflection versus x and t')
312:    pause
313:    print -deps endpos1
314:    xc=linspace(0,1,nc); zmax=1.2*max(abs(x(:)));
315: 
316:    clf, surf(xc,t,x), view(30,35)
317:    colormap([1 1 1])
318:    axis([0,1,0,tmax,-zmax,zmax])
319:    xlabel('x axis'); ylabel('time')
320:    zlabel('deflection')
321:    title(['Cantilever Beam Deflection ' ...
322:           'for Varying Position and Time'])
323:    figure(gcf);
324:    print -deps endpos2
325:    disp(' '), disp(['Press [Enter] to animate',...
326:          ' the beam motion'])
327:    pause
328: 
329:    titl='Cantilever Beam Animation';
330:    xlab='x axis'; ylab='displacement';
331:    beamanim(hdist,x,0.1,titl,xlab,ylab), close
332: end
333: 
334: %===============================================
335: 
336: % function beamanim(x,u,tpause,titl,xlabl,ylabl)
337: % See Appendix B
338: 
339: %===============================================
340: 
341: function plotsave(wex,wfd,pefd,wfe,pefem)
342: %
343: % function plotsave(wex,wfd,pefd,wfe,pefem)
344: % ~~~~~~~~~~~~~~~~~~~~~~~~~~~~~~~~~~~~~~~~~
345: % This function plots errors in frequencies
346: % computed by two approximate methods.
347: %
```

```
348: % wex         - exact frequencies
349: % wfd         - finite difference frequencies
350: % wfe         - finite element frequencies
351: % pefd,pefem - percent errors by both methods
352: %
353: % User m functions called:  none
354: %----------------------------------------------
355:
356: % plot results comparing accuracy
357: % of both frequency methods
358: w=[wex(:);wfd(:);wfd];
359: wmin=min(w); wmax=max(w);
360: n=length(wex); wht=wmin+.001*(wmax-wmin);
361: j=1:n;
362:
363: semilogy(j,wex,'k-',j,wfe,'k--',j,wfd,'k:')
364: title('Cantilever Beam Frequencies')
365: xlabel('frequency number')
366: ylabel('frequency values')
367: legend('Exact freq.','Felt. freq.', ...
368:        'Fdif. freq.',2); figure(gcf)
369: disp(['Press [Enter] for a frequency ',...
370:       'error plot']), pause
371: print -deps beamfrq1
372:
373: plot(j,abs(pefd),'k--',j,abs(pefem),'k-')
374: title(['Cantilever Beam Frequency ' ...
375:        'Error Percentages'])
376: xlabel('frequency number')
377: ylabel('percent frequency error')
378: legend('Fdif. pct. error','Felt. pct. error',4)
379: figure(gcf)
380: disp(['Press [Enter] for a transient ',...
381: 'response calculation'])
382: pause
383: print -deps beamfrq2
384:
385: %=============================================
386:
387: % function varargout=inputv(prompt)
388: % See Appendix B
```

10.7 Vibration Modes of an Elliptic Membrane

10.7.1 Analytical Formulation

Examples using eigenvalues and modal functions of rectangular or circular membranes were presented in chapter 9. In this section we analyze modal vibrations of an elliptic membrane. In this case the natural frequencies and modal functions cannot be obtained easily in explicit form. The problem can be formulated in elliptical coordinates leading to Mathieu type differential equations [74]. Library routines to compute these functions are not widely available; so, a different approach is employed using least squares approximation and the MATLAB function **eig**. Consider a membrane with major and minor semi-diameters a and b. The analytic function $z = h\,\cosh(\varsigma)$ where $h = \sqrt{a^2 - b^2}$ and $\zeta = \xi + i\,\eta$ maps the rectangle defined by $0 \leq \xi \leq R = \tanh^{-1}(b/a), \quad -\pi \leq \eta \leq \pi$ onto the interior of the ellipse. This transformation takes lines of constant ξ into a system of confocal ellipses and lines of constant η into hyperbolas intersecting the ellipses orthogonally. The following function was used to produce the elliptic coordinate plot in Figure 10.13.

```
function z = elipmap(a,b,neta,nxi)
h=sqrt(a^2-b^2); R=atanh(b/a);
[xi,eta]=meshgrid(...
linspace(0,R,nxi),linspace(-pi,pi,neta));
z=h*cosh(xi+i*eta); x=real(z); y=imag(z);
plot(x,y,'k',x',y','k')
title('ELLIPTICAL COORDINATE SYSTEM')
xlabel('x axis'), ylabel('y axis')
axis equal, grid off, shg
```

Transforming the wave equation to (ξ, η) coordinates gives

$$U_{\xi\xi} + U_{\eta\eta} = \frac{h^2}{2}[\cosh(2\xi) - \cos(2\eta)]\,U_{tt},$$

and assuming separable solutions of the form

$$U = f(\eta)g(\xi)\sin(\Omega\,t)$$

leads to

$$\frac{f''(\eta)}{f(\eta)} + \frac{g''(\xi)}{g(\xi)} = -\lambda\,[\cosh(2\xi) - \cos(2\eta)],$$

where $\lambda = \Omega^2 h^2/2$. So f and g are found to satisfy the following two Mathieu type differential equations:

$$f''(\eta) + [\alpha - \lambda\cos(2\eta)]f(\eta) = 0, \quad -\pi \leq \eta \leq \pi$$

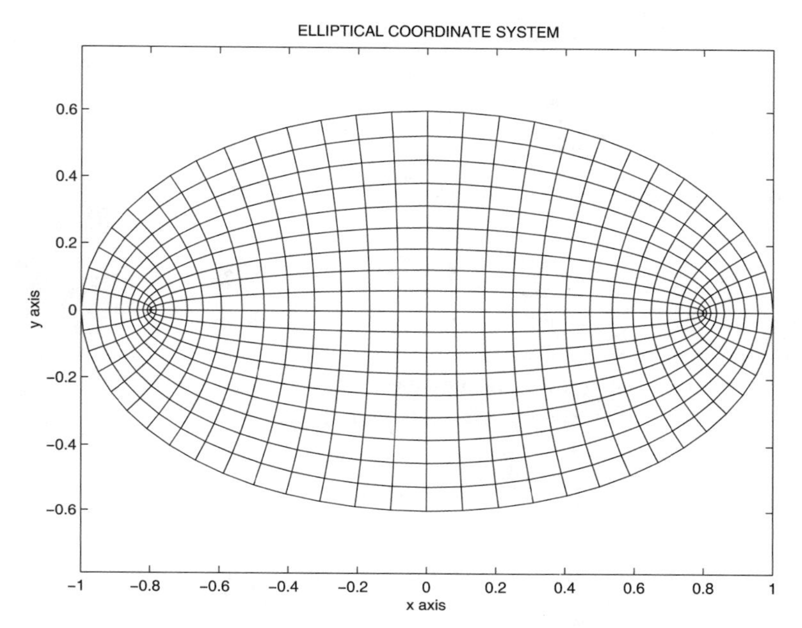

Figure 10.13: Elliptic Coordinate Grid

and

$$g''(\xi) - [\alpha - \lambda \cosh(2\xi)]g(\xi) = 0, \quad 0 \leq \xi \leq R$$

where the eigenvalue parameters α and λ are determined to make $f(\eta)$ have period 2π and make $g(\xi)$ vanish at $\xi = R$. The modal functions can be written in terms of Mathieu functions as products of the form

$$ce(\eta, q)Ce(\xi, q)$$

for modes symmetric about the x-axis and

$$se(\eta, \overline{q})Se(\xi, \overline{q})$$

for modes anti-symmetric about the x-axis. The functions ce and se are periodic Mathieu functions pertaining to the circumferential direction, while Ce and Se are modified Mathieu functions pertaining to the radial direction. The structure of these functions motivates using the following series approximation for the functions for even modes:

$$f(\eta) = \sum_{k=1}^{N} \cos(\eta(k-1))\, a_k, \quad g(\xi) = \sum_{l=1}^{M} \cos(\frac{\pi\xi}{R}(l - 1/2))\, b_l.$$

The analogous approximations for the modes anti-symmetric about the x-axis are:

$$f(\eta) = \sum_{k=1}^{N} \sin(\eta k)\, a_k, \quad g(\xi) = \sum_{l=1}^{M} \sin(\frac{\pi\xi}{R}\, l)\, b_l.$$

Thus the expressions for both cases take the form:

$$f(\eta) = \sum_{k=1}^{N} f_k(\eta)\, a_k \text{ and } g(\xi) = \sum_{l=1}^{M} g_l(\xi)\, b_l.$$

Let us choose a set of collocation points $\eta_i,\ i = 1, \ldots, n$, and $\xi_j,\ j = 1, \ldots, m$. Then substituting the series approximation for $f(\eta)$ into the differential equation gives the following over-determined system of equations:

$$\sum_{k=1}^{N} f_k^{''}(\eta_i) a_k + \alpha \sum_{k=1}^{N} f_k(n_i) a_k - \lambda \cos(2\eta_i) \sum_{k=1}^{N} f_k(\eta_i) a_k = 0, \quad i = 1, \ldots, n.$$

Denote F as the matrix having $f_k(\eta_i)$ as the element in row i and column k. Then multiplying the last equation on the left by the generalized inverse of F gives a matrix equation of the form

$$C\,A + \alpha\,A - \lambda\,D\,A = 0,$$

where A is a column matrix consisting of the coefficients a_k. A similar equation results when the series for $g(\xi)$ is substituted into the differential equation for the radial direction. It reduces to

$$E\,B - \alpha\,B + \lambda\,G\,B = 0.$$

The parameter α can be eliminated from the last two equations to yield a single eigenvalue equation

$$W\,E' + C\,W = \lambda\,(-W\,G' + D\,W)$$

where $W = A\,B'$, and the tic mark indicates matrix transposition. By addressing the two-dimensional array W in terms of a single index, the eigenvalues λ and the modal multipliers defined by W can be computed using the function **eig**. Then the values of the other eigenvalue parameter α can also be obtained using the known λ, W combinations. The mathematical developments just given are implemented below in a program which animates the various natural frequency vibration modes for an elliptic membrane.

10.7.2 Computer Formulation

The program **elipfreq** was written to compute frequencies and mode shapes for an elliptic membrane. The primary data input includes the ellipse semi-diameters, a flag indicating whether even modes, odd modes, or both are desired, the number of least squares points used, and the number of terms used in the approximation series. Natural frequencies and data needed to produce modal surfaces are returned. The program also animates the various mode shapes arranged in the order of increasing frequency. The modules employed are described in the following table.

elipfreq	reads data, calls other computational modules, and outputs modal plots
frqsimpl	forms the matrix approximations of the Mathieu equations and calls **eigenrec** to generate frequencies and mode shapes
eigenrec	solves the rectangular eigenvalue problem
plotmode	generates animated plots of the modal functions
modeshap	computes modal function shapes using the approximating function series
funcxi	approximating series functions in the xi variable
funceta	approximating series functions in the eta variable

The accuracy of the formulation developed above was assessed by 1) comparison with circular membrane frequencies known in terms of Bessel function roots and 2) results obtained from the commercial **PDE toolbox** from MathWorks employing triangular finite element analysis. The elliptic coordinate formulation is singular for a circular shape, but a nearly circular shape with $a = 1$ and $b = 0.9999$ causes no numerical difficulty. Figure 10.14 shows how well frequencies from **elipfreq** with nlsq=[200,200] and nfuns=[30,30] compare with the roots of $J_n(r)$. The first fifty frequencies were accurate to within 0.8 percent and the first one hundred frequencies were accurate to within 5 percent. The function **pdetool** from the **PDE toolbox** was also used to compute circular membrane frequencies with a quarter circular shape and 2233 node points. The first two hundred even mode frequencies from this model were accurate to within 1 percent for the first one hundred frequencies and to within 7 percent for the first 200 frequencies. Since the function **pdetool** would probably give comparable accuracy for an elliptic membrane, results from **elipfreq** were compared with those from **pdetool** using an ellipse with $a = 1$ and $b = 0.5$. The percent difference between the frequencies from the two methods appears in Figure 10.15. This comparison suggests that the first fifty frequencies produced by **elipfreq** for the elliptic membrane are probably accurate to within about 2 percent.

The various modal surfaces of an elliptic membrane have interesting shapes. The program **elipfreq** allows a sequence of modes to be exhibited by selecting vectors of frequency numbers such as 1:10 or 10:2:20. Two typical shapes are shown in Figures 10.16 and 10.17. The particular modes shown have no special significance besides their esthetic appeal. A listing of some interactive computer output and the source code for **elipfreq** follows.

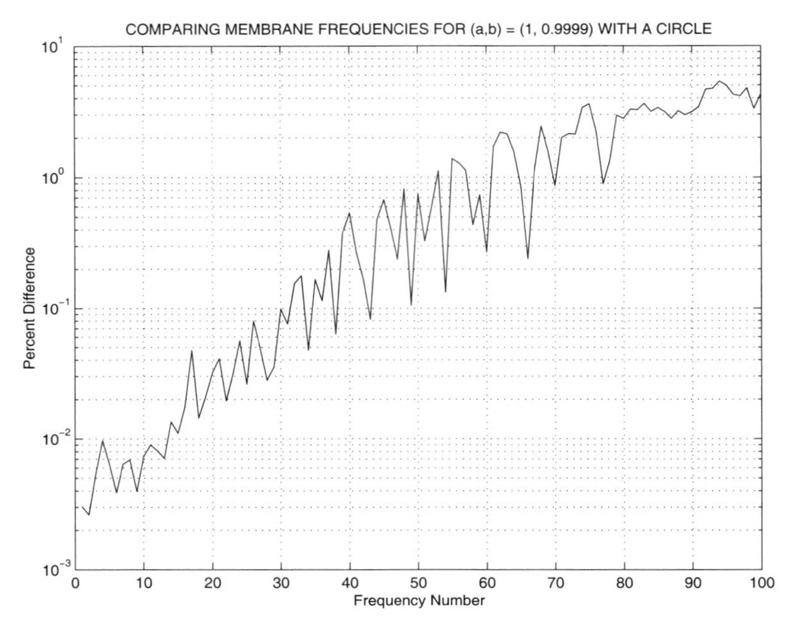

Figure 10.14: **Comparing Elipfreq Results with Bessel Function Roots**

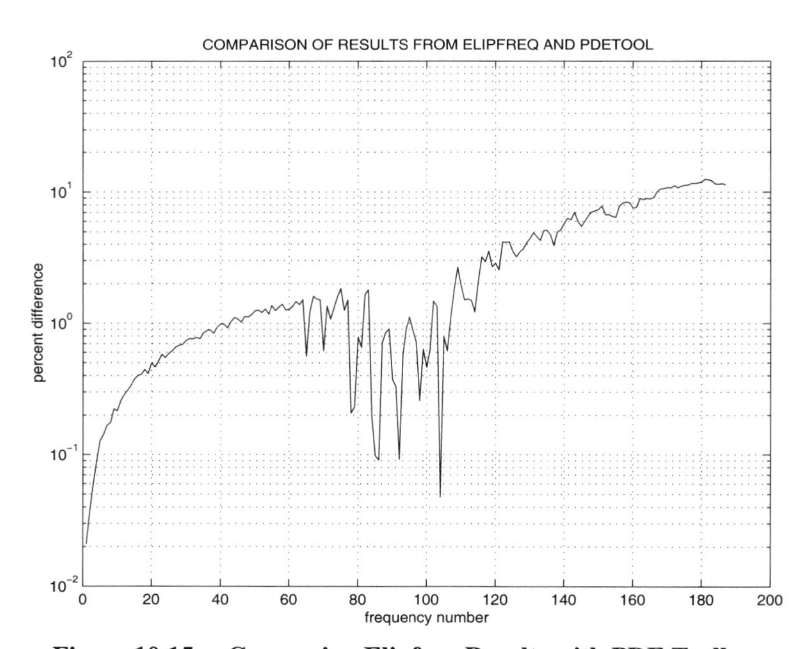

Figure 10.15: **Comparing Elipfreq Results with PDE Toolbox**

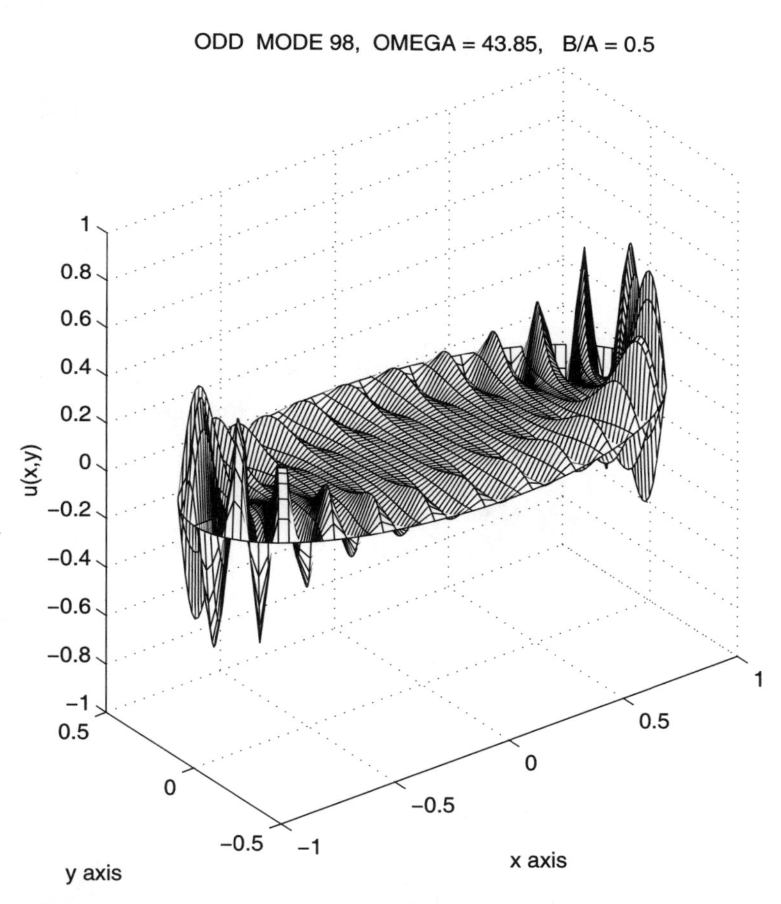

Figure 10.16: Surface for Anti-Symmetric Mode Number 98

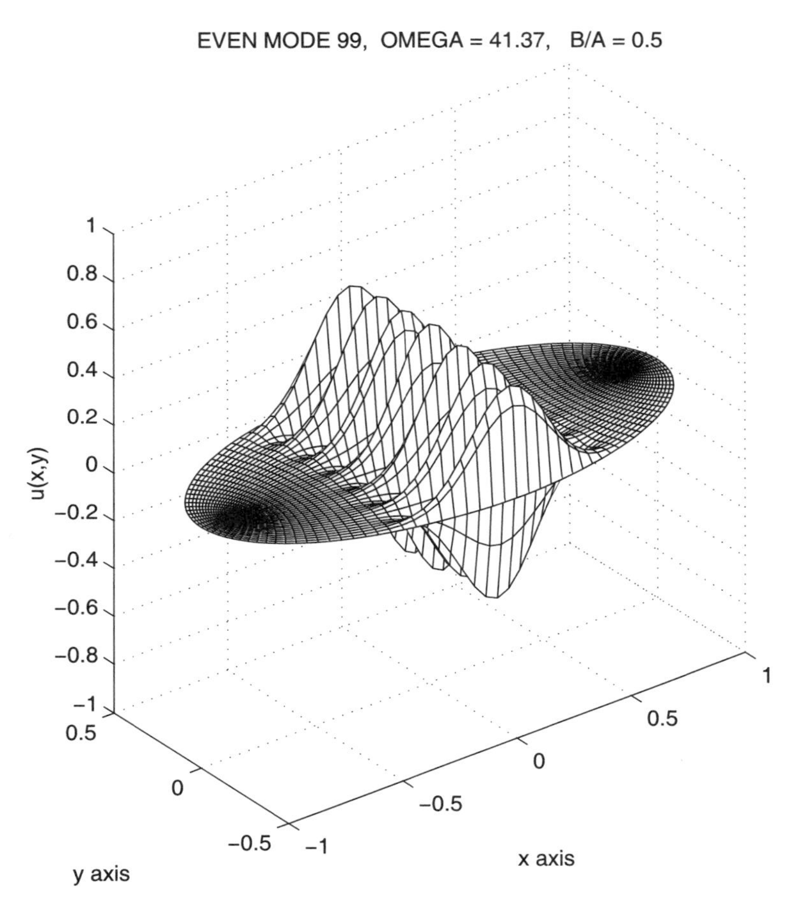

Figure 10.17: **Surface for Symmetric Mode Number 99**

Interactive Input-Output for Program elipfreq

```
>> elipfreq;

VIBRATION MODE SHAPES AND FREQUENCIES
       OF AN ELLIPTIC MEMBRANE

Input the major and minor semi-diameters > ? 1,.5

Select the modal form option
1<=>even, 2<=>odd, 3<=>both > ? 1

The computation takes awhile. Please wait.

Computation time  = 44.1 seconds.
Number of modes   = 312
Highest frequency = 116.979

Press return to see modal plots.

Give a vector of mode indices (try 10:2:20)
enter 0 to stop > ? 1

Give a vector of mode indices (try 10:2:20)
enter 0 to stop > ? 2:6

Give a vector of mode indices (try 10:2:20)
enter 0 to stop > ? [20 25 30]

Give a vector of mode indices (try 10:2:20)
enter 0 to stop > ? 0
>>
```

Elliptic Membrane Program

```
1: function [frqs,modes,indx,x,y,alpha,cptim]=elipfreq(...
2:                               a,b,type,nlsq,nfuns,noplot)
3: % [frqs,modes,indx,x,y,alpha,cptim]=elipfreq(...
4: %                             a,b,type,nlsq,nfuns,noplot)
5: % ~~~~~~~~~~~~~~~~~~~~~~~~~~~~~~~~~~~~~~~~~~~~~~~~~~~~~~~
6: % This function computes natural frequencies and mode
7: % shapes for an elliptical membrane. Modes that are
8: % symmetrical or anti-symmetrical about the x axis are
9: % included. An approximate solution is obtained using
```

```
10: % a separation of variables formulation in elliptical
11: % coordinates.
12: %
13: % a,b       - the ellipse major and minor semi-
14: %             diameters along the x and y axes
15: % nlsq      - two-component vector giving the number
16: %             of least square points in the eta and
17: %             xi directions
18: % nfuns     - two-component vector giving the number of
19: %             functions used to solve the differential
20: %             equations for the eta and xi directions.
21: % type      - use 1 for even modes symmetric about the
22: %             x-axis. Use 2 for odd modes anti-
23: %             symmetric about the x-axis. Use 3 to
24: %             combine both even and odd modes.
25: %
26: % frqs      - a vector of natural frequencies
27: %             arranged in increasing order.
28: % modes     - a three dimensional array in which
29: %             modes(:,:,j) defines the modal
30: %             deflection surface for frequency
31: %             frqs(j).
32: % indx      - a vector telling whether each
33: %             mode is even (1) or odd (2)
34: % x,y       - curvilinear coordinate arrays of
35: %             points in the membrane where modal
36: %             function values are computed.
37: % alpha     - a vector of eigenvalue parameters in
38: %             the Mathieu equation: u''(eta)+...
39: %             (alpha-lambda*cos(2*eta))*u(eta)=0
40: %             where lambda=(h*freq)^2/2 and
41: %             h=atanh(b/a)
42: % cptim     - the cpu time in seconds used to
43: %             form the equations and solve for
44: %             eigenvalues and eigenvectors
45: % noplot    - enter any value to skip mode plots
46: %
47: % User m functions called:
48: %                   frqsimpl eigenrec plotmode
49: %                   modeshap funcxi funceta
50:
51: if nargin==0
52:    disp(' ')
53:    disp('VIBRATION MODE SHAPES AND FREQUENCIES')
54:    disp('      OF AN ELLIPTIC MEMBRANE       ')
```

```
55:     disp(' ')
56:
57:     nlsq=[300,300]; nfuns=[25,25];
58:
59:     v=input(['Input the major and minor ',...
60:           'semi-diameters > ? '],'s');
61:     v=eval(['[',v,']']); a=v(1); b=v(2); disp(' ')
62:     disp('Select the modal form option')
63:     type=input(...
64:        '1<=>even, 2<=>odd, 3<=>both > ? ');
65:     disp(' ')
66:     disp(['The computation takes awhile.',...
67:           ' PLEASE WAIT.'])
68: end
69:
70: if type ==1 | type==2 % Even or odd modes
71: [frqs,modes,x,y,alpha,cptim]=frqsimpl(...
72: a,b,type,nlsq,nfuns);
73:   indx=ones(length(frqs),1)*type;
74: else % Both modes
75:     [frqs,modes,x,y,alpha,cptim]=frqsimpl(...
76:  a,b,1,nlsq,nfuns);
77:     indx=ones(length(frqs),1);
78:     [frqso,modeso,x,y,alphao,cpto]=frqsimpl(...
79:  a,b,2,nlsq,nfuns);
80:     frqs=[frqs;frqso]; alpha=[alpha;alphao];
81:  modes=cat(3,modes,modeso);
82:     indx=[indx;2*ones(length(frqso),1)];
83:     [frqs,k]=sort(frqs); modes=modes(:,:,k);
84:     indx=indx(k); cptim=cptim+cpto;
85: end
86:
87: if nargin==6, return, end
88:
89: % Plot a sequence of modal functions
90: neig=length(frqs);
91: disp(' '), disp(['Computation time  = ',...
92:        num2str(sum(cptim)),' seconds.'])
93: disp(['Number of modes   = ',num2str(neig)]);
94: disp(['Highest frequency = ',...
95:        num2str(frqs(end))]), disp(' ')
96: disp('Press return to see modal plots.')
97: pause, plotmode(a,b,x,y,frqs,modes,indx)
98:
99: %=========================================
```

```
100:
101: function [frqs,Modes,x,y,alpha,cptim]=frqsimpl(...
102:                               a,b,type,nlsq,nfuns)
103: % [frqs,Modes,x,y,alpha,cptim]=frqsimpl(...
104: %                                a,b,type,nlsq,nfuns)
105: % ~~~~~~~~~~~~~~~~~~~~~~~~~~~~~~~~~~~~~~~~~~~~~~~~~~~~~~~
106:
107: % a,b   - ellipse major and minor semi-diameters
108: % type  - numerical values of one or two for modes
109: %         symmetric or anti-symmetric about the x axis
110: % nlsq  - vector [neta,nxi] giving the number of least
111: %         square points used for the eta and xi
112: %         directions
113: % nfuns - vector [meta,mxi] giving the number of
114: %         approximating functions used for the eta and
115: %         xi directions
116: % frqs  - natural frequencies arranged in increasing
117: %         order
118: % Modes - modal surface shapes in the ellipse
119: % x,y   - coordinate points in the ellipse
120: % alpha - vector of values for the eigenvalues in the
121: %         Mathieu differential equation:
122: %         u''(eta)+(alpha-lambda*cos(2*eta))*u(eta)=0
123: % cptim - vector of computation times
124: %
125: % User m functions called: funceta  funcxi
126: %                          eigenrec modeshap
127: if nargin==0
128: a=cosh(2); b=sinh(2); type=1;
129:   nlsq=[200,200]; nfuns=[30,30];
130: end
131: h=sqrt(a^2-b^2); R=atanh(b/a); neta=nlsq(1); alpha=[];
132: nxi=nlsq(2); meta=nfuns(1); mxi=nfuns(2);
133: eta=linspace(0,pi,neta)'; xi=linspace(0,R,nxi)';
134: [Xi,Eta]=meshgrid(xi,eta); z=h*cosh(Xi+i*Eta);
135: x=real(z); y=imag(z); cptim=zeros(1,3);
136:
137: % Form the Mathieu equation for the circumferential
138: % direction as: A*E+alpha*E-lambda*B*E=0
139: tic; [Veta,A]=funceta(meta,type,eta);
140: A=Veta\[A,repmat(cos(2*eta),1,meta).*Veta];
141: B=A(:,meta+1:end); A=A(:,1:meta);
142:
143: % Form the modified Mathieu equation for the radial
144: % direction as: P*F-alpha*F+lambda*Q*F=0
```

```
145: [Vxi,P]=funcxi(a,b,mxi,type,xi);
146: P=Vxi\[P,repmat(cosh(2*xi),1,mxi).*Vxi];
147: Q=P(:,mxi+1:end); P=P(:,1:mxi);
148: cptim(1)=toc; tic
149:
150: % Solve the eigenvalue problem. This takes most
151: % of the computation time
152: [frqs,modes]=eigenrec(P',A,-Q',B);
153: % Keep only half of the modes and frequencies
154: nmax=fix(length(frqs)/2); frqs=frqs(1:nmax);
155: modes=modes(:,:,1:nmax); cptim(2)=toc;
156:
157: % Compute values of the second eigenvalue
158: % parameter in Mathieu's equation
159: alpha=zeros(1,nmax); tic;
160: s=size(modes); s=s(1:2); Vxi=Vxi';
161:
162: % Obtain the modal surface shapes
163: Neta=91; Nxi=25; Modes=zeros(Neta,Nxi,nmax);
164: for k=1:nmax
165: Mk=modes(:,:,k); [dmk,K]=max(abs(Mk(:)));
166: [I,J]=ind2sub(s,K); Ej=Mk(:,J);
167: alpha(k)=(B(I,:)*Ej*frqs(k)-A(I,:)*Ej)/Mk(K);
168: [Modes(:,:,k),x,y]=modeshap(a,b,type,Mk,Nxi,Neta);
169: end
170: frqs=sqrt(2*frqs)/h; cptim(3)=toc;
171:
172: %==============================================
173:
174: function [eigs,vecs,Amat,Bmat]=eigenrec(A,B,C,D)
175: % [eigs,vecs,Amat,Bmat]=eigenrec(A,B,C,D)
176: % Solve a rectangular eigenvalue problem of the
177: % form: X*A+B*X=lambda*(X*C+D*X)
178: %
179: % A,B,C,D - square matrices defining the problem.
180: %           A and C have the same size. B and D
181: %           have the same size.
182: % eigs    - vector of eigenvalues
183: % vecs    - array of eigenvectors where vecs(:,:,j)
184: %           contains the rectangular eigenvector
185: %           for eigenvalue eigs(j)
186: % Amat,
187: % Bmat    - matrices that express the eigenvalue
188: %           problem as Amat*V=lambda*Bmat*V
189: %
```

```
190: n=size(B,1); m=size(A,2); s=[n,m]; N=n*m;
191: Amat=zeros(N,N); Bmat=Amat; kn=1:n; km=1:m;
192: for i=1:n
193:   IK=sub2ind(s,i*ones(1,m),km);
194:   Bikn=B(i,kn); Dikn=D(i,kn);
195:   for j=1:m
196:     I=sub2ind(s,i,j);
197:     Amat(I,IK)=A(km,j)'; Bmat(I,IK)=C(km,j)';
198:     KJ=sub2ind(s,kn,j*ones(1,n));
199:     Amat(I,KJ)=Amat(I,KJ)+ Bikn;
200:     Bmat(I,KJ)=Bmat(I,KJ)+ Dikn;
201:    end
202: end
203: [vecs,eigs]=eig(Bmat\Amat);
204: [eigs,k]=sort(diag(eigs));
205: vecs=reshape(vecs(:,k),n,m,N);
206:
207: %=============================================
208:
209: function plotmode(a,b,x,y,eigs,modes,indx)
210: %
211: % plotdmode(a,b,x,y,eigs,modes,indx)
212: % ~~~~~~~~~~~~~~~~~~~~~~~~~~~~~~~~~~~
213: % This function makes animated plots of the
214: % mode shapes of an elliptic membrane for
215: % various frequencies
216: % a,b    - major and minor semi-diameters
217: % x,y    - arrays of points defining the
218: %          curvilinear coordinate grid
219: % eigs   - vector of sorted frequencies
220: % modes  - array of modal surfaces for
221: %          the corresponding frequencies
222: % indx   - vector of indices designating
223: %          each mode as even (1) or odd (2)
224:
225: range=[-a,a,-b,b,-a,a];
226: nf=25; ft=cos(linspace(0,4*pi,nf));
227: boa=[',   B/A = ',num2str(b/a,4)];
228: while 1
229:    jlim=[];
230:    while isempty(jlim), disp(' ')
231:      disp(['Give a vector of mode ',...
232:            'indices (try 10:2:20) > ? ']);
233:      jlim=input('(input 0 to stop > ? ');
234:    end
```

```
235:     if any(jlim==0)
236:       disp(' '), disp('All done'), break, end
237:     for j=jlim
238:        if indx(j)==1, type='EVEN'; f=1;
239:        else, type ='ODD '; f=-1; end
240:        u=a/2*modes(:,:,j);
241:
242:        for kk=1:nf
243:           surf(x,y,ft(kk)*u)
244:           axis equal, axis(range)
245:           xlabel('x axis'), ylabel('y axis')
246:           zlabel('u(x,y)')
247:           title([type,' MODE ',num2str(j),...
248:           ',  OMEGA = ',num2str(eigs(j),4),boa])
249:           %colormap([127/255 1 212/255])
250:           colormap([1 1 0])
251:           drawnow, shg
252:        end
253:        pause(1);
254:     end
255: end
256:
257: %==================================================
258:
259: function [u,x,y]=modeshap(...
260:                       a,b,type,modemat,nxi,neta,H)
261: %
262: % [u,x,y]=modeshap(a,b,type,modemat,nxi,neta,H)
263: % ~~~~~~~~~~~~~~~~~~~~~~~~~~~~~~~~~~~~~~~~~~~~~~
264: % This function uses the eigenvectors produced by
265: % the rectangular eigenvalue solver to form modal
266: % surface shapes in cartesian coordinates.
267: % a,b       - major and minor semi-diameters
268: % type      - 1 for even, 2 for odd
269: % modemat   - eigenvector matrix output by eigenrec
270: % nxi,neta - number of radial and circumferential
271: %             coordinate values
272: % H         - maximum height of the modal surfaces.
273: %             The default value is one.
274: % u,x,y     - modal surface array and corresponding
275: %             cartesian coordinate matrices. u(:,:j)
276: %             gives the modal surface for the j'th
277: %             natural frequency.
278:
279: if nargin<7, H=1; end
```

```
280: if nargin<6, neta=81; end; if nargin<5, nxi=22; end
281: h=sqrt(a^2-b^2); r=atanh(b/a); x=[]; y=[];
282: xi=linspace(0,r,nxi); eta=linspace(-pi,pi,neta);
283: if nargout>1
284:   [Xi,Eta]=meshgrid(xi,eta); z=h*cosh(Xi+i*Eta);
285:   x=real(z); y=imag(z);
286: end
287: [Neta,Nxi]=size(modemat);
288: mateta=funceta(Neta,type,eta);
289: matxi=funcxi(a,b,Nxi,type,xi);
290: u=mateta*modemat*matxi'; [umax,k]=max(abs(u(:)));
291: u=H/u(k)*u;
292:
293: %=============================================================
294:
295: function [f,f2]=funcxi(a,b,n,type,xi)
296: %
297: % [f,f2]=funcxi(a,b,n,type,xi)
298: % ~~~~~~~~~~~~~~~~~~~~~~~~~~~~~
299: % This function defines the approximating functions
300: % for the radial direction
301: % a,b  - ellipse major and minor half-diameters
302: % n    - number of series terms used
303: % type - 1 for even valued, 2 for odd valued
304: % xi   - vector of radial coordinate values
305: % f,f2 - matrix of function and second derivative
306: %        values
307:
308: xi=xi(:); nxi=length(xi); R=atanh(b/a);
309: if type==1, N=pi/R*(1/2:n); f=cos(xi*N);
310: else, N=pi/R*(1:n); f=sin(xi*N); end
311: f2=-repmat(N.^2,nxi,1).*f;
312:
313: %=============================================================
314:
315: function [f,f2]=funceta(n,type,eta)
316: %
317: % [f,f2]=funceta(n,type,eta)
318: % ~~~~~~~~~~~~~~~~~~~~~~~~~~
319: % This function defines the approximating functions
320: % for the circumferential direction
321: % n    - number of series terms used
322: % type - 1 for even valued, 2 for odd valued
323: % xi   - vector of circumferential coordinate values
324: % f,f2 - matrix of function and second derivative
```

```
325: %           values
326:
327: eta=eta(:); neta=length(eta);
328: if type==1, N=0:n-1; f=cos(eta*N);
329: else, N=1:n; f=sin(eta*N); end
330: f2=-repmat(N.^2,neta,1).*f;
```

Chapter 11

Bending Analysis of Beams of General Cross Section

11.1 Introduction

Elastic beams are important components in many types of structures. Consequently methods to analyze the shear, moment, slope, and deflection in beams with complex loading and general cross section variation are of significant interest. A typical beam of the type considered is shown in Figure 11.1. The study of Euler beam theory is generally regarded as an elementary topic dealt with in undergraduate engineering courses. However, simple analyses presented in standard textbooks usually do not reveal difficulties encountered with statically indeterminate problems and general geometries [115]. Finite element approximations intended to handle arbitrary problems typically assume a piecewise constant depth profile and a piecewise cubic transverse deflection curve. This contradicts even simple instances such as a constant depth beam subjected to a linearly varying distributed load which actually leads to a deflection curve which is a fifth order polynomial. Exact solutions of more involved problems where the beam depth changes linearly, for example, are more complicated. Therefore, an exact analysis of the beam problem is desirable to handle depth variation, a combination of concentrated and distributed loads, and static indeterminacy providing for general end conditions and multiple in-span supports. The current formulation considers a beam carrying any number of concentrated loads and linearly varying distributed loads. The equations for the shear and moment in the beam are obtained explicitly. Expressions for slope and deflection are formulated for evaluation by numerical integration allowing as many integration steps as necessary to achieve high accuracy. A set of simultaneous equations imposing desired constraints at the beam ends and at supports is solved for support reactions and any unknown end conditions. Knowledge of these quantities then allows evaluation of internal load and deformation quantities throughout the beam. The analytical formulation is implemented in a program using a concise problem definition specifying all loading, geometry, and constraint conditions without reference to beam elements or nodal points as might be typical in a finite element formulation. The program and example problem are discussed next.

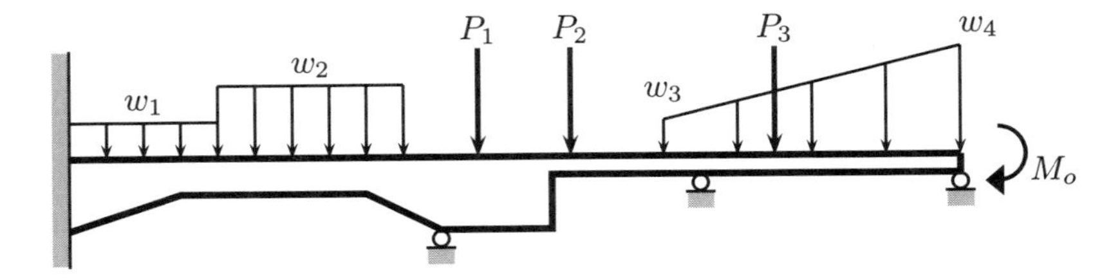

Figure 11.1: General Beam

11.1.1 Analytical Formulation

Solution of beam problems utilizes some mathematical idealizations such as a concentrated load, which implies infinite load intensity acting over an infinitesimal area. Also of importance are linearly varying distributed loads, or ramp loads. Treatment of these entities is facilitated by use of singularity functions [9]. The singularity function of order n is denoted by $< x - x_0 >^n$ and is defined as

$$< x - x_0 >^n = \begin{cases} 0, & x < x_0 \\ (x - x_0)^n & x \geq x_0. \end{cases}$$

For $n \geq 0$, the function satisfies

$$\int_0^x < x - x_0 >^n \, dx = \frac{< x - x_0 >^{n+1}}{n+1}.$$

The special case where $n = -1$ is appropriate for describing a concentrated load. The term $< x - x_0 >^{-1}$ means the limit as $\epsilon \rightarrow 0$ of the following function

$$< x - x_0 >^{-1} = \begin{cases} 0 & x < x_0, \\ \frac{1}{\epsilon} & x_0 \leq x \leq (x_0 + \epsilon), \\ 0 & x > (x_0 + \epsilon). \end{cases}$$

Consequently, in the limit as ϵ approaches zero the integral becomes

$$\int_0^x < x - x_0 >^{-1} \, dx = \; < x - x_0 >^0 .$$

Analyzing the loads and deformations in the beam requires computation of the shear, moment, slope, and deflection designated as $v(x)$, $m(x)$, $y'(x)$, and $y(x)$. The beam lies in the range $0 \leq x \leq L$. A total of four end conditions are imposed at $x = 0$ and $x = L$. Normally, two conditions will be specified at each end; so, two unknown conditions applicable at $x = 0$ need to be found during the solution process. Along with the end conditions, interior supports may exist at $x = r_\jmath$, $1 \leq \jmath \leq N_s$. Displacements $y_\jmath$ will occur at supports, and the reactions $R_\jmath$, as well as four end conditions, needed to cause the deflections will have to be determined during the analysis. Within the beam span, the applied loading will consist of known external loads described as $w_e(x)$ and the support reactions. Fundamentals of Euler beam theory developed in standard textbooks [9, 102] imply the following differential and integral relations:

I) **Load**

$$v'(x) = w_e(x) + \sum_{j=1}^{N_s} R_j < x - r_j >^{-1};$$

II) **Shear**

$$v(x) = v_0 + v_e(x) + \sum_{j=1}^{N_s} R_j < x - r_j >^0,$$

$$v_e(x) = \int_0^x w_e(x)\, dx;$$

III) **Moment and Second Derivative**

$$m'(x) = v,$$

$$m(x) = m_0 + v_0 x + m_e(x) + \sum_{j=1}^{N_s} R_j < x - r_j >^1;$$

$$m_e(x) = \int_0^x v_e(x)\, dx,$$

$$y''(x) = k(x) \left[m_0 + v_0 x + m_e(x) + \sum_{j=1}^{N_s} R_j < x - r_j >^1 \right],$$

$$k(x) = \frac{1}{E(x)\, I(x)};$$

IV) **Slope**

$$y'(x) = y_0' + m_0 \int_0^x k(x)\, dx + v_0 \int_0^x x\, k(x)\, dx +$$

$$\int_0^x k(x)\, m_e(x)\, dx + \sum_{j=1}^{N_s} R_j \int_0^x < x - r_j >^1 k(x)\, dx;$$

V) **Deflection**

$$y(x) = y_0 + y_0' x + m_0 \int_0^x \int_0^x k(x)\, dx\, dx +$$
$$v_0 \int_0^x \int_0^x x\, k(x)\, dx\, dx + \int_0^x \int_0^x k(x)\, m_e(x)\, dx\, dx +$$
$$\sum_{j=1}^{N_s} R_j \int_0^x \int_0^x < x - r_j >^1 k(x)\, dx\, dx$$

where $E(x)I(x)$ is the product of the Young's modulus and the cross section moment of inertia, y_0, y_0', v_0, m_0, are the left-end values of the deflection, slope, shear and moment respectively. The property $k(x)$ will be spatially variable unless EI is constant, which yields the following simple formulas

$$EIy'(x) = EIy_0' + m_0 x + \frac{v_0 x^2}{2} + \int_0^x m_e(x)\, dx + \frac{1}{2} \sum_{j=1}^{N_s} R_j < x - r_j >^2,$$

$$EIy(x) = EI\,(y_0 + y_0' x) + \frac{m_0 x^2}{2} + \frac{v_0 x^3}{6} + \int_0^x \int_0^x m_e(x)\, dx\, dx +$$
$$\frac{1}{6} \sum_{j=1}^{N_s} R_j < x - r_j >^3 .$$

The external loading conditions employed here can handle most practical situations. It is assumed that several concentrated loads F_j act at positions f_j, $1 \le j \le N_f$. Distributed loads are described by linearly varying ramp loads. A typical ramp load starts at position p_j with intensity P_j and varies linearly to magnitude Q_j at position q_j. The ramp load is zero unless $p_j \le x \le q_j$. A total of N_r ramp loads may be present. Instances where $P_j = Q_j$ can also occur, implying a uniformly distributed load. The general external loading chosen can be represented as

$$w_e(x) = \sum_{j=1}^{N_f} F_j < x - f_j >^{-1} +$$
$$\sum_{j=1}^{N_r} \left[P_j < x - p_j >^0 - Q_j < x - q_j >^0 + \right.$$
$$\left. S_j \left(< x - p_j >^1 - < x - q_j >^1 \right) \right]$$

where

$$S_j = \frac{Q_j - P_j}{q_j - p_j}$$

and each summation extends over the complete range of pertinent values. Similarly, integration using the properties of singularity functions yields

$$v_e(x) = \sum_{j=1}^{N_f} F_j < x - f_j >^0 + \sum_{j=1}^{N_r} [P_j < x - p_j >^1 - Q_j < x - q_j >^1 + \frac{S_j}{2} (< x - p_j >^2 - < x - q_j >^2)]$$

and

$$m_e(x) = \sum_{j=1}^{N_f} F_j < x - f_j >^1 + \sum_{j=1}^{N_r} \left[\frac{P_j}{2} < x - p_j >^2 - \frac{Q_j}{2} < x - q_j >^2 + \frac{S_j}{6} (< x - p_j >^3 - < x - q_j >^3) \right].$$

The single and double integrals given earlier involving $m_e(x)$ and $k(x)$ can easily be evaluated exactly when EI is constant, but these are not needed here. Since $k(x)$ will generally be spatially variable in the target problem set, the integrations to compute $y'(x)$ and $y(x)$ are best performed numerically. Leaving the number of integration increments as an independent parameter allows high accuracy evaluation of all integrals whenever this is desirable. Typically, problems using several hundred integration points only require a few seconds to solve using a personal computer.

Completing the problem solution requires formulations and solution of a system of simultaneous equations involving v_0, m_0, y_0', y_0, R_1, ..., R_{N_s}. The desired equations are created by specifying the displacement constraints at the supports, as well as four of eight possible end conditions. To present the equations more concisely the following notation is adopted:

$$\int_0^x k(x)\,dx = K_1(x)\ , \int_0^x \int_0^x k(x)\,dx\,dx = K_2(x),$$

$$\int_0^x x\,k(x)\,dx = L_1(x)\ , \int_0^x \int_0^x x\,k(x)\,dx\,dx = L_2(x),$$

$$\int_0^x m_e(x)\,k(x)\,dx = I_1(x)\ , \int_0^x \int_0^x m_e(x)\,k(x)\,dx\,dx = I_2(x),$$

$$\int_0^x < x - r_j >^1 k(x)\,dx = J_1(x, r_j)$$

$$\int_0^x \int_0^x < x - r_j >^1 k(x)\,dx\,dx = J_2(x, r_j),$$

and it is evident from their definitions that both $J_1(x, r_j)$ and $J_2(x, r_j)$ both equal zero for $x \leq r_j$.

At a typical support location r_i, the deflection will have an imposed value y_i. Consequently, the displacement constraints require

$$y_0 + r_i y_0' + K_2(r_i)\, m_0 + L_2(r_i)\, v_0 + \sum_{j=i+1}^{N_s} J_2(r_i, r_j)\, R_j = y_i - I_2(r_i)$$

for $1 \leq i \leq N_s$. The remaining four end conditions can specify any legitimate combination of conditions yielding a unique solution. For example, a beam cantilevered at $x = 0$ and pin supported at $x = L$ would require $y(0) = 0$, $y'(0) = 0$, $m(L) = 0$, and $y(L) = 0$. In general, conditions imposed at $x = 0$ have an obvious form since only v_0, m_0, y_0, or y_0' are explicitly involved. To illustrate a typical right end condition, let us choose slope, for example. This yields

$$y_0 + y_0' + K_1(L)\, m_0 + L_1(L)\, v_0 + \sum_{j=1}^{N_s} J_1(L, r_j)\, R_j = y'(L) - I_1(L).$$

Equations for other end conditions have similar form, and all eight possibilities are implemented in the computer program listed at the end of the chapter. Once the reactions and any initially unknown left-end conditions have been determined, load and deformation quantities anywhere in the beam can be readily found.

11.1.2 Program to Analyze Beams of General Cross Section

A program to solve general beam problems was written which tabulates and plots the shear, moment, slope, and deflection. The driver program **vdb** defines the data, calls the analysis functions, and outputs the results. Six functions that implement the methods given in this section were written. Understanding the program details can best be achieved by studying the code closely. The program was checked extensively using examples from several texts and reference books. The three span beam having parabolically tapered haunches shown in Figure 11.2 was analyzed previously by Arbabi and Li [5]. The program **vdb** was used to analyze the same problem and produces the results in Figure 11.3, which agree well with the paper.

We believe that the computer program is general enough to handle a wide variety of practical problems. Some readers may want to extend the program by adding interactive input or input from a data file. Such a modification is straightforward.

11.1.3 Program Output and Code

Output from Arbabi and Li Example

```
Analysis of a Variable Depth Elastic Beam
-----------------------------------------

Title: Problem from Arbabi and Li
```

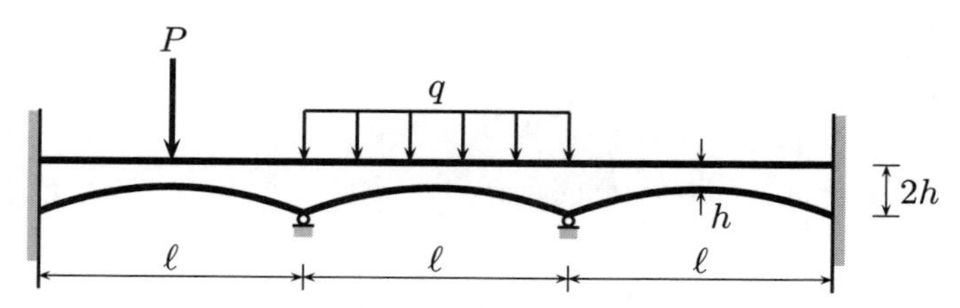

Figure 11.2: **Parabolic Beam from Arbabi and Li**

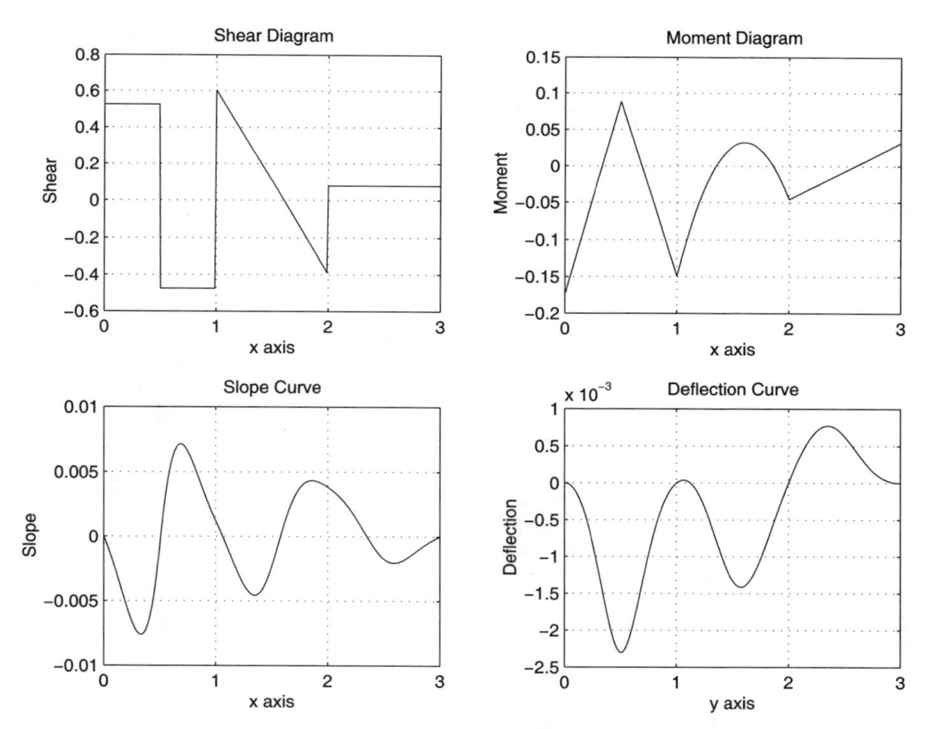

Figure 11.3: **Results for Arbabi and Li Example**

```
Beam Length:                        3
Number of integration segments: 301
Print frequency for results:      10

Interior Supports: (2)
  |   #    X-location   Deflection
  | --- ------------ ------------
  |   1  1.0000e+000  0.0000e+000
  |   2  2.0000e+000  0.0000e+000

Concentrated Forces: (1)
  |   #    X-location        Force
  | --- ------------ ------------
  |   1  5.0000e-001 -1.0000e+000

Ramp loads: (1)
  |   #       X-start         Load        X-end         Load
  | --- ------------ ------------ ------------ ------------
  |   1  1.0000e+000 -1.0000e+000  2.0000e+000 -1.0000e+000

End conditions:
  | End    Function            Value
  | ------ ----------  ------------
  | left   slope         0.0000e+000
  | left   deflection    0.0000e+000
  | right  slope         0.0000e+000
  | right  deflection    0.0000e+000

EI values are specified
  |   #       X-start     EI-value
  | --- ------------ ------------
  |   1  0.0000e+000  7.9976e+000
  |   2  1.0101e-002  7.5273e+000
  |   3  2.0202e-002  7.0848e+000
  |   4  3.0303e-002  6.6688e+000
  |   5  4.0404e-002  6.2776e+000

  Material deleted for publication

  | 296  2.9596e+000  6.2776e+000
  | 297  2.9697e+000  6.6688e+000
  | 298  2.9798e+000  7.0848e+000
  | 299  2.9899e+000  7.5273e+000
  | 300  3.0000e+000  7.9976e+000

Solution time was 0.55 secs.


Reactions at Internal Supports:
  |   X-location     Reaction
  | ------------ ------------
  |            1  1.0782e+000
  |            2  4.7506e-001

Table of Results:
```

```
| X-location        Shear        Moment         Theta         Delta
| ----------- ------------ ------------ ------------ ------------
|           0  5.2494e-001 -1.7415e-001  0.0000e+000  0.0000e+000
|         0.1  5.2494e-001 -1.2166e-001 -2.4859e-003 -1.1943e-004
|         0.2  5.2494e-001 -6.9164e-002 -5.3262e-003 -5.0996e-004
|         0.3  5.2494e-001 -1.6670e-002 -7.4251e-003 -1.1612e-003
|         0.4  5.2494e-001  3.5824e-002 -6.5761e-003 -1.8965e-003
|         0.5 -4.7506e-001  8.8318e-002 -5.5680e-004 -2.3003e-003
|         0.6 -4.7506e-001  4.0812e-002  5.6976e-003 -1.9998e-003
|         0.7 -4.7506e-001 -6.6940e-003  7.1119e-003 -1.3258e-003
|         0.8 -4.7506e-001 -5.4200e-002  5.6441e-003 -6.7385e-004
|         0.9 -4.7506e-001 -1.0171e-001  3.3302e-003 -2.2402e-004
|           1  6.0309e-001 -1.4921e-001  1.2242e-003 -2.4286e-017
|         1.1  5.0309e-001 -9.3903e-002 -7.9439e-004  2.3707e-005
|         1.2  4.0309e-001 -4.8593e-002 -2.8814e-003 -1.6165e-004
|         1.3  3.0309e-001 -1.3284e-002 -4.3574e-003 -5.3250e-004
|         1.4  2.0309e-001  1.2025e-002 -4.2883e-003 -9.8078e-004
|         1.5  1.0309e-001  2.7334e-002 -2.3015e-003 -1.3242e-003
|         1.6  3.0918e-003  3.2643e-002  6.5694e-004 -1.4078e-003
|         1.7 -9.6908e-002  2.7953e-002  3.0625e-003 -1.2125e-003
|         1.8 -1.9691e-001  1.3262e-002  4.1954e-003 -8.3907e-004
|         1.9 -2.9691e-001 -1.1429e-002  4.2843e-003 -4.0860e-004
|           2  7.8151e-002 -4.6120e-002  3.8358e-003 -1.1102e-016
|         2.1  7.8151e-002 -3.8305e-002  3.1202e-003  3.5021e-004
|         2.2  7.8151e-002 -3.0490e-002  2.0801e-003  6.1308e-004
|         2.3  7.8151e-002 -2.2675e-002  7.2881e-004  7.5555e-004
|         2.4  7.8151e-002 -1.4860e-002 -6.9898e-004  7.5597e-004
|         2.5  7.8151e-002 -7.0445e-003 -1.7447e-003  6.2865e-004
|         2.6  7.8151e-002  7.7058e-004 -2.0539e-003  4.3228e-004
|         2.7  7.8151e-002  8.5857e-003 -1.7105e-003  2.4008e-004
|         2.8  7.8151e-002  1.6401e-002 -1.0840e-003  9.9549e-005
|         2.9  7.8151e-002  2.4216e-002 -4.7454e-004  2.2493e-005
|           3  7.8151e-002  3.2031e-002 -4.4409e-016 -2.2204e-016
```

Variable Depth Beam Program

```
1: function vdb
2: % Example: vdb
3: % ~~~~~~~~~~~~
4: %
5: % This program calculates the shear, moment,
6: % slope, and deflection of a variable depth
7: % indeterminate beam subjected to complex
8: % loading and general end conditions. The
9: % input data are defined in the program
10: % statements below.
11: %
12: % User m functions required:
13: %   bmvardep, extload, lintrp, oneovrei,
14: %   sngf, trapsum
```

```
15:
16: clear all; Problem=1;
17: if Problem == 1
18:   Title=['Problem from Arbabi and Li'];
19:   Printout=10;  % Output frequency
20:   BeamLength=3; % Beam length
21:   NoSegs=301;   % # of beam divisions for
22:                 % integration
23:   % External concentrated loads and location
24:   ExtForce= [-1]; ExtForceX=[.5];
25:   % External ramp loads and range
26:   %        q1  q2  x1  x2
27:   ExtRamp=[-1  -1   1   2];
28:   % Interior supports: initial displacement
29:   % and location
30:   IntSupX=    [1; 2]; IntSupDelta=[0; 0];
31:   % End (left and right) conditions
32:   EndCondVal= [0; 0; 0; 0];   % magnitude
33:   % 1=shear,2=moment,3=slope,4=delta
34:   EndCondFunc=[3; 4; 3; 4];
35:   % 1=left end,2=right end
36:   EndCondEnd= [1; 1; 2; 2];
37:   % EI or beam depth specification
38:   EIorDepth=1;  % 1=EI values specified
39:                 % 2=depth values specified
40:   if EIorDepth == 1
41:     % Discretize the parabolic haunch for the
42:     % three spans
43:     Width=1; E=1; a=0.5^2; Npts=100;
44:     h1=0.5; k1=1; x1=linspace(0,1,Npts);
45:     h2=1.5; k2=1; x2=linspace(1,2,Npts);
46:     h3=2.5; k3=1; x3=linspace(2,3,Npts);
47:     y1=(x1-h1).^2/a+k1; y2=(x2-h2).^2/a+k2;
48:     y3=(x3-h3).^2/a+k3;
49:     EIx=[x1 x2 x3]'; h=[y1 y2 y3]';
50:     EIvalue=E*Width/12*h.^3;
51:     mn=min(EIvalue); EIvalue=EIvalue./mn;
52:   else
53:     % Beam width and Young's modulus
54:     BeamWidth=[]; BeamE=[]; Depth=[]; DepthX=[];
55:   end
56: elseif Problem == 2
57:   Title=['From Timoshenko and Young,', ...
58:          ' p 434, haunch beam'];
59:   Printout=12; NoSegs=144*4+1; BeamLength=144;
```

```
60:     ExtForce=[]; ExtForceX=[];
61:     ExtRamp=[-1 -1 0 108];
62:     IntSupX=[36; 108]; IntSupDelta=[0; 0];
63:     EndCondVal=[0; 0; 0; 0];
64:     EndCondFunc=[2; 4; 2; 4];
65:     EndCondEnd= [1; 1; 2; 2]; EIorDepth=2;
66:     if EIorDepth == 1
67:       EIvalue=[]; EIx=[];
68:     else
69:       BeamWidth=[1]; BeamE=[1];
70:       % Discretize the parabolic sections
71:       a=36^2/5; k=2.5; h1=0; h2=72; h3=144;
72:       N1=36; N2=72; N3=36;
73:       x1=linspace(  0, 36,N1); y1=(x1-h1).^2/a+k;
74:       x2=linspace( 36,108,N2); y2=(x2-h2).^2/a+k;
75:       x3=linspace(108,144,N3); y3=(x3-h3).^2/a+k;
76:       Depth=[y1 y2 y3]'; DepthX=[x1 x2 x3]';
77:       % Comparison values
78:       I=BeamWidth*Depth.^3/12; Imin=min(I); L1=36;
79:       k1=BeamE*Imin/L1; k2=k1/2; k3=k1;
80:       t0=10.46/k1; t1=15.33/k1; t2=22.24/k1;
81:       t3=27.95/k1;
82:       fprintf('\n\nValues from reference');
83:       fprintf('\n  Theta (x=  0): %12.4e',t0);
84:       fprintf('\n  Theta (x= 36): %12.4e',t1);
85:       fprintf('\n  Theta (x=108): %12.4e',t2);
86:       fprintf('\n  Theta (x=144): %12.4e\n',t3);
87:     end
88: end
89:
90: % Load input parameters into matrices
91: Force=[ExtForce,ExtForceX];
92: NoExtForce=length(ExtForce);
93: [NoExtRamp,ncol]=size(ExtRamp);
94: IntSup=[IntSupDelta,IntSupX];
95: NoIntSup=length(IntSupX);
96: EndCond=[EndCondVal,EndCondFunc,EndCondEnd];
97: if EIorDepth == 1
98:   BeamProp=[]; NoEIorDepths=length(EIx);
99:   EIdata=[EIvalue EIx];
100: else
101:   BeamProp=[BeamWidth BeamE];
102:   NoEIorDepths=length(DepthX);
103:   EIdata=[Depth DepthX];
104: end
```

```
105:
106: % more on
107:
108: % Output input data
109: label1=['shear     ';'moment    '; ...
110:         'slope     ';'deflection'];
111: label2=['left   ';'right  '];
112: fprintf('\n\nAnalysis of a Variable Depth ');
113: fprintf('Elastic Beam');
114: fprintf('\n-------------------------------');
115: fprintf('---------');
116: fprintf('\n\n');
117: disp(['Title: ' Title]);
118: fprintf...
119:   ('\nBeam Length:                     %g', ...
120:   BeamLength);
121: fprintf...
122:   ('\nNumber of integration segments: %g', ...
123:   NoSegs);
124: fprintf...
125:   ('\nPrint frequency for results:    %g', ...
126:   Printout);
127: fprintf('\n\nInterior Supports: (%g)', ...
128:   NoIntSup);
129: if NoIntSup > 0
130:   fprintf('\n  |   #   X-location   Deflection');
131:   fprintf('\n  | --- ------------ ------------');
132:   for i=1:NoIntSup
133:     fprintf('\n  |%4.0f %12.4e %12.4e', ...
134:             i,IntSup(i,2),IntSup(i,1));
135:   end
136: end
137: fprintf('\n\nConcentrated Forces: (%g)', ...
138:   NoExtForce);
139: if NoExtForce > 0
140:   fprintf('\n  |   #   X-location        Force');
141:   fprintf('\n  | --- ------------ ------------');
142:   for i=1:NoExtForce
143:     fprintf('\n  |%4.0f %12.4e %12.4e', ...
144:             i,Force(i,2),Force(i,1));
145:   end
146: end
147: fprintf('\n\nRamp loads: (%g)', NoExtRamp);
148: if NoExtRamp > 0
149:   fprintf('\n  |   #      X-start         Load');
```

```
150:   fprintf('          X-end          Load');
151:   fprintf('\n  | --- ----------- -----------');
152:   fprintf(' ----------- -----------');
153:   for i=1:NoExtRamp
154:     fprintf('\n  |%4.0f %12.4e %12.4e ', ...
155:             i,ExtRamp(i,3),ExtRamp(i,1));
156:     fprintf('%12.4e %12.4e', ...
157:             ExtRamp(i,4),ExtRamp(i,2));
158:   end
159: end
160: fprintf('\n\nEnd conditions:');
161: fprintf('\n  | End   Function            Value');
162: fprintf('\n  ');
163: fprintf('| ------ ----------  -----------\n');
164: for i=1:4
165:   j=EndCond(i,3); k=EndCond(i,2);
166:   strg=sprintf('  %12.4e',EndCond(i,1));
167:   disp(['  | ' label2(j,:) label1(k,:) strg]);
168: end
169: if EIorDepth == 1
170:   fprintf('\nEI values are specified');
171:   fprintf('\n  |   #      X-start    EI-value')
172:   fprintf('\n  | --- ----------- -----------');
173:   for i=1:NoEIorDepths
174:     fprintf('\n  |%4.0f %12.4e %12.4e', ...
175:             i,EIdata(i,2),EIdata(i,1));
176:   end
177: else
178:   fprintf('\nDepth values are specified for ');
179:   fprintf('rectangular cross section');
180:   fprintf('\n  | Beam width:       %12.4e', ...
181:           BeamProp(1));
182:   fprintf('\n  | Young''s modulus: %12.4e', ...
183:           BeamProp(2));
184:   fprintf('\n  |');
185:   fprintf('\n  |   #      X-start         Depth')
186:   fprintf('\n  | --- ----------- -----------');
187:   for i=1:NoEIorDepths
188:     fprintf('\n  |%4.0f %12.4e %12.4e', ...
189:             i,EIdata(i,2),EIdata(i,1));
190:   end
191: end
192: disp(' ');
193:
194: % Begin analysis
```

```
195: x=linspace(0,BeamLength,NoSegs)'; t=clock;
196: [V,M,Theta,Delta,Reactions]= ...
197:   bmvardep(NoSegs,BeamLength,Force,ExtRamp, ...
198:            EndCond,IntSup,EIdata,BeamProp);
199: t=etime(clock,t);
200:
201: % Output results
202: disp(' ');
203: disp(['Solution time was ',num2str(t),' secs.']);
204: if NoIntSup > 0
205:   fprintf('\nReactions at Internal Supports:');
206:   fprintf('\n  |   X-location     Reaction');
207:   fprintf('\n  | ------------ ------------');
208:   for i=1:NoIntSup
209:     fprintf('\n  | %12.8g %12.4e', ...
210:             IntSup(i,2),Reactions(i));
211:   end
212: end
213: fprintf('\n\nTable of Results:');
214: fprintf('\n  |  X-location        Shear');
215: fprintf('       Moment');
216: fprintf('        Theta        Delta');
217: fprintf('\n  | ----------- ------------ ');
218: fprintf('------------');
219: fprintf(' ------------ ------------');
220: if Printout > 0
221:   for i=1:Printout:NoSegs
222:     fprintf('\n  |%12.4g %12.4e %12.4e', ...
223:             x(i),V(i),M(i));
224:     fprintf(' %12.4e %12.4e',Theta(i),Delta(i));
225:   end
226:   disp(' ');
227: else
228:   i=1; j=NoSegs;
229:   fprintf('\n  |%12.4g %12.4e %12.4e', ...
230:           x(i),V(i),M(i));
231:   fprintf(' %12.4e %12.4e',Theta(i),Delta(i));
232:   fprintf('\n  |%12.8g %12.4e %12.4e', ...
233:           x(j),V(j),M(j));
234:   fprintf(' %12.4e %12.4e',Theta(j),Delta(j));
235: end
236: fprintf('\n\n');
237: subplot(2,2,1);
238:   plot(x,V,'k-'); grid; xlabel('x axis');
239:   ylabel('Shear'); title('Shear Diagram');
```

```
240: subplot(2,2,2);
241:   plot(x,M,'k-'); grid; xlabel('x axis');
242:   ylabel('Moment'); title('Moment Diagram')
243: subplot(2,2,3);
244:   plot(x,Theta,'k-'); grid; xlabel('x axis');
245:   ylabel('Slope'); title('Slope Curve');
246: subplot(2,2,4);
247:   plot(x,Delta,'k-'); grid; xlabel('y axis');
248:   ylabel('Deflection');
249:   title('Deflection Curve'); subplot
250: drawnow; figure(gcf)
251: %print -deps vdb
252:
253: % more off
254:
255: %==============================================
256:
257: function [V,M,Theta,Delta,Reactions]= ...
258:   bmvardep(NoSegs,BeamLength,Force,ExtRamp, ...
259:   EndCond,IntSup,EIdata,BeamProp)
260: % [V,M,Theta,Delta,Reactions]=bmvardep ...
261: % (NoSegs,BeamLength,Force,ExtRamp,EndCond, ...
262: % IntSup,EIdata,BeamProp)
263: % ~~~~~~~~~~~~~~~~~~~~~~~~~~~~~~~~~~~~~~~~~~~~~
264: %
265: % This function computes the shear, moment,
266: % slope, and deflection in a variable depth
267: % elastic beam having specified end conditions,
268: % intermediate supports with given
269: % displacements, and general applied loading,
270: % allowing concentrated loads and linearly
271: % varying ramp loads.
272: %
273: % NoSegs     - number of beam divisions for
274: %              integration
275: % BeamLength - beam length
276: % Force      - matrix containing the magnitudes
277: %              and locations for concentrated
278: %              loads
279: % ExtRamp    - matrix containing the end
280: %              magnitudes and end locations
281: %              for ramp loads
282: % EndCond    - matrix containing the type of
283: %              end conditions, the magnitudes,
284: %              and whether values are for the
```

```
285: %                left or right ends
286: % IntSup     - matrix containing the location
287: %                and delta for interior supports
288: % EIdata     - either EI or depth values
289: % BeamProp   - either null or beam widths
290: %
291: % V          - vector of shear values
292: % M          - vector of moment values
293: % Theta      - vector of slope values
294: % Delta      - vector of deflection values
295: % Reactions  - reactions at interior supports
296: %
297: % User m functions required:
298: %    oneovrei, extload, sngf, trapsum
299: %----------------------------------------------
300:
301: if nargin < 8, BeamProp=[]; end
302: % Evaluate function value coordinates and 1/EI
303: x=linspace(0,BeamLength,NoSegs)';
304: kk=oneovrei(x,EIdata,BeamProp);
305:
306: % External load contributions to shear and
307: % moment interior to span and at right end
308: [ve,me]=extload(x,Force,ExtRamp);
309: [vv,mm]=extload(BeamLength,Force,ExtRamp);
310:
311: % Deflections and position of interior supports
312: ns=size(IntSup,1);
313: if ns > 0
314:   ysprt=IntSup(:,1); r=IntSup(:,2);
315:   snf=sngf(x,r,1);
316: else
317:   ysprt=[]; r=[]; snf=zeros(NoSegs,0);
318: end
319:
320: % Form matrix governing y''(x)
321: smat=kk(:,ones(1,ns+3)).* ...
322:      [x,ones(NoSegs,1),snf,me];
323:
324: % Integrate twice to get slope and deflection
325: % matrices
326: smat=trapsum(0,BeamLength,smat);
327: ymat=trapsum(0,BeamLength,smat);
328:
329: % External load contributions to
```

```
330: % slope/deflection at the right end
331: ss=smat(NoSegs,ns+3); yy=ymat(NoSegs,ns+3);
332:
333: % Equations to solve for left end conditions
334: % and internal reactions
335: ns4=ns+4; j=1:4; a=zeros(ns4,ns4);
336: b=zeros(ns4,1); js=1:ns; js4=js+4;
337:
338: % Account for four independent boundary
339: % conditions.  Usually two conditions will be
340: % imposed at each end.
341: for k=1:4
342:   val=EndCond(k,1); typ=EndCond(k,2);
343:   wchend=EndCond(k,3);
344:   if wchend==1
345:     b(k)=val; row=zeros(1,4); row(typ)=1;
346:     a(k,j)=row;
347:   else
348:     if typ==1      % Shear
349:       a(k,j)=[1,0,0,0]; b(k)=val-vv;
350:       if ns>0
351:         a(k,js4)=sngf(BeamLength,r,0);
352:       end
353:     elseif typ==2 % Moment
354:       a(k,j)=[BeamLength,1,0,0]; b(k)=val-mm;
355:       if ns>0
356:         a(k,js4)=sngf(BeamLength,r,1);
357:       end
358:     elseif typ==3 % Slope
359:       a(k,j)=[smat(NoSegs,1:2),1,0];
360:       b(k)=val-ss;
361:       if ns>0
362:         a(k,js4)=smat(NoSegs,3:ns+2);
363:       end
364:     else           % Deflection
365:       a(k,j)=[ymat(NoSegs,1:2),BeamLength,1];
366:       b(k)=val-yy;
367:       if ns>0
368:         a(k,js4)=ymat(NoSegs,3:ns+2);
369:       end
370:     end
371:   end
372: end
373:
374: % Interpolate to assess how support deflections
```

```
375: % are affected by end conditions, external
376: % loads, and support reactions.
377: if ns>0
378:   a(js4,1)=interp1(x,ymat(:,1),r);
379:   a(js4,2)=interp1(x,ymat(:,2),r);
380:   a(js4,3)=r; a(js4,4)=ones(ns,1);
381:   for j=1:ns-1
382:     a(j+5:ns+4,j+4)= ...
383:       interp1(x,ymat(:,j+2),r(j+1:ns));
384:   end
385: end
386: b(js4)=ysprt-interp1(x,ymat(:,ns+3),r);
387:
388: % Solve for unknown reactions and end conditions
389: c=a\b; v0=c(1); m0=c(2); s0=c(3); y0=c(4);
390: Reactions=c(5:ns+4);
391:
392: % Compute the shear, moment, slope, deflection
393: % for all x
394: if ns > 0
395:   V=v0+ve+sngf(x,r,0)*Reactions;
396:   M=m0+v0*x+me+sngf(x,r,1)*Reactions;
397:   Theta=s0+smat(:,ns+3)+smat(:,1:ns+2)* ...
398:         [v0;m0;Reactions];
399:   Delta=y0+s0*x+ymat(:,ns+3)+ ...
400:         ymat(:,1:ns+2)*[v0;m0;Reactions];
401: else
402:   Reactions=[]; V=v0+ve; M=m0+v0*x+me;
403:   Theta=s0+smat(:,ns+3)+smat(:,1:2)*[v0;m0];
404:   Delta=y0+s0*x+ymat(:,ns+3)+ ...
405:         ymat(:,1:2)*[v0;m0];
406: end
407:
408: %=============================================
409:
410: function [V,M,EITheta,EIDelta]=extload ...
411:           (x,Force,ExtRamp)
412: % [V,M,EITheta,EIDelta]=extload ...
413: %                         (x,Force,ExtRamp)
414: % ~~~~~~~~~~~~~~~~~~~~~~~~~~~~~~~~~~~~~~~~
415: %
416: % This function computes the shear, moment,
417: % slope, and deflection in a uniform depth
418: % Euler beam which is loaded by a series of
419: % concentrated loads and ramp loads. The values
```

```
420: % of shear, moment, slope and deflection all
421: % equal zero when x=0.
422: %
423: % x       - location along beam
424: % Force   - concentrated force matrix
425: % ExtRamp - distributed load matrix
426: %
427: % V       - shear
428: % M       - moment
429: % EITheta - slope
430: % EIDelta - deflection
431: %
432: % User m functions required: sngf
433: %----------------------------------------------
434:
435: nf=size(Force,1); nr=size(ExtRamp,1);
436: nx=length(x); V=zeros(nx,1); M=V;
437: EITheta=V; EIDelta=V;
438: % Concentrated load contributions
439: if nf > 0
440:   F=Force(:,1); f=Force(:,2); V=V+sngf(x,f,0)*F;
441:   M=M+sngf(x,f,1)*F;
442:   if nargout > 2
443:     EITheta=EITheta+sngf(x,f,2)*(F/2);
444:     EIDelta=EIDelta+sngf(x,f,3)*(F/6);
445:   end
446: end
447: % Ramp load contributions
448: if nr > 0
449:   P=ExtRamp(:,1); Q=ExtRamp(:,2);
450:   p=ExtRamp(:,3); q=ExtRamp(:,4);
451:   S=(Q-P)./(q-p); sp2=sngf(x,p,2);
452:   sq2=sngf(x,q,2); sp3=sngf(x,p,3);
453:   sq3=sngf(x,q,3); sp4=sngf(x,p,4);
454:   sq4=sngf(x,q,4);
455:   V=V+sngf(x,p,1)*P-sngf(x,q,1)* ... % Shear
456:     Q+(sp2-sq2)*(S/2);
457:   M=M+sp2*(P/2)-sq2*(Q/2)+ ...       % Moment
458:     (sp3-sq3)*(S/6);
459:   if nargout > 2
460:     EITheta=EITheta+sp3*(P/6)- ...  % EI*Theta
461:             sq3*(Q/6)+(sp4-sq4)*(S/24);
462:     EIDelta=EIDelta+sp4*(P/24)- ... % EI*Delta
463:             sq4*(Q/24)+(sngf(x,p,5)- ...
464:             sngf(x,q,5))*(S/120);
```

```
465:   end
466: end
467:
468: %==============================================
469:
470: function val=oneovrei(x,EIdata,BeamProp)
471: % [val]=oneovrei(x,EIdata,BeamProp)
472: % ~~~~~~~~~~~~~~~~~~~~~~~~~~~~~~~~~~
473: %
474: % This function computes 1/EI by piecewise
475: % linear interpolation through a set of data
476: % values.
477: %
478: % x        - location along beam
479: % EIdata   - EI or depth values
480: % BeamProp - null or width values
481: %
482: % val      - computed value for 1/EI
483: %
484: % User m functions required: none
485: %----------------------------------------------
486:
487: if size(EIdata,1) < 2  % uniform depth case
488:   v=EIdata(1,1);
489:   EIdata=[v,min(x);v,max(x)];
490: end
491: if ( nargin > 2 ) & ( sum(size(BeamProp)) > 0)
492:   % Compute properties assuming the cross
493:   % section is rectangular and EIdata(:,1)
494:   % contains depth values
495:   width=BeamProp(1); E=BeamProp(2);
496:   EIdata(:,1)=E*width/12*EIdata(:,1).^3;
497: end
498: val=1./lintrp(EIdata(:,2),EIdata(:,1),x);
499:
500: %==============================================
501:
502: function y=sngf(x,x0,n)
503: % y=sngf(x,x0,n)
504: % ~~~~~~~~~~~~~~
505: %
506: % This function computes the singularity
507: % function defined by
508: %    y=<x-x0>^n for n=0,1,2,...
509: %
```

```
510: % User m functions required: none
511: %----------------------------------------------
512:
513: if nargin < 3, n=0; end
514: x=x(:); nx=length(x); x0=x0(:)'; n0=length(x0);
515: x=x(:,ones(1,n0)); x0=x0(ones(nx,1),:); d=x-x0;
516: s=(d>=zeros(size(d))); v=d.*s;
517: if n==0
518:   y=s;
519: else
520:   y=v;
521:   for j=1:n-1; y=y.*v; end
522: end
523:
524: %=============================================
525:
526: function v=trapsum(a,b,y,n)
527: %
528: % v=trapsum(a,b,y,n)
529: % ~~~~~~~~~~~~~~~~~~
530: %
531: % This function evaluates:
532: %
533: %   integral(a=>x, y(x)*dx) for a<=x<=b
534: %
535: % by the trapezoidal rule (which assumes linear
536: % function variation between succesive function
537: % values).
538: %
539: % a,b - limits of integration
540: % y   - integrand that can be a vector-valued
541: %       function returning a matrix such that
542: %       function values vary from row to row.
543: %       It can also be input as a matrix with
544: %       the row size being the number of
545: %       function values and the column size
546: %       being the number of components in the
547: %       vector function.
548: % n   - the number of function values used to
549: %       perform the integration.  When y is a
550: %       matrix then n is computed as the number
551: %       of rows in matrix y.
552: %
553: % v   - integral value
554: %
```

```
555: % User m functions called:  none
556: %----------------------------------------------
557:
558: if isstr(y)
559:   % y is an externally defined function
560:   x=linspace(a,b,n)'; h=x(2)-x(1);
561:   Y=feval(y,x); % Function values must vary in
562:                 % row order rather than column
563:                 % order or computed results
564:                 % will be wrong.
565:   m=size(Y,2);
566: else
567:   % y is column vector or a matrix
568:   Y=y; [n,m]=size(Y); h=(b-a)/(n-1);
569: end
570: v=[zeros(1,m); ...
571:   h/2*cumsum(Y(1:n-1,:)+Y(2:n,:))];
572:
573: %=============================================
574:
575: % function y=lintrp(xd,yd,x)
576: % See Appendix B
```

Chapter 12

Applications of Analytic Functions

12.1 Properties of Analytic Functions

Complex valued functions of a single complex variable are useful in various disciplines such as physics and numerical approximation theory. The current chapter summarizes a number of attractive properties of analytic functions and presents some applications in which MATLAB is helpful. Excellent textbooks presenting the theory of analytic functions [18, 75, 119] are available which fully develop various theoretical concepts employed in this chapter. Therefore, only the properties which may be helpful in subsequent discussions are included.

12.2 Definition of Analyticity

We consider a complex valued function

$$F(z) = u(x, y) + iv(x, y) \quad , \quad z = x + iy$$

which depends on the complex variable z. The function $F(z)$ is analytic at point z if it is differentiable in the neighborhood of z. Differentiability requires that the limit

$$\lim_{|\Delta z| \to 0} \left[\frac{F(z + \Delta z) - F(z)}{\Delta z} \right] = F'(z)$$

exists independent of how $|\Delta z|$ approaches zero. Necessary and sufficient conditions for analyticity are continuity of the first partial derivatives of u and v and satisfaction of the Cauchy-Riemann conditions (CRC)

$$\frac{\partial u}{\partial x} = \frac{\partial v}{\partial y} \quad , \quad \frac{\partial u}{\partial y} = -\frac{\partial v}{\partial x}$$

These conditions can be put in more general form as follows. Let n denote an arbitrary direction in the z-plane and let s be the direction obtained by a $90\,^\circ$ counterclockwise rotation from the direction of n. The generalized CRC are:

$$\frac{\partial u}{\partial n} = \frac{\partial v}{\partial s} \quad , \quad \frac{\partial u}{\partial s} = -\frac{\partial v}{\partial n}$$

Satisfaction of the CRC implies that both u and v are solutions of Laplace's equation

$$\nabla^2 u = \frac{\partial^2 u}{\partial x^2} + \frac{\partial^2 u}{\partial y^2} = 0$$

and

$$\frac{\partial^2 v}{\partial x^2} + \frac{\partial^2 v}{\partial y^2} = 0$$

These functions are called harmonic. Functions related by the CRC are also said to be harmonic conjugates. When one function u is known, its harmonic conjugate v can be found within an additive constant by using

$$v = \int dv = \int \frac{\partial v}{\partial x}\,dx + \int \frac{\partial v}{\partial y}\,dy = \int \left(-\frac{\partial u}{\partial y}\,dx + \frac{\partial u}{\partial x}\,dy \right) + \text{constant}$$

Harmonic conjugates also have the properties that curves $u = \text{constant}$ and $v = \text{constant}$ intersect orthogonally. This follows because $u = \text{constant}$ implies $\frac{\partial u}{\partial n}$ is zero in a direction tangent to the curve. However $\frac{\partial u}{\partial n} = \frac{\partial v}{\partial s}$ so $v = \text{constant}$ along a curve intersecting $u = \text{constant}$ orthogonally.

Sometimes it is helpful to regard a function of x and y as a function of $z = x + iy$ and $\bar{z} = x - iy$. The inverse is $x = (z + \bar{z})/2$ and $y = (z - \bar{z})/(2i)$. Chain rule differentiation applied to a general function ϕ yields

$$\frac{\partial \phi}{\partial x} = \frac{\partial \phi}{\partial z} + \frac{\partial \phi}{\partial \bar{z}} \quad , \quad \frac{\partial \phi}{\partial y} = i\frac{\partial \phi}{\partial z} - i\frac{\partial \phi}{\partial \bar{z}}$$

so that

$$\left(\frac{\partial}{\partial x} - i\frac{\partial}{\partial y} \right) \phi = 2\frac{\partial \phi}{\partial z} \quad , \quad \left(\frac{\partial}{\partial x} + i\frac{\partial}{\partial y} \right) \phi = 2\frac{\partial \phi}{\partial \bar{z}}$$

So Laplace's equation becomes

$$\frac{\partial^2 \phi}{\partial x^2} + \frac{\partial^2 \phi}{\partial y^2} = 4\frac{\partial^2 \phi}{\partial z \partial \bar{z}} = 0$$

It is straightforward to show the condition that a function F be an analytic function of z is expressible as

$$\frac{\partial F}{\partial \bar{z}} = 0$$

It is important to note that most of the functions routinely employed with real arguments are analytic in some part of the z-plane. These include:

$$z^n,\ \sqrt{z},\ \log(z),\ e^z,\ \sin(z),\ \cos(z),\ \arctan(z),$$

to mention a few. The real and imaginary parts of these functions are harmonic and they arise in various physical applications. The integral powers of z are especially significant. We can write

$$z = re^{i\theta} \quad , \quad r = \sqrt{x^2 + y^2} \quad , \quad \theta = \tan^{-1}\left(\frac{y}{x}\right)$$

and get

$$z^n = u + iv \quad , \quad u = r^n \cos(n\theta) \quad , \quad v = r^n \sin(n\theta)$$

The reader can verify by direct differentiation that both u and v are harmonic.

Points where $F(z)$ is nondifferentiable are called singular points and these are categorized as isolated or nonisolated. Isolated singularities are termed either poles or essential singularities. Branch points are the most common type of nonisolated singularity. Singular points and their significance are discussed further below.

12.3 Series Expansions

If $F(z)$ is analytic inside and on the boundary of an annulus defined by $a \leq |z - z_0| \leq b$ then $F(z)$ is representable in a Laurent series of the form

$$F(z) = \sum_{n=-\infty}^{\infty} a_n (z - z_0)^n \quad , \quad a \leq |z - z_0| \leq b$$

where

$$F(z) = \frac{1}{2\pi i} \int_L \frac{F(t)\, dt}{(t - z_0)^{n+1}}$$

and L represents any closed curve encircling z_0 and lying between the inner circle $|z - z_0| = a$ and the outer circle $|z - z_0| = b$. The direction of integration along the curve is counterclockwise. If $F(z)$ is also analytic for $|z - z_0| < a$, the negative powers in the Laurent series drop out to give Taylor's series

$$F(z) = \sum_{n=0}^{\infty} a_n (z - z_0)^n \quad , \quad |z - z_0| \leq b$$

Special cases of the Laurent series lead to classification of isolated singularities as poles or essential singularities. Suppose the inner radius can be made arbitrarily small but nonzero. If the coefficients below some order, say $-m$, vanish but $a_{-m} \neq 0$, we classify z_0 as a pole of order m. Otherwise, we say z_0 is an essential singularity.

Another term of importance in connection with Laurent series is a_{-1}, the coefficient of $(z - z_0)^{-1}$. This coefficient, called the residue at z_0, is sometimes useful for evaluating integrals.

12.4 Integral Properties

Analytic functions have many useful integral properties. One of these properties that concerns integrals around closed curves is:

Cauchy-Goursat Theorem: If $F(z)$ is analytic at all points in a simply connected region R, then

$$\int_L F(z)\,dz = 0$$

for every closed curve L in the region.

An immediate consequence of this theorem is that the integral of $F(z)$ along any path between two end points z_1 and z_2 is independent of the path (this only applies for simply connected regions).

12.4.1 Cauchy Integral Formula

If $F(z)$ is analytic inside and on a closed curve L bounding a simply connected region R then

$$F(z) = \frac{1}{2\pi\imath}\int_L \frac{F(t)\,dt}{t-z} \qquad \text{for } z \text{ inside } L$$

$$F(z) = 0 \qquad \text{for } z \text{ outside } L$$

The Cauchy integral formula provides a simple means for computing $F(z)$ at interior points when its boundary values are known. We refer to any integral of the form

$$I(z) = \frac{1}{2\pi i}\int_L \frac{F(t)\,dt}{t-z}$$

as a Cauchy integral, regardless of whether $F(t)$ is the boundary value of an analytic function. $I(z)$ defines a function analytic in the complex plane cut along the curve L. When $F(t)$ is the boundary value of a function analytic inside a closed curve L, $I(z)$ is evidently discontinuous across L since $I(z)$ approaches $F(z)$ as z approaches L from the inside but gives zero for an approach from the outside. The theory of Cauchy integrals for both open and closed curves is extensively developed in Muskhelishvili's texts [72, 73] and is used to solve many practical problems.

12.4.2 Residue Theorem

If $F(z)$ is analytic inside and on a closed curve L except at isolated singularities $z_1, z_2, \ldots, z_n$ where it has Laurent expansions, then

$$\int_L F(z)\,dz = 2\pi\imath\sum_{j=1}^{j=n} B_j$$

where B_j is the residue of $F(z)$ at $z = z_j$. In the instance where $z_\imath$ is a pole of order m, the residue can be computed as

$$a_{-1} = \frac{1}{(m-1)!}\left\{\frac{d^{m-1}}{dz^{m-1}}\left[F(z)(z-z_\imath)^m\right]\right\}_{\lim\limits_{z\to z_\imath}}$$

12.5 Physical Problems Leading to Analytic Functions

Several physical phenomena require solutions involving real valued functions satisfying Laplace's equation. Since an analytic function has harmonic real and imaginary parts, a harmonic function can often be expressed concisely as the real part of an analytic function. Useful tools such as Taylor series can yield effective computational devices. One of the simplest practical examples involves determining a function u harmonic inside the unit disk $|z| \leq 1$ and having boundary values described by a Fourier series. In the following equations, and in subsequent articles, we will often refer to a function defined inside and on the unit circle in terms of polar coordinates as $u(r, \theta)$ while we may, simultaneously, think of it as a function of the complex variable $z = r\sigma$ where $\sigma = e^{i\theta}$. Hence we write the boundary condition for the circular disk as

$$u(1, \theta) = \sum_{n=-\infty}^{\infty} c_n \sigma^n \ , \ \sigma = e^{\imath\theta}$$

with $c_{-n} = \bar{c}_n$ because u is real. The desired function can be found as

$$u(r, \theta) = mbox\mathbf{real}(\ F(z)\ )$$

where

$$F(z) = c_0 + 2\sum_{n=1}^{\infty} c_n z^n \quad , \quad |z| \leq 1$$

This solution is useful because the Fast Fourier Transform (FFT) can be employed to generate Fourier coefficients for quite general boundary conditions, and the series for $F(z)$ converges rapidly when $|z| < 1$. This series will be employed below to solve both the problem where boundary values are given (the Dirichlet problem) and where normal derivative values are known on the boundary (the Neumann problem). Several applications where analytic functions occur are mentioned below.

12.5.1 Steady-State Heat Conduction

The steady-state temperature distribution in a homogeneous two-dimensional body is harmonic. We can take $u = \mathbf{Real}[F(z)]$. Boundary curves where $u =$ constant lead to conditions

$$F(z) + \overline{F(z)} = \text{constant}$$

in the complex plane. Boundary curves insulated to prevent transverse heat flow lead to $\frac{\partial u}{\partial n} = 0$, which implies

$$F(z) - \overline{F(z)} = \text{constant}$$

12.5.2 Incompressible Inviscid Fluid Flow

Some flow problems for incompressible, nonviscous fluids involve velocity components obtainable in terms of the first derivative of an analytic function. A complex velocity potential $F(z)$ exists such that

$$u - iv = F'(z)$$

At impermeable boundaries the flow normal to the boundary must vanish which implies

$$F(z) - \overline{F(z)} = \text{constant.}$$

Furthermore, a uniform flow field with $u = U$, $v = V$ is easily described by

$$F(z) = (U - iV)z$$

12.5.3 Torsion and Flexure of Elastic Beams

The distribution of stresses in a cylindrical elastic beam subjected to torsion or bending can be computed using analytic functions [90]. For example, in the torsion problem shear stresses τ_{XZ} and τ_{YZ} can be sought as

$$\tau_{XZ} - i\tau_{YZ} = \mu\,\varepsilon[f'(z) - i\bar{z}]$$

and the condition of zero traction on the lateral faces of the beam is described by

$$f(z) - \overline{f(z)} = iz\bar{z}$$

If the function $z = \omega(\zeta)$ which maps $|\zeta| \le 1$ onto the beam cross section is known, then an explicit integral formula solution can be written as

$$f(\zeta) = \frac{1}{2\pi} \int_{|\sigma|=1} \frac{\omega(\sigma)\overline{\omega(\sigma)}d\sigma}{\sigma - \zeta}$$

Consequently, the torsion problem for a beam of simply connected cross section is represented concisely in terms of the function which maps a circular disk onto the cross section.

12.5.4 Plane Elastostatics

Analyzing the elastic equilibrium of two-dimensional bodies satisfying conditions of plane stress or plane strain can be reduced to determining two analytic functions. The formulas to find three stress components and two displacement components are more involved than the ones just stated. They will be investigated later when stress concentrations in a plate having a circular or elliptic hole are discussed.

12.5.5 Electric Field Intensity

Electromagnetic field theory is concerned with the field intensity ϵ which is described in terms of the electrostatic potential $\mathcal{E}$ [92] such that

$$\mathcal{E} = E_x + iE_y = -\frac{\partial \phi}{\partial x} - i\frac{\partial \phi}{\partial y}$$

where ϕ is a harmonic function at all points not occupied by charge. Consequently a complex electrostatic potential $\Omega(z)$ exists such that

$$\mathcal{E} = -\overline{\Omega'(z)}$$

The electromagnetic problem is analogous to inviscid incompressible fluid flow problems. We will also find that harmonic functions remain harmonic under the geometry change of a conformal transformation, which will be discussed later. This produces interesting situations where solutions for new problems can sometimes be derived by simple geometry changes.

12.6 Branch Points and Multivalued Behavior

Before specific types of maps are examined, we need to consider the concept of branch points. A type of singular point quite different from isolated singularities such as poles arises when a singular point of $F(z)$ cannot be made the interior of a small circle on which $F(z)$ is single valued. Such singularities are called branch points and the related behavior is typified by functions such as $\sqrt{z - z_0}$ and $\log(z - z_0)$. To define $p = \log(z - z_0)$, we accept any value p such that e^p produces the value $z - z_0$. Using polar form we can write

$$(z - z_0) = |z - z_0|e^{i(\theta + 2\pi k)} \quad \text{where} \quad \theta = \arg(z - z_0)$$

with k being any integer. Taking

$$p = \log|z - z_0| + i(\theta + 2\pi k)$$

yields an infinity of values all satisfying $e^p = z - z_0$. Furthermore, if z traverses a counterclockwise circuit around a circle $|z - z_0| = \delta$, θ increases by 2π and $\log(z - z_0)$ does not return to its initial value. This shows that $\log(z - z_0)$ is discontinuous on a path containing z_0. A similar behavior is exhibited by $\sqrt{z - z_0}$, which changes sign for a circuit about $|z - z_0| = \delta$.

Functions with branch points have the characteristic behavior that the relevant functions are discontinuous on contours enclosing the branch points. Computing the function involves selection among a multiplicity of possible values. Hence $\sqrt{4}$ can equal $+2$ or -2, and choosing the proper value depends on the functions involved.

For sake of definiteness MATLAB uses what are called principal branch definitions such that

$$\sqrt{z} = |z|^{1/2}\, e^{\imath\theta/2} \quad , \quad -\pi < \theta = \tan^{-1}\left(\frac{y}{x}\right) \le \pi$$

and

$$\log(z) = \log|z| + i\theta$$

The functions defined this way have discontinuities across the negative real axis. Futhermore, $\log(z)$ becomes infinite at $z = 0$.

Dealing carelessly with multivalued functions can produce strange results. Consider the function

$$p = \sqrt{z^2 - 1}$$

which will have discontinuities on lines such that $z^2 - 1 = -|h|$, where h is a general parameter. Discontinuity trouble occurs when

$$z = \pm\sqrt{1 - |h|}$$

Taking $0 \le |h| \le 1$ gives a discontinuity line on the real axis between -1 and $+1$, and taking $|h| > 1$ leads to a discontinuity on the imaginary axis. Figure 12.1 illustrates the odd behavior exhibited by `sqrt(z.^2-1)`. The reader can easily verify that using

```
sqrt(z-1).*sqrt(z+1)
```

defines a different function that is continuous in the plane cut along a straight line between -1 and $+1$.

Multivalued functions arise quite naturally in solutions of boundary value problems, and the choices of branch cuts and branch values are usually evident from physical circumstances. For instance, consider a steady-state temperature problem for the region $|z| < 1$ with boundary conditions requiring

$$u(1,\theta) = 1 \quad , \; 0 < \theta < \pi \;\; \text{and} \;\; \frac{\partial u(1,\theta)}{\partial r} = 0 \quad , \; \pi < \theta < 2\pi .$$

It can be shown that the desired solution is

$$u = \textbf{real}\left\{\frac{1}{\pi i}\left[\log(z+1) - \log(z-1)\right]\right\} + \frac{3}{2}$$

where the logarithms must be defined so u is continuous inside the unit circle and u equals $1/2$ at $z = 0$. Appropriate definitions result by taking

$$-\pi < \arg(z+1) \le \pi \quad , \quad 0 \le \arg(z-1) \le 2\pi$$

MATLAB does not provide this definition intrinsically; so, the user must handle each problem individually when branch points arise.

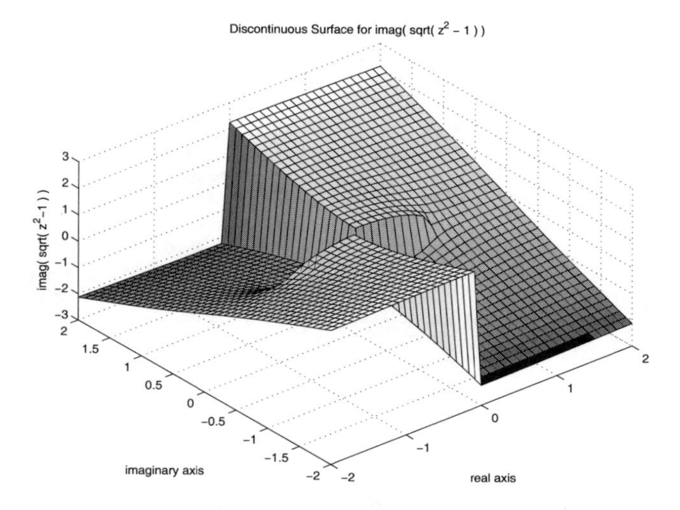

Figure 12.1: Discontinuous Surface for imag (sqrt $(z^2 - 1)^{1/2}$)

12.7 Conformal Mapping and Harmonic Functions

A transformation of the form

$$x = x(\xi, \eta) \ , \ \ y = y(\xi, \eta)$$

is said to be conformal if the angle between intersecting curves in the (ξ, η) plane remains the same for corresponding mapped curves in the (x, y) plane. Consider the transformation implied by $z = \omega(\zeta)$ where ω is an analytic function of ζ. Since

$$dz = \omega'(\zeta)\, d\zeta$$

it follows that

$$|dz| = |\omega'(\zeta)|\, |d\zeta| \qquad \text{and} \qquad \arg(dz) = \arg(\, \omega'(\zeta)\,) + \arg(d\zeta)$$

This implies that the element of length $|d\zeta|$ is stretched by a factor of $|\omega'(\zeta)|$ and the line element $d\zeta$ is rotated by an angle $\arg[\omega'(\zeta)]$. The transformation is conformal at all points where $\omega'(\zeta)$ exists and is nonzero.

Much of the interest in conformal mapping results from the fact that harmonic functions remain harmonic under a conformal transformation. To see why this is true, examine Laplace's equation written in the form

$$\nabla^2_{xy} u = 4 \frac{\partial^2 u}{\partial z \partial \bar{z}} = 0$$

For a conformal map we have

$$z = \omega(\zeta) \quad , \quad \bar{z} = \overline{\omega(\zeta)}$$

$$\frac{\partial u}{\partial z} = \frac{1}{\omega'(\zeta)} \frac{\partial u}{\partial \zeta} \quad , \quad \frac{\partial u}{\partial \bar{z}} = \frac{1}{\overline{\omega'(\zeta)}} \frac{\partial u}{\partial \bar{\zeta}}$$

Since z depends only on ζ and $\bar{z}$ depends only on $\bar{\zeta}$ we find that

$$\nabla_{xy}^2 u = 4 \frac{1}{\omega'(\zeta)\overline{\omega'(\zeta)}} \frac{\partial^2 u}{\partial \zeta \partial \bar{\zeta}} = \frac{1}{|\omega'(\zeta)|^2} \nabla_{\xi\eta}^2 u$$

It follows that

$$\frac{\partial^2 u}{\partial x^2} + \frac{\partial^2 u}{\partial y^2} = 0 \quad \text{implies} \quad \frac{\partial^2 u}{\partial \xi^2} + \frac{\partial^2 u}{\partial \eta^2} = 0$$

wherever $\omega'(\zeta) \neq 0$. The transformed differential equation in the new variables is identical to that of the original differential equation. Hence, when $u(x, y)$ is a harmonic function of (x, y), then $u(x(\xi, \eta), y(\xi, \eta))$ is a harmonic function of (ξ, η), provided $\omega(\zeta)$ is an analytic function. This is a remarkable and highly useful property. Normally, changing the independent variables in a differential equation changes the form of the equation greatly. For instance, with the polar coordinate transformation

$$x = r\cos(\theta), \quad y = r\sin(\theta)$$

the Laplace equation becomes

$$\nabla^2 u = \frac{\partial^2 u}{\partial r^2} + \frac{1}{r}\frac{\partial u}{\partial r} + \frac{1}{r^2}\frac{\partial^2 u}{\partial \theta^2} = 0$$

The appearance of this equation is very different from the Cartesian form because $x+iy$ is not an analytic function of $r+i\theta$. On the other hand, using the transformation

$$z = \log(\zeta) = \log(|\zeta|) + i\arg(\zeta)$$

gives

$$\nabla_{xy}^2 u = (\zeta\bar{\zeta})\nabla_{\xi\eta}^2 u$$

and $\nabla_{xy}^2 u = 0$ implies $\nabla_{\xi\eta}^2 u = 0$ at points other than $\zeta = 0$ or $\zeta = \infty$.

Because solutions to Laplace's equation are important in physical applications, and such functions remain harmonic under a conformal map, an analogy between problems in two regions often can be useful. This is particularly attractive for problems where the harmonic function has constant values or zero normal gradient on critical boundaries. An instance pertaining to inviscid fluid flow about an elliptic cylinder will be used later to illustrate the harmonic function analogy. In the subsequent sections we discuss several transformations and their relevant geometrical interpretation.

12.8 Mapping onto the Exterior or the Interior of an Ellipse

We will examine in some detail the transformation

$$z = \left(\frac{a+b}{2}\right)\zeta + \left(\frac{a-b}{2}\right)\zeta^{-1} = R(\zeta + m\zeta^{-1}) \quad , \quad \zeta \geq 1$$

where $R = (a+b)/2$ and $m = (a-b)/(a+b)$. The derivative

$$z'(\zeta) = R(1 - m\zeta^{-2})$$

becomes nonconformal when $z'(\zeta) = 0$ or $\zeta = \pm\sqrt{m}$. For sake of discussion, we temporarily assume $a \geq b$ to make $\sqrt{m}$ real rather than purely imaginary. A circle $\zeta = \rho_0 e^{\imath\theta}$ transforms into

$$x + \imath y = R(\rho_0 + m\rho_0^{-1})\cos(\theta) + iR(\rho_0 - m\rho_0^{-1})\sin(\theta)$$

yielding an ellipse. When $\rho_0 = 1$ we get $x = a\cos(\theta)$, $y = b\sin(\theta)$. This mapping function is useful in problems such as inviscid flow around an elliptic cylinder or stress concentration around an elliptic hole in a plate. Furthermore, the mapping function is easy to invert by solving a quadratic equation to give

$$\zeta = \frac{z + \sqrt{(z-\alpha)(z+\alpha)}}{a+b} \quad , \quad \alpha = \sqrt{a^2 - b^2}$$

The radical should be defined to have a branch cut on the x-axis from $-\alpha$ to α and to behave like $+z$ for large $|z|$. Computing the radical in MATLAB as

```
sqrt(z-alpha).*sqrt(z+alpha)
```

works fine when α is real because MATLAB uses

$$-\pi < \arg(z \pm \alpha) \leq \pi$$

and the sign change discontinuities experienced by both factors on the negative real axis cancel to make the product of radicals continuous. However, when $a < b$ the branch points occur at $\pm z_0$ where $z_0 = i\sqrt{b^2 - a^2}$, and a branch cut is needed along the imaginary axis. We can give a satisfactory definition by requiring

$$-\frac{\pi}{2} < \arg(z \pm z_0) \leq \frac{3\pi}{2}$$

The function **elipinvr** provided below handles general a and b.

Before leaving the problem of ellipse mapping we mention the fact that mapping the interior of a circle onto the interior of an ellipse is rather complicated but can be formulated by use of elliptic functions [75]. However, a simple solution to compute boundary point correspondence between points on the circle and points on the ellipse appears in [52]. This can be used to obtain mapping functions in rational form which are quite accurate. The function **elipdplt** produces the mapping. Results showing how a polar coordinate grid in the ζ-plane maps onto a two to one ellipse appears in Figure 12.2. In these examples and other similar ones, grid networks in polar coordinates always use constant radial increments and constant angular increments. Only the region corresponding to $0.3 \le |\zeta| \le 1$ and $0 \le \arg(\zeta) \le \frac{\pi}{2}$ is shown. Note that the distortion of line elements at different points of the grid is surprisingly large. This implies that the stretching effect, depending on $|\omega'(\zeta)|$,varies more than might at first be expected.

Often it is desirable to see how a rectangular or polar coordinate grid distorts under a mapping transformation. This is accomplished by taking the point arrays and simultaneously plotting rows against rows and columns against columns as computed by the following function **gridview** which works for general input arrays x, y. If the input data are vectors instead of arrays, then the routine draws a single curve instead of a surface. When **gridview** is executed with no input, it generates the plot in Figure 12.3 which shows how a polar coordinate grid in the ζ-plane maps under the transformation

$$z = R\left(\zeta + \frac{m}{\zeta}\right)$$

The new grid consists of a system of confocal ellipses orthogonally intersecting a system of hyperbolas.

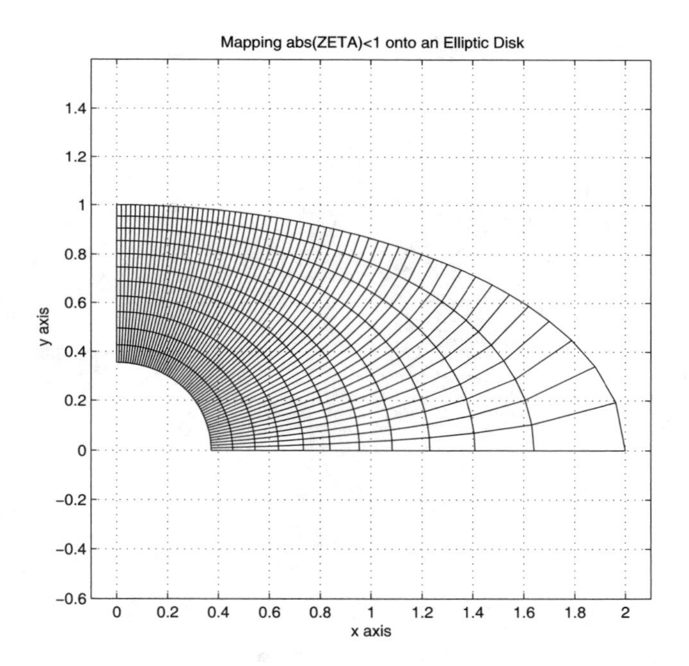

Figure 12.2: Mapping $|z| < 1$ onto an Elliptic Disk

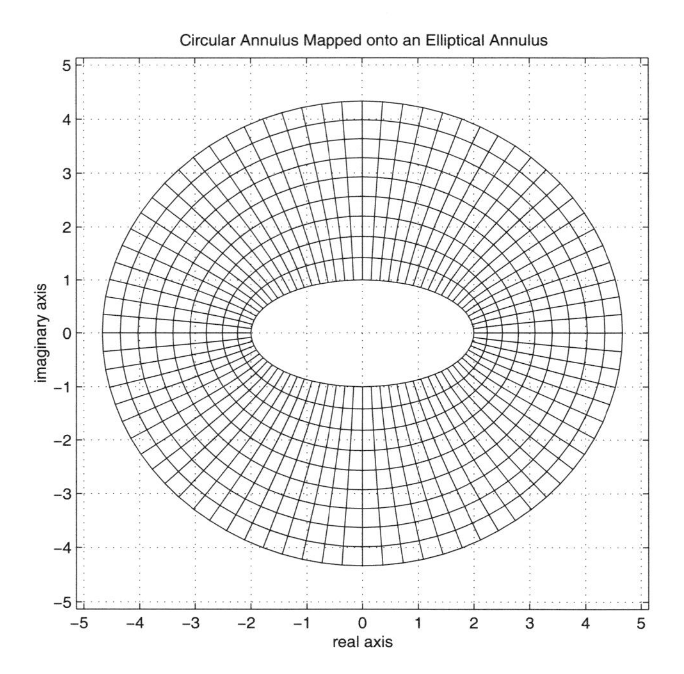

Figure 12.3: Circular Annulus Mapped onto an Elliptic Annulus

12.8.1 Program Output and Code

Function sqrtsurf

```
1: function sqrtsurf
2: %
3: % sqrtsurf
4: % ~~~~~~~~
5: %
6: % This function illustrates the discontinuity
7: % in the function w=sqrt(z*z-1).
8:
9: xx=linspace(-2,2,41); [x,y]=meshgrid(xx,xx);
10: z=x+i*y; w=sqrt(z.*z-1); close
11: surf(x,y,imag(w)); view(-40,50);
12: xlabel('real axis'); ylabel('imaginary axis');
13: zlabel('imag( sqrt( z^2-1 ) )');
14: title(['Discontinuous Surface for imag( sqrt', ...
15:        '( z^2 - 1 ) )']);
16: grid on; figure(gcf);
17: %print -deps sqrtsurf
```

Function elipinvr

```
1: function zeta=elipinvr(a,b,z)
2: %
3: % zeta=elipinvr(a,b,z)
4: % ~~~~~~~~~~~~~~~~~~~~
5: %
6: % This function inverts the transformation
7: % z=(a+b)/2*zeta+(a-b)/2/zeta which maps
8: % abs(zeta)>=1 onto (x/a).^2+(y/b).^2 >= 1
9: %
10: %  a    - semi-diameter on x-axis
11: %  b    - semi-diameter on y-axis
12: %  z    - array of complex values
13: %
14: %  zeta - array of complex values for the
15: %         inverse mapping function
16: %
17: % User m functions called:  none
18:
```

```
19: z0=sqrt(a^2-b^2); ab=a+b;
20: if a==b
21:   zeta=z/a;
22: elseif a>b  % branch cut along the real axis
23:   zeta=(z+sqrt(z-z0).*sqrt(z+z0))/ab;
24: else       % branch cut along the imaginary axis
25:   ap=angle(z+z0); ap=ap+2*pi*(ap<=-pi/2);
26:   am=angle(z-z0); am=am+2*pi*(am<=-pi/2);
27:   zeta=(z+sqrt(abs(z.^2-z0.^2)).*exp(...
28:         i/2*(ap+am)))/ab;
29: end
```

Function elipdplt

```
1: function [z,a,b]=elipdplt(rx,ry)
2: % [z,a,b]=elipdplt(rx,ry)
3: % ~~~~~~~~~~~~~~~~~~~~~~~~
4: % This function plots contour lines showing
5: % how a polar coordinate grid in a circular
6: % disk maps onto an elliptic disk.
7: %
8: % User m functions called: elipdisk, gridview
9:
10: if nargin==0, rx=2; ry=1; end
11: zeta=linspace(.3,1,12)'* ...
12:      exp(i*linspace(0,pi/2,61));
13: [z,a,b]=elipdisk(zeta,rx,ry);
14: x=real(z); y=imag(z);
15: gridview(x,y,'x axis','y axis',...
16:   'Mapping abs(ZETA)<1 onto an Elliptic Disk');
17: colormap([1 1 1]); shg
18: print -deps elipdisk
19:
20: %============================================
21:
22: function gridview(x,y,xlabl,ylabl,titl)
23: %
24: % gridview(x,y,xlabl,ylabl,titl)
25: % ~~~~~~~~~~~~~~~~~~~~~~~~~~~~~~
26: %
27: % This function views a surface from the top
```

```
28: % to show the coordinate lines of the surface.
29: % It is useful for illustrating how coordinate
30: % lines distort under a conformal transformation.
31: % Calling gridview with no arguments depicts the
32: % mapping of a polar coordinate grid map under
33: % a transformation of the form
34: % z=R*(zeta+m/zeta).
35: %
36: %  x,y           - real matrices defining a
37: %                  curvilinear coordinate system
38: %  xlabl,ylabl - labels for x and y axes
39: %  titl          - title for the graph
40: %
41: % User m functions called:  cubrange
42: %----------------------------------------------------
43:
44: % close
45: if nargin<5
46:   xlabl='real axis'; ylabl='imaginary axis';
47:   titl='';
48: end
49:
50: % Default example using z=R*(zeta+m/zeta)
51: if nargin==0
52:   zeta=linspace(1,3,10)'* ...
53:        exp(i*linspace(0,2*pi,81));
54:   a=2; b=1; R=(a+b)/2; m=(a-b)/(a+b);
55:   z=R*(zeta+m./zeta); x=real(z); y=imag(z);
56:   titl=['Circular Annulus Mapped onto an ', ...
57:        'Elliptical Annulus'];
58: end
59:
60: range=cubrange([x(:),y(:)],1.1);
61:
62: % The data defin a curve
63: if size(x,1)==1 | size(x,2)==1
64:   plot(x,y,'-k'); xlabel(xlabl); ylabel(ylabl);
65:   title(titl); axis('equal'); axis(range);
66:   grid on; figure(gcf);
67:   if nargin==0
68:     print -deps gridviewl
69:   end
70: % The data defin a surface
71: else
72:   plot(x,y,'k-',x',y','k-')
```

```
73:   xlabel(xlabl); ylabel(ylabl); title(titl);
74:   axis('equal'); axis(range); grid on;
75:   figure(gcf);
76:   if nargin==0
77:     print -deps gridview
78:   end
79: end
80:
81: %==============================================
82:
83: function [z,a,b]=elipdisk(zeta,rx,ry)
84: %
85: % [z,a,b]=elipdisk(zeta,rx,ry)
86: % ~~~~~~~~~~~~~~~~~~~~~~~~~~~~~~
87: %
88: % This function computes a rational function
89: % mapping abs(zeta)<=1 onto an elliptical disk
90: % defined by (x/rx)^2+(y/ry)^2<=1. Boundary
91: % points are computed using theory from
92: %   P. Henrici, Applied Complex Analysis,
93: %   Vol 3, p391.
94: % The rational function approximation has the
95: % form:
96: %        z=sum(a(j)*zeta^(2*j-1)) /
97: %          (1+sum(b(j)*zeta^(2*j));
98: %
99: %  zeta   - matrix of points with abs(zeta)<=1
100: %  rx,ry - ellipse semidiameters on x and y
101: %          axes
102: %
103: %  z     - points into which zeta maps
104: %  a,b   - coefficients in the rational
105: %          function defining the map
106: %
107: % User m functions called: ratcof
108: %----------------------------------------------
109:
110: ntrms=100; ntheta=251;
111: tau=(0:2*pi/ntheta:2*pi)';
112: ep=(rx-ry)/(rx+ry);
113: z=exp(i*tau); z=z+ep*conj(z);
114: j=1:ntrms;  ep=ep.^j; ep=ep./(j.*(1+ep.*ep));
115: theta=tau+2*( sin((2*tau+pi)*j)*ep');
116: zta=exp(i*theta); z=rx/max(real(z))*z;
117: [a,b]=ratcof(zta.^2,z./zta,8);
```

```
118: a=fix(real(1e8*a))/1e8; b=fix(real(1e8*b))/1e8;
119: af=flipud(a(:)); bf=flipud([1;b(:)]);
120: zta2=zeta.^2;
121: z=zeta.*polyval(af,zta2)./polyval(bf,zta2);
122:
123: %===============================================
124:
125: function [a,b]=ratcof(xdata,ydata,ntop,nbot)
126: %
127: % [a,b]=ratcof(xdata,ydata,ntop,nbot)
128: % ~~~~~~~~~~~~~~~~~~~~~~~~~~~~~~~~~~~~~~
129: %
130: % Determine a and b to approximate ydata as
131: % a rational function of the variable xdata.
132: % The function has the form:
133: %
134: %    y(x) = sum(1=>ntop) ( a(j)*x^(j-1) ) /
135: %         ( 1 + sum(1=>nbot) ( b(j)*x^(j)) )
136: %
137: % xdata,ydata - input data vectors (real or
138: %               complex)
139: % ntop,nbot   - number of series terms used in
140: %               the numerator and the
141: %               denominator.
142: %
143: % User m functions called: none
144: %-----------------------------------------------
145:
146: ydata=ydata(:); xdata=xdata(:);
147: m=length(ydata);
148: if nargin==3, nbot=ntop; end;
149: x=ones(m,ntop+nbot); x(:,ntop+1)=-ydata.*xdata;
150: for i=2:ntop, x(:,i)=xdata.*x(:,i-1); end
151: for i=2:nbot
152:   x(:,i+ntop)=xdata.*x(:,i+ntop-1);
153: end
154: ab=x\ydata;
155: a=ab(1:ntop); b=ab(ntop+1:ntop+nbot);
156:
157: %===============================================
158:
159: % function range=cubrange(xyz,ovrsiz)
160: % See Appendix B
```

12.9 Linear Fractional Transformations

The mapping function defined by

$$w = \frac{az + b}{cz + d}$$

is called a linear fractional, or bilinear, transformation where a, b, c, and d are constants. It can be inverted to yield

$$z = \frac{-dw + b}{cw - a}$$

If c is zero the transformation is linear. Otherwise, we can divide out c to get

$$w = \frac{Az + B}{z + D}$$

The three remaining constants can be found by making three points in the z-plane map to three given points in the w-plane. Note that $z = \infty$ maps to $w = A$ and $z = -D$ maps to $w = \infty$.

The transformation has the attractive property that circles or straight lines map into circles or straight lines. An equation defining a circle or straight line in the z-plane has the form

$$Pz\bar{z} + Qz + \bar{Q}\bar{z} + S = 0$$

where P and S are real. A straight line is obtained when P is zero. Expressing z in terms of w and clearing fractions leads to an equation of the form

$$P_0 w\bar{w} + Q_0 w + \bar{Q}_0 \bar{w}_0 + S_0 = 0$$

which defines a circle in the w-plane when P_0 is nonzero. Otherwise, a straight line in the w-plane results.

Determining the bilinear transformation to take three z-points to three w-points is straightforward except for special cases. Let

`Z=[z1;z2;z3]` and `W=[w1;w2;w3]`

If `det([Z,W,ones(3,1)])` vanishes then a linear transformation with $c = 0$ and $d = 1$ applies. If $z = \infty$ maps to w_1 we take $a = w_1$, $c = 1$. If $z = z_1$ maps to $w = \infty$ we take $c = 1$, $d = -z_1$. In the usual situation we simply write $w(z + D) = Az + B$ and solve the system

```
[Z,ones(3,1),-W]*[A;B;D]=W.*Z
```

Function **linfrac**, used to compute the coefficients in the transformation, is provided at the end of this section. Points at infinity are handled by including ∞ (represented in MATLAB by `inf`) as a legitimate value in the components of z or w. For example, the transformation $w = (2z+3)/(z-1)$ takes $z = \infty$ to $w = 2$, $z = 1$ to $w = \infty$, and $z = 1 + \imath$ to $w = 2 - 5\imath$. The expression

```
cz=linfrac([inf,1,1+i],[2,inf,2-5i]);
```

produces the coefficients in the transformation. Similarly, the transformation is inverted by

```
cw=linfrac([2,inf,2-5i],[inf,1,1+i]);
```

or equivalently by

```
cw=linfrac([0,1,2i],[-1.5,-4,-0.25-1.25i]);
```

Another type of problem of interest in connection with a known bilinear transformation is to find the circle or straight line into which a given circle or straight line maps. Function **crc2crc** performs this task. The coefficients c are given along with three points lying on a circle or a straight line. Then parameters w_0, r_0 pertaining to the w-plane are computed. If parameter `type` equals 1, then w_0 and r_0 specify the center and radius of a circle. Otherwise, w_0 and r_0 are two points defining a straight line.

The linear fractional transformation can be used to map an eccentric annulus such as that in Figure 12.4 onto a concentric annulus. Suppose a region $1 \le |z| \le R$ is to be mapped onto the region defined by

$$|w| \ge R_1 \quad , \quad |w - w_0| \le R_0$$

The radius R and mapping coefficients c can be obtained by solving a system of nonlinear simultaneous equations. Function **ecentric** accomplishes the task. A function call of

```
[c,r]=ecentric(0.25,-0.25,1);
```

produces

$$w = \frac{3.4821z + 0.25}{z + 13.9282} \quad , \quad R = 3.7321$$

and the plot in Figure 12.4 shows the mapped image of a polar coordinate grid using

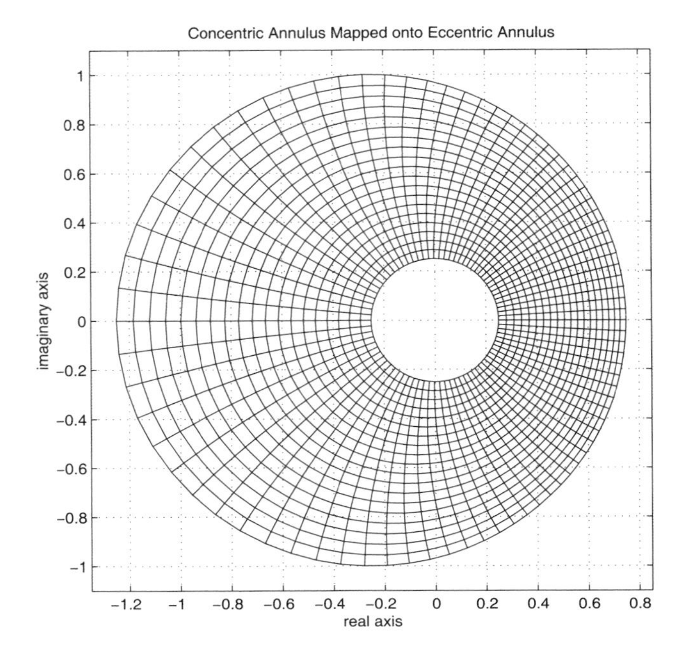

Figure 12.4: Concentric Annulus Mapped onto Eccentric Annulus

constant radial and angular increment in the z plane.

To demonstrate the utility of the transformation just discussed, consider the problem of determining the steady-state temperature field in an eccentric annulus with the inner and outer boundaries held at u_1 and u_0, respectively. The temperature field will be a harmonic function that remains harmonic under a conformal transformation. The related problem for the concentric annulus has the simple form

$$u = u_1 + \frac{(u_0 - u_1)\ln(r)}{\ln(R)} \quad , \quad 1 \leq r \leq R$$

By analogy, expressing $r = |z|$ in terms of w gives the temperature distribution at points in the w-plane.

12.9.1 Program Output and Code

Function linfrac

```
1: function c=linfrac(z,w)
2: %
3: % c=linfrac(z,w)
4: % ~~~~~~~~~~~~~~
5: %
6: % This function determines the linear
7: % fractional transformation to map any three
8: % points in the z-plane into any three points
9: % in the w plane. Not more than one point in
10: % either the z or w plane may be located at
11: % infinity.
12: %
13: % z  - vector of complex values [z1,z2,z3]
14: % w  - vector of complex values [w1,w2,w3]
15: %
16: % c  - vector defining the bilinear
17: %      transformation
18: %        w=(c(1)*z + c(2))/(c(3)*z + c(4))
19: %
20: % User m functions called:  none
21: %----------------------------------------------
22:
23: z=z(:); w=w(:); c=ones(4,1);
24: k=find(z==inf); j=find(w==inf); kj=[k;j];
25:
26: % z and w both contain points at infinity
27: if length(kj)==2
28:   c(1)=w(k); c(4)=-z(j); w(kj)=[]; z(kj)=[];
29:   c(2)=(w-c(1))*z+w*c(4);
30:   return
31: end
32:
33: % z=infinity maps to a finite w point
34: if ~isempty(k) & isempty(j)
35:   c(1)=w(k); z(k)=[]; w(k)=[];
36:   c([2 4])=[[1;1],-w]\[(w-c(1)).*z];
37:   return
38: end
39:
40: % a finite z point maps to w = infinity
```

```
41: if ~isempty(j) & isempty(k)
42:   c(4)=-z(j); z(j)=[]; w(j)=[];
43:   c([1 2])=[z,[1;1]]\[w.*(z+c(4))];
44:   return
45: end
46:
47: % case where all  points are finite
48: mat=[z,ones(3,1),-w];
49:
50: % case for a general transformation
51: if det(mat)~=0
52:   c([1 2 4])=mat\[w.*z];
53: % case where transformation is linear
54: else
55:   c(3)=0; c([1 2])=[z,ones(3,1)]\w;
56: end
```

Function crc2crc

```
1: function [w0,r0,type]=crc2crc(c,z)
2: %
3: % [w0,r0,type]=crc2crc(c,z)
4: % ~~~~~~~~~~~~~~~~~~~~~~~~~~~~
5: %
6: % This function determines the circle or
7: % straight line into which a circle or straight
8: % line maps under a linear fractional
9: % transformation.
10: %
11: % c    - coefficients defining a linear
12: %        fractional transformation
13: %          w=(c(1)*z+c(2))/(c(3)*z*c(4))
14: %        where c(2)*c(3)-c(1)*c(4) is nonzero
15: % z    - a vector of three complex values
16: %        lying on a circle or a straight line
17: %
18: % w0   - center of a circle in the w plane
19: %        if type=1, or a point on a straight
20: %        line if type=2
21: % r0   - radius of a circle in the w plane
22: %        if type=1, or a point on a straight
23: %        line if type=2
24: % type - equals 1 to denote a circle or 2 to
```

```
25: %          denote a straight line in the w plane
26: %
27: % User m functions called:  none
28: %----------------------------------------------
29:
30: % check for degenerate transformation
31: if c(2)*c(3)==c(1)*c(4)
32:   disp(['Degenerate transformation in ', ...
33:         'function crc2crc']);
34:   w0=[]; r0=[]; type=[]; return;
35: end
36:
37: % evaluate the mapping of the z points
38: w=(c(1)*z(:)+c(2))./(c(3)*z(:)+c(4));
39:
40: % check whether a point passes to infinity or
41: % the three z points define a straight line
42: k=find(w==inf);
43: dt=det([real(w),imag(w),ones(3,1)]);
44: if ~isempty(k); w(k)=[]; end
45:
46: % case for a straight line in the w plane
47: % defined by two points on the line
48: if dt==0 | ~isempty(k)
49:   type=2; w0=w(1); r0=w(2);
50: % case for a circle in the w plane defined by
51: % a center point and the circle radius
52: else
53:   type =1;
54:   v=[2*real(w),2*imag(w),ones(3,1)]\abs(w).^2;
55:   w0=v(1)+i*v(2); r0=sqrt(v(3)+abs(w0)^2);
56: end
```

Function ecentric

```
1: function [c,r]=ecentric(ri,wo,ro,nopl)
2: %
3: % [c,r]=ecentric(ri,wo,ro,nopl)
4: % ~~~~~~~~~~~~~~~~~~~~~~~~~~~~
5: %
6: % This function determines the bilinear
7: % transformation which maps the region
8: % 1<=abs(z)<=r onto an eccentric annulus
```

```
 9: % defined by
10: %      abs(w)>=ri & abs(w-wo)<=ro
11: %
12: % The coefficients c in the transformation
13: %      w=(c(1)*z+c(2))/(c(3)*z+c(4))
14: % must be found as well as the outer radius r
15: % of the annulus in the z plane.
16: %
17: % ri   - radius of inner circle abs(w)=ri
18: % wo   - center of outer circle abs(w-wo)=ro
19: % ro   - radius of outer circle
20: %
21: % c    - coefficients in the mapping function
22: % r    - radius of outer circle abs(z)=r
23: % nopl- no plot is given if nopl is input
24: %
25: % User m functions called: gridview
26:
27: if nargin==0, ri=.25; wo=-.25; ro=1; end
28:
29: if wo~=0
30:   c1=(wo+ro)/ri; c2=(wo-ro)/ri; c3=2/(c1+c2);
31:   c4=(c2-c1)/(c1+c2); c5=c3-c1-c1*c4; c6=1-c1*c3;
32:   rt=sqrt(c5^2-4*c4*c6);
33:   r1=(-c5+rt)/(2*c4); r2=(-c5-rt)/(2*c4);
34:   r=max([r1,r2]); d=c3+c4*r; c=[ri*d;ri;1;d];
35: else
36:   c=[ri;0;0;1]; r=ro/ri;
37: end
38: if nargin > 3, return, end
39:
40: % Show the region onto which a polar coordinate
41: % grid in the z-plane maps.
42: z=linspace(1,r,20)'*exp(i*linspace(0,2*pi,81));
43: w=(c(1)*z+c(2))./(c(3)*z+c(4));
44: titl=['Concentric Annulus Mapped onto ', ...
45:       'Eccentric Annulus'];
46: gridview(real(w),imag(w),...
47:   'real axis','imaginary axis',titl); shg
48: % print -deps ecentric
```

12.10 Schwarz-Christoffel Mapping onto a Square

The Schwarz-Christoffel transformation [75] provides integral formulas defining transformations to map the interior of a circle onto the interior or exterior of a polygon. Special cases obtained by allowing selected vertices to pass to infinity lead to a variety of results [58]. In general situations, evaluating the parameters and integrals in the Schwarz-Christoffel transformation is difficult and requires use of special software [35]. We will examine only two cases: a) where the interior of a circle is mapped onto the interior of a square, and b) where the exterior of a circle is mapped onto the exterior of a square. The function

$$z = C \int_0^{\zeta} (1 + t^4)^{-1/2} \, dt,$$

where C is a scaling constant, maps $|\zeta| \leq 1$ inside the square defined by

$$(|x| \leq 1) \cap (|y| \leq 1).$$

Expanding this radical by the binominal expansion and integrating gives

$$z = c \sum_{n=0}^{\infty} (-1)^n \left[\frac{\Gamma(n + \frac{1}{2})}{n!(4n + 1)} \right] \zeta^{1+4n} \quad , \quad |\zeta| \leq 1$$

A reasonably good approximation to the mapping function can be obtained by taking several hundred terms in the mapping function and adjusting the constant c to make $\zeta = 1$ match $z = 1$. This series expansion converges slowly and rounds the corners of the square because the derivative of the mapping function behaves like $(\zeta - \zeta_o)^{-1/2}$ at $\zeta_o = \pm e^{\pm \imath \pi/4}$.

The transformation to map $|\zeta| \geq 1$ onto the square exterior defined by

$$(|x| \geq 1) \cup (|y| \geq 1)$$

has the form

$$z = c_0 \int_1^{\zeta} (1 + t^{-4})^{1/2} \, dt + c_1,$$

where c_0 and c_1 are arbitrary constants. Using the binomial expansion again and term by term integration leads to

$$z = c \sum_{n=0}^{\infty} (-1)^n \left[\frac{\Gamma(n - \frac{1}{2})}{n!(4n - 1)} \right] \zeta^{1-4n} \quad , \quad |\zeta| \geq 1$$

The function **swcsqmap** provides both interior and exterior polynomial maps. Once again, truncating the series after a specified number of terms and making $\zeta = 1$ map to $z = 1$ gives an approximate mapping function which converges much more rapidly than the series for the interior problem. Rounding of the square corners is greatly reduced because the mapping function derivative behaves like $(\zeta - \zeta_o)^{1/2}$ at $\zeta_o = \pm e^{\pm \imath\pi/4}$. Figure 12.5 illustrates results produced by the ten term series for both interior and exterior regions. Using rational functions to produce better results than polynomials was discussed earlier in Chapter 3. The function **squarat**, which provides both interior and exterior maps, appears below.

It should be noted that inverting a mapping function $z = \omega(\zeta)$ to get $\zeta = g(z)$ explicitly is often difficult, if not impossible. For example, consider the form

$$z = \frac{\zeta(a + b\zeta^4 + c\zeta^8)}{1 + d\zeta^4 + e\zeta^8} \quad , \quad |\zeta| \leq 1$$

which requires solving the polynomial

$$c\zeta^9 - ez\zeta^8 + b\zeta^5 - dz\zeta^4 + a\zeta - z = 0$$

and picking the root inside or on the unit circle. Although the MATLAB function **roots** efficiently factors polynomials with complex coefficients, inverting the mapping function for hundreds or thousands of values can be time consuming.

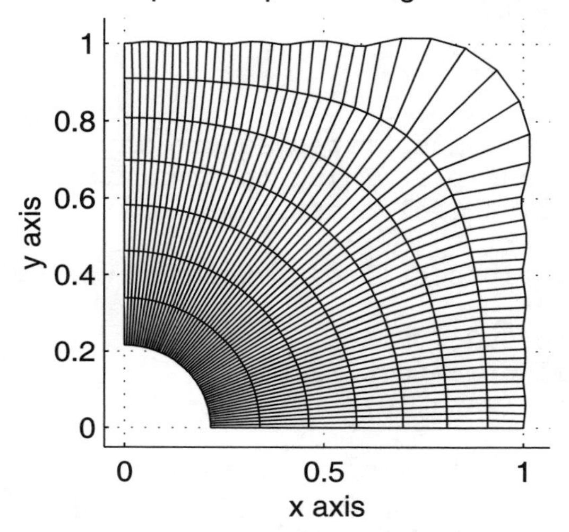

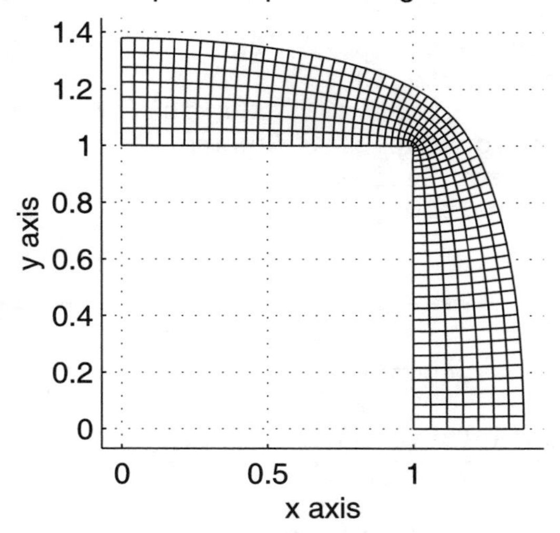

Figure 12.5: **Square Maps Using a 10-term Series**

12.10.1 Program Output and Code

Function swcsq10

```
1: function swcsq10
2: % Example: swcsq10
3: % ~~~~~~~~~~~~~~~~
4: %
5: % This example demonstrates square map
6: % approximations pertaining to truncated
7: % Schwarz-Christoffel transformations.
8: %
9: % User m functions called:  swcsqmap, gridview
10:
11: zeta=linspace(0.2,1,8)'* ...
12:      exp(i*linspace(0,pi/2,61));
13: [z,a]=swcsqmap(zeta,10);
14: subplot(211)
15: gridview(real(z),imag(z),'x axis','y axis', ...
16:         ['Interior Map of a Square Using', ...
17:          ' a 10-term Series']);
18: subplot(212)
19: zeta=linspace(1,1.25,8)'* ...
20:      exp(i*linspace(0,pi/2,61));
21: [z0,a]=swcsqmap(zeta,10,1);
22: gridview(real(z0),imag(z0),'x axis','y axis', ...
23:         ['Exterior Map of a Square Using ', ...
24:         'a 10-term Series']);
25: print -deps sqrplt10
26: subplot
27:
28: %=============================================
29:
30: function [z,a]=swcsqmap(zeta,ntrms,ifout)
31: %
32: % [z,a]=swcsqmap(zeta,ntrms,ifout)
33: % ~~~~~~~~~~~~~~~~~~~~~~~~~~~~~~~~
34: %
35: % This function evaluates power series
36: % approximations for mapping either the inside
37: % of a circle onto the inside of a square, or
38: % mapping the outside of a circle onto the
39: % outside of a square. The Schwarz-Christoffel
40: % integrals defining the mapping functions are
```

```
41: % expanded in Taylor series and are truncated
42: % to produce approximations in the following
43: % polynomial forms:
44: %
45: %   For the interior problem:
46: %     z=sum(a(n)*zeta^(4*n-3),n=1:ntrms)
47: %
48: %   For the exterior problem:
49: %     z=sum(a(n)*zeta^(-4*n+5),n=1:ntrms)
50: %
51: % The side length of the square is adjusted
52: % to equal 2.
53: %
54: % zeta  - complex values where the mapping
55: %         function is evaluated
56: % ntrms - number of terms used in the
57: %         truncated series
58: % ifout - a parameter omitted if an interior
59: %         map applies.  ifout can have any
60: %         value (such as 1) to show that an
61: %         exterior map is to be performed.
62: %
63: % z     -  values of the mapping function
64: % a     -  coefficients in the mapping series
65: %
66: % User m functions called:  none
67: %----------------------------------------------
68:
69: n=0:ntrms-2;
70: if nargin==2   % recursion formula for mapping
71:                % interior on interior
72:    p1=(n+1/2)./(n+1); p2=(n+1/4)./(n+5/4);
73: else           % recursion formula for mapping
74:                %exterior on exterior
75:    p1=(n-1/2)./(n+1); p2=(n-1/4)./(n+3/4);
76: end
77: a=[1,cumprod(-p1.*p2)]; a=a(:)/sum(a);
78: z4=zeta.^4;
79: if nargin ==3, z4=1./z4; end;
80: z=zeta.*polyval(flipud(a(:)),z4);
81:
82: %=============================================
83:
84: % function gridview(x,y,xlabl,ylabl,titl)
```

```
85: % See Appendix B
```

Function squarat

```
 1: function [z,a,b]=squarat(zeta,ifout)
 2: %
 3: % [z,a,b]=squarat(zeta,ifout)
 4: % ~~~~~~~~~~~~~~~~~~~~~~~~~~~~~
 5: %
 6: % This function maps either the interior of a
 7: % circle onto the interior of a square, or maps
 8: % the exterior of a circle onto the exterior of
 9: % a square using a rational function having the
10: % approximate form:
11: %
12: %  z(zeta) = zeta *
13: %
14: %      Sum(a(j)*zeta4^j)/(1+Sum(b(j)*zeta4^j),
15: %
16: % where zeta4=zeta^4 for an interior problem,
17: % or zeta4=zeta^(-4) for an exterior problem.
18: %
19: % zeta  - matrix of complex values such that
20: %         abs(zeta)<=1 for an interior map,
21: %         or abs(zeta)>=1 for an exterior map
22: % ifout - parameter present in the call list
23: %         only when an exterior mapping is
24: %         required
25: %
26: % z     - matrix of values of the mapping
27: %         function
28: % a,b   - coefficients of the polynomials
29: %         defining the rational mapping
30: %         function
31: %
32: % User m functions called: none
33: %----------------------------------------------
34:
35: zeta4=zeta.^4;
36:
37: if nargin==1 % map interior on interior
38:   a=[ 1.07835, 1.37751,-0.02642, -0.09129, ...
39:       0.13460,-0.15763, 0.07430,  0.14858, ...
```

```
40:          0.01878,-0.00354 ]';
41:    b=[ 1.37743, 0.07157,-0.11085,  0.12778, ...
42:        -0.13750, 0.05313, 0.14931,  0.02683, ...
43:        -0.00350,-0.000120 ]';
44: else          % map exterior on exterior
45:    a = [1.18038, 1.10892, 0.13365, -0.02910]';
46:    b = [1.10612, 0.27972, 0.00788]';
47:    zeta4=1./zeta4;
48: end
49:
50: % Evaluate the mapping function
51: af=flipud(a); bf=flipud([1;b]);
52: z=zeta.*polyval(af,zeta4)./polyval(bf,zeta4);
```

12.11 Determining Harmonic Functions in a Circular Disk

The problem of determining a function that is harmonic for $|z| < 1$ and satisfies certain boundary conditions can be analyzed effectively using series methods. In problems pertaining to the unit circle, it is often convenient to consider a function u, in polar cordinates, and write $u(r, \theta)$. Simultaneously, we may wish to think in terms of the related complex variable $z = r\,\sigma$ where $\sigma = e^{i\theta}$. Three basic problems will be considered.

I) **Dirichlet Problem**

$$\nabla^2 u = 0 \quad , \quad |z| < 1$$

$$u(1, \theta) = f(\theta) \quad , \quad 0 \leq \theta \leq 2\pi$$

We assume $f(\theta)$ is a real piecewise continuous function expandable in a Fourier series as

$$f(\theta) = \sum_{n=-\infty}^{\infty} f_n \sigma^n \quad , \quad f_{-n} = \overline{f}_n$$

Then u is given by the series

$$u = f_0 + 2\,\mathbf{real}(\sum_{n=1}^{\infty} f_n z^n) \quad , \quad |z| \leq 1$$

II) **Neumann Problem**

$$\nabla^2 u = 0 \quad , \quad |z| < 1$$

$$\frac{\partial u(1, \theta)}{\partial r} = g(\theta) \quad , \quad 0 \leq \theta \leq 2\pi$$

We assume that the gradient function g is expandable in a Fourier series as

$$g(\theta) = \sum_{n=-\infty}^{\infty} g_n \sigma^n \quad , \quad g_{-n} = \overline{g}_n.$$

The solution only exists if the integral of $g(\theta)$ with respect to arc length around the boundary is zero. Hence, when

$$g_0 = \frac{1}{2\pi} \int_0^{2\pi} g(\theta)\, d\theta = 0,$$

then the series solution is

$$u = 2\,\mathbf{real}\left(\sum_{n=1}^{\infty} \left(\frac{g_n}{n}\right) z^n \right) + c \quad , \quad |z| \leq 1$$

where c is an arbitrary real constant.

III) Mixed Problem

In the third type of problem the function value is specified on one part of the boundary and the normal gradient is specified on the remainder. In the general situation a solution can be constructed by methods using Cauchy integrals [73]. Only a simple case will be examined here. We require

$$\nabla^2 u = 0 \qquad |z| < 1$$

$$u(1,\theta) = f(\theta) \quad , \quad \theta_1 < \theta < \theta_2$$

$$\frac{\partial u(1,\theta)}{\partial r} = g(\theta) \quad , \quad \theta_2 < \theta < (2\pi + \theta_1)$$

For convenience use the notation

$$\begin{array}{lll} L: & z = e^{\imath\theta}\,, & \theta_1 < \theta < \theta_2 \\ L': & z = e^{\imath\theta}\,, & \theta_2 < \theta < (2\pi + \theta_1) \end{array}$$

The mixed problem can be reduced to a case where g is zero by first solving a Neumann problem for a harmonic function v such that

$$\frac{\partial v}{\partial r} = g(\theta) \quad , \quad z \in L'$$

$$\frac{\partial v}{\partial r} = -\frac{\int_{\theta_2}^{2\pi+\theta_1} g(\theta)\, d\theta}{\theta_2 - \theta_1} \quad , \quad z \in L$$

Then we replace $f(\theta)$ by $f(\theta) - v(1,\theta)$ to get a problem where

$$u = f(\theta) - v(1,\theta) \quad , \quad z \in L$$

$$\frac{\partial u}{\partial r} = 0 \quad , \quad z \in L'$$

The complete solution then equals the sum of u and v. Consequently, no loss of generality results in dealing with the problem

$$u = f \quad , \quad z \in L$$

$$\frac{\partial u}{\partial r} = 0 \quad , \quad z \in L'$$

Consider the function

$$R(z) = \sqrt{(z-a)(z-b)} \quad , \quad a = e^{\imath\theta_1} \quad , \quad b = e^{\imath\theta_2}$$

defined in the complex plane cut along L. We choose the branch of R satisfying

$$R(0) = e^{i(\theta_1+\theta_2)/2}$$

The solution to the mixed boundary value problem can be expressed as

$$u = \textbf{real}(\frac{R(z)}{\pi i} \int_L \frac{f(t)\, dt}{R^+(t)(t-z)}) \quad , \quad t = e^{\imath\theta} \quad , \quad \theta_1 < \theta < \theta_2$$

where $R^+(t)$ means the boundary value of $R(z)$ on the inside of the arc. As an example take

$$\theta_1 = -\frac{\pi}{2} \quad , \quad \theta_2 = \frac{\pi}{2}$$

$$R(z) = \sqrt{z^2+1} \quad , \quad R(0) = 1$$

$$u = \cos(\theta) \quad , \quad -\frac{\pi}{2} \le \theta \le \frac{\pi}{2}$$

Carrying out the integration gives

$$u = \textbf{real}(\, F(z)\,)$$

where

$$F(z) = \frac{z + z^{-1} + (1 - z^{-1})\sqrt{z^2+1}}{2} \quad , \quad |z| \le 1$$

and the square root equals $+1$ at $z = 0$. This function is employed as a test case in subsequent calculations. The exact solution is evaluated in function **mbvtest**.

12.11.1 Numerical Results

The function **lapcrcl** solves either Dirichlet or Neumann problems for the unit disk. The boundary values are specified as piecewise linear functions of the polar angle. Then function **lintrp** is used to obtain a dense set of boundary values which are transformed by the FFT to produce coefficients in the series solution. When

lapcrcl is executed with no input data, a Dirichlet problem is solved having the boundary condition

$$u(1,\theta) = 1 + \frac{\cos(16\theta)}{10} \quad , \quad -\frac{\pi}{2} < \theta < \frac{\pi}{2}$$

$$u(1,\theta) = \frac{\cos(16\theta)}{10} \quad , \quad \frac{\pi}{2} < \theta < \frac{3\pi}{2}$$

This chosen boundary condition produces the interesting surface plot shown in Figure 12.6 where the solution was evaluated on a polar coordinate grid employing constant radial and angular increments.

The mixed boundary value problem is more difficult to handle than the Dirichlet or Neumann problems because numerical evaluation of the Cauchy integral must be performed cautiously. As z approaches a point on L, the integrand becomes singular. Theoretical developments involving Cauchy principal value integrals and the Plemelj formulas are needed to handle this situation thoroughly [73]. Even when z is close to the boundary, large integrand magnitude may cause inaccurate numerical integration. Furthermore, the integrand will have square root type singularities at the ends of L unless $f(a) = f(b) = 0$. Regularization procedures that can cope fully with these difficulties [26] will not be investigated in this text. Instead a simplified approach is presented.

The function **cauchint** was written to evaluate a contour integral involving a general density function $f(\zeta)$ defined on a curve L of general shape.We consider

$$F(z) = \frac{1}{2\pi i} \int_L \frac{f(\zeta)\, d\zeta}{\zeta - z}$$

with both the density function f and the shape of L being defined using cubic spline interpolation. A set of points

$$[\zeta_1, \zeta_2, \ldots, \zeta_m] \quad , \quad \zeta = \xi + i\eta$$

lying on L, along with boundary values

$$[f(\zeta_1), f(\zeta_2), \ldots, f(\zeta_m)] = [f_1, f_2, \ldots, f_m]$$

are given. Spline functions $\zeta(t)$, $f(t)$ are defined for $1 \le t \le m$ such that

$$\zeta(j) = \zeta_j \qquad \text{and} \qquad f(j) = f_j \qquad j = 1, 2, \ldots, n$$

The integrand in parametric form becomes

$$F(z) = \frac{1}{2\pi i} \int_1^n \frac{f(t)\,[\xi'(t) + i\eta'(t)]\, dt}{\zeta(t) - z}$$

and this integral is evaluated using function **gcquad** which computes Gaussian base points and weight factors using eigenvalue methods. It should be remembered that

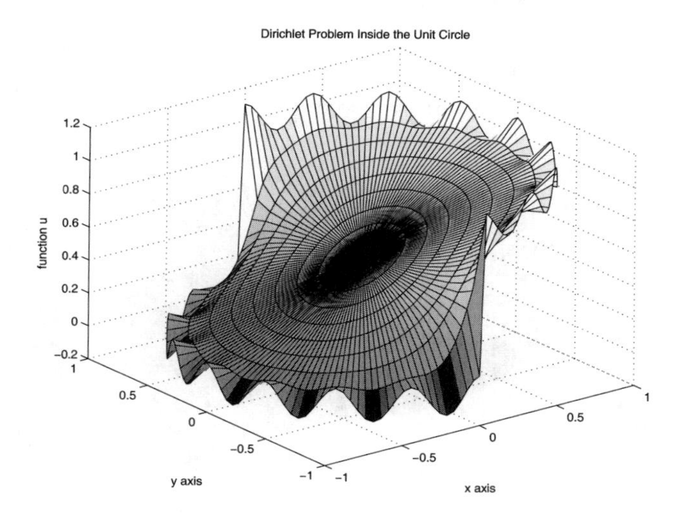

Figure 12.6: Dirichlet Problem Inside the Unit Circle

when z is a point on the contour of integration, the integrand has a first order singularity. Hence, procedures to regularize the integrand would be needed to achieve accurate numerical integration in such cases.

Function **cauchtst** was employed to produce an approximate solution of the problem cited above. A surface plot of the exact solution appears in Figure 12.7. A plot of the difference between the exact and approximate solutions for $0 \leq r \leq 0.99$ is shown in Figure 12.8. This error is about three orders-of-magnitude smaller than the maximum function values in the solution. The reader can verify that using $r = 0.999$ and $-\pi/2 < \theta < \pi/2$ leads to much larger errors. The authors have found function **cauchint** to be helpful if proper caution is exercised for results involving points near the boundary.

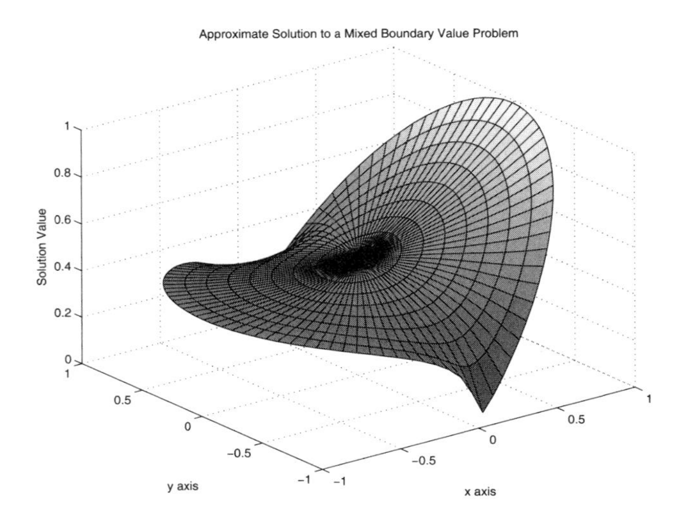

Figure 12.7: Approximate Solution to a Mixed Boundary Value Problem

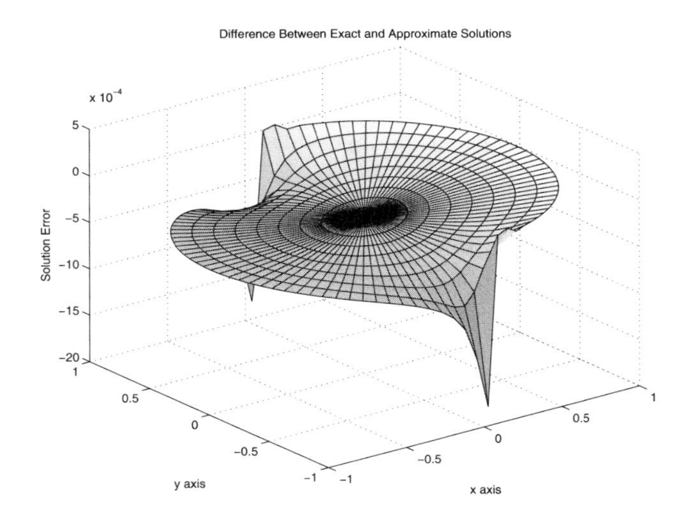

Figure 12.8: Difference Between Exact and Approximate Solutions

12.11.2 Program Output and Code

Function lapcrcl

```
1: function [u,r,th]=lapcrcl ...
2:                     (bvtyp,bvdat,rvec,thvec,nsum)
3: %
4: % [u,r,th]=lapcrcl(bvtyp,bvdat,rvec,thvec,nsum)
5: % ~~~~~~~~~~~~~~~~~~~~~~~~~~~~~~~~~~~~~~~~~~~~~~
6: %
7: % This function solves Laplace's equation
8: % inside a circle of unit radius. Either a
9: % Dirichlet problem or a Neumann problem can be
10: % analyzed using boundary values defined by
11: % piecewise linear interpolation of data
12: % specified in terms of the polar angle.
13: %
14: % bvtyp      - parameter determining what type
15: %              of boundary value problem is
16: %              solved. If bvtyp equals one,
17: %              boundary data specify function
18: %              values and a Dirichlet problem
19: %              is solved. Otherwise, the
20: %              boundary data specify values
21: %              of normal gradient, and a Neumann
22: %              problem is solved if, in accord
23: %              with the existence conditions for
24: %              this problem, the average value
25: %              of gradient on the boundary is
26: %              zero (negligibly small in an
27: %              approximate solution).
28: % bvdat      - a matrix of boundary data. Each
29: %              bvdat(j,:) gives a function value
30: %              and polar angle (in degrees) of
31: %              a data point used by function
32: %              lintrp to linearly interpolate
33: %              for all other boundary values
34: %              needed to generate the solution.
35: % rvec,thvec - vectors of radii and polar
36: %              coordinate values used to form a
37: %              polar coordinate grid of points
38: %              inside the unit circle. No values
39: %              of r exceeding unity are allowed.
40: % nsum       - the number of terms summed in the
```

```
41: %                   series expansion of the analytic
42: %                   function which has u as its real
43: %                   part. Typically, no more than one
44: %                   hundred terms are needed to
45: %                   produce a good solution.
46: %
47: % u             - values of the harmonic function
48: %                   evaluated at a set of points on
49: %                   a polar coordinate grid inside
50: %                   the unit circle.
51: % r,th          - the grid of polar coordinate
52: %                   values in which the function is
53: %                   evaluated
54: %
55: % User m functions called:  lintrp
56:
57: % Default test case solves a Dirichlet problem
58: % for a function having the following exact
59: % solution:
60: %
61: % -1/2+imag(log((z-i)/(z+i))/pi)+real(z^16)/10
62: %
63: if nargin ==0
64:   bvtyp=1; th=linspace(0,2*pi,201)';
65:   bv=1-(th>pi/2)+(th>3*pi/2)+cos(16*th)/10;
66:   bvdat=[bv,180/pi*th];
67:   rvec=linspace(1,0,10);
68:   thvec=linspace(0,360,161); nsum=200;
69: end
70:
71: nft=512;
72: thfft=linspace(0,2*pi*(nft-1)/nft,nft);
73: if nargin<5, nsum=200; end;
74: nsum=min(nsum,nft/2-1);
75: fbv=bvdat(:,1); thbv=pi/180*bvdat(:,2);
76: nev=size(bvdat,1); nr=length(rvec);
77: nth=length(thvec); neval=nr*nth;
78: [R,Th]=meshgrid(rvec,pi/180*thvec);
79: r=R(:); th=Th(:);
80:
81: % Check for any erroneous points outside the
82: % unit circle
83: rvec=rvec(:);
84: kout=find(rvec>1); nout=length(kout);
85: if length(kout)>0
```

```
86:     print('Input data are incorrect. The ');
87:     print('following r values lie outside the ');
88:     print('unit circle:'); disp(rvec(kout)');
89:     return
90:  end
91:
92:  if bvtyp==1 % Solve a Dirichlet problem
93:    % Check for points on the boundary where
94:    % function values are known. Interpolate
95:    % these directly
96:    konbd=find(r==1); onbndry=length(konbd);
97:    if onbndry > 0
98:      u(konbd)=lintrp(thbv,fbv,th(konbd));
99:    end
100:
101:   % Evaluate the series solution
102:   kinsid=find(r<1); inside=length(kinsid);
103:
104:   if inside > 0
105:     a=fft(lintrp(thbv,fbv,thfft));
106:     a=a(1:nsum)/(nft/2);
107:     a(1)=a(1)/2; Z=r(kinsid).*exp(i*th(kinsid));
108:     u(kinsid)=real(polyval(flipud(a(:)),Z));
109:   end
110:
111:   titl= ...
112:   'Dirichlet Problem Inside the Unit Circle';
113:
114: else % Solve a Neumann problem
115:   gbv=lintrp(thbv,fbv,thfft);
116:   a=fft(gbv)/(nft/2);
117:   erchek=abs(a(1))/sum(abs(gbv));
118:   if erchek>1e-3
119:     disp(' ');
120:     disp('ERROR DUE TO NONZERO AVERAGE VALUE');
121:     disp('OF NORMAL GRADIENT ON THE BOUNDARY.');
122:     disp('CORRECT THE INPUT DATA AND RERUN.');
123:     disp(' '); u=[]; r=[]; th=[]; return;
124:   end
125:   a=a(2:nsum)./(1:nsum-1)'; z=r.*exp(i*th);
126:   u=real(polyval(flipud([0;a(:)]),z));
127:   titl='Neumann Problem Inside the Unit Circle';
128: end
129:
130: u=reshape(u,nth,nr); r=R; th=Th;
```

```
131: surf(r.*cos(th),r.*sin(th),u);
132: xlabel('x axis'); ylabel('y axis');
133: zlabel('function u'); title(titl);
134: colormap('default');
135: grid on; figure(gcf);
136: % print -deps dirich
137:
138: %==============================================
139:
140: % function y=lintrp(xd,yd,x)
141: % See Appendix B
```

Function cauchtst

```
1: function u=cauchtst(z,nquad)
2: %
3: % u=cauchtst(z,nquad)
4: % ~~~~~~~~~~~~~~~~~~~
5: %
6: % This function solves a mixed boundary
7: % value problem for the interior of a circle
8: % by numerically evaluating a Cauchy integral.
9: %
10: %  z     - matrix of complex coordinates where
11: %          function values are computed
12: %  nquad - order of Gauss quadrature used to
13: %          perform numerical integration
14: %
15: %  u     - computed values of the approximate
16: %          solution
17: %
18: % User m functions called: cauchint, mbvtest,
19: %                gcquad, splined
20:
21: if nargin<2, nquad=50; end; nbdat=61;
22: if nargin==0
23:   z=linspace(0,.99,10)'* ...
24:     exp(i*linspace(0,2*pi,91));
25: end
26: th=linspace(-pi/2,pi/2,nbdat); zb=exp(i*th);
27: fb=sqrt(zb-i).*sqrt(zb+i); fb(1)=1; fb(nbdat)=1;
```

```
28: fb=cos(th)./fb; fb(1)=0; fb(end)=0;
29: F=cauchint(fb,zb,z,nquad);
30: F=F.*sqrt(z-i).*sqrt(z+i); u=2*real(F);
31:
32: surf(real(z),imag(z),u); xlabel('x axis');
33: ylabel('y axis'); zlabel('Solution Value')
34: title(['Approximate Solution to ', ...
35:        'a Mixed Boundary Value Problem']);
36: grid on; figure(gcf); %gra(.4);
37: fprintf('\nPress [Enter] to solution error\n');
38: pause
39: %print -deps caucher1
40: uexact=mbvtest(z,1); udif=u-uexact;
41: clf; surf(real(z),imag(z),udif);
42: title(['Difference Between Exact and ', ...
43:        'Approximate Solutions']);
44: xlabel('x axis'); ylabel('y axis');
45: zlabel('Solution Error')
46: grid on; figure(gcf); %gra(.4)
47: %print -deps caucher2
48:
49: %==============================================
50:
51: function u=mbvtest(z,noplot)
52: %
53: % u=mbvtest(z,noplot)
54: % ~~~~~~~~~~~~~~~~~~~~
55: %
56: % This function determines a function which is
57: % harmonic for abs(z)<1 and satisfies at r=1,
58: %   u=cos(theta), -pi/2<theta<pi/2
59: %   du/dr=0,       pi/2<theta<3*pi/2
60: % The solution only applies for points inside
61: % or on the unit circle.
62: %
63: % z      - matrix of complex values where the
64: %          solution is computed.
65: % noplot - option set to one if no plot is
66: %          requested, otherwise option is not
67: %          required.
68: %
69: % u      - values of the harmonic function
70: %          defined inside the unit circle
71: %
72: % User m functions called:  none
```

```
73: %----------------------------------------------
74:
75: if nargin==0
76:   noplot=0;
77:   z=linspace(0,1,10)'* ...
78:     exp(i*linspace(0,2*pi,81));
79: end
80: [n,m]=size(z); z=z(:); u=1/2*ones(size(z));
81: k=find(abs(z)>0); Z=z(k);
82: U=(Z+1./Z+(1-1./Z).*sqrt(Z-i).*sqrt(Z+i))/2;
83: u(k)=real(U); u=reshape(u,n,m);
84: if nargin==1 | noplot==0
85:   z=reshape(z,n,m);
86:   surf(real(z),imag(z),u); xlabel('x axis');
87:   ylabel('y axis');
88:   title(['Mixed Boundary Value Problem ', ...
89:          'for a Circular Disk']);
90:   grid; figure(gcf); %gra(.4), pause
91:   %print -deps mbvtest
92: end
93:
94: %=============================================
95:
96: function F=cauchint(fb,zb,z,nquad)
97: %
98: % F=cauchint(fb,zb,z,nquad)
99: % ~~~~~~~~~~~~~~~~~~~~~~~~~
100: %
101: % This function numerically evaluates a Cauchy
102: % integral of the form:
103: %
104: %   F(z)=1/(2*pi*i)*Integral(f(t)/(t-z)*dt)
105: %
106: % where t denotes points on a curve in the
107: % complex plane. The boundary curve is defined
108: % by spline interpolation through data points
109: % zb lying on the curve. The values of f(t)
110: % are also specified by spline interpolation
111: % through values fb corresponding to the
112: % points zb. Numerical evaluation of the
113: % integral is performed using a composite
114: % Gauss formula of arbitrary order.
115: %
116: % fb    - values of density function f
117: %         at point on the curve
```

```
118: % zb    - points where fb is given. The
119: %          number of values of zb must be
120: %          adequate to define the curve
121: %          accurately.
122: % z     - a matrix of values at which the
123: %          Cauchy integral is to be evaluated.
124: %          If any of the z-values lie on path
125: %          of integration or too close to the
126: %          path of integration, incorrect
127: %          results will be obtained.
128: % nquad - the order of Gauss quadrature
129: %          formula used to perform numerical
130: %          integration
131: %
132: % F     - The value of the Cauchy integral
133: %          corresponding to matrix argument z
134: %
135: % User m functions called:  gcquad splined
136: %----------------------------------------------
137:
138: n=length(fb); [nr,nc]=size(z); z=z(:).';
139: nz=length(z); t=1:n;
140: [dummy,bp,wf]=gcquad('',1,n,nquad,n-1);
141: fq=spline(t,fb,bp); zq=spline(t,zb,bp);
142: zqd=splined(t,zb,bp); nq=length(fq);
143: fq=fq(:).*zqd(:);
144:
145: bdrylen=sum(abs(zq(2:nq)-zq(1:nq-1)));
146:
147: closnes=1e100; bigz=max(abs(z));
148: for j=1:nq
149:   closnes=min([closnes,abs(zq(j)-z)]);
150: end
151: if closnes/bdrylen<.01 | closnes/bigz<.01
152:   disp(' ')
153:   disp(['WARNING! SOME DATA VALUES ARE ', ...
154:         'EITHER NEAR OR ON']);
155:   disp(['THE BOUNDARY. COMPUTED RESULTS ', ...
156:         'MAY BE INACCURATE']); disp(' ')
157: end
158: F=wf(:)'*(fq(:,ones(1,nz))./(zq(:,ones(1,nz))...
159:                               -z(ones(nq,1),:)));
160: F=reshape(F,nr,nc)/(2*pi*i);
161:
162: %=============================================
```

```
163:
164: % function [val,bp,wf]=gcquad(func,xlow,...
165: %                    xhigh,nquad,mparts,varargin)
166: % See Appendix B
167:
168: %=============================================
169:
170: % function val=splined(xd,yd,x,if2)
171: % See Appendix B
```

12.12 Inviscid Fluid Flow around an Elliptic Cylinder

This section analyzes inviscid flow around an elliptic cylinder in an infinite field. Flow around a circular cylinder is treated first. Then the function conformally mapping the exterior of a circle onto the exterior of an ellipse is used in conjunction with the invariance of harmonic functions under a conformal transformation. Results describing the elliptic cylinder flow field for uniform velocity components at infinity are presented.

Let us solve for the flow around a circular cylinder in the region $|\zeta| \geq 1$, $\zeta = \xi + i\eta$ with the requirement that the velocity components at infinity have constant values

$$u = U \quad , \quad v = V$$

where (u, v) are the horizontal and vertical components of velocity. These components are derivable from a potential function ϕ such that

$$u = \frac{\partial \phi}{\partial \xi} \quad , \quad v = \frac{\partial \phi}{\partial \eta}$$

where ϕ is a harmonic function. The velocity normal to the cylinder boundary must be zero. This requires that the function ψ, the harmonic conjugate of ϕ, must be constant on the boundary. The constant can be taken as zero without loss of generality. In terms of the complex velocity potential

$$f(\zeta) = \phi + i\psi$$

we need

$$f(\zeta) - \overline{f(\zeta)} = 0 \quad \text{on} \quad |\zeta| = 1$$

The velocity field is related to the complex velocity potential by

$$u - iv = f'(\zeta)$$

so the flow condition at infinity is satisfied by

$$f(\zeta) = p\zeta + O(1) \quad \text{where} \quad p = U - iV$$

A Laurent series can be used to represent $f(\zeta)$ in the form

$$f(\zeta) = p\zeta + a_0 + \sum_{n=1}^{\infty} a_n \zeta^{-n}$$

Imposition of the boundary condition on the cylinder surface requiring

$$f(\sigma) - \overline{f(\sigma)} = 0 \quad \text{where} \quad \sigma = e^{\imath\theta}$$

leads to

$$p\sigma + a_0 + \sum_{n=1}^{\infty} a_n \sigma^{-n} - \bar{p}\sigma^{-1} - \overline{a_0} - \sum_{n=1}^{\infty} \overline{a_n} \sigma^{n} = 0$$

Taking $a_0 = 0$, $a_1 = \bar{p}$, and $a_n = 0$, $n \geq 2$ satisfies all conditions of the problem and yields

$$f(\zeta) = p\zeta + \bar{p}\zeta^{-1}$$

as the desired complex potential function giving the velocity field as

$$u - iv = f'(\zeta) = p - \bar{p}\zeta^{-2} \quad , \quad |\zeta| \geq 1$$

Now consider flow about an elliptic cylinder lying in the z-plane. If the velocity at infinity has components (U, V) then we need a velocity potential $F(z)$ such that $F'(\infty) = U - iV$ and

$$F(z) - \overline{F(z)} = 0 \quad \text{for} \quad \left(\frac{x}{a}\right)^2 + \left(\frac{y}{b}\right)^2 = 1$$

This is nearly the same problem as was already solved in the ζ-plane except that

$$\frac{dF}{dz} = \frac{d\zeta}{dz}\frac{dF}{d\zeta} = \frac{1}{\omega'(\zeta)}\frac{dF}{d\zeta}$$

where $\omega(\zeta)$ is the mapping function

$$z = \omega(\zeta) = R(\zeta + m\zeta^{-1}) \quad , \quad R = \frac{a+b}{2} \quad , \quad m = \frac{a-b}{a+b}$$

In terms of ζ we would need

$$\frac{dF}{d\zeta} = \omega'(\infty)[U - iV] = R(U - iV) \quad \text{at} \quad \zeta = \infty$$

Consequently, the velocity potential for the elliptic cylinder problem expressed in terms of ζ is

$$F = p\zeta + \bar{p}\zeta^{-1} \quad , \quad p = R(U - \imath V)$$

and the velocity components in the z-plane are given by

$$u - iv = \frac{1}{\omega'(\zeta)}\left[p - \bar{p}\zeta^{-2}\right] = \frac{(U - iV) - (U - iV)\zeta^{-2}}{1 - m\zeta^{-2}}.$$

To get values for a particular choice of z we can use the inverse mapping function

$$\zeta = \frac{z + \sqrt{z^2 - 4mR^2}}{2R}$$

to eliminate ζ or we can compute results in terms of ζ.

To complete our discussion of this flow problem we will graph the lines characterizing the directions of flow. The velocity potential $F = \phi + i\psi$ satisfies

$$u = \frac{\partial \phi}{\partial x} = \frac{\partial \psi}{\partial y} \quad , \quad v = \frac{\partial \phi}{\partial y} = -\frac{\partial \psi}{\partial x}$$

so a curve tangent to the velocity field obeys

$$\frac{dy}{dx} = \frac{v}{u} = -\frac{\partial \psi / \partial x}{\partial \psi / \partial y}$$

or

$$\frac{\partial \psi}{\partial x} dx + \frac{\partial \psi}{\partial y} dy = 0 \quad , \quad \psi = \text{constant}$$

Consequently, the flow lines are the contours of function ψ, which is called the stream function. The function we want to contour does not exist inside the ellipse, but we can circumvent this problem by computing ψ in the ellipse exterior and then setting ψ to zero inside the ellipse. The function **elipcyl** analyzes the cylinder flow and produces the accompanying contour plot shown in Figure 12.9.

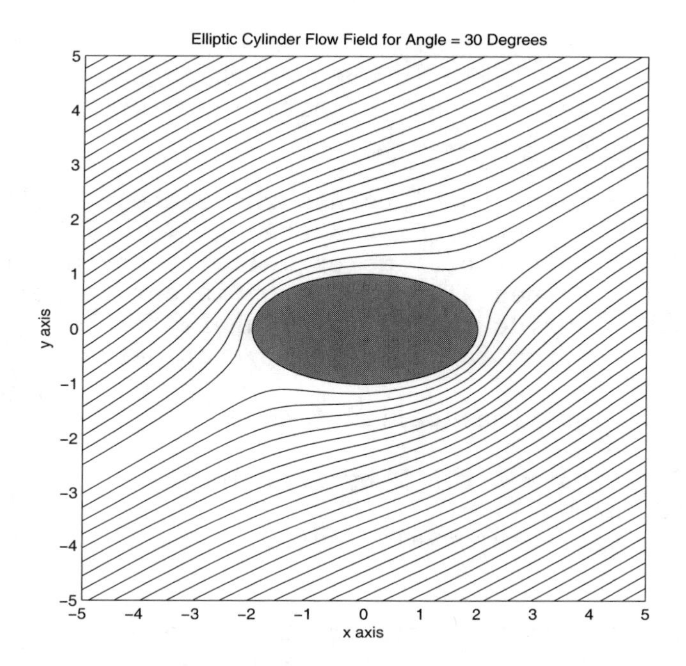

Figure 12.9: **Elliptic Cylinder Flow Field for Angle =** 30°

12.12.1 Program Output and Code

Function elipcyl

```
1: function [x,y,F]=elipcyl(a,n,rx,ry,ang)
2: %
3: % [x,y,F]=elipcyl(a,n,rx,ry,ang)
4: % ~~~~~~~~~~~~~~~~~~~~~~~~~~~~~~
5: %
6: % This function computes the flow field around
7: % an elliptic cylinder. The velocity direction
8: % at infinity is arbitrary.
9: %
10: % a      - defines the region -a<x<a, -a<y<a
11: %          within which the flow field is
12: %          computed
13: % n      - this determines the grid size which
14: %          uses n by n points
15: % rx,ry - major and minor semi-diameters af the
16: %          ellipse lying on the x and y axes,
```

```
17: %           respectively
18: % ang -     the angle in degrees which the
19: %           velocity at infinity makes with the
20: %           x axis
21: %
22: % x,y   - matrices of points where the velocity
23: %           potential is computed
24: % F     - matrix of complex velocity potential
25: %           values. This function is set to zero
26: %           inside the ellipse, where the
27: %           potential is actually not defined
28: %
29: % User m functions called:  none
30:
31: % default data for a 2 by 1 ellipse
32: if nargin==0
33:  a=5; n=81; rx=2; ry=1; ang=30;
34: end
35:
36: % Compute a square grid in the z plane.
37: ar=pi/180*ang; p=(rx+ry)/2*exp(-i*ar);
38: cp=conj(p); d=linspace(-a,a,n);
39: [x,y]=meshgrid(d,d); m=sqrt(rx^2-ry^2);
40:
41: % Obtain points in the zeta plane outside
42: % the ellipse
43: z=x(:)+i*y(:); k=find((x/rx).^2+(y/ry).^2>=1);
44: Z=z(k); zeta=(Z+sqrt(Z-m).*sqrt(Z+m))/(rx+ry);
45: F=zeros(n*n,1);
46:
47: % Evaluate the potential for a circular
48: % cylinder
49: F(k)=p*zeta+cp./zeta; F=reshape(F,n,n);
50:
51: % Contour the stream function to show the
52: % direction of flow
53:
54: clf; contourf(x(1,:),y(:,1),abs(imag(F)),30);
55: axis('square'); zb=exp(i*linspace(0,2*pi,101));
56: xb=rx*real(zb); yb=ry*imag(zb);
57: xb(end)=xb(1); yb(end)=yb(1);
58: hold on; fill(xb,yb,[127/255 1 212/255]);
59: xlabel('x axis'); ylabel('y axis');
60: title(['Elliptic Cylinder Flow Field for ', ...
61:        'Angle = ',num2str(ang),' Degrees']);
```

```
62: colormap hsv; figure(gcf); hold off;
63: %print -deps elipcyl
```

12.13 Torsional Stresses in a Beam Mapped onto a Unit Disk

Torsional stresses in a cylindrical beam can be computed from an integral formula when the function $z = \omega(\zeta)$ mapping the unit disk, $|\zeta| \leq 1$, onto the beam cross section is known [90]. The complex stress function

$$f(\zeta) = \frac{1}{2\pi} \int_{\gamma} \frac{\omega(\sigma)\overline{\omega(\sigma)}\, d\sigma}{\sigma - \zeta} + \text{constant},$$

where γ denotes the unit circle, can be evaluated exactly by contour integration in some cases. However, an approach employing series methods is easy to implement and gives satisfactory results if enough series terms are taken. When $\omega(\zeta)$ is a polynomial, $f(\zeta)$ is a polynomial of the same order as $\omega(\zeta)$. Furthermore, when $\omega(\zeta)$ is a rational function, residue calculus can be employed to compute $f(\zeta)$ exactly, provided the poles of $\overline{\omega}(1/\zeta)$ can be found. A much simpler approach is to use the FFT to expand $\omega(\sigma)\overline{\omega(\sigma)}$ in a complex Fourier series and write

$$\omega(\sigma)\overline{\omega(\sigma)} = \sum_{n=-\infty}^{\infty} c_n \sigma^n \quad , \quad \sigma = e^{\imath\theta}$$

Then the complex stress function is

$$f(\zeta) = i \sum_{n=1}^{\infty} c_n \zeta^n + \text{constant}$$

where the constant has no influence on the stress state. The shear stresses relative to the curvilinear coordinate system are obtainable from the formula

$$\frac{\tau_{\rho Z} - i\tau_{\alpha Z}}{\mu\varepsilon} = \frac{\left[f'(\zeta) - i\overline{\omega(\zeta)}\omega'(\zeta)\right]\zeta}{|\zeta\omega'(\zeta)|}$$

where μ is the shear modulus and ε is the angle of twist per unit length. The capital Z subscript on shear stresses refers to the direction of the beam axis normal to the xy plane rather than the complex variable $z = x + \imath y$. The series expansion gives

$$f'(\zeta) = i \sum_{n=1}^{\infty} n c_n \zeta^{n-1}$$

and this can be used to compute stresses. Differentiated series expansions often converge slowly or may even be divergent. To test the series expansion solution, a

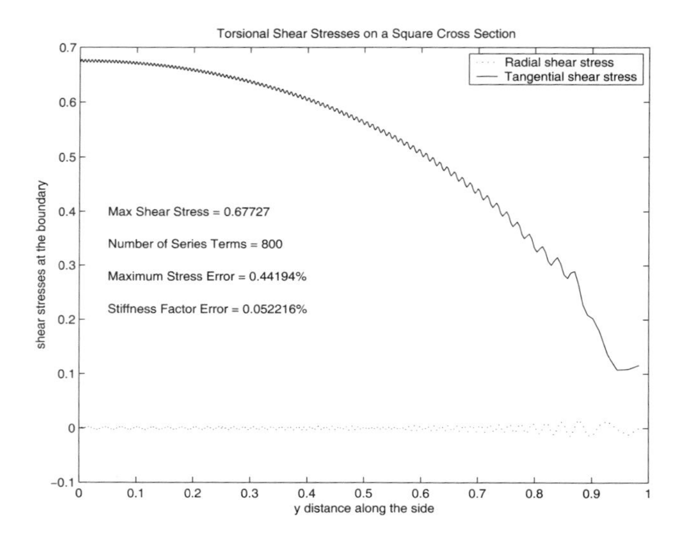

Figure 12.10: Torsional Shear Stresses on a Square Cross Section

rational function mapping $|\zeta| < 1$ onto a square defined by $|x| \leq 1$ and $|y| \leq 1$ was employed. Function **mapsqr** which computes $z(\zeta)$ and $z'(\zeta)$ is used by function **torstres** to evaluate stresses in terms of ζ. A short driver program **runtors** evaluates stresses on the boundary for $x = 1, 0 \leq y \leq 1$. Stresses divided by the side length of 2 are plotted and results produced from a highly accurate solution [90] are compared with values produced using 800 terms in $f(\zeta)$. Results depicted in Figure 12.10 show that the error in maximum shear stress was only 0.44% and the torsional stiffness was accurate within 0.05%. The numerical solution gives a nonzero stress value for $y = 1$, which disagree with the exact solution. This error is probably due more to the mapping function giving slightly rounded corners than to slow convergence of the series solution. Even though the differentiated series converges slowly, computation time is still small. The reader can verify that using 1500 terms reduces the boundary stress oscillations to negligible magnitude and produces a maximum stress error of 0.03%. Although taking 1500 terms to achieve accurate results seems excessive, less than 400 nonzero terms are actually involved because geometrical symmetry implies a series increasing in powers of four. For simplicity and generality, no attempt was made to account for geometrical symmetry exhibited by a particular mapping function. It appears that a series solution employing a mapping function is a viable computational tool to deal with torsion problems.

12.13.1 Program Output and Code

Program runtors

```
1: function runtors(ntrms)
2: % Example: runtors(ntrms)
3: % ~~~~~~~~~~~~~~~~~
4: %
5: % Example showing torsional stress computation
6: % for a beam of square cross section using
7: % conformal mapping and a complex stress
8: % function.
9: %
10: % ntrms - number of series terms used to
11: %         represent abs(w(zeta))^2
12: %
13: % User m functions called: torstres, mapsqr
14:
15: % Generate zeta values defining half of a side
16: theta=linspace(0,pi/4,501); zeta=exp(i*theta);
17: if nargin==0, ntrms=800; end
18:
19: % Compute stresses using an approximate rational
20: % function mapping function for the square
21: [tr,ta,z,c,C]= ...
22:   torstres('mapsqr',zeta,ntrms,4*1024);
23:
24: % Results from the exact solution
25: n=1:2:13;
26: tmexact=1-8/pi^2*sum(1./(n.^2.*cosh(n*pi/2)));
27: err=abs(ta(1)/2-tmexact)*100/tmexact;
28: stfexct=16/3-1024/pi^5*sum(tanh(pi/2*n)./n.^5);
29: stfaprx=8/3-pi*sum((1:ntrms)'.* ...
30:         abs(C(2:ntrms+1)).^2);
31: ster=100*abs(stfaprx-stfexct)/stfexct;
32:
33: % Plot circumferential and normal stresses at
34: % the boundary
35: th=180/pi*theta;
36: clf; plot(imag(z),tr/2,'k:',imag(z),ta/2,'k-')
37: xlabel('y distance along the side');
38: ylabel('shear stresses at the boundary');
39: title(['Torsional Shear Stresses on a ', ...
40:        'Square Cross Section']);
```

```
41: text(.05,.40, ...
42:   ['Max Shear Stress = ',num2str(max(ta)/2)]);
43: text(.05,.34, ...
44:   ['Number of Series Terms = ',num2str(ntrms)]);
45: text(.05,.28, ...
46:   ['Maximum Stress Error = ',num2str(err),'%']);
47: text(.05,.22,['Stiffness Factor Error = ', ...
48:   num2str(ster),'%']);
49: legend('Radial shear stress',...
50:   'Tangential shear stress');
51: figure(gcf);
52: %disp('Use mouse to locate legend block');
53: %disp('Press [Enter] when finished');
54: %print -deps torsion
55:
56: %==============================================
57:
58: function [trho,talpha,z,c,C]= ...
59:                 torstres(mapfun,zeta,ntrms,nft)
60: %
61: % [trho,talpha,z,c,C]= ...
62: %                 torstres(mapfun,zeta,ntrms,nft)
63: % ~~~~~~~~~~~~~~~~~~~~~~~~~~~~~~~~~~~~~~~~~~~~~
64: %
65: % This function computes torsional stresses in
66: % a beam such that abs(zeta)<=1 is mapped onto
67: % the beam cross section by a function named
68: % mapfun.
69: %
70: % mapfun - a character string giving the name
71: %          of the mapping function
72: % zeta   - values in the zeta plane
73: %          corresponding to which torsional
74: %          stresses are computed
75: % ntrms  - the number of terms used in the
76: %          series expansion of the mapping
77: %          function
78: % nft    - the number of function values
79: %          employed to compute Fourier
80: %          coefficients of the complex stress
81: %          function
82: %
83: % trho   - torsional stresses in directions
84: %          normal to the lines into which
85: %          abs(zeta)=const map. These values
```

```
86: %              should be zero at the boundary
87: %              corresponding to abs(zeta)=1.
88: % talpha  - torsional stresses in directions
89: %              tangent to the curves into which
90: %              abs(zeta)=const map. The maximum
91: %              value of shear stress always occurs
92: %              at some point on the boundary defined
93: %              by abs(zeta)=1.
94: % z        - values of z where stresses are
95: %              computed
96: % c        - coefficients in the series expansion
97: %              of the complex stress function
98: % C        - complex Fourier coefficients of
99: %              z.*conj(z) on the boundary of the
100: %             beam cross section
101: %
102: % User m functions called:  none
103: %----------------------------------------------
104:
105: if nargin<4, nft=4096; end;
106: if nargin<3, ntrms=800; end
107:
108: % Compute boundary values of the mapping
109: % function needed to construct the complex
110: % stress function
111: zetab=exp(i*linspace(0,2*pi*(nft-1)/nft,nft));
112: zb=feval(mapfun,zetab); zb=zb(:);
113:
114: % Evaluate z and z'(zeta) at other
115: % desired points
116: [z,zp]=feval(mapfun,zeta);
117:
118: % Compute Fourier coefficients for the complex
119: % stress function and its derivative
120: C=fft(zb.*conj(zb))/nft;
121: c=i*C(2:ntrms+1).*(1:ntrms)';
122: fp=polyval(flipud(c),zeta);
123:
124: % Evaluate stresses relative to the curvilinear
125: % coordinate system
126: tcplx=zeta./abs(zeta.*zp).*(fp-i*conj(z).*zp);
127:
128: % trho is the radial shear stress that should
129: % vanish at the boundary
130: trho=real(tcplx);
```

```
131:
132: % talpha is the circumferential stress which
133: % gives the maximum stress of interest at the
134: % boundary
135: talpha=-imag(tcplx);
136:
137: %==============================================
138:
139: function [z,zp]=mapsqr(zeta);
140: %
141: % [z,zp]=mapsqr(zeta)
142: % ~~~~~~~~~~~~~~~~~~~~
143: %
144: % This function maps the interior of a circle
145: % onto the interior of a square using a rational
146: % function of the approximate form:
147: %
148: % z(zeta)=zeta*Sum(a(j)* ...
149: %         zeta4^(j-1)/(1+Sum(b(j)*zeta4^(j-1))
150: %
151: % where zeta4=zeta^4
152: %
153: % zeta - matrix of complex values such that
154: %        abs(zeta)<=1
155: % z,zp - matrices of values of the mapping
156: %        function and its first derivative
157: %
158: % User m functions called:  none
159: %----------------------------------------------
160:
161: a=[ 1.07835,  1.37751, -0.02642, -0.09129, ...
162:     0.13460, -0.15763,  0.07430,  0.14858, ...
163:     0.01878, -0.00354 ]';
164: b=[ 1.37743,  0.07157, -0.11085,  0.12778, ...
165:    -0.13750,  0.05313,  0.14931,  0.02683, ...
166:    -0.00350, -0.000120 ]';
167:
168: % Evaluate the mapping function
169: zeta4=zeta.^4; p=zeta.*polyval(flipud(a),zeta4);
170: q=polyval(flipud([1;b]),zeta4); z=p./q;
171:
172: % Exit if the derivative of z is not needed
173: if nargout==1, return, end
174:
175: % evaluate z'(zeta)
```

```
176: na=length(a); nb=length(b);
177: pp=polyval(flipud((4*(1:na)'-3).*a),zeta4);
178: qp=4*zeta.^3.*polyval(flipud((1:nb)'.*b),zeta4);
179: zp=(q.*pp-p.*qp)./q.^2;
```

12.14 Stress Analysis by the Kolosov-Muskhelishvili Method

Two-dimensional problems in linear elastostatics of homogeneous bodies can be analyzed with the use of analytic functions. The primary quantities of interest are cartesian stress components τ_{xx}, τ_{yy}, and τ_{xy} and displacement components u and v. These can be expressed as

$$\tau_{xx} + \tau_{yy} = 2[\Phi(z) + \overline{\Phi(z)}]$$

$$-\tau_{xx} + \tau_{yy} + 2i\tau_{xy} = 2[\bar{z}\Phi'(z) + \Psi(z)]$$

$$2\mu(u + iv) = \kappa\phi(z) - z\overline{\Phi(z)} - \overline{\psi(z)}$$

$$\phi(z) = \int \Phi(z)\, dz \quad , \quad \psi(z) = \int \Psi(z)\, dz$$

where μ is the shear modulus and κ depends on Poisson's ratio ν according to $\kappa = 3 - 4\nu$ for plane strain or $\kappa = (3 - \nu)/(1 + \nu)$ for plane stress. The above relations are known as the Kolosov-Muskhelishvili formulas [73] and they have been used to solve many practical problems employing series or integral methods. Bodies such as a circular disk, a plate with a circular hole, and a circular annulus can be handled for quite general boundary conditions. Solutions can also be developed for geometries where a rational function is known that maps the interior of a circle onto the desired geometry. Futhermore, complex variable methods provide the most general techniques available for solving a meaningful class of mixed boundary value problems such as contact problems typified by pressing a rigid punch into a half plane.

Fully understanding all of the analyses presented in [72, 73] requires familiarity with contour integration, conformal mapping, and multivalued functions. However, some of the closed form solutions given in these texts can be used without extensive background in complex variable methods or the physical concepts of elasticity theory. With that perspective let us examine the problem of computing stresses in an infinite plate uniformly stressed at infinity and having a general normal stress $N(\theta)$ and tangential shear $T(\theta)$ applied to the hole. We will use the general solution of Muskhelishvili[1] [72] to evaluate stresses anywhere in the plate with particular interest on stress concentrations occurring around the hole. The stress functions Ψ and Φ

[1] Chapter 20.

can be represented as follows

$$\Phi(z) = -\frac{1}{2\pi\imath}\int_{\gamma}\frac{(N+\imath T)d\sigma}{\sigma - z} + \alpha + \beta z^{-1} + \delta z^{-2} \quad , \quad \sigma = e^{\imath\theta}$$

where γ denotes counterclockwise contour integration around the boundary of the hole and the other constants are given by

$$\alpha = \frac{\tau_{xx}^{\infty} + \tau_{yy}^{\infty}}{4} \quad , \quad \delta = \frac{-\tau_{xx}^{\infty} + \tau_{yy}^{\infty} + 2\imath\tau_{xy}^{\infty}}{2}$$

$$\beta = -\frac{\kappa}{1+\kappa}\frac{1}{2\pi}\int_0^{2\pi}(N+\imath T)e^{i\theta}\,d\theta$$

Parameters α and δ depend only on the components of stress at infinity, while β is determined by the force resultant on the hole caused by the applied loading. The quantity $N + \imath T$ is the boundary value of radial stress τ_{rr} and shear stress $\tau_{r\theta}$ in polar coordinates. Hence

$$N + \imath T = \tau_{rr} + i\tau_{r\theta} \quad , \quad |z| = 1$$

The transformation formulas relating Cartesian stresses τ_{xx}, τ_{yy}, τ_{xy} and polar coordinate stresses τ_{rr}, $\tau_{\theta\theta}$, $\tau_{r\theta}$ are

$$\tau_{rr} + \tau_{\theta\theta} = \tau_{xx} + \tau_{yy} \quad , \quad -\tau_{rr} + \tau_{\theta\theta} + 2\imath\tau_{r\theta} = (-\tau_{xx} + \tau_{yy} + 2i\tau_{xy})e^{2\imath\theta}$$

Let us assume that $N + \imath T$ is expandable in a Fourier series of the form

$$N + \imath T = \sum_{n=-\infty}^{\infty} c_n \sigma^n \quad , \quad \sigma = e^{i\theta}$$

where c_n can be obtained by integration as

$$c_n = \frac{1}{2\pi}\int_0^{2\pi}(N + iT)\sigma^{-n}\,d\theta$$

or we can compute the approximate coefficients more readily using the FFT.

The stress function $\Psi(z)$ is related to $\Phi(z)$ according to

$$\Psi = \frac{1}{z^2}\overline{\Phi\left(\frac{1}{\bar{z}}\right)} - \frac{d}{dz}\left[\frac{1}{z}\Phi(z)\right] \quad , \quad |z| \geq 1$$

Substituting the complex Fourier series into the integral formula for Φ gives

$$\Phi = -\sum_{n=0}^{\infty} c_n z^n + \alpha + \beta z^{-1} + \delta z^{-2} \quad , \quad |z| \leq 1$$

$$\Phi = \sum_{n=1}^{\infty} c_{-n} z^{-n} + \alpha + \beta z^{-1} + \delta z^{-2} \quad , \quad |z| \geq 1$$

which has the form

$$\Phi = \sum_{n=0}^{\infty} a_n z^{-n} \quad , \quad |z| \geq 1$$

These two relations then determine Ψ as

$$\Psi = \bar{\delta} + \bar{\beta} z^{-1} + (\alpha + a_0 - \overline{c_0}) z^{-2} + \sum_{n=3}^{\infty} \left[(n-1) a_{n-2} - \overline{c_{n-2}} \right] z^{-n}$$

The last equation has the form

$$\Psi = \sum_{n=0}^{\infty} b_n z^{-n} \quad , \quad |z| \geq 1$$

where the coefficients b_n are obtainable by comparing coefficients of corresponding powers in the two series. Hence, the series expansions of functions $\Phi(z)$ and $\Psi(z)$ can be generated in terms of the coefficients c_n and the stress components at infinity. The stresses can be evaluated by using the stress functions. Displacements can also be obtained by integrating Φ and Ψ, but this straightforward calculation is not discussed here.

The program **runplate** was written to evaluate the above formulas by expanding $N + iT$ using the FFT. Truncating the series for harmonics above some specified order, say np, gives approximations for $\Phi(z)$ and $\Psi(z)$, which exactly represent the solution corresponding to the boundary loading defined by the truncated Fourier series. Using the same approach employed in Chapter 6 we can define N and T as piecewise linear functions of the polar angle θ.

The program utilizes several routines described in the table below.

runplate	define N, T, stresses at infinity, z-points where results are requested, and the number of series terms used.
platecrc	computes series coefficients defining the stress functions.
strfun	evaluates Φ, Ψ, and Φ'.
cartstrs	evaluates Cartesian stresses for given values of z and the stress functions.
rec2polr	transforms from Cartesian stresses to polar coordinate stresses.
polflip	simplified interface to function **polyval**.

The program solves two sample problems. The first one analyzes a plate having no loading on the hole, and stresses at infinity given by $\tau_{yy}^{\infty} = 1$, $\tau_{xx}^{\infty} = \tau_{xy}^{\infty} = 0$. Figure 12.11 shows that the circumferential stress on the hole varies between -1 and

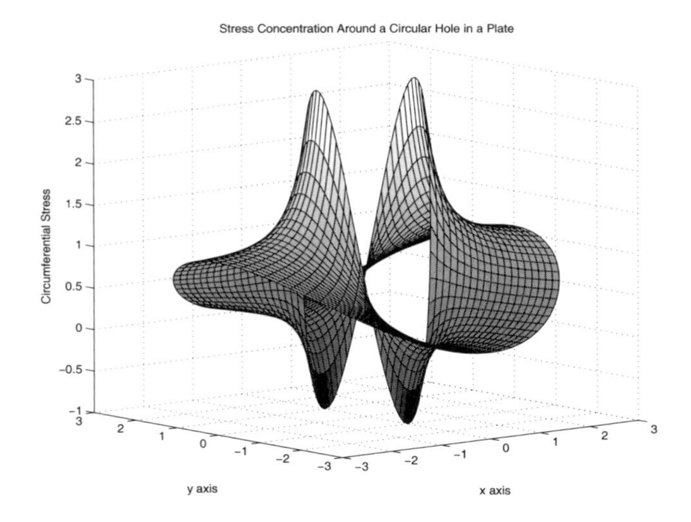

Figure 12.11: **Stress Concentration around a Circular Hole in a Plate**

3, producing a stress concentration factor of three due to the presence of the hole. The second problem applies a sinusoidally varying normal stress on the hole while the stresses at infinity are zero. Taking

```
T=0; ti=[0,0,0];
th=linspace(0,2*pi,81);
N=[cos(4*th), 180/pi*th];
```

gives the results depicted in Figure 12.12. Readers may find it interesting to investigate how stresses around the hole change with different combinations of stress at infinity and normal stress distributions on the hole.

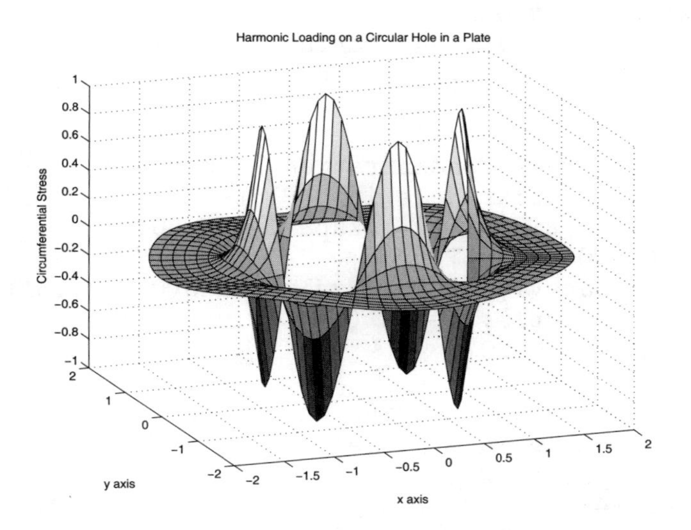

Figure 12.12: **Harmonic Loading on a Circular Hole in a Plate**

12.14.1 Program Output and Code

Program runplate

```
1: function runplate(WhichProblem)
2: % Example: runplate(WhichProblem)
3: % ~~~~~~~~~~~~~~~~~~~
4: %
5: % Example to compute stresses around a
6: % circular hole in a plate using the
7: % Kolosov-Muskhelishvili method.
8: %
9: % User m functions required:
10: %    platecrc, strfun, cartstrs,
11: %    rec2polr, polflip, lintrp
12:
13: if nargin==0
14:   titl=['Stress Concentration Around a ', ...
15:         'Circular Hole in a Plate'];
16:   N=0; T=0; ti=[0,1,0]; kapa=2; np=50;
17:   Nn='N = 0'; Tt='T = 0';
18:   rz=linspace(1,3,20)'; tz=linspace(0,2*pi,81);
19:   z=rz*exp(i*tz); x=real(z); y=imag(z);
20:   viewpnt=[-40,10];
21: else
22:   titl=['Harmonic Loading on a Circular', ...
23:         ' Hole in a Plate'];
24:   th=linspace(0,2*pi,81)';
25:   N=[cos(4*th),180/pi*th];
26:   Nn='N = cos(4*theta)'; Tt='T = 0';
27:   T=0; ti=[0,0,0]; kapa=2; np=10;
28:   rz=linspace(1,2,10)'; tz=linspace(0,2*pi,81);
29:   z=rz*exp(i*tz); x=real(z); y=imag(z);
30:   viewpnt=[-20,20];
31: end
32:
33: fprintf('\nSTRESSES IN A PLATE WITH A ')
34: fprintf('CIRCULAR HOLE')
35: fprintf('\n\nStress components at infinity ')
36: fprintf('are: '); fprintf('%g ',ti);
37: fprintf('\nNormal stresses on the hole are ')
38: fprintf(['defined by ',Nn]);
39: fprintf('\nTangential stresses on the hole ')
40: fprintf(['are defined by ',Tt])
```

```
41: fprintf('\nElastic constant kappa equals: ')
42: fprintf('%s',num2str(kapa));
43: fprintf('\nHighest harmonic order used is: ')
44: fprintf('%s',num2str(np));
45:
46: [a,b,c]=platecrc(N,T,ti,kapa,np);
47:
48: fprintf('\n');
49: fprintf('\nThe Kolosov-Muskhelishvili stress ');
50: fprintf('functions have\nthe series forms:');
51: fprintf('\nPhi=sum(a(k)*z^(-k+1), k=1:np+1)');
52: fprintf('\nPsi=sum(b(k)*z^(-k+1), k=1:np+3)');
53: fprintf('\n');
54: fprintf('\nCoefficients defining stress ');
55: fprintf('function Phi are:\n');
56: disp(a(:));
57: fprintf('Coefficients defining stress ');
58: fprintf('function Psi are:\n');
59: disp(b(:));
60:
61: % Evaluate the stress functions
62: [Phi,Psi,Phip]=strfun(a,b,z);
63:
64: % Compute the Cartesian stresses and the
65: % principal stresses
66: [tx,ty,txy,pt1,pt2]=cartstrs(z,Phi,Psi,Phip);
67: theta=angle(z./abs(z)); x=real(z); y=imag(z);
68: [tr,tt,trt]=rec2polr(tx,ty,txy,theta);
69: pmin=num2str(min([pt1(:);pt2(:)]));
70: pmax=num2str(max([pt1(:);pt2(:)]));
71:
72: disp(...
73: ['Minimum Principal Stress = ',num2str(pmin)]);
74: disp(...
75: ['Maximum Principal Stress = ',num2str(pmax)]);
76: fprintf('\nPress [Enter] for a surface ');
77: fprintf('plot of the\ncircumferential stress ');
78: fprintf('in the plate\n'); input('','s'); clf;
79: close; colormap('hsv');
80: surf(x,y,tt); xlabel('x axis'); ylabel('y axis');
81: zlabel('Circumferential Stress');
82: title(titl); grid on; view(viewpnt); figure(gcf);
83: %if nargin==0, print -deps strconc1
84: %else, print -deps strconc2; end
85: fprintf('All Done\n');
```

```
86:
87: %=============================================
88:
89: function [a,b,c]=platecrc(N,T,ti,kapa,np)
90: %
91: % [a,b,c]=platecrc(N,T,ti,kapa,np)
92: % ~~~~~~~~~~~~~~~~~~~~~~~~~~~~~~~~~~
93: %
94: % This function computes coefficients in the
95: % series expansions that define the Kolosov-
96: % Muskhelishvili stress functions for a plate
97: % having a circular hole of unit radius. The
98: % plate is uniformly stressed at infinity. On
99: % the surface of the hole, normal and tangential
100: % stress distributions N and T defined as
101: % piecewise linear functions are applied.
102: %
103: % N    - a two column matrix with each row
104: %        containing a value of normal stress
105: %        and polar angle in degrees used to
106: %        specify N as a piecewise linear
107: %        function of the polar angle. Step
108: %        discontinuities can be included by
109: %        using successive values of N with the
110: %        same polar angle values.  The data
111: %        should cover the range of theta from
112: %        0 to 360.  N represents boundary values
113: %        of the polar coordinate radial stress.
114: %        A single constant value can be input
115: %        when N is constant (including zero
116: %        if desired).
117: % T    - a two column matrix defining values of
118: %        the polar coordinate shear stress on
119: %        the hole defined as a piecewise linear
120: %        function. The points where function
121: %        values of T are specified do not need
122: %        to be the same as as those used to
123: %        specify N. Input a single constant
124: %        when T is constant on the boundary.
125: % ti   - vector of Cartesian stress components
126: %        [tx,ty,txy] at infinity.
127: % kapa - a constant depending on Poisson's ratio
128: %        nu.
129: %            kapa=3-4*nu for plane strain
130: %            kapa=(3-nu)/(1+nu) for plane stress
```

```
131: %          When the resultant force on the hole
132: %          is zero, then kapa has no effect on
133: %          the solution.
134: % np    - the highest power of exp(i*theta) used
135: %          in the series expansion of N+i*T. This
136: %          should not exceed 255.
137: %
138: % a     - coefficients in the series expansion
139: %          defining the stress function
140: %              Phi=sum(a(k)*z^(-k+1), k=1:np+1)
141: % b     - coefficients in the series expansion
142: %          defining the stress function
143: %              Psi=sum(b(k)*z^(-k+1), k=1:np+3)
144: %
145: % User m functions called:  lintrp
146: %----------------------------------------------
147:
148: % Handle case of constant boundary stresses
149: if length(N(:))==1; N=[N,0;N,360]; end
150: if length(T(:))==1; T=[T,0;T,360]; end
151:
152: % Expand the boundary stresses in a Fourier
153: % series
154: f=pi/180; nft=512; np=min(np,nft/2-1);
155: thta=linspace(0,2*pi*(nft-1)/nft,nft);
156:
157: % Interpolate linearly for values at the
158: % Fourier points
159: Nft=lintrp(f*N(:,2),N(:,1),thta);
160: Tft=lintrp(f*T(:,2),T(:,1),thta);
161: c=fft(Nft(:)+i*Tft(:))/nft;
162:
163: % Evaluate auxiliary parameters in the
164: % series solutions
165: alp=(ti(1)+ti(2))/4; bet=-kapa*c(nft)/(1+kapa);
166: sig=(-ti(1)+ti(2)-2*i*ti(3))/2;
167:
168: % Generate a and b coefficients using the
169: % Fourier coefficients of N+i*T.
170: a=zeros(np+1,1); b=zeros(np+3,1); j=(1:np)';
171: a(j+1)=c(nft+1-j); a(1)=alp;
172: a(2)=bet+c(nft); a(3)=sig+c(nft-1);
173: j=(3:np+2)'; b(j+1)=(j-1).*a(j-1)-conj(c(j-1));
174: b(1)=conj(sig); b(2)=conj(bet);
175: b(3)=alp+a(1)-conj(c(1));
```

```
176:
177: % Discard any negligibly small high order
178: % coefficients.
179: tol=max(abs([N(:);T(:);ti(:)]))/1e4;
180: ka=max(find(abs(a)>tol));
181: if isempty(ka), a=0; else, a(ka+1:np+1)=[]; end
182: kb=max(find(abs(b)>tol));
183: if isempty(kb), b=0; else, b(kb+1:np+3)=[]; end
184:
185: %==============================================
186:
187: function [Phi,Psi,Phip]=strfun(a,b,z)
188: %
189: % [Phi,Psi,Phip]=strfun(a,b,z)
190: % ~~~~~~~~~~~~~~~~~~~~~~~~~~~~
191: %
192: % This function evaluates the complex
193: % stress functions Phi(z) and Psi(z)
194: % as well as the derivative function Phi'(z)
195: % using series coefficients determined from
196: % function platecrc. The calculation also
197: % uses a function polflip defined such that
198: % polflip(a,z)=polyval(flipud(a(:)),z).
199: %
200: % a,b     - series coefficients defining Phi
201: %           and Psi
202: % z       - matrix of complex values
203: %
204: % Phi,Psi - complex stress function values
205: % Phip    - derivative Phi'(z)
206: %
207: % User m functions called: polflip
208: %----------------------------------------------
209:
210: zi=1./z; np=length(a); a=a(:);
211: Phi=polflip(a,zi); Psi=polflip(b,zi);
212: Phip=-polflip((1:np-1)'.*a(2:np),zi)./z.^2;
213:
214: %==============================================
215:
216: function [tx,ty,txy,tp1,tp2]= ...
217:                        cartstrs(z,Phi,Psi,Phip)
218: %
219: % [tx,ty,txy,tp1,tp2]=cartstrs(z,Phi,Psi,Phip)
220: % ~~~~~~~~~~~~~~~~~~~~~~~~~~~~~~~~~~~~~~~~~~~~
```

```
221: %
222: % This function uses values of the complex
223: % stress functions to evaluate Cartesian stress
224: % components relative to the x,y axes.
225: %
226: % z           - matrix of complex values where
227: %               stresses are required
228: % Phi,Psi     - matrices containing complex stress
229: %               function values
230: % Phip        - values of  Phi'(z)
231: %
232: % tx,ty,txy - values of the Cartesian stress
233: %               components for the x,y axes
234: % tp1,tp2     - values of maximum and minimum
235: %               principal stresses
236: %
237: % User m functions called:  none
238: %----------------------------------------------------
239:
240: A=2*real(Phi); B=conj(z).*Phip+Psi;
241: C=A-B; R=abs(B);
242: tx=real(C); ty=2*A-tx; txy=-imag(C);
243: tp1=A+R; tp2=A-R;
244:
245: %====================================================
246:
247: function [tr,tt,trt]=rec2polr(tx,ty,txy,theta)
248: %
249: % [tr,tt,trt]=rec2polr(tx,ty,txy,theta)
250: % ~~~~~~~~~~~~~~~~~~~~~~~~~~~~~~~~~~~~~~~~
251: %
252: % This function transforms Cartesian stress
253: % components tx,ty,txy to polar coordinate
254: % stresses tr,tt,trt.
255: %
256: % tx,ty,txy - matrices of Cartesian stress
257: %               components
258: % theta       - a matrix of polar coordinate
259: %               values.  This can also be a
260: %               single value if all stress
261: %               components are rotated by the
262: %               same angle.
263: %
264: % tr,tt,trt - matrices of polar coordinate
265: %               stresses
```

```
266: %
267: % User m functions called:  none
268: %----------------------------------------------
269:
270: if length(theta(:))==1
271:   theta=theta*ones(size(tx)); end
272: a=(tx+ty)/2;
273: b=((tx-ty)/2-i*txy).*exp(2*i*theta);
274: c=a+b; tr=real(c); tt=2*a-tr; trt=-imag(c);
275:
276: %==============================================
277:
278: function y=polflip(a,x)
279: %
280: % y=polflip(a,x)
281: % ~~~~~~~~~~~~~~~
282: %
283: % This function evaluates polyval(a,x) with
284: % the order of the elements reversed.
285: %
286: %----------------------------------------------
287:
288: y=polyval(a(end:-1:1),x);
289:
290: %==============================================
291:
292: % function y=lintrp(xd,yd,x)
293: % See Appendix B
```

12.14.2 Stressed Plate with an Elliptic Hole

This chapter is concluded with an example using conformal mapping in elasticity theory. We discussed earlier the useful property that harmonic functions remain harmonic under a conformal transformation. However, linear elasticity leads to the biharmonic Airy stress function which satisfies

$$\left[\frac{\partial^2}{\partial x^2} + \frac{\partial^2}{\partial y^2}\right]^2 U = 0$$

Unfortunately, a conformal transformation $x + iy = \omega(\xi + i\eta)$ does not imply

$$\left[\frac{\partial^2}{\partial \xi^2} + \frac{\partial^2}{\partial \eta^2}\right]^2 U = 0$$

except when the mapping function has the trivial linear form $z = c_1\zeta + c_0$. Consequently, the analogy employed in the ideal flow problem is not applicable in linear elasticity. This does not preclude use of conformal mapping in elasticity, but we encounter equations of very different structure in the mapped variables. We will examine that problem enough to illustrate the kind of differences involved. Let a mapping function $z = \omega(\zeta)$ define curvilinear coordinate lines in the z-plane. A polar coordinate grid corresponding to $\arg(\zeta) = $ constant and $|\zeta| = $ constant maps into curves we term ρ lines and α lines, respectively. Plotting of such lines was demonstrated previously with function **gridview** (mapping the exterior of a circle onto the exterior of an ellipse). It can be shown that curvilinear coordinate stresses $\tau_{\rho\rho}$, $\tau_{\alpha\alpha}$, $\tau_{\rho\alpha}$ are related to cartesian stresses according to

$$\tau_{\rho\rho} + \tau_{\alpha\alpha} = \tau_{xx} + \tau_{yy} \quad , \quad -\tau_{\rho\rho} + \tau_{\alpha\alpha} + 2i\tau_{\rho\alpha} = h(-\tau_{xx} + \tau_{yy} + 2i\tau_{xy})$$

where

$$h = \frac{\zeta\omega'(\zeta)}{\overline{\zeta\omega'(\zeta)}}$$

Muskhelishvili [72] has developed a general solution for a plate with an elliptic hole allowing general boundary tractions. Here we use one solution from his text which employs the mapping function

$$z = \omega(\zeta) = R\left(\zeta + \frac{m}{\zeta}\right)$$

and the stress functions

$$\phi(z) = \int \Phi(z)\, dz \qquad \psi(z) = \int \Psi(z)\, dz$$

When ζ is selected as the primary reference variable, we have to perform chain rule differentiation and write

$$\Phi(z) = \frac{\phi'(\zeta)}{\omega'(\zeta)} \qquad \Psi(z) = \frac{\psi'(\zeta)}{\omega'(\zeta)}$$

$$\Phi'(z) = \frac{\omega'(\zeta)\phi''(\zeta) - \omega''(\zeta)\phi'(\zeta)}{\omega'(\zeta)^3}$$

in order to compute stresses in terms of the ζ-variable. Readers unaccustomed to using conformal mapping in this context should remember that there is no stress state in the ζ-plane comparable to the analogous velocity components which can be envisioned in a potential flow problem. We are simply using ζ as a convenient reference variable to analyze physical stress and displacement quantities existing only in the z-plane.

Suppose the infinite plate has an elliptic hole defined by

$$\left(\frac{x}{r_x}\right)^2 + \left(\frac{y}{r_y}\right)^2 = 1$$

and the hole is free of applied tractions. The stress state at infinity consists of a tension p inclined at angle λ with the x-axis. The stress functions relating to that problem are found to be ([72], page 338)

$$\phi(\zeta) = b\zeta + \frac{c}{\zeta}$$

$$\psi(\zeta) = d\zeta + \frac{e}{\zeta} + \frac{f\zeta}{\zeta^2 - m} \quad , \quad |\zeta| \geq 1$$

$$a = e^{2\imath\lambda} \quad , \quad b = \frac{pr}{4} \quad , \quad c = b(2a - m)$$

$$d = -\frac{pr\bar{a}}{2} \quad , \quad e = -\frac{pra}{2m} \quad , \quad f = \frac{pr(m + \frac{1}{m})}{2}$$

Clearly these functions have no obvious relation to the simpler results shown earlier for a plate with a circular hole. The function **eliphole** computes curvilinear coordinate stresses in the z-plane expressed in terms of the ζ-variable. When $\lambda = \pi/2$, the plate tension acts along the y-axis and the maximum circumferential stress occurs at $z = r_x$ corresponding to $\zeta = 1$. A surface plot produced by **eliphole** for the default data case using $r_x = 2$ and $r_y = 1$ is shown in Figure 12.13. It is also interesting to graph $\tau_{\alpha\alpha}^{\max}/\tau_{yy}^{\infty}$ as a function of r_x/r_y. The program **elpmaxst** produces the plot in Figure 12.14 showing that the circumferential stress concentration increases linearly according to

$$\frac{\tau_{\alpha\alpha}^{\max}}{p} = 1 + 2\left(\frac{r_x}{r_y}\right)$$

which can also be verified directly from the stress functions.

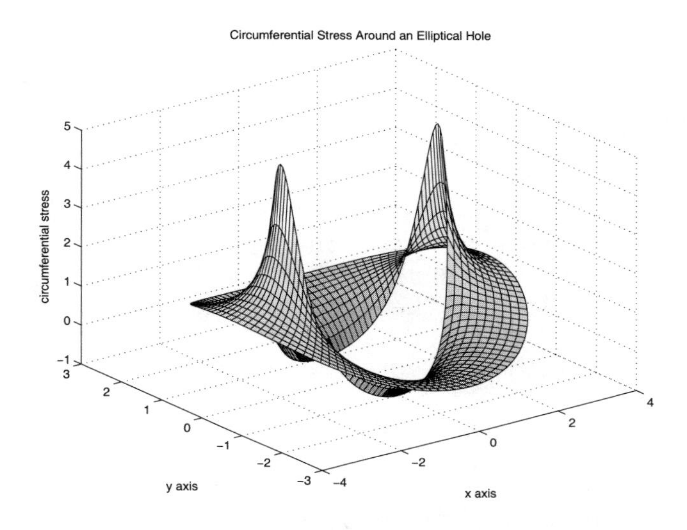

Figure 12.13: Circumferential Stress around an Elliptical Hole

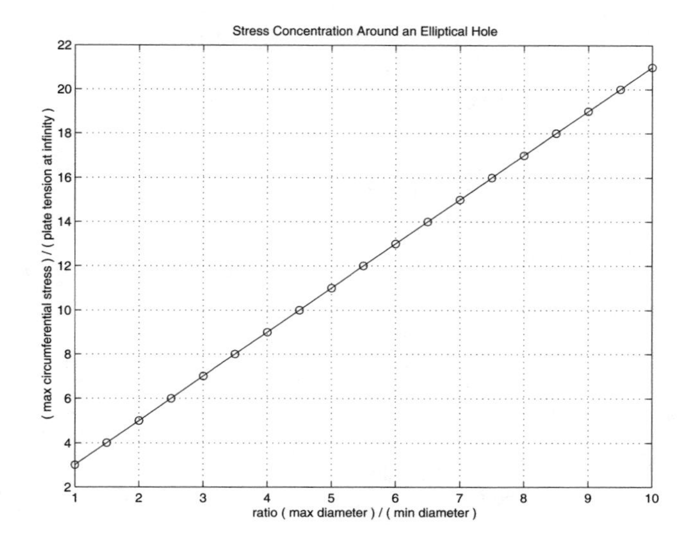

Figure 12.14: Stress Concentration around an Elliptical Hole

12.14.3 Program Output and Code

Program elpmaxst

```
1: function elpmaxst
2: % Example: elpmaxst
3: % ~~~~~~~~~~~~~~~~~
4: %
5: % MATLAB example to plot the stress
6: % concentration around an elliptic hole
7: % as a function of the semi-diameter ratio.
8: %
9: % User m functions required: eliphole
10:
11: rx=2; ry=1; p=1; ang=90; ifplot=1;
12: zeta=linspace(1,2,11)'* ...
13:        exp(i*linspace(0,2*pi,121));
14: eliphole(rx,ry,p,ang,zeta,1);
15:
16: r=linspace(1.001,10,19); tamax=zeros(size(r));
17: for j=1:19
18:   [tr,tamax(j)]=eliphole(r(j),1,1,90,1);
19: end
20: plot(r,tamax,'-',r,tamax,'o');
21: title(['Stress Concentration Around an ', ...
22:        'Elliptical Hole']);
23: xlabel(['ratio ( max diameter ) / ', ...
24:         '( min diameter )']);
25: ylabel(['( max circumferential stress ) / ',...
26:         '( plate tension at infinity )']);
27: grid on; figure(gcf);
28: %print -deps elpmaxst
29:
30: %=============================================
31:
32: function [tr,ta,tra,z]=eliphole...
33:                 (rx,ry,p,ang,zeta,ifplot)
34: %
35: % [tr,ta,tra,z]=eliphole(rx,ry,p,ang,...
36: %                             zeta,ifplot)
37: % ~~~~~~~~~~~~~~~~~~~~~~~~~~~~~~~~~~~~~~~~
38: %
39: % This function determines curvilinear
40: % coordinate stresses around an elliptic hole
```

```
41: % in a plate uniformly stressed at infinity.
42: %
43: % rx,ry    - ellipse semidiameters on the x and
44: %            y axes
45: % p        - values of uniaxial tension at
46: %            infinity
47: % ang      - angle of inclination in degrees
48: %            of the tensile stress at infinity
49: % zeta     - curvilinear coordinate values for
50: %            which stresses are evaluated
51: % ifplot   - optional parameter that is given
52: %            a value if a surface plot of the
53: %            circumferential stress is desired
54: %
55: % tr       - tensile stress normal to an
56: %            elliptical coordinate line
57: % ta       - tensile stress in a direction
58: %            tangential to the elliptical
59: %            coordinate line
60: % tra      - shear stress complementary to the
61: %            normal stresses
62: % z        - points in the z plane where
63: %            stresses are computed
64: %
65: % User m functions called: none
66: %----------------------------------------------
67:
68: if nargin<6, ifplot=0; end
69: if nargin==0
70:   rx=2; ry=1; p=1; ang=90; ifplot=1;
71:   zeta=linspace(1,2,11)'* ...
72:        exp(i*linspace(0,2*pi,121));
73: end
74:
75: % The complex stress functions and mapping
76: % function have the form
77: %   phi(zeta)=b*zeta+c/zeta
78: %   psi(zeta)=d*zeta+e/zeta+f*zeta/(zeta^2-m)
79: %   z=w(zeta)=r(zeta+m/zeta)
80: %   Phi(zeta)=phi'(zeta)/w'(zeta)
81: %   Psi(zeta)=psi'(zeta)/w'(zeta)
82: %   d(Phi)/dz=(w'(zeta)*phi''(zeta)-...
83: %              w''(zeta)*phi'(zeta))/w'(zeta)^3
84:
85: r=(rx+ry)/2; m=(rx-ry)/(rx+ry);
```

```
86: z=r*(zeta+m./zeta); zeta2=zeta.^2;
87: zeta3=zeta.^3; wp=r*(1-m./zeta2);
88: wpp=2*r*m./zeta3; a=exp(2*i*pi/180*ang);
89: b=p*r/4; c=b*(2*a-m); d=-p*r/2*conj(a);
90: e=-p*r/2*a/m; f=p*r/2*(m+1/m)*(a-m);
91: phip=b-c./zeta2; phipp=2*c./zeta3;
92: h=wp.*zeta; h=h./conj(h);
93: Phi=phip./wp; Phipz=(wp.*phipp-wpp.*phip)./wp.^3;
94: Psi=(d-e./zeta2-f*(zeta2+m)./(zeta2-m).^2)./wp;
95: A=2*real(Phi); B=(conj(z).*Phipz+Psi).*h;
96: C=A-B; tr=real(C); ta=2*A-tr; tra=imag(B);
97: if ifplot>0
98:   %colormap('gray'); brighten(.95);
99:   surf(real(z),imag(z),ta);
100:   xlabel('x axis'); ylabel('y axis');
101:   zlabel('circumferential stress');
102:   title(['Circumferential Stress Around ', ...
103:          'an Elliptical Hole']);
104:   grid on; figure(gcf); input('','s');
105:   %print -deps eliphole
106: end
```

Chapter 13

Nonlinear Optimization Applications

13.1 Basic Concepts

Optimization problems occur for a diverse range of topics. Perhaps the simplest type of optimization problem involves a scalar function of several variables. For example, the cost of a product having several ingredients may need to be minimized. This problem can be represented by a function $F(x)$ which depends on the vector $x = [x_1; x_2; \ldots; x_n]$ in n-dimensional space. Function F is called the *objective function* and cases where the independent variables $x_\imath$ can vary arbitrarily are considered unconstrained. Most problems have constraints requiring $x_\imath$ to remain within given bounds or satisfy other functional equations. Different analysis procedures exist for solving problems depending on whether they are linear or nonlinear, constrained or unconstrained. General solutions are available to handle linear objective functions with linear equality and inequality constraints. The discipline devoted to such problems is known as *linear programming* [41] and applications involving thousands of independent variables can be analyzed.[1] Although this class of linear problems is important, it does not offer the versatility of methods used to address nonlinear problems (which are more compute intensive for problems of similar dimensionality).[2] The material in this chapter addresses nonlinear problems with a few independent variables which are either constrained or restricted to lie within bounds of the form

$$a_\imath \leq x_\imath \leq b_\imath.$$

This type of constraint can be satisfied by taking

$$x_\imath = a_\imath + (b_\imath - a_\imath)\sin^2(z_\imath)$$

and letting $z_\imath$ vary arbitrarily. The MATLAB intrinsic functions **fminbnd** and **fminsearch** are employed for solving this class of problems. The following five examples are presented to illustrate the nature of nonlinear optimization methods:

1. Computing the inclination angle necessary to cause a projectile to strike a stationary distant object;

[1] High dimensionality linear problems should always be solved using the appropriate specialized software.

[2] The MathWorks markets an "Optimization Toolbox" intended to satisfy a number of specialized optimization needs.

2. Finding parameters of a nonlinear equation to closely fit a set of data values;

3. Determining components of end force on a statically loaded cable necessary to make the endpoint assume a desired position;

4. Computing the shape of a curve between two points such that a smooth particle slides from one end to the other in the minimum time;

5. Determining the closest points on two surfaces.

Before addressing specific problems, some of the general concepts of optimization will be discussed.

The minimum of an unconstrained differentiable function

$$F(x_1, x_2, \ldots, x_n)$$

will occur at a point where the function has a zero gradient. Thus the condition

$$\frac{\partial F}{\partial x_\imath} = 0 \ , \ 1 \leq \imath \leq n$$

leads to n nonlinear simultaneous equations. Such systems often have multiple solutions, and a zero gradient indicates either a maximum, or a minimum, or a saddle point. No reliable general methods currently exist to obtain all solutions to a general system of nonlinear equations. However, practical situations do occur where one unique point providing a relative minimum is expected. In such cases $F(x)$ is called *unimodal* and we seek x_0 which makes

$$F(x_0) < F(x_0 + \Delta) \text{ for } |\Delta| > 0.$$

Most unconstrained nonlinear programming software starts from an initial point and searches iteratively for a point where the gradient vanishes. *Multimodal*, or non-unimodal, functions can sometimes be solved by initiating searches from multiple starting points and using the best result obtained among all the searches. Since situations such as false convergence are fairly common with nonlinear optimization methods, results obtained warrant greater scrutiny than might be necessary for linear problems.

The intrinsic MATLAB functions **fminbnd** and **fminsearch** are adequate to address many optimization problems. Readers should study the documentation available for **fminbnd**, which performs a one-dimensional search within specified limits, and **fminsearch**, which performs an unconstrained multi-dimensional search starting from a user selected point. Both functions require objective functions of acceptable syntactical form. Various options controlling convergence tolerances and function evaluation counts should be studied to insure that the parameter choices are appropriately defined.

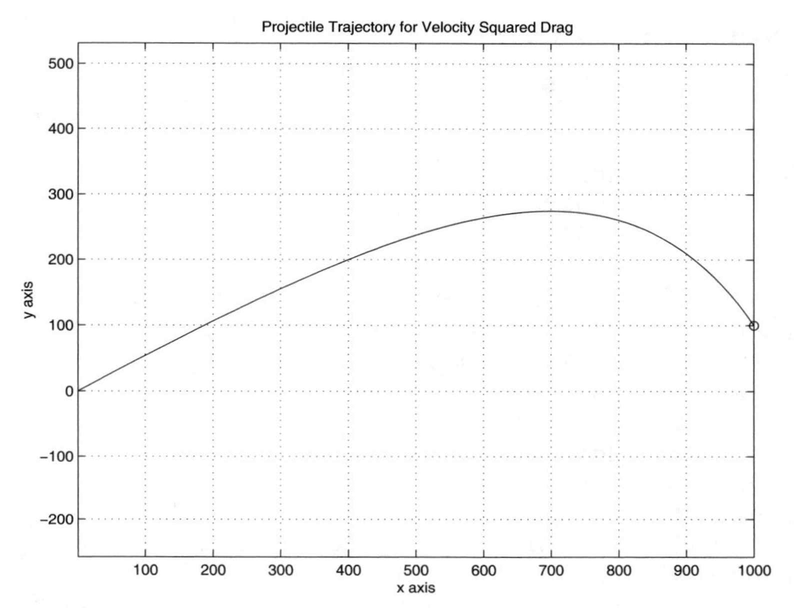

Figure 13.1: Projectile Trajectory for v^2 Drag Condition

13.2 Initial Angle for a Projectile

In Chapter 8, equations of motion for motion of a projectile with atmospheric drag were formulated and a function **traject** producing a solution $y(x)$ passing through $(x, y) = (0, 0)$ with arbitrary inclination was developed. The solution is generated for $0 \leq x \leq x_f$ assuming the initial velocity is large enough for the projectile to reach x_f. Therefore, program execution terminates if dx/dt goes to zero. In order to hit a target at position (x_f, y_f), the starting angle of the trajectory must be selected iteratively because the equations of motion cannot be solved exactly (except for the undamped case). With the aid of an optimization method we calculate $|y(x_f) - y_f)|$ and minimize this quantity (described in function **missdis** which has the firing angle as its argument). Function **fminbnd** seeks the angle to minimize the "miss" distance. Program **runtraj** illustrates the solution to the problem described and Figure 13.1 shows the trajectory required for the projectile to strike the object.

Depending on the starting conditions, zero, one, or two solutions exist to cause the "miss" distance to approach zero. Function **fminbnd** terminates at either a local minimum or at one of the search limits. The reader will need to examine how the initial data correlate to the final answers. For example, if the projectile misses the target by a significant amount, the initial projectile velocity was not large enough to reach the target.

Program Output and Code

Trajectory Analysis Program

```
1: function runtraj
2: % Example: runtraj
3: % ~~~~~~~~~~~~~~~~
4: %
5: % This program integrates the differential
6: % equations governing two-dimensional motion
7: % of a projectile subjected to gravity loading
8: % and atmospheric drag proportional to the
9: % velocity squared. The initial inclination
10: % angle needed to hit a distant target is
11: % computed repeatedly and function fmin is
12: % employed to minimize the square of the
13: % distance by which the target is missed. The
14: % optimal value of the miss distance is zero
15: % and the optimum angle will typically be found
16: % unless the initial velocity is too small
17: % and the horizontal velocity becomes zero
18: % before the target is passed. The initial
19: % velocity of the projectile must be large
20: % enough to make the problem well posed.
21: % Otherwise, the program will terminate with
22: % an error message.
23: %
24: % User m functions called: missdis, traject,
25: %                          projcteq
26:
27: clear all;
28: global Vinit Gravty Cdrag Xfinl Yfinl
29:
30: vinit=600; gravty=32.2; cdrag=0.002;
31: xfinl=1000; yfinl=100;
32:
33: disp(' ');
34: disp('SEARCH FOR INITIAL INCLINATION ANGLE ');
35: disp('TO MAKE A PROJECTILE STRIKE A DISTANT');
36: disp('OBJECT'); disp(' ');
37: disp(['Initial velocity = ',num2str(vinit)]);
38: disp(['Gravity constant = ',num2str(gravty)]);
39: disp(['Drag coefficient = ',num2str(cdrag)]);
40: disp(['Coordinates of target = (', ...
```

```
41:        num2str(xfinl),',',...
42:        num2str(yfinl),')']); disp(' ');
43:
44: % Replicate input data as global variables
45: Vinit=vinit; Gravty=gravty; Cdrag=cdrag;
46: Xfinl=xfinl; Yfinl=yfinl;
47:
48: % Perform the minimization search
49: fstart=180/pi*atan(yfinl/xfinl); fend=75;
50: disp('Please wait for completion of the')
51: disp('minimization search');
52: bestang=fminbnd(@missdis,fstart,fend);
53:
54: % Display final results
55: [y,x,t]=traject ...
56:          (bestang,vinit,gravty,cdrag,xfinl);
57: dmiss=abs(yfinl-y(length(y))); disp(' ')
58: disp(['Final miss distance is ', ...
59:      num2str(dmiss),' when the']);
60: disp(['initial inclination angle is ', ...
61:      num2str(bestang),...
62:      ' degrees']);
63:
64: %==============================================
65:
66: function [dsq,x,y]=missdis(angle)
67: %
68: % [dsq,x,y]=missdis(angle)
69: % ~~~~~~~~~~~~~~~~~~~~~~~~~~
70: %
71: % This function is used by fminbnd. It returns
72: % an error measure indicating how much the
73: % target is missed for a particular initial
74: % inclination angle of the projectile.
75: %
76: % angle - the initial inclination angle of
77: %         the projectile in degrees
78: %
79: % dsq   - the square of the difference between
80: %         Yfinal and the final value of y found
81: %         using function traject.
82: % x,y   - points on the trajectory.
83: %
84: % Several global parameters (Vinit, Gravty,
85: % Cdrag, Xfinl) are passed to missdis by the
```

```
86: % driver program runtraj.
87: %
88: % User m functions called: traject
89: %-----------------------------------------------
90: 
91: global Vinit Gravty Cdrag Xfinl Yfinl
92: [y,x,t]=traject ...
93:        (angle,Vinit,Gravty,Cdrag,Xfinl,1);
94: dsq=(y(length(y))-Yfinl)^2;
95: 
96: %===============================================
97: 
98: function [y,x,t]=traject ...
99:        (angle,vinit,gravty,cdrag,xfinl,noplot)
100: %
101: % [y,x,t]=traject ...
102: %       (angle,vinit,gravty,cdrag,xfinl,noplot)
103: % ~~~~~~~~~~~~~~~~~~~~~~~~~~~~~~~~~~~~~~~~~~~~~~~
104: %
105: % This function integrates the dynamical
106: % equations for a projectile subjected to
107: % gravity loading and atmospheric drag
108: % proportional to the square of the velocity.
109: %
110: % angle  - initial inclination of the
111: %          projectile in degrees
112: % vinit  - initial velocity of the projectile
113: %          (muzzle velocity)
114: % gravty - the gravitational constant
115: % cdrag  - drag coefficient specifying the
116: %          drag force per unit mass which
117: %          equals cdrag*velocity^2.
118: % xfinl  - the projectile is fired toward the
119: %          right from x=0.  xfinl is the
120: %          largest x value for which the
121: %          solution is computed. The initial
122: %          velocity must be large enough that
123: %          atmospheric damping does not reduce
124: %          the horizontal velocity to zero
125: %          before xfinl is reached.  Otherwise
126: %          an error termination will occur.
127: % noplot - plotting of the trajectory is
128: %          omitted when this parameter is
129: %          given an input value
130: %
```

```
131: % y,x,t  - the y, x and time vectors produced
132: %          by integrating the equations of
133: %          motion
134: %
135: % Global variables:
136: %
137: % grav,   - two constants replicating gravty and
138: % dragc     cdrag, for use in function projcteq
139: % vtol    - equal to vinit/1e6, used in projcteq
140: %           to check whether the horizontal
141: %           velocity has been reduced to zero
142: %
143: % User m functions called: projcteq
144: %----------------------------------------------
145:
146: global grav dragc vtol
147:
148: % Default data case generated when input is null
149: if nargin ==0
150:   angle=45; vinit=600; gravty=32.2;
151:   cdrag=0.002; xfinl=1000;
152: end;
153:
154: % Assign global variables and evaluate
155: % initial velocity
156: grav=gravty; dragc=cdrag; ang=pi/180*angle;
157: vtol=vinit/1e6;
158: z0=[vinit*cos(ang); vinit*sin(ang); 0; 0];
159:
160: % Integrate the equations of motion defined
161: % in function projcteq
162: deoptn=odeset('RelTol',1e-6);
163: [x,z]=ode45('projcteq',[0,xfinl],z0,deoptn);
164:
165: y=z(:,3); t=z(:,4); n=length(x);
166: xf=x(n); yf=y(n);
167:
168: % Plot the trajectory curve
169: if nargin < 6
170:   plot(x,y,'k-',xf,yf,'ko');
171:   xlabel('x axis'); ylabel('y axis');
172:   title(['Projectile Trajectory for ', ...
173:          'Velocity Squared Drag']);
174:   axis('equal'); grid on; figure(gcf);
175:   %print -deps trajplot
```

```
176: end
177:
178: %==========================================
179:
180: function zp=projcteq(x,z)
181: %
182: % zp=projcteq(x,z)
183: % ~~~~~~~~~~~~~~~~
184: %
185: % This function defines the equation of motion
186: % for a projectile loaded by gravity and
187: % atmospheric drag proportional to the square
188: % of the velocity.
189: %
190: % x     -  the horizontal spatial variable
191: % z     -  a vector containing [vx; vy; y; t];
192: %
193: % zp    -  the derivative dz/dx which equals
194: %          [vx'(x); vy'(x); y'(x); t'(x)];
195: %
196: % Global variables:
197: %
198: % grav  -  the gravity constant
199: % dragc -  the drag coefficient divided by
200: %          gravity
201: % vtol  -  a global variable used to check
202: %          whether vx is zero
203: %
204: % User m functions called:  none
205: %-------------------------------------------
206:
207: global grav dragc vtol
208: vx=z(1); vy=z(2); v=sqrt(vx^2+vy^2);
209:
210: % Check to see whether drag reduced the
211: % horizontal velocity to zero before the
212: % xfinl was reached.
213: if abs(vx) < vtol
214:   disp(' ');
215:   disp('************************************');
216:   disp('ERROR in function projcteq. The ');
217:   disp('  initial velocity of the projectile');
218:   disp('  was not large enough for xfinal to');
219:   disp('  be reached.');
220:   disp('EXECUTION IS TERMINATED.');
```

```
221:     disp('*************************************');
222:     disp(' '),error(' ');
223: end
224: zp=[-dragc*v; -(grav+dragc*v*vy)/vx; ...
225:       vy/vx; 1/vx];
```

13.3 Fitting Nonlinear Equations to Data

Often an equation of known form is needed to approximately fit some given data values. An equation $y(t)$ to fit m data values (t_i, y_i) might be sought from an equation expressible as

$$y = f(a_1, a_2, \ldots, a_n, t)$$

where n parameters $a_1, a_2, \ldots, a_n$ are needed to minimize the least squares error

$$\epsilon(a_1, a_2, \ldots, a_n) = \sum_{j=1}^{n} [y_j - f(a_1, a_2, \ldots, a_n, t_j)]^2 .$$

The smallest possible error would be zero when the equation passes exactly through all the data values. Function ϵ can be minimized with an optimizer such as **fminsearch**, or conditions seeking a zero gradient of ϵ which require

$$\frac{\partial \epsilon}{\partial a_i} = 2\sum_{j=1}^{n} [f(a_1, a_2, \ldots, a_n, t_j) - y_j] \left(\frac{\partial f}{\partial a_i}\right)$$

can be written. Note that the problem of minimizing a function and the problem of solving a set of nonlinear simultaneous equations are closely related. Solving large systems of nonlinear equations is difficult. Therefore, data fitting by use of function minimization procedures is typically more effective.

The formulation assuming y depends on a single independent variable could just as easily have involved several independent variables $x_1, x_2, \ldots, x_N$, which would yield an equation of the form

$$y = f(a_1, a_2, \ldots, a_n, x_1, x_2, \ldots, x_N).$$

For instance, we might choose the simplest useful equation depending linearly on the independent variables

$$y = \sum_{k=0}^{N} x_k a_k$$

where $x_0 = 1$. The least squares error can be expressed as

$$\epsilon(a_0, a_1, \ldots, a_n) = \sum_{j=1}^{n} \left[y_j - \sum_{k=0}^{N} X_{jk} a_k \right]^2$$

where X_{jk} means the value of the kth independent variable at the jth data point. The condition that ϵ have a zero gradient gives

$$\sum_{k=0}^{N}\left[\sum_{j=1}^{n} X_{ji}X_{jk}\right] a_k = \sum_{j=1}^{n} X_{ji}y_j \;,\; 1 \le i \le N.$$

This linear system can be solved using traditional methods. Since the multiple indices in the equation are slightly cryptic, expressing the relationship in matrix notation is helpful. We get

$$Y \approx XA$$

where

$$Y = \begin{bmatrix} y_1 \\ y_2 \\ \vdots \\ y_n \end{bmatrix} \;,\; X = [1, X_1, X_2, \ldots, X_N] \;,\; A = \begin{bmatrix} a_0 \\ a_1 \\ \vdots \\ a_N \end{bmatrix}$$

with X_i being the column matrix $[x_{i1}, x_{i2}, \ldots, x_{in}]$ and the first column of X contains all ones. The requirement to minimize ϵ is simply

$$(X^T X)A = X^T Y$$

and MATLAB produces the desired solution using

```
A=X\Y;
```

Although taking y as a linear function of parameters $a_0, a_1, \ldots, a_N$ produces solvable linear equations, the general situation yields nonlinear equations, and a minimization search procedure has greater appeal. We conclude this section with an example employing a minimization search.

Consider an experiment where data values (t_i, y_i) are expected to conform to the transient response of a linear harmonic oscillator governed by the differential equation

$$m_0\ddot{y} + c_0\dot{y} + k_0 y = 0.$$

This equation has a solution representable as

$$y = a_1 e^{-|a_2|t}\cos(|a_3|t + a_4)$$

where $|a_2|$ makes the response decay exponentially and $|a_3|$ assures that the damped natural frequency is positive. Minimizing the error function

$$\epsilon(a_1, a_2, a_3, a_4) = \sum_{j=1}^{n}\left[y_j - a_1 e^{-1|a_2|t_j}\cos(|a_3|t_j + a_4)\right]^2$$

requires a four-dimensional search.

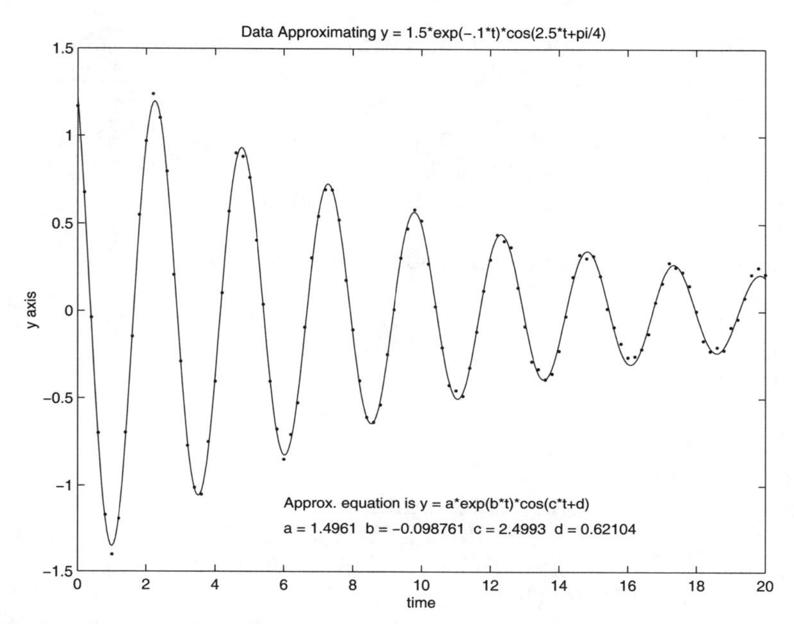

Figure 13.2: Data Approximating $y = 1.5\exp(-0.1t)\cos(2.5t + \pi/4)$

The program **vibfit** tests data deviating slightly from an equation employing specific values of a_1, a_2, a_3, a_4. Then function **fminsearch** is used to verify whether the coefficients can be recovered from the data points. Figure 13.2 shows the data values and the equation resulting from the nonlinear least square fit. The results produced are quite acceptable.

Program Output and Code

Program vibfit

```
1: function vibfit
2: %
3: % Example: vibfit
4: % ~~~~~~~~~~~~~~~~
5: %
6: % This program illustrates use of the Nelder
7: % and Mead multi-dimensional function
8: % minimization method to determine an equation
9: % for y(t) which depends nonlinearly on several
10: % parameters chosen to closely fit known data
```

```
11: % values. The program minimizes the sum of the
12: % squares of error deviations between the data
13: % values and results produced by the chosen
14: % equation. The example pertains to the time
15: % response curve characterizing free vibrations
16: % of a damped linear harmonic oscillator.
17: %
18: % User m functions called: vibfun
19: %
20: % Make the data vectors global to allow
21: % access from function vibfun
22: global timdat ydat
23:
24: echo off;
25: disp(' ');
26: disp('          CHOOSING PARAMETERS');
27: disp('    IN THE THE NONLINEAR EQUATION');
28: disp('      Y = A*EXP(B*T)*COS(C*T+D)');
29: disp('TO OBTAIN THE BEST FIT TO GIVEN DATA');
30: fprintf('\nPress [Enter] to list function\n');
31: fprintf('vibfun which is to be minimized\n');
32: pause;
33:
34: % Generate a set of data to be fitted by a
35: % chosen equation.
36: a=1.5; b=-.1; c=2.5; d=pi/5;
37: timdat=0:.2:20;
38: ydat=a*exp(b*timdat).*cos(c*timdat+d);
39:
40: % Add some random noise to the data
41: ydat=ydat+.1*(-.5+rand(size(ydat)));
42:
43: % Function vibfun defines the quantity to be
44: % minimized by a search using function fmins.
45: disp(' ');
46: disp('The function to be minimized is:');
47: type vibfun.m; disp(' ');
48: disp('The input data will be plotted next.');
49: disp('Press [Enter] to continue'); pause;
50: plot(timdat,ydat,'k.');
51: title('Input Data'); xlabel('time');
52: ylabel('y axis'); grid off; figure(gcf);
53: input('','s');
54:
55: % Initiate the four-dimensional search
```

```
56: x=fminsearch(@vibfun,[1 1 1 1]);
57:
58: % Check how well the computed parameters
59: % fit the data.
60: aa=x(1); bb=-abs(x(2)); cc=abs(x(3)); dd=x(4);
61: as=num2str(aa); bs=num2str(bb);
62: cs=num2str(cc); ds=num2str(dd);
63: ttrp=0:.05:20;
64: ytrp=aa*exp(bb*ttrp).*cos(cc*ttrp+dd);
65: disp(' ');
66: disp('Press [Enter] to see how well');
67: disp('the equation fits the data'); pause;
68: plot(ttrp,ytrp,'k-',timdat,ydat,'k.');
69: str1=['Approx. equation is y = ', ...
70:       'a*exp(b*t)*cos(c*t+d)'];
71: str2=['a = ',as,'  b = ',bs,'  c = ', ...
72:       cs,'  d = ',ds];
73: text(6,-1.1,str1); text(6,-1.25,str2);
74: xlabel('time'); ylabel('y axis');
75: title(['Data Approximating ', ...
76:        'y = 1.5*exp(-.1*t)*cos(2.5*t+pi/4)']);
77: grid off; figure(gcf);
78: print -deps apprxdat
79:
80: %===============================================
81:
82: function z=vibfun(x)
83: %
84: % z=vibfun(x)
85: % ~~~~~~~~~~~
86: %
87: % This function evalautes the least square
88: % error for a set of vibration data. The data
89: % vectors timdat and ydat are passed as global
90: % variables. The function to be fitted is:
91: %
92: %   y=a*exp(b*t)*cos(c*t+d)
93: %
94: % x - a vector defining a,b,c and d
95: %
96: % z - the square of the norm for the vector
97: %     of error deviations between the data and
98: %     results the equation gives for current
99: %     parameter values
100: %
```

```
101: % User m functions called:  none
102: %----------------------------------------
103:
104: global timdat ydat
105: a=x(1); b=-abs(x(2)); c=abs(x(3)); d=x(4);
106: z=a*exp(b*timdat).*cos(c*timdat+d);
107: z=norm(z-ydat)^2;
```

13.4 Nonlinear Deflections of a Cable

We will now present an optimization procedure to determine the static equilibrium position of a perfectly flexible inextensible cable having given end positions and a known distributed load per unit length. If $\boldsymbol{R}(s)$ is the position of any point on the cable as a function of arc length $0 \leq s \leq L$, then the internal tension at position s is

$$\boldsymbol{T}(s) = \boldsymbol{F}_e + \int_s^L \boldsymbol{q}(s)\, ds$$

with $\boldsymbol{q}(s)$ being the applied force per unit length and $\boldsymbol{F}_e$ being the support force at $s = L$. The end force to produce a desired end deflection has to be determined in the analysis. However, the end deflection resulting from any particular choice of end force can be computed by observing that the tangent to the deflection curve will point along the direction of the cable tension. This means

$$\frac{d\boldsymbol{R}}{ds} = \frac{\boldsymbol{T}(s)}{|\boldsymbol{T}(s)|}$$

and

$$\boldsymbol{R}(s) = \int_0^s \frac{\boldsymbol{T}(s) ds}{|\boldsymbol{T}(s)|} = \int_0^s \frac{\left(\boldsymbol{F}_e + \int_s^L \boldsymbol{q}\, ds\right) ds}{|\boldsymbol{F}_e + \int_s^L \boldsymbol{q}\, ds|}$$

where $\boldsymbol{R}(0) = 0$ is taken as the position at the starting end. The deflection at $s = L$ will have some specified position $\boldsymbol{R}_e$ so that requiring $\boldsymbol{R}(L) = \boldsymbol{R}_e$ gives a vector equation depending parametrically on $\boldsymbol{F}_e$. Thus, we need to solve three nonlinear simultaneous equations in the Cartesian components of force $\boldsymbol{F}_e$. A reasonable analytical approach is to employ an optimization search to minimize $|\boldsymbol{R}(L) - \boldsymbol{R}_e|$ in terms of the components of $\boldsymbol{F}_e$.

The procedure described for a cable with continuous loading extends easily to a cable having several rigid links connected at frictionless joints where arbitrary concentrated forces are applied. The function **cabldefl** evaluates the position of each joint when the joint forces and outer end force are given. With the end force on the last link treated as a parameter, function **endfl** computes an error measure $|\boldsymbol{F}(L) - \boldsymbol{R}_E|^2$ to be minimized using function **fminsearch**. The optimization search seeks the components of $\boldsymbol{F}_e$ needed to reduce the error measure to zero. Specifying a sensible problem obviously requires that $|\boldsymbol{R}_e|$ must not exceed the total length of all members

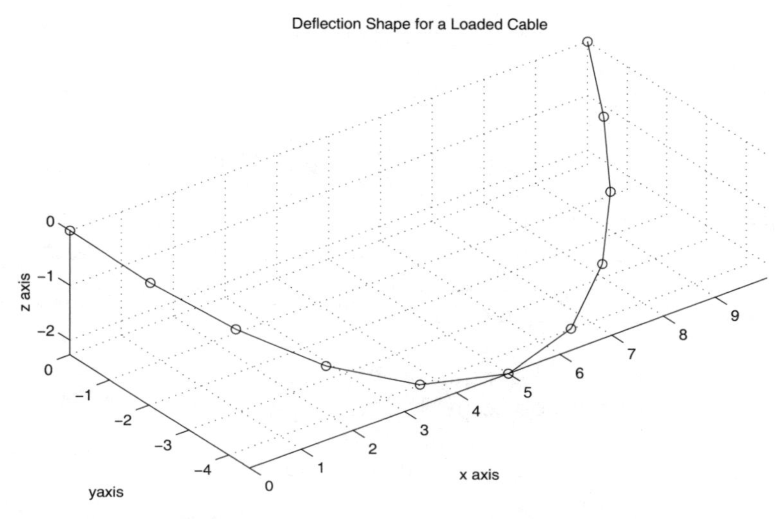

Figure 13.3: Deflected Shape for a Loaded Cable

in the chain. Initiating the search with a randomly generated starting force leads to a final force produced by **fminsearch**, which is then employed in another call to **cabldefl** to determine and plot the final deflection position as shown in Figure 13.3. Using a random initial guess for the end force was done to show that choosing bad starting data, insufficiently stringent convergence tolerances, or too few allowable function iterations can sometimes produce erroneous results. This emphasizes the need to always examine the results from nonlinear search procedures to assure that satisfactory answers are obtained.

Program Output and Code

Program cablsolv

```
1: function [r,t,pends]=cablsolv(Len,P,Rend)
2: %
3: % [r,t,pends]=cablsolv(Len,P,Rend)
4: % ~~~~~~~~~~~~~~~~~~~~~~~~~~~~~~~~~
5: %
6: % This function computes the equilibrium
7: % position for a cable composed of rigid
8: % weightless links with loads applied at the
9: % frictionless joints. The ends of the cable
10: % are assumed to have a known position.
11: %
```

```
12: % Len   - a vector containing the lengths
13: %         Len(1), ..., Len(n)
14: % P     - matrix of force components applied
15: %         at the interior joints. P(:,i)
16: %         contains the Cartesian components of
17: %         the force at joint i.
18: % Rend  - the coordinate vector giving the
19: %         position of the outer end of the last
20: %         link, assuming the outer end of the
21: %         first link is at [0,0,0].
22: %
23: % r     - a matrix with rows giving the
24: %         computed equilibrium positions of all
25: %         ends
26: % t     - a vector of tension values in the
27: %         links
28: % pends - a matrix having two rows which
29: %         contain the force components acting
30: %         at both ends of the chain to maintain
31: %         equilibrium
32: %
33: % User m functions called: endfl, cabldefl
34:
35: if nargin < 3
36:   % Example for a ten link cable with vertical
37:   % and lateral loads
38:   Len=1.5*ones(10,1); Rend=[10,0,0];
39:   P=ones(9,1)*[0,-2,-1];
40: end
41:
42: global len p rend
43: len=Len; rend=Rend; p=P; tol=sum(Len)/1e8;
44:
45: % Start the search with a random force applied
46: % at the far end
47:
48: % Perform several searches to minimize the
49: % length of the vector from the outer end of
50: % the last link to the desired position Rend
51: % where the link is to be connected to a
52: % support. The final end force should reduce
53: % the deflection error to zero if the search
54: % is successful.
55:
56: opts=optimset('tolx',tol,'tolfun',tol,...
```

```
57:                'maxfunevals',2000);
58: endval=realmax;
59:
60: % Iterate several times to avoid false
61: % convergence
62: for k=1:5
63:   p0=10*max(abs(p(:)))*rand(size(p,2),1);
64:   [pendk,endvalk,exitf]=...
65:   fminsearch(@endfl,p0,opts);
66:     if endvalk < endval
67:       pend=pendk(:); endval=endvalk;
68:     end
69: end
70:
71: % Use the computed end force to obtain the
72: % final deflection. Also return the
73: % support forces.
74: [r,t,pstart]=cabldefl(len,[p;pend']);
75: x=r(:,1); y=r(:,2); z=r(:,3);
76: pends=[pstart(:)';pend(:)'];
77:
78: % Plot the deflection curve of the cable
79: plot3(x,y,z,'k-',x,y,z,'ko'); xlabel('x axis');
80: ylabel('yaxis'); zlabel('z axis');
81: title('Deflection Shape for a Loaded Cable');
82: axis('equal'); grid on; figure(gcf);
83: print -deps defcable
84:
85: %==============================================
86:
87: function enderr=endfl(pend)
88: %
89: % enderr=endfl(pend)
90: % ~~~~~~~~~~~~~~~~~~
91: %
92: % This function computes how much the
93: % position of the outer end of the last link
94: % deviates from the desired position when an
95: % arbitrary force pend acts at the cable end.
96: %
97: % pend   - vector of force components applied
98: %          at the outer end of the last link
99: %
100: % enderr - the deflection error defined by the
101: %          square of the norm of the vector
```

```
102: %           from the computed end position and
103: %           the desired end position. This error
104: %           should be zero for the final
105: %           equilibrium position
106: %
107: % User m functions called: cabldefl
108: %----------------------------------------------
109:
110: % Pass the lengths, the interior forces and the
111: % desired position of the outer end of the last
112: % link as global variables.
113: global len p rend
114:
115: % use function cabldefl to compute the
116: % desired error
117: r=cabldefl(len,[p;pend(:)']);
118: rlast=r(size(r,1),:);
119: d=rlast(:)-rend(:); enderr=d'*d;
120:
121: %==============================================
122:
123: function [r,t,pbegin]=cabldefl(len,p)
124: %
125: % [r,t,pbegin]=cabldefl(len,p)
126: % ~~~~~~~~~~~~~~~~~~~~~~~~~~~~
127: %
128: % This function computes the static equilibrium
129: % position for a cable of rigid weightless
130: % links having concentrated loads applied at
131: % the joints and the outside of the last link.
132: % The outside of the first link is positioned
133: % at the origin.
134: %
135: % len    - a vector of link lengths
136: %          len(1), ..., len(n)
137: % p      - a matrix with rows giving the
138: %          force components acting at the
139: %          interior joints and at the outer
140: %          end of the last link
141: %
142: % r      - matrix having rows which give the
143: %          final positions of each node
144: % t      - vector of member tensions
145: % pbegin - force acting at the outer end of
146: %          the first link to achieve
```

```
147: %              equilibrium
148: %
149: % User m functions called:  none
150: %----------------------------------------------
151:
152: n=length(len); len=len(:); nd=size(p,2);
153:
154: % Compute the forces in the links
155: T=flipud(cumsum(flipud(p)));
156: t=sqrt(sum((T.^2)')');
157:
158: % Obtain the deflections of the outer ends
159: % and the interior joints
160: r=cumsum(T./t(:,ones(1,nd)).*len(:,ones(1,nd)));
161: r=[zeros(1,nd);r]; pbegin=-t(1)*r(2,:)/len(1);
```

13.5 Quickest Time Descent Curve (the Brachistochrone)

The subject of variational calculus addresses methods to find a function producing the minimum value for an integral depending parametrically on the function. Typically, we have a relationship of the form

$$I(y) = \int_{x_1}^{x_2} G(x, y, y'(x))\, dx$$

where values of y at $x = x_1$ and $x = x_2$ are known, and $y(x)$ for $x_1 < x < x_2$ is sought to minimize I. A classical example in this subject is determining a curve starting at $(0, 0)$ and ending at (a, b) so that a smooth particle will slide from one end to the other in the shortest possible time. Let X and Y be measured positive to the right and downward. Then the descent time for frictionless movement along the curve will be

$$t = \frac{1}{\sqrt{2g}} \int_0^a \sqrt{\frac{1 + Y'(X)^2}{Y(X)}}\, dX \ , \ Y(0) = 0 \ , \ Y(a) = b.$$

This problem is solved in various advanced calculus books.[3] The curve is a cycloid expressed in parametric form as

$$X = k[\theta - \sin(\theta)] \ , \ Y = k[1 - \cos(\theta)]$$

[3]Weinstock [105] provides an excellent discussion of the brachistochrone problem using calculus of variation methods.

where $0 < \theta < \theta_f$. Values of θ_f and k are found to make $x(\theta_f) = a$ and $Y(\theta_f) = b$. The exact descent time is

$$t_{\text{best}} = \theta_f \sqrt{\frac{k}{g}}$$

which is significantly smaller than the descent time for a straight line, which is

$$t_{\text{line}} = \sqrt{\frac{2(a^2 + b^2)}{gb}}.$$

Two functions, **brfaltim** and **bracifun**, are used to compute points on the brachistochrone curve and evaluate the descent time.

The main purpose of this section is to illustrate how optimization search can be used to minimize an integral depending parametrically on a function. The method used chooses a set of base points through which an interpolation curve is constructed to specify the function. Using numerical integration gives a value for the integral. Holding the x values for the interpolation points constant and allowing the y values to vary expresses the integral as a function of the y values at a finite number of points. Then a multi-dimensional search function such as **fminsearch** can optimize the choice of Y values. Before carrying out this process for the brachistochrone problem it is convenient to change variables so that $x = X/a$ and

$$Y(X) = b[x + y(x)] \ , \ 0 \leq x \leq 1,$$

with

$$y(0) = y(1) = 0.$$

Then the descent integral becomes

$$t = \frac{a}{\sqrt{2gb}} \int_0^1 \sqrt{\frac{1 + (b/a)^2[1 + y'(x)]^2}{x + y}} \, dx.$$

For any selected set of interpolation points, functions **spline** and **splined** can evaluate $y(x)$ and $y'(x)$ needed in the integrand, and function **gcquad** can be used to perform Gaussian integration. An optimization search employing **fminsearch** produces the curve heights yielding an approximation to the brachistochrone as shown in Figure 13.4.

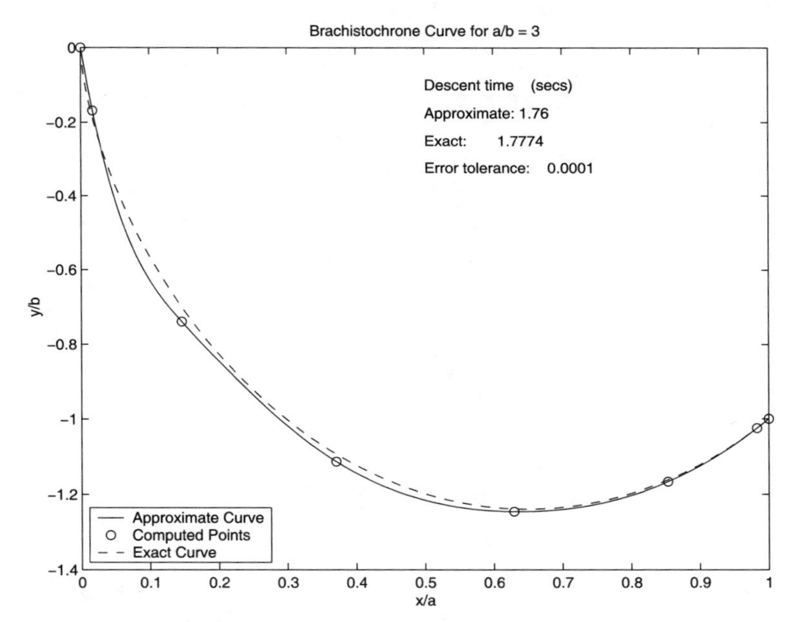

Figure 13.4: Brachistochrone Curve for $\frac{a}{b} = 3$

Program Output and Code

Program brachist

```
1: function brachist
2: % Example: brachist
3: % ~~~~~~~~~~~~~~~~~~
4: % This program determines the shape of a
5: % smooth curve down which a particle can slide
6: % in minimum possible time. The analysis
7: % employs a piecewise cubic spline to define
8: % the curve by interpolation using a fixed set
9: % of base point positions. The curve shape
10: % becomes a function of the heights chosen at
11: % the base points. These heights are determined
12: % by computing the descent time as a function
13: % of the heights and minimizing the descent
14: % time by use of an optimization program. The
15: % Nelder and Mead unconstrained search
16: % procedure is used to compute the minimum.
17: %
18: % User m functions called:
19: %       chbpts, brfaltim, fltim, gcquad,
20: %       bracifun, splined
21:
22: global cbp cwf cofs n xc yc a b b_over_a ...
23:        grav nparts nquad nfcls
24:
25: fprintf(...
26: '\nBRACHISTOCHRONE DETERMINATION BY NONLINEAR');
27: fprintf('\n         OPTIMIZATION SEARCH \n');
28: fprintf(['\nPlease wait. The ',...
29:   'calculation takes a while.\n']);
30:
31: % Initialize
32: a=30; b=10; grav=32.2; nparts=1; nquad=50;
33: tol=1e-4; n=6; b_over_a = b/a;
34:
35: [dummy,cbp,cwf]=gcquad('',0,1,nquad,nparts);
36: xc=chbpts(0,1,n); xc=xc(:);
37: y0=5*sin(pi*xc); xc=[0;xc;1];
38:
39: % Calculate results from the exact solution
40: [texact,xexact,yexact]=brfaltim(a,b,grav,100);
```

```
41:
42: % Perform minimization search for
43: % approximate solution
44: opts=optimset('tolx',tol,'tolfun',tol);
45: [yfmin,fmin,flag,outp] =...
46:                   fminsearch(@fltim,y0,opts);
47:
48: % Evaluate final position and approximate
49: % descent time
50: Xfmin=xc; Yfmin=Xfmin+[0;yfmin(:);0];
51: % tfmin=a/sqrt(2*grav*b)*fltim(yfmin(:));
52: tfmin=a/sqrt(2*grav*b)*fmin;
53: nfcls=1+outp.funcCount;
54:
55: % Summary of calculations
56: fprintf('\nBrachistochrone Summary');
57: fprintf('\n----------------------');
58: fprintf('\nNumber of function calls:   ');
59: fprintf('%g',nfcls);
60: fprintf('\nDescent time:   ');
61: fprintf('%g',tfmin), fprintf('\n')
62:
63: % Plot results comparing the approximate
64: % and exact solutions
65: xplot=linspace(0,1,100);
66: yplot=spline(Xfmin,Yfmin,xplot);
67: plot(xplot,-yplot,'-',Xfmin,-Yfmin,'o', ...
68:      xexact/a,-yexact/b,'--');
69: xlabel('x/a'); ylabel('y/b'); % grid
70: title(['Brachistochrone Curve for ', ...
71:        'a/b = ',num2str(a/b)]);
72: text(.5,-.1,   'Descent time    (secs)')
73: text(.5,-.175,['Approximate: ',num2str(tfmin)])
74: text(.5,-.25, ['Exact:         ',num2str(texact)]);
75: text(.5,-.325, ...
76:   sprintf('Error tolerance:    %g',tol));
77: legend('Approximate Curve', ...
78:        'Computed Points','Exact Curve',3);
79: figure(gcf);
80: print -deps brachist
81:
82: %============================================
83:
84: function [tfall,xbrac,ybrac]=brfaltim ...
85:                              (a,b,grav,npts)
```

```
86: %
87: %
88: % [tfall,xbrac,ybrac]=brfaltim(a,b,grav,npts)
89: % ~~~~~~~~~~~~~~~~~~~~~~~~~~~~~~~~~~~~~~~~~~~~
90: %
91: % This function determines the descent time
92: % and a set of points on the brachistochrone
93: % curve passing through (0,0) and (a,b).
94: % The curve is a cycloid expressible in
95: % parametric form as
96: %
97: %    x=k*(th-sin(th)),
98: %    y=k*(1-cos(th))    for 0<=th<=thf
99: %
100: % where thf is found by solving the equation
101: %
102: %    b/a=(1-cos(thf))/(thf-sin(thf)).
103: %
104: % Once thf is known then k is found from
105: %
106: %    k=a/(th-sin(th)).
107: %
108: % The exact value of the descent time is given
109: % by
110: %
111: %    tfall=sqrt(k/g)*thf
112: %
113: % a,b   - final values of (x,y) on the curve
114: % grav  - the gravity constant
115: % npts  - the number of points computed on
116: %         the curve
117: %
118: % tfall - the time required for a smooth
119: %         particle to slide along the curve
120: %         from (0,0) to (a,b)
121: % xbrac - x points on the curve with x
122: %         increasing to the right
123: % ybrac - y points on the curve with y
124: %         increasing downward
125: %
126: % User m functions called: none
127: %----------------------------------------------
128:
129: brfn=inline('cos(th)-1+cof*(th-sin(th))','th','cof');
130:
```

```
131: ba=b/a; [th,fval,flag]=fzero(...
132:                  brfn,[.01,10],optimset('fzero'),ba);
133:
134: k=a/(th-sin(th)); tfall=sqrt(k/grav)*th;
135: if nargin==4
136:   thvec=(0:npts-1)'*(th/(npts-1));
137:   xbrac=k*(thvec-sin(thvec));
138:   ybrac=k*(1-cos(thvec));
139: end
140:
141: %==========================================
142:
143: function x=chbpts(xmin,xmax,n)
144: %
145: % x=chbpts(xmin,xmax,n)
146: % ~~~~~~~~~~~~~~~~~~~~~
147: % Determine n points with Chebyshev spacing
148: % between xmin and xmax.
149: %
150: % User m functions called:  none
151: %-------------------------------------------
152:
153: x=(xmin+xmax)/2+((xmin-xmax)/2)* ...
154:   cos(pi/n*((0:n-1)'+.5));
155:
156: %==========================================
157:
158: function t=fltim(y)
159: %
160: % t=fltim(y)
161: % ~~~~~~~~~~
162: %
163: % This function evaluates the time descent
164: % integral for a spline curve having heights
165: % stored in y.
166: %
167: % y - vector defining the curve heights at
168: %     interior points corresponding to base
169: %     positions in xc
170: %
171: % t - the numerically integrated time descent
172: %     integral evaluated by use of base points
173: %     cbp and weight factors cwf passed as
174: %     global variables
175: %
```

```
176: % User m functions called: splined
177: %----------------------------------------------
178:
179: global xc cofs nparts bp wf nfcls cbp cwf ...
180:        b_over_a
181:
182: nfcls=nfcls+1; x=cbp;
183:
184: % Generate coefficients used in spline
185: % interpolation
186: yc=[0;y(:);0];
187: y=spline(xc,yc,x); yp=splined(xc,yc,x);
188:
189: % Evaluate the integrand
190: f=(1+(b_over_a*(1+yp)).^2)./(x+y); f=sqrt(f);
191:
192: % Evaluate the integral
193: t=cwf(:)'*f(:);
194:
195: %==============================================
196:
197: % function [val,bp,wf]=gcquad(func,xlow,...
198: %                 xhigh,nquad,mparts,varargin)
199: % See Appendix B
200:
201: %==============================================
202:
203: % function val=splined(xd,yd,x,if2)
204: % See Appendix B
```

13.6 Determining the Closest Points on Two Surfaces

Determining the closest points on two surfaces arises in applications such as robotic collision avoidance and container packing. Many types of surfaces can be parameterized using two curvilinear coordinates; so, the problem reduces to a four dimensional search to minimize the length of a line from a point on one surface to a point on the other surface. We call this the proximity problem and will consider typical instances involving two circular cylinders arbitrarily positioned in space as illustrated by the test examples of Figure 13.5. This application illustrates that, despite the apparent simplicity of this problem, convergence difficulties can occur with minimization search procedures, and several runs may be needed to get correct results.

An elementary way to analyze the proximity of two surfaces is to describe each surface by a grid of points and find the smallest element in a matrix describing the distance from point **i** of the first surface to a point **j** of the second surface. Large array

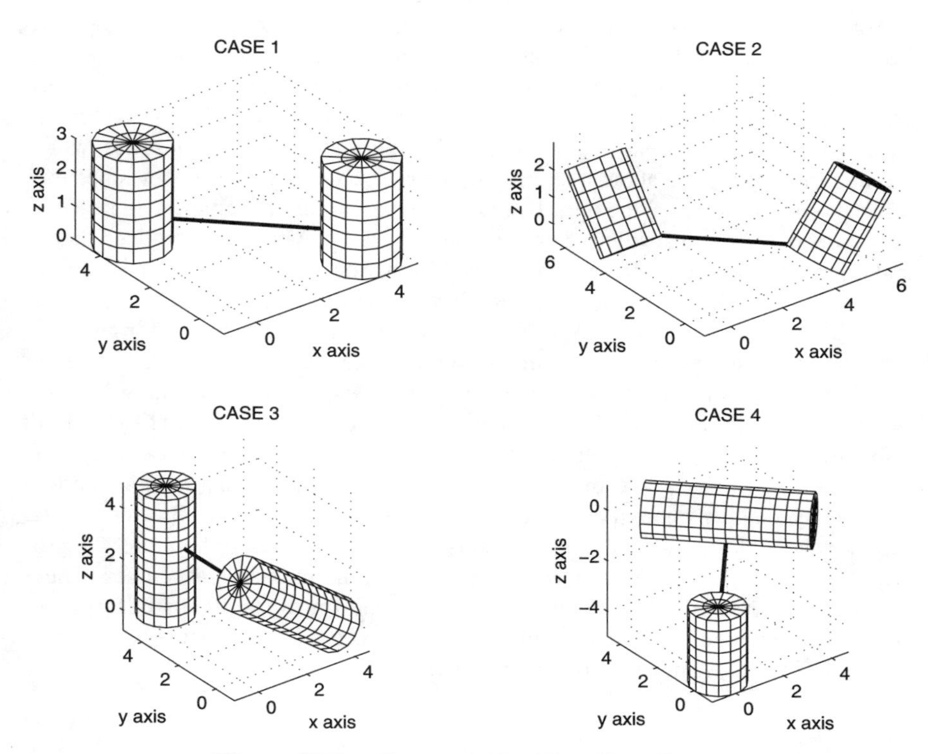

Figure 13.5: **Geometry for Four Test Cases**

dimensions can occur since a typical 100 by 100 surface grid involves 10,000 points and 30,000 coordinate values. The adjacency matrix for two surfaces, each using 10,000 points, has one hundred million points and would consume 2400 megabytes of memory when stored unpartitioned. However, memory limitations can be overcome by processing a few points at a time. In the program given below, a function **surf2surf** is presented to perform exhaustive search. It works well for the cylinder to cylinder problem and also handles some special cases. Since points and space curves are degenerate examples of surfaces, **surf2surf** can solve problems like obtaining the point on a curve closest to an arbitrary point in space.

For surfaces described by equations of the form $\boldsymbol{r}(s_1, s_2)$ and $\boldsymbol{R}(s_3, s_4)$, the proximity problem can be treated by minimizing $\mathbf{norm}(\boldsymbol{r} - \boldsymbol{R})^2$as a function of $[s_1 , s_2, s_3 , s_4]$. In this context, let us discuss briefly the concepts used in function **fminsearch** based on the flexible polyhedron search procedure developed by Nelder and Mead [ ]. The search employs a polyhedron having $n + 1$ corners in n space, which are initially aggregated about a starting point x_0. A sequence of moves repeatedly replaces corners at which the objective function has maximum values, with new corners corresponding to smaller values. Ultimately, the polyhedron is reduced in size and contracts to a point where the objective function is perceived to have a relative minimum. The algorithm embodied in **fminsearch** is useful but it sometimes gives false convergence. This experience led the authors to implement, for comparison purposes, a somewhat shorter version of the Nelder and Mead algorithm given in function **nelmed** shown in the following program **cylclose.** This program is designed to solve four test problems using functions **fminsearch**, **nelmed**, or **surf2surf**. Both implementations of the flexible polyhedron search are vulnerable to false convergence; so, it is necessary to initiate the search several times using random starting points. By making enough trials so that the same best result is obtained several times, reasonable confidence in the answers can be achieved. Furthermore, the program shows images of the cylinders and connecting minimum distance lines. These images can be rotated interactively to observe the validity of results. In the test cases considered, about eight trials was sufficient to produce the same best results at least twice. Some results showing computer output for case 4 are typical.

```
cylclose(1);

CASE 4 USING FUNCTION NELMED
Trial     Minimum    Function
Number    Distance  Evaluations
   1       1.915        163
   2       1.916        161
   3       1.710        207
   4       2.293        156
   5       1.710        154
   6       2.165        139
   7       2.165        122
   8       1.710        182
```

```
The analysis used FUNCTION NELMED
Shortest Distance     =     1.710
Function Evaluations =       1284
Compute Time          =     4.450 secs

cylclose(1);

CASE 4 USING FUNCTION FMINSEARCH
Trial     Minimum    Function
Number    Distance  Evaluations
   1        1.710        223
   2        2.293        472
   3        2.293        693
   4        2.293        295
   5        2.165        286
   6        2.165        585
   7        1.710        265
   8        1.915        231

The analysis used FUNCTION FMINSEARCH
Shortest Distance     =     1.710
Function Evaluations =       3050
Compute Time          =    10.930 secs

cylclose(3);

CASE 4 USING EXHAUSTIVE SEARCH
Shortest Distance     =     1.729
Function Evaluations =        546
Compute Time          =     0.440 secs
```

Note that incorrect answers were obtained repeatedly by **fminsearch** and **nelmed**, whereas exhaustive search gave the fastest and most reliable solution. Readers interested in exploring the convergence problems occurring with the Nelder and Mead search will find it instructive to run program **cylclose** to observe the variations in results produced from randomly chosen starting points. This example problem shows clearly that, unless the best result among a number of trials is taken, an incorrect answer may occur.

13.6.1 Discussion of the Computer Code

Program **cylclose** uses minimization search to determine the closest points on two arbitrarily positioned circular cylinders. Three solution methods are provided using functions **fminsearch**, **nelmed**, or **surf2surf**. Four test cases are included, and other geometries can be analyzed by modifying data lines in function **cylclose.** The various modules in the program are listed in the following table.

Routine	Line	Operation
cylclose	1-155	several functions are called to plot the geometry and perform the minimization search
cylpoint	159-178	gives the position of a point on a cylinder surface
dcyl2cyl	182-197	computes the distance between points on two cylinders
cylfigs	201 -244	plots the geometries for four data cases
plot2cyls	248-276	plots the geometry for two cylinders
cylpts	280-300	generates a grid of points on a cylinder surface
crnrpts	304-321	generates a dense set of points in an increasing set of data set
ortbas	325-332	creates orthonormal base vectors needed to define cylinder geometry
nelmed	336-475	function which performs the Nelder-Mead search
surf2surf	479-513	uses discrete search to compute closest points on two surfaces defined by coordinate grids. Large grids can be handled by calling function **srf2srf**
srf2srf	517-534	uses discrete search to compute closest points on two surfaces defined by coordinate grids
rads	538-550	gives base radii for example problems

Program cylclose

```
1: function [dbest,r,R]= cylclose(srchtype,...
2:                       ntrials,sidlen,tolx,tolf)
3: % [dbest,r,R]= cylclose(srchtype,ntrials,...
4: %                              sidlen,tolx,tolf)
5: % ~~~~~~~~~~~~~~~~~~~~~~~~~~~~~~~~~~~~~~~~~~~~~
6: %
7: % This program locates the points closest
8: % together on the surfaces of two circular
9: % cylinders arbitrarily positioned in space.
10: % A four-dimensional unconstrained search
11: % is performed using functions NELMED,
12: % FMINSEARCH, or SURF2SURF. The quantity
```

```
13: % minimized is the square of the distance
14: % from an arbitrary point on one cylinder
15: % to an arbitrary point on the other cylinder.
16: % The search parameters specify axial and
17: % circumferential coordinates of points on
18: % the cylinder surfaces.
19: %
20: % srchtype - selects the solution method. Use
21: %            1,2, or 3 for NELMED, FMINSEARCH,
22: %            or SURF2SURF
23: % ntrials  - Number of times the solution is
24: %            repeated to avoid false
25: %            convergence
26: % sidlen   - initial polyhedron side length
27: % tolx     - Convergence tolerance on solution
28: %            vector
29: % tolf     - Convergence tolerance on function
30: %            value
31: %
32: % User m functions called:
33: %      cylpoint, dcyl2cyl, cylfigs, plot2cyls
34: %      cylpts, cornrpts, ortbasis, nelmed,
35: %      surfmany, surf2surf, srf2srf, rads
36:
37: if nargin<5, tolf=1e-4;  end
38: if nargin<4, tolx=1e-2;  end
39: if nargin<3, sidlen=.5;  end
40: if nargin<2, ntrials=8;  end
41: if nargin<1, srchtype=1; end
42:
43: if srchtype==1
44:   fname='FUNCTION NELMED';
45: elseif srchtype==2
46:   fname='FUNCTION FMINSEARCH';
47: else
48:   fname='EXHAUSTIVE SEARCH';
49: end
50:
51: disp(' '),
52: disp(' CYLINDER PROXIMITY ANALYSIS')
53: disp('USING A FOUR-DIMENSIONAL SEARCH')
54:
55: cylfigs, drawnow, disp(' '), dumy=input(...
56: 'Press return to begin the search','s');
57: close; ncases=4;
```

```
58:
59: for jcase=1:ncases
60:   disp(' '), disp(['CASE ',...
61:       num2str(jcase),' USING ',fname])
62:
63:   % Define several data cases
64:   switch jcase
65:   case 1
66:     rad=1; len=3; r0=[4,0,0]; v=[0,0,1];
67:     Rad=1; Len=3; R0=[0,4,0]; V=[0,0,1];
68:   case 2
69:     rad=1; len=3; r0=[4,0,0]; v=[3,0,4];
70:     Rad=1; Len=3; R0=[0,4,0]; V=[0,3,4];
71:   case 3
72:     rad=1; len=5; r0=[4,0,0]; v=[-4,0,3];
73:     Rad=1; Len=5; R0=[0,4,0]; V=[0,0,1];
74:   case 4
75:     rad=1; len=4*sqrt(2); r0=[4,0,0];
76:     v=[-1,1,0];
77:     Rad=1; Len=3; R0=[0,0,-2]; V=[0,0,-1];
78:   end
79:
80:   % Create data parameters used repeatedly
81:   % during the search process
82:
83:   % First cylinder
84:   dat=cumsum([0;rad;len;rad]);
85:   dat=dat/max(dat); zdat=[dat,[0;0;len;len]];
86:   rdat=[dat,[0;rad;rad;0]]; m=ortbasis(v);
87:
88:   % Second cylinder
89:   dat=cumsum([0;Rad;Len;Rad]);
90:   dat=dat/max(dat); Zdat=[dat,[0;0;Len;Len]];
91:   Rdat=[dat,[0;Rad;Rad;0]]; M=ortbasis(V);
92:
93:   % Make several searches starting from
94:   % randomly chosen points and keep
95:   % the best answer obtained
96:   ntotal=0; ntype=zeros(1,5); disp(' ')
97:   tic; dbest=realmax; opts=optimset;
98:   if srchtype<3
99:     disp('Trial    Minimum    Function')
100:    disp('Number   Distance  Evaluations')
101:    for k=1:ntrials
102:      winit=2*pi*rand(4,1);
```

```
103:       if srchtype==1 % Search using nelmed
104:         [w,fmin,nvals,ntyp]=nelmed(@dcyl2cyl,...
105:              winit,sidlen,tolx,tolf,2000,0,...
106:              r0,m,rdat,zdat,R0,M,Rdat,Zdat);
107:       elseif srchtype==2 % Search using fminsearch
108:         [w,fmin,xflag,outp]=fminsearch(@dcyl2cyl,...
109:         winit,opts,r0,m,rdat,zdat,R0,M,Rdat,Zdat);
110:         nvals=outp.funcCount; ntyp=zeros(1,5);
111:       end
112:       dk=sqrt(dcyl2cyl(w,r0,m,rdat,zdat,...
113:                                  R0,M,Rdat,Zdat));
114:       fprintf('%4i   %8.3f   %7i\n',k,dk,nvals)
115:       if dk<dbest, dbest=dk; W=w; end
116:       ntotal=ntotal+nvals; ntype=ntype+ntyp;
117:     end
118:     w=W; r=cylpoint(w(1),w(2),r0,m,rdat,zdat);
119:     R=cylpoint(w(3),w(4),R0,M,Rdat,Zdat);
120:     t=toc;
121:     fprintf(['\nThe analysis used ',fname,'\n'])
122:     %if srchtype==1
123:     %  fprintf(['\nReflect  Expand  Contract  ',...
124:     %  'Shrink \n%4i %7i %9i %7i\n'],ntype(2),...
125:     %  ntype(3),ntype(4),ntype(5))
126:     %end
127:   else
128:     dplot=0.3; tic;
129:     [x,y,z,X,Y,Z]=plot2cyls(rad,len,r0,v,...
130:     Rad,Len,R0,V,dplot,' '); close;
131:     [dbest,r,R]=surf2surf(x,y,z,X,Y,Z);
132:     ntotal=length(x)*length(X); t=toc;
133:   end
134:   fprintf(...
135:   ['Shortest Distance      = %8.3f\n',...
136:    'Function Evaluations = %8i\n',...
137:    'Compute Time           = %8.3f secs\n'],...
138:    dbest,ntotal,t)
139:
140:   n=1; Rr=repmat(R,1,n+1)+(r-R)*(0:n)/n;
141:   hold off; clf,
142:   titl=['CASE ',num2str(jcase),' USING ',fname];
143:   dplot=0.3; plot2cyls(...
144:         rad,len,r0,v,Rad,Len,R0,V,dplot,titl);
145:   colormap([1 1 0]), hold on,
146:   plot3(Rr(1,:),Rr(2,:),Rr(3,:),'linewidth',2)
147:   title([titl,' : DISTANCE = ',...
```

```
148:   num2str(dbest),',  CPU TIME = ',...
149:   num2str(t),' SECS'])
150:   rotate3d on, shg, disp(' ')
151:   disp('Rotate the figure or press')
152:   disp('return to continue')
153:   dumy=input(' ','s'); close
154:
155: end
156:
157: %===========================================
158:
159: function r=cylpoint(w1,w2,r0,m,rdat,zdat)
160: % r=cylpoint(w1,w2,v,r0,m,rdat,zdat)
161: % ~~~~~~~~~~~~~~~~~~~~~~~~~~~~~~~~~~~
162: % This function computes the position of a
163: % point on the surface of a circular cylinder
164: % arbitrarily positioned in space. The argument
165: % list parameters have the following form,
166: % where rad means cylinder radius, and len
167: % means cylinder length.
168: % b=2*rad+len;
169: % zdat=[[0,0]; [rad/b, 0];
170: %       [(rad+len)/b, len];[ 1, len]];
171: % rdat=zdat; rdat(2,2)=rad;
172: % rdat(3,2)=rad; rdat(4,2)=0;
173:
174: u=2*pi*sin(w1)^2; v=sin(w2)^2;
175: z=interp1(zdat(:,1),zdat(:,2),v);
176: rho=interp1(rdat(:,1),rdat(:,2),v);
177: x=rho*cos(u); y=rho*sin(u);
178: r=r0(:)+m*[x;y;z];
179:
180: %===========================================
181:
182: function dsqr=dcyl2cyl(...
183:               w,r0,m,rdat,zdat,R0,M,Rdat,Zdat)
184: %~~~~~~~~~~~~~~~~~~~~~~~~~~~~~~~~~~~~~~~~~~~
185: % dsqr=dcyl2cyl(w,r0,m,rdat,zdat,R0,M,Rdat,Zdat)
186: % This function computes the square of the
187: % distance between generic points on the
188: % surfaces of two circular cylinders in three
189: % dimensions.
190: %
191: % User m functions called: cylpoint
192:
```

```
193: global fcount
194: fcount=fcount+1;
195: r=cylpoint(w(1),w(2),r0,m,rdat,zdat);
196: R=cylpoint(w(3),w(4),R0,M,Rdat,Zdat);
197: dsqr=norm(r-R)^2;
198:
199: %=============================================
200:
201: function cylfigs
202: % cylfigs
203: % ~~~~~~~
204: % This function plots the geometries
205: % pertaining to four data cases used
206: % to test closest proximity problems
207: % involving two circular cylinders
208: %
209: % User m functions called: plot2cyls
210:
211: w=rads; p=1:2; q=3:4; s=5:6; t=7:8;
212:
213: rad=1; len=3; r0=[4,0,0]; v=[0,0,1];
214: Rad=1; Len=3; R0=[0,4,0]; V=[0,0,1];
215: d=.4; subplot(2,2,1)
216: [x,y,z,X,Y,Z]=plot2cyls(rad,len,r0,v,Rad,Len,...
217:                  R0,V,d,'CASE 1'); hold on
218: plot3(w(p,1),w(p,2),w(p,3),'linewidth',2')
219: hold off
220:
221: rad=1; len=3; r0=[4,0,0]; v=[3,0,4];
222: Rad=1; Len=3; R0=[0,4,0]; V=[0,3,4];
223: d=.4; subplot(2,2,2);
224: [x,y,z,X,Y,Z]=plot2cyls(rad,len,r0,v,Rad,Len,...
225:                  R0,V,d,'CASE 2'); hold on
226: plot3(w(q,1),w(q,2),w(q,3),'linewidth',2')
227: hold off
228:
229: rad=1; len=5; r0=[4,0,0]; v=[-4,0,3];
230: Rad=1; Len=5; R0=[0,4,0]; V=[0,0,1];
231: d=.4; subplot(2,2,3)
232: [x,y,z,X,Y,Z]=plot2cyls(rad,len,r0,v,Rad,Len,...
233:                  R0,V,d,'CASE 3'); hold on
234: plot3(w(s,1),w(s,2),w(s,3),'linewidth',2')
235: hold off
236:
237: rad=1; len=4*sqrt(2);  r0=[4,0,0]; v=[-1,1,0];
```

```
238: Rad=1; Len=3; R0=[0,0,-2]; V=[0,0,-1];
239: d=.4; subplot(2,2,4);
240: [x,y,z,X,Y,Z]=plot2cyls(rad,len,r0,v,Rad,Len,...
241:                R0,V,d,'CASE 4'); hold on
242: plot3(w(t,1),w(t,2),w(t,3),'linewidth',2')
243: hold off, subplot
244: % print -deps cylclose
245:
246: %==========================================
247:
248: function [x,y,z,X,Y,Z]=plot2cyls(...
249:              rad,len,r0,vc,Rad,Len,R0,Vc,d,titl)
250: % [x,y,z,X,Y,Z]=plot2cyls(rad,len,r0,vc,Rad,...
251: %                          Len,R0,Vc,d,titl)
252: % ~~~~~~~~~~~~~~~~~~~~~~~~~~~~~~~~~~~~~~~~~~~~
253: % This function generates point grids on the
254: % surfaces of two circular cylinders and plots
255: % both cylinders together
256: %
257: % User m functions called: cornrpts surfmany
258: %                          cylpts
259: if nargin==0
260:    titl='TWO CYLINDERS';
261:    rad=1; len=3; r0=[4,0,0]; vc=[3,0,4];
262:    Rad=1; Len=3; R0=[0,4,0]; Vc=[0,3,4]; d=.2;
263: end
264: if isempty(titl), titl=' '; end
265: u=2*rad+len; v=2*pi*rad;
266: nu=ceil(u/d); nv=ceil(v/d);
267: u=cornrpts([0,rad,rad+len,u],nu)/u;
268: v=linspace(0,1,nv);
269: [x,y,z]=cylpts(u,v,rad,len,r0,vc);
270: U=2*Rad+Len; V=2*pi*Rad;
271: Nu=ceil(U/d); Nv=ceil(V/d);
272: U=cornrpts([0,Rad,Rad+Len,U],Nu)/U;
273: V=linspace(0,1,Nv);
274: [X,Y,Z]=cylpts(U,V,Rad,Len,R0,Vc);
275: surfmany(x,y,z,X,Y,Z), title(titl)
276: colormap([1 1 0]), shg
277:
278: %==========================================
279:
280: function [x,y,z]=cylpts(...
281:                 axial,circum,rad,len,r0,vectax)
282: % [x,y,z]=cylpts(axial,circum,rad,len,r0,vectax)
```

```
283: % ~~~~~~~~~~~~~~~~~~~~~~~~~~~~~~~~~~~~~~~~~~~~~
284: % This function computes a grid of points on the
285: % surface of a circular cylinder
286: %
287: % User m functions called: ortbasis
288:
289: U=2*rad+len; u=U*axial(:); n=length(u);
290: v=2*pi*circum(:)'; m=length(v);
291: ud=[0,rad,rad+len,U];
292: r=interp1(ud,[0,rad,rad,0],u);
293: z=interp1(ud,[0,0,len,len],u);
294: x=r*cos(v); y=r*sin(v); z=repmat(z,1,m);
295: % w=basis(vectax)*[x(:),y(:),z(:)]';
296: w=ortbasis(vectax)*[x(:),y(:),z(:)]';
297:
298: x=r0(1)+reshape(w(1,:),n,m);
299: y=r0(2)+reshape(w(2,:),n,m);
300: z=r0(3)+reshape(w(3,:),n,m);
301:
302: %==============================================
303:
304: function v=cornrpts(u,N)
305: % v=cornrpts(u,N)
306: % ~~~~~~~~~~~~~~
307: % This function generates approximately N
308: % points between min(u) and max(u) including
309: % all points in u plus additional points evenly
310: % spaced in each successive interval.
311: % u   -  vector of points
312: % N   -  approximate number of output points
313: %        between min(u(:)) and max(u(:))
314: % v   -  vector of points in increasing order
315:
316: u=sort(u(:))'; np=length(u);
317: d=u(np)-u(1); v=u(1);
318: for j=1:np-1
319:   dj=u(j+1)-u(j); nj=max(1,fix(N*dj/d));
320:   v=[v,[u(j)+dj/nj*(1:nj)]];
321: end
322:
323: %==============================================
324:
325: function mat=ortbasis(v)
326: % mat=ortbasis(v)
327: % ~~~~~~~~~~~~~~
```

```
328: % This function generates a rotation matrix
329: % having v(:)/norm(v) as the third column
330:
331: v=v(:)/norm(v); mat=[null(v'),v];
332: if det(mat)<0, mat(:,1)=-mat(:,1); end
333:
334: %===========================================
335:
336: function [xmin,fmin,m,ntype]=nelmed(...
337:               F,x0,dx,epsx,epsf,M,ifpr,varargin)
338: % [xmin,fmin,m,ntype]=nelmed(...
339: %              F,x0,dx,epsx,epsf,M,ifpr,varargin)
340: % ~~~~~~~~~~~~~~~~~~~~~~~~~~~~~~~~~~~~~~~~~~~~~~
341: % This function performs multidimensional
342: % unconstrained function minimization using the
343: % direct search procedure developed by
344: % J. A. Nelder and R. Mead. The method is
345: % described in various books such as:
346: % 'Nonlinear Optimization', by M. Avriel
347: %
348: % F        - objective function of the form
349: %            F(x,p1,p2,...) where x is vector
350: %            in n space and p1,p2,... are any
351: %            auxiliary parameters needed to
352: %            define F
353: % x0       - starting vector to initiate
354: %            the search
355: % dx       - initial polyhedron side length
356: % epsx     - convergence tolerance on x
357: % epsf     - convergence tolerance on
358: %            function values
359: % M        - function evaluation limit to
360: %            terminate search
361: % ifpr     - when this parameter equals one,
362: %            different stages in the search
363: %            are printed
364: % varargin - variable length list of parameters
365: %            which can be passed to function F
366: % xmin     - coordinates of the smallest
367: %            function value
368: % fmin     - smallest function value found
369: % m        - total number of function
370: %            evaluations made
371: % ntype    - a vector containing
372: %            [ninit,nrefl,nexpn,ncontr,nshrnk]
```

```
373: %               which tells the number of reflect-
374: %               ions, expansions, contractions,and
375: %               shrinkages performed
376: %
377: % User m functions called: objective function
378: %                     named in the argument list
379:
380: if isempty(ifpr), ifpr=0; end
381: if isempty(M), M=500; end;
382: if isempty(epsf), epsf=1e-5; end
383: if isempty(epsx), epsx=1e-5; end
384:
385: % Initialize the simplex array
386: x0=x0(:); n=length(x0); N=n+1; f=zeros(1,N);
387: x=repmat(x0,1,N)+[zeros(n,1),dx*eye(n,n)];
388: for k=1:N
389:   f(k)=feval(F,x(:,k),varargin{:});
390: end
391:
392: ninit=N; nrefl=0; nexpn=0; ncontr=0;
393: nshrnk=0; m=N;
394:
395: Erx=realmax; Erf=realmax;
396: alpha=1.0; % Reflection coefficient
397: beta= 0.5; % Contraction coefficient
398: gamma=2.0; % Expansion coefficient
399:
400: % Top of the minimization loop
401:
402: while Erx>epsx | Erf>epsf
403:
404:   [f,k]=sort(f); x=x(:,k);
405:
406:   % Exit if maximum allowable number of
407:   % function values is exceeded
408:   if m>M, xmin=x(:,1); fmin=f(1); return; end
409:
410:   % Generate the reflected point and
411:   % function value
412:   c=sum(x(:,1:n),2)/n; xr=c+alpha*(c-x(:,N));
413:   fr=feval(F,xr,varargin{:}); m=m+1;
414:   nrefl=nrefl+1;
415:   if ifpr==1, fprintf(' :RFL \n'); end
416:
417:   if fr<f(1)
```

```
418:      % Expand and take best from expansion
419:      % or reflection
420:      xe=c+gamma*(xr-c);
421:      fe=feval(F,xe,varargin{:});
422:      m=m+1; nexpn=nexpn+1;
423:      if ifpr==1, fprintf(' :EXP \n'); end
424:
425:      if fr<fe
426:        % The reflected point was best
427:        f(N)=fr; x(:,N)=xr;
428:      else
429:        % The expanded point was best
430:        f(N)=fe; x(:,N)=xe;
431:      end
432:
433:    elseif fr<=f(n)  % In the middle zone
434:      f(N)=fr; x(:,N)=xr;
435:
436:    else
437:      % Reflected point exceeds the second
438:      % highest value so either use contraction
439:      % or shrinkage
440:      if fr<f(N)
441:        xx=xr; ff=fr;
442:      else
443:        xx=x(:,N); ff=f(N);
444:      end
445:
446:      xc=c+beta*(xx-c);
447:      fc=feval(F,xc,varargin{:});
448:      m=m+1; ncontr=ncontr+1;
449:
450:      if fc<=ff
451:        % Accept the contracted value
452:        x(:,N)=xc; f(N)=fc;
453:        if ifpr==1, fprintf(' :CNT \n'); end
454:
455:      else
456:        % Shrink the simplex toward
457:        % the best point
458:        x=(x+repmat(x(:,1),1,N))/2;
459:        for j=2:N
460:          f(j)=feval(F,x(:,j),varargin{:});
461:        end
462:        m=m+n; nshrnk=nshrnk+n;
```

```
463:       if ifpr==1, fprintf(' :SHR \n'); end
464:     end
465:   end
466:
467:   % Evaluate parameters to check convergence
468:   favg=sum(f)/N; Erf=sqrt(sum((f-favg).^2)/n);
469:   xcent=sum(x,2)/N; xdif=x-repmat(xcent,1,N);
470:   Erx=max(sqrt(sum(xdif.^2)));
471:
472: end % Bottom of the optimization loop
473:
474: xmin=x(:,1); fmin=f(1);
475: ntype=[ninit,nrefl,nexpn,ncontr,nshrnk];
476:
477: %=================================================
478:
479: function [d,r,R]=surf2surf(x,y,z,X,Y,Z,n)
480: % [d,r,R]=surf2surf(x,y,z,X,Y,Z,n)
481: % ~~~~~~~~~~~~~~~~~~~~~~~~~~~~~~~~~~
482: % This function determines the closest points on two
483: % surfaces and the distance between these points. It
484: % is similar to function srf2srf except that large
485: % arrays can be processed.
486: %
487: % x,y,z  -  arrays of points on the first surface
488: % X,Y,Z  -  arrays of points on the second surface
489: % d      -  the minimum distance between the surfaces
490: % r,R    -  vectors containing the coordinates of the
491: %           nearest points on the first and the
492: %           second surface
493: % n      -  length of subvectors used to process the
494: %           data arrays. Sending vectors of length
495: %           n to srf2srf and taking the best of the
496: %           subresults allows processing of large
497: %           arrays of data points
498: %
499: % User m functions used: srf2srf
500:
501: if nargin<7, n=500; end
502: N=prod(size(x)); M=prod(size(X)); d=realmax;
503: kN=max(1,floor(N/n)); kM=max(1,floor(M/n));
504: for i=1:kN
505:   i1=1+(i-1)*n; i2=min(i1+n,N); i12=i1:i2;
506:   xi=x(i12); yi=y(i12); zi=z(i12);
507:   for j=1:kM
```

```
508:     j1=1+(j-1)*n; j2=min(j1+n,M); j12=j1:j2;
509:     [dij,rij,Rij]=srf2srf(...
510:                   xi,yi,zi,X(j12),Y(j12),Z(j12));
511:     if dij<d, d=dij; r=rij; R=Rij; end
512:   end
513: end
514: 
515: %==============================================
516: 
517: function [d,r,R]=srf2srf(x,y,z,X,Y,Z)
518: % [d,r,R]=srf2srf(x,y,z,X,Y,Z)
519: % ~~~~~~~~~~~~~~~~~~~~~~~~~~~~
520: % This function determines the closest points on two
521: % surfaces and the distance between these points.
522: % x,y,z  -  arrays of points on the first surface
523: % X,Y,Z  -  arrays of points on the second surface
524: % d      -  the minimum distance between the surfaces
525: % r,R    -  vectors containing the coordinates of the
526: %           nearest points on the first and the
527: %           second surface
528: 
529: x=x(:); y=y(:); z=z(:); n=length(x); v=ones(n,1);
530: X=X(:)'; Y=Y(:)'; Z=Z(:)'; N=length(X); h=ones(1,N);
531: d2=(x(:,h)-X(v,:)).^2; d2=d2+(y(:,h)-Y(v,:)).^2;
532: d2=d2+(z(:,h)-Z(v,:)).^2;
533: [u,i]=min(d2); [d,j]=min(u); i=i(j); d=sqrt(d);
534: r=[x(i);y(i);z(i)]; R=[X(j);Y(j);Z(j)];
535: 
536: %==============================================
537: 
538: function R=rads
539: % R=rads
540: % Radii for the problem solutions
541: 
542: R=[...
543: 0.7045    3.2903    0.8263
544: 3.2932    0.7074    0.8295
545: 0.7783    3.4977    0.3767
546: 3.4994    0.7800    0.3755
547: 0.0026    3.0000    2.9934
548: 0.0028    1.0000    3.0001
549: 0.7034    0.7107   -2.0000
550: 1.5139    1.5320   -0.7382];
551: 
552: %==============================================
```

```
553:
554: % surfmany(x1,y1,z1,x2,y2,z2,x3,y3,z3,...
555: %          xn,yn,zn)
556: % See Appendix B
```

Appendix A

List of MATLAB Routines with Descriptions

Table A.1: Description of MATLAB Programs and Selected Functions

Routine	Chapter	Description
finance	1	Financial analysis program illustrating programming methods.
inputv	1	Function to read several data items on one line.
polyplot	2	Program comparing polynomial and spline Interpolation.
squarrun	2	Program illustrating conformal mapping of a square.
squarmap	2	Function for Schwarz-Christoffel mapping of a circular disk inside a square.
cubrange	2	Function to compute data range limits for 2D or 3D data.
pendulum	2	Program showing animated large oscillations of a pendulum.
animpen	2	Function showing pendulum animation.
smdplot	2	Program to animate forced motion of a spring-mass-damper system.
smdsolve	2	Function to solve a constant coefficient linear second order differential equation with a harmonic forcing function.
strngrun	2	Program animating wave motion in a string with given initial deflection.
strngwav	2	Function to compute deflections of a vibrating string.
animate	2	Function to show animation of a vibrating string.

continued on next page

Routine	Chapter	Description
splinerr	2	Program showing differential geometry properties of a space curve.
curvprpsp	2	Function using spline interpolation to compute differential properties of a space curve.
splined	2	Function to compute first or second derivatives of a cubic spline.
srfex	2	Program illustrating combined plotting of several surfaces.
frus	2	Function to compute points on a frustum.
surfmany	2	Function to plot several functions together without distortion.
rgdbodmo	2	Program illustrating 3D rigid body rotation and translation.
rotatran	2	Function to perform coordinate rotation.
membran	3	Program illustrating static deflection of a membrane.
mbvprun	3	Program to solve a mixed boundary value problem for a circular disk.
makratsq	3	Program showing conformal mapping of a square using rational functions.
ratcof	3	Function to compute coefficients for rational function interpolation.
raterp	3	Function to evaluate a rational function using coefficients from function **raterp**.
strdyneq	3	Program to solve the structural dynamics equation using eigenvalue-eigenvector methods.
fhrmck	3	Function to solve a linear second order matrix differential equation having a harmonic forcing function.
recmemfr	3	Program illustrating use of functions **null** and **eig** to compute rectangular membrane frequencies.

continued on next page

Routine	Chapter	Description
multimer	3	Program comparing execution of intrinsic MATLAB matrix multiplication and slow Fortran style using loops.
lintrp	4	Function for piecewise linear interpolation allowing finite jump discontinuities.
curvprop	4	Program to compute the length and area bounded by a curve defined by spline interpolation.
spcof	4	Function to compute spline interpolation coefficients used by function **spterp**.
spterp	4	Function to interpolate, differentiate, and integrate a cubic spline having general end conditions.
powermat	4	Function used by functions **spcof** and **spterp**.
splineq	4	Function to interpolate, integrate, and differentiate using the intrinsic function **spline**.
splincof	4	Function that computes coefficients used by **splineg** to handle general end conditions.
matlbdat	4	Program that draws the word MATLAB using a spline.
finitdif	4	Program to compute finite difference formulas.
findifco	4	Function to compute finite difference formulas for derivatives of arbitrary order.
simpson	5	Function using Simpson's rule to integrate an exact function or one defined by spline interpolation.
gcquad	5	Function to perform composite Gauss integration of arbitrary order, and return the base points and weight factors.
quadtest	5	Program comparing the performance of **gcquad** and **quadl** for several test functions.

continued on next page

Routine	Chapter	Description
areaprog	5	Program to compute area, centroidal coordinates and inertial properties of general areas bounded by spline curves.
aprop	5	Function to compute geometrical properties of general areas.
volrevol	5	Program to compute geometrical properties of partial volumes of revolution bounded by spline curves.
volrev	5	Function to compute geometrical properties of partial volumes of revolution.
rotasurf	5	Function to plot a partial surface of revolution.
ropesymu	5	Program using numerical and symbolic computation to evaluate geometrical properties of a rope shaped solid.
ropedraw	5	Function to draw a twisted rope shaped surface.
twistprop	5	Function using symbolic computation to obtain geometrical properties.
srfv	5	Function to compute geometrical properties of a solid specified by general surface coordinates.
polhdrun	5	Program to produce geometrical properties and a surface plot of an arbitrary polyhedron.
polhedron	5	Function for geometrical properties of a polyhedron.
polyxy	5	Function for geometrical properties of a polygon.
sqrtquadtest	5	Program using **quadl** and **gcquad** to evaluate integrals having square root type singularities at the integration end points.
quadqsqrt	5	Function applying **gcquad** to integrals having square root type singularities.
quadlsqrt	5	Function applying **quadl** to integrals having square root type singularities.

continued on next page

Routine	Chapter	Description
triplint	5	Program applying Gauss quadrature to evaluate a triple integral with variable integration limits.
plotjrun	6	Program to compute and plot integer order Bessel functions using the FFT.
runimpv	6	Program using the FFT to analyze earthquake data.
fouapprox	6	Function for Fourier series approximation of a general function.
fouseris	6	Program to plot truncated Fourier series expansions of general functions.
fousum	6	Function to sum a Fourier series and include coefficient smoothing.
cablinea	7	Program showing modal superposition analysis of a swinging cable.
udfrevib	7	Function computing undamped response of a second order matrix differential equation with general initial conditions.
strdynrk	7	Function using **ode45** to solve a second order matrix differential equation.
deislner	7	Program comparing implicit second and fourth order integrators which use fixed stepsize.
mckde2i	7	Function to solve a matrix ODE using a second order fixed stepsize integrator.
mckde4i	7	Function to solve a matrix ODE using a fourth order fixed stepsize integrator.
rkdestab	8	Program to plot stability zones for Runge-Kutta integrators.
prun	8	Program illustrating **ode45** response calculation of an inverted pendulum.
toprun	8	Program for dynamic response of a spinning top.
traject	8	Program for a projectile trajectory.
cablenl	8	Program illustrating animated nonlinear dynamic response for a multi-link cable of rigid links.

continued on next page

Routine	Chapter	Description
plotmotn	8	Function to animate the dynamic response of a cable.
sprchan	8	Program for animated nonlinear dynamics of an elastic cable shaken at both ends.
laplarec	9	Program using Fourier series to solve the Laplace equation in a rectangle having general boundary conditions.
recseris	9	Function to compute a harmonic function and gradient components in a rectangular region.
stringft	9	Program for Fourier series solution and animated response for a string with given initial displacement.
forcmove	9	Program for response of a string subjected to a moving concentrated load.
membwave	9	Program animating the response of a rectangular or circular membrane subjected to an oscillating concentrated force.
memrecwv	9	Function for dynamic response of a rectangular membrane.
memcirwv	9	Function for dynamic response of a circular membrane.
besjroot	9	Function to compute a table of integer order Bessel function roots.
membanim	9	Function to show animated membrane response.
bemimpac	9	Program showing wave propagation in a simply supported beam subjected to an oscillating end moment.
beamanim	9	Function to animate the motion of a vibrating beam.
pilevibs	9	Program illustrating the response of a pile embedded in an oscillating elastic foundation.
slabheat	9	Program for heat conduction in a slab having sinusoidally varying end temperature.

continued on next page

Routine	Chapter	Description
heatcyln	9	Program analyzing transient heat conduction in a circular cylinder.
tempstdy	9	Function for the steady-state temperature in a circular cylinder with general boundary conditions.
foubesco	9	Function to compute coefficients in a Fourier-Bessel series.
besjtabl	9	Function giving a table of integer order Bessel function roots.
rector	9	Program to compute torsional stresses in a beam of rectangular cross section.
eigverr	10	Program comparing eigenvalues of a second order differential equation computed using finite difference methods and using collocation with spline interpolation.
prnstres	10	Function to compute principal stresses and principal directions for a symmetric second order stress tensor.
trusvibs	10	Program to compute and show animation of the natural vibration modes of a general pin connected truss.
drawtruss	10	Function to draw the deflection modes of a truss.
eigsym	10	Function solving the constrained eigenvalue problem associated with an elastic structure fixed as selected points.
elmstf	10	Function to form mass and stiffness matrices of a pin connected truss.
colbuc	10	Program to compute buckling loads of a variable depth column with general end conditions.
cbfreq	10	Program comparing cantilever beam natural frequencies computed by exact, finite difference, and finite element methods.

continued on next page

Routine	Chapter	Description
cbfrqnwm	10	Function to compute exact cantilever beam frequencies.
cbfrqfdm	10	Function to compute cantilever beam frequencies using finite difference methods.
cbfrqfem	10	Function to compute cantilever beam frequencies using the finite element method.
elipfreq	10	Program for natural frequencies and animation of the mode shapes of an elliptic membrane.
frqsimpl	10	Function to compute elliptic membrane natural frequencies and mode shapes.
eigenrec	10	Function to solve a rectangular eigenvalue problem of the form: $XA + BX = \lambda(XC + DX)$.
plotmode	10	Function to plot the mode shapes of the membrane.
vdb	11	Program to compute shear, moment, slope, and deflection in a variable depth multi-support beam with general external loading conditions.
extload	11	Function to compute load and deformation quantities for distributed and concentrated loading on a beam.
sngf	11	Singularity function used to describe beam loads.
trapsum	11	Trapezoidal rule function used to integrate beam functions.
sqrtsurf	12	Function used to illustrate branch cut discontinuities for an analytic function.
elipinvr	12	Function to invert the function mapping the exterior of a circle onto the exterior of an ellipse.
elipdplt	12	Program showing grid lines for conformal mapping of a circular disk onto an elliptic disk.

continued on next page

Routine	Chapter	Description
elipdisk	12	Function mapping an elliptic disk onto a circular disk.
gridview	12	Function to plot a curvilinear coordinate grid.
linfrac	12	Function to perform linear fractional transformations.
crc2crc	12	Function analyzing mapping of circles and straight lines under a linear fractional transformation.
ecentric	12	Function to determine a concentric annulus which maps onto a given eccentric annulus.
swcsq10	12	Program illustrating both interior and exterior maps regarding a circle and a square.
squarat	12	Rational function map taking the inside of a circle onto the interior of a square or the exterior of a square onto the exterior of a square.
swcsqmap	12	Function using truncated series expansions in relation to circle to square maps.
lapcrcl	12	Program solving the Laplace equation in a circular disk for either Dirichlet or Neumann boundary conditions.
cauchtst	12	Program using a Cauchy integral to solve a mixed boundary value problem for a circular disk.
cauchint	12	Function to numerically evaluate a Cauchy integral.
elipcyl	12	Program illustrating inviscid fluid flow about an elliptic cylinder in an infinite stream.
runtors	12	Program using a Cauchy integral and conformal mapping to compute torsional stresses in a beam.
runplate	12	Program using complex stress functions to compute stresses in a plate with a circular hole.

continued on next page

Routine	Chapter	Description
platecrc	12	Function computing series coefficients for complex stress functions pertaining to a plate with a circular hole.
strfun	12	Function to evaluate stress functions phi and psi.
cartstrs	12	Function using complex stress functions to evaluate Cartesian stress components.
rec2polr	12	Function transforming stress components from Cartesian to polar coordinates.
elipmaxst	12	Program using conformal mapping and complex stress functions to compute stress in a plate with an elliptic hole.
runtraj	13	Program using one-dimensional search to optimize a projectile trajectory.
vibfit	13	Program using multi-dimensional search to fit a nonlinear equation to vibration response data.
cablsolv	13	Program to compute large deflection static equilibrium of a loaded cable.
brachist	13	Program to determine a minimum time descent curve (brachistochrone).
cylclose	13	Program using multi-dimensional search to find the closest points on two adjacent circular cylinders.
surf2surf	13	Function using exhaustive search to find the closest points on two surfaces.
nelmed	13	Function similar to **fminsearch** which implements the Nelder and Mead algorithm for multi-dimensional search.

Appendix B

Selected Utility and Application Functions

Function animate

```
1: function animate(x,y,titl,tim,trace)
2: %
3: % animate(x,y,titl,tim,trace)
4: % ~~~~~~~~~~~~~~~~~~~~~~~~~~~~~
5: % This function performs animation of a 2D curve
6: % x,y - arrays with columns containing curve positions
7: %       for successive times. x can also be a single
8: %       vector if x values do not change. The animation
9: %       is done by plotting (x(:,j),y(:,j)) for
10: %      j=1:size(y,2).
11: % titl- title for the graph
12: % tim - the time in seconds between successive plots
13:
14: if nargin<5, trace=0; else, trace=1; end;
15: if nargin<4, tim=.05; end
16: if nargin<3, trac=''; end; [np,nt]=size(y);
17: if min(size(x))==1, j=ones(1,nt); x=x(:);
18: else, j=1:nt; end; ax=newplot;
19: if trace, XOR='none'; else, XOR='xor'; end
20: r=[min(x(:)),max(x(:)),min(y(:)),max(y(:))];
21: %axis('equal') % Needed for an undistorted plot
22: axis(r), % axis('off')
23: curve = line('color','k','linestyle','-',...
24: 'erase',XOR, 'xdata',[],'ydata',[]);
25: xlabel('x axis'), ylabel('y axis'), title(titl)
26: for k = 1:nt
27:    set(curve,'xdata',x(:,j(k)),'ydata',y(:,k))
28:    if tim>0, pause(tim), end, drawnow, shg
29: end
```

Function aprop

```
1: function [p,zplot]=aprop(xd,yd,kn)
2: %
3: % [p,zplot]=aprop(xd,yd,kn)
4: % ~~~~~~~~~~~~~~~~~~~~~~~~~~~~
5: % This function determines geometrical properties
6: % of a general plane area bounded by a spline
7: % curve
8: %
9: % xd,yd - data points for spline interpolation
10: %         with the boundary traversed in counter-
11: %         clockwise direction. The first and last
12: %         points must match for boundary closure.
13: % kn    - vector of indices of points where the
14: %         slope is discontinuous to handle corners
15: %         like those needed for shapes such as a
16: %         rectangle.
17: % p     - the vector [a,xcg,ycg,axx,axy,ayy]
18: %         containing the area, centroid coordinates,
19: %         moment of inertia about the y-axis,
20: %         product of inertia, and moment of inertia
21: %         about the x-axis.
22: % zplot - complex vector of boundary points for
23: %         plotting the spline interpolated geometry.
24: %         The points include the numerical quadrature
25: %         points interspersed with data values.
26: %
27: % User functions called: gcquad, curve2d
28: if nargin==0
29:   td=linspace(0,2*pi,13); kn=[1,13];
30:   xd=cos(td)+1; yd=sin(td)+1;
31: end
32: nd=length(xd); nseg=nd-1;
33: [dum,bp,wf]=gcquad([],1,nd,6,nseg);
34: [z,zplot,zp]=curve2d(xd,yd,kn,bp);
35: w=[ones(size(z)), z, z.*conj(z), z.^2].*...
36:    repmat(imag(conj(z).*zp),1,4);
37: v=(wf'*w)./[2,3,8,8]; vr=real(v); vi=imag(v);
38: p=[vr(1:2),vi(2),vr(3)+vr(4),vi(4),vr(3)-vr(4)];
39: p(2)=p(2)/p(1); p(3)=p(3)/p(1);
```

Function besjroot

```
1: function rts=besjroot(norder,nrts,tol)
2: %
3: % rts=besjroot(norder,nrts,tol)
4: % ~~~~~~~~~~~~~~~~~~~~~~~~~~~~~
5: % This function computes an array of positive roots
6: % of the integer order Bessel functions besselj of
7: % the first kind for various orders. A chosen number
8: % of roots is computed for each order
9: % norder - a vector of function orders for which
10: %          roots are to be computed. Taking 3:5
11: %          for norder would use orders 3,4, and 5.
12: % nrts   - the number of positive roots computed for
13: %          each order. Roots at x=0 are ignored.
14: % rts    - an array of roots having length(norder)
15: %          rows and nrts columns. The element in
16: %          column k and row i is the k'th root of
17: %          the function besselj(norder(i),x).
18: % tol    - error tolerance for root computation.
19:
20: if nargin<3, tol=1e-5; end
21: jn=inline('besselj(n,x)','x','n');
22: N=length(norder); rts=ones(N,nrts)*nan;
23: opt=optimset('TolFun',tol,'TolX',tol);
24: for k=1:N
25:    n=norder(k); xmax=1.25*pi*(nrts-1/4+n/2);
26:    xsrch=.1:pi/4:xmax; fb=besselj(n,xsrch);
27:    nf=length(fb); K=find(fb(1:nf-1).*fb(2:nf)<=0);
28:    if length(K)<nrts
29:       disp('Search error in function besjroot')
30:       rts=nan; return
31:    else
32:       K=K(1:nrts);
33:       for i=1:nrts
34:          interval=xsrch(K(i):K(i)+1);
35:          rts(k,i)=fzero(jn,interval,opt,n);
36:       end
37:    end
38: end
```

Function cubrange

```
 1: function range=cubrange(xyz,ovrsiz)
 2: %
 3: % range=cubrange(xyz,ovrsiz)
 4: % ~~~~~~~~~~~~~~~~~~~~~~~~~~~
 5: % This function determines limits for a square
 6: % or cube shaped region for plotting data values
 7: % in the columns of array xyz to an undistorted
 8: % scale
 9: %
10: % xyz    - a matrix of the form [x,y] or [x,y,z]
11: %          where x,y,z are vectors of coordinate
12: %          points
13: % ovrsiz - a scale factor for increasing the
14: %          window size. This parameter is set to
15: %          one if only one input is given.
16: %
17: % range  - a vector used by function axis to set
18: %          window limits to plot x,y,z points
19: %          undistorted. This vector has the form
20: %          [xmin,xmax,ymin,ymax] when xyz has
21: %          only two columns or the form
22: %          [xmin,xmax,ymin,ymax,zmin,zmax]
23: %          when xyz has three columns.
24: %
25: % User m functions called:  none
26: %----------------------------------------------
27: 
28: if nargin==1, ovrsiz=1; end
29: pmin=min(xyz); pmax=max(xyz); pm=(pmin+pmax)/2;
30: pd=max(ovrsiz/2*(pmax-pmin));
31: if length(pmin)==2
32:   range=pm([1,1,2,2])+pd*[-1,1,-1,1];
33: else
34:   range=pm([1 1 2 2 3 3])+pd*[-1,1,-1,1,-1,1];
35: end
```

Function curve2d

```
 1: function [z,zplot,zp]=curve2d(xd,yd,kn,t)
 2: %
```

```
 3: % [z,zplot,zp]=curve2d(xd,yd,kn,t)
 4: %~~~~~~~~~~~~~~~~~~~~~~~~~~~~~~~~~
 5: % This function generates a spline curve through
 6: % given data points with corners (slope dis-
 7: % continuities) allowed as selected points.
 8: % xd,yd - real data vectors of length nd
 9: %         defining the curve traversed in
10: %         counterclockwise order.
11: % kn    - vectors of point indices, between one
12: %         and nd, where slope discontinuities
13: %         occur
14: % t     - a vector of parameter values at which
15: %         points on the spline curve are
16: %         computed. The components of t normally
17: %         range from one to nd, except when t is
18: %         a negative integer,-m. Then t is
19: %         replaced by a vector of equally spaced
20: %         values using m steps between each
21: %         successive pair of points.
22: % z     - vector of points on the spline curve
23: %         corresponding to the vector t
24: % zplot - a complex vector of points suitable
25: %         for plotting the geometry
26: % zp    - first derivative of z with respect to
27: %         t for the same values of t as is used
28: %         to compute z
29: %
30: % User m functions called:  splined
31: %----------------------------------------------
32:
33: nd=length(xd); zd=xd(:)+i*yd(:); td=(1:nd)';
34: if isempty(kn), kn=[1;nd]; end
35: kn=sort(kn(:)); if kn(1)~=1, kn=[1;kn]; end
36: if kn(end)~=nd, kn=[kn;nd]; end
37: N=length(kn)-1; m=round(abs(t(1)));
38: if -t(1)==m, t=linspace(1,nd,1+N*m)'; end
39: z=[]; zp=[]; zplot=[];
40: for j=1:N
41:   k1=kn(j); k2=kn(j+1); K=k1:k2;
42:   k=find(k1<=t & t<k2);
43:   if j==N, k=find(k1<=t & t<=k2); end
44:   if ~isempty(k)
45:     zk=spline(K,zd(K),t(k)); z=[z;zk];
46:     zplot=[zplot;zd(k1);zk];
47:     if nargout==3
```

```
48:       zp=[zp;splined(K,zd(K),t(k))];
49:     end
50:   end
51: end
52: zplot=[zplot;zd(end)];
```

Function eigenrec

```
1: function [eigs,vecs,Amat,Bmat]=eigenrec(A,B,C,D)
2: % [eigs,vecs,Amat,Bmat]=eigenrec(A,B,C,D)
3: % Solve a rectangular eigenvalue problem of the
4: % form: X*A+B*X=lambda*(X*C+D*X)
5: n=size(B,1); m=size(A,2); s=[n,m]; N=n*m;
6: Amat=zeros(N,N); Bmat=Amat; kn=1:n; km=1:m;
7: for i=1:n
8:   IK=sub2ind(s,i*ones(1,m),km);
9:   Bikn=B(i,kn); Dikn=D(i,kn);
10:   for j=1:m
11:     I=sub2ind(s,i,j);
12:     Amat(I,IK)=A(km,j)'; Bmat(I,IK)=C(km,j)';
13:     KJ=sub2ind(s,kn,j*ones(1,n));
14:     Amat(I,KJ)=Amat(I,KJ)+ Bikn;
15:     Bmat(I,KJ)=Bmat(I,KJ)+ Dikn;
16:    end
17: end
18: [vecs,eigs]=eig(Bmat\Amat);
19: [eigs,k]=sort(diag(eigs));
20: vecs=reshape(vecs(:,k),n,m,N);
```

Function eigsym

```
1: function [evecs,eigvals]=eigsym(k,m,c)
2: %
3: % [evecs,eigvals]=eigsym(k,m,c)
4: % ~~~~~~~~~~~~~~~~~~~~~~~~~~~~~~~~
5: % This function solves the eigenvalue of the
6: % constrained eigenvalue problem
7: %    k*x=(lambda)*m*x, with c*x=0.
8: % Matrix k must be real symmetric and matrix
9: % m must be symmetric and positive definite;
10: % otherwise, computed results will be wrong.
```

```
11: %
12: % k       - a real symmetric matrix
13: % m       - a real symmetric positive
14: %           definite matrix
15: % c       - a matrix defining the constraint
16: %           condition c*x=0. This matrix is
17: %           omitted if no constraint exists.
18: %
19: % evecs   - matrix of eigenvectors orthogonal
20: %           with respect to k and m. The
21: %           following relations apply:
22: %           evecs'*m*evecs=identity_matrix
23: %           evecs'*k*evecs=diag(eigvals).
24: % eigvals - a vector of the eigenvalues
25: %           sorted in increasing order
26: %
27: % User m functions called: trifacsm
28: %----------------------------------------------
29:
30: if nargin==3
31:   q=null(c); m=q'*m*q; k=q'*k*q;
32: end
33: u=trifacsm(m); k=u'\k/u; k=(k+k')/2;
34: [evecs,eigvals]=eig(k);
35: [eigvals,j]=sort(diag(eigvals));
36: evecs=evecs(:,j); evecs=u\evecs;
37: if nargin==3, evecs=q*evecs; end
```

Function fhrmck

```
1: function [t,y,lam]=fhrmck(m,c,k,f1,f2,w,tlim,nt,y0,v0)
2: %
3: % [t,y,lam]=fhrmck(m,c,k,f1,f2,w,tlim,nt,y0,v0)
4: % ~~~~~~~~~~~~~~~~~~~~~~~~~~~~~~~~~~~~~~~~~~~~~~
5: % This function uses eigenfunction analysis to solve
6: % the matrix differential equation
7: %   m*y''(t)+c*y'(t)+k*y(t)=f1*cos(w*t)+f2*sin(w*t)
8: % with initial conditions of y(0)=y0, y'(0)=v0
9: % The solution is general unless 1) a zero or repeated
10: % eigenvalue occurs or 2) the system is undamped and
11: % the forcing function matches a natural frequency.
12: % If either error condition occurs, program execution
13: % terminates with t and y set to nan.
```

```
14: %
15: % m,c,k    - mass, damping, and stiffness matrices
16: % f1,f2    - amplitude vectors for the sine and cosine
17: %            forcing function components
18: % w        - frequency of the forcing function
19: % tlim     - a vector containing the minimum and
20: %            maximum time limits for evaluation of
21: %            the solution
22: % nt       - the number of times at which the solution
23: %            is evaluated within the chosen limits
24: %            for which y(t) is computed
25: % y0,v0    - initial position and velocity vectors
26: %
27: % t        - vector of time values for the solution
28: % y        - matrix of solution values where y(i,j)
29: %            is the value of component j at time t(i)
30: % lam      - the complex natural frequencies arranged
31: %            in order of increasing absolute value
32:
33: if nargin==0 % Generate default data using 2 masses
34:   m=eye(2,2); k=[2,-1;-1,1]; c=.3*k;
35:   f1=[0;1]; f2=[0;0]; w=0.6; tlim=[0,100]; nt=400;
36: end
37: n=size(m,1); t=linspace(tlim(1),tlim(2),nt);
38: if nargin<10, y0=zeros(n,1); v0=y0; end
39:
40: % Determine eigenvalues and eigenvectors for
41: % the homogeneous solution
42: A=[zeros(n,n), eye(n,n); -m\[k, c]];
43: [U,lam]=eig(A); [lam,j]=sort(diag(lam)); U=U(:,j);
44:
45: % Check for zero or repeated eigenvalues and
46: % for undamped resonance
47: wmin=abs(lam(1)); tol=wmin/1e6;
48: [dif,J]=min(abs(lam-i*w)); lj=num2str(lam(J));
49: if wmin==0, disp(' ')
50:   disp('The homogeneous equation has a zero')
51:   disp('eigenvalue which is not allowed.')
52:   disp('Execution is terminated'), disp(' ')
53:   t=nan; y=nan; return
54: elseif any(abs(diff(lam))<tol)
55:   disp('A repeated eigenvalue occurred.')
56:   disp('Execution is terminated'),disp(' ')
57:   t=nan; y=nan; return
58: elseif dif<tol & sum(abs(c(:)))==0
```

```
59:   disp('The system is undamped and the forcing')
60:   disp(['function resonates with ',...
61:         'eigenvalue ',lj])
62:   disp('Execution is terminated.')
63:   disp(' '), t=nan; y=nan; return
64: else
65:   % Determine the particular solution
66:   a=(-w^2*m+k+i*w*c)\(f1-i*f2);
67:   yp=real(a*exp(i*w*t));
68:   yp0=real(a); vp0=real(i*w*a);
69: end
70:
71: % Scale the homogeneous solution to satisfy the
72: % initial conditions
73: U=U*diag(U\[y0-yp0; v0-vp0]);
74: yh=real(U(1:n,:)*exp(lam*t));
75:
76: % Combine results to obtain the total solution
77: t=t(:); y=[yp+yh]';
78:
79: % Show data graphically only for default case
80: if nargin==0
81:   waterfall(t,(1:n),y'), xlabel('time axis')
82:   ylabel('mass index'), zlabel('Displacements')
83:   title(['DISPLACEMENT HISTORY FOR A ',...
84:          int2str(n),'-MASS SYSTEM'])
85:   colormap([1,0,0]), shg
86: end
```

Function findifco

```
1: function [c,e,m,crat]=findifco(k,a)
2: %
3: % [c,e,m,crat]=findifco(k,a)
4: % ~~~~~~~~~~~~~~~~~~~~~~~~~~
5: % This function approximates the k'th derivative
6: % of a function using function values at n
7: % interpolation points. Let f(x) be a general
8: % function having its k'th derivative denoted
9: % by F(x,k). The finite difference approximation
10: % for the k'th derivative employing a stepsize h
11: % is given by:
12: % F(x,k)=Sum(c(j)*f(x+a(j)*h), j=1:n)/h^k +
```

```
13: %         TruncationError
14: % with m=n-k being the order of truncation
15: % error which decreases like h^m and
16: % TruncationError=(h^m)*(e(1)*F(x,n)+...
17: % e(2)*F(x,n+1)*h+e(3)*F(x,n+2)*h^2+O(h^3))
18: %
19: % a    - a vector of length n defining the
20: %        interpolation points x+a(j)*h where
21: %        x is an arbitrary parameter point
22: % k    - order of derivative evaluated at x
23: % c    - the weighting coeffients in the
24: %        difference formula above. c(j) is
25: %        the multiplier for value f(x+a(j)*h)
26: % e    - error component vector in the above
27: %        difference formula
28: % m    - order of truncation order in the
29: %        formula. The relation m=n-k applies.
30: % crat - a matrix of integers such that c is
31: %        approximated by crat(1,:)./crat(2,:)
32:
33: a=a(:); n=length(a); m=n-k; mat=ones(n,n+4);
34: for j=2:n+4; mat(:,j)=a/(j-1).*mat(:,j-1); end
35: A=pinv(mat(:,1:n)); ec=-A*mat(:,n+1:n+4);
36: c=A(k+1,:); e=-ec(k+1,:);
37: [ctop,cbot]=rat(c,1e-8); crat=[ctop(:)';cbot(:)'];
```

Function gcquad

```
1: function [val,bp,wf]=gcquad(func,xlow,...
2:                      xhigh,nquad,mparts,varargin)
3: %
4: % [val,bp,wf]=gcquad(func,xlow,...
5: %     xhigh,nquad,mparts,varargin)
6:
7: % ~~~~~~~~~~~~~~~~~~~~~~~~~~~~~~~~~
8: %
9: % This function integrates a general function using
10: % a composite Gauss formula of arbitrary order. The
11: % integral value is returned along with base points
12: % and weight factors obtained by an eigenvalue based
13: % method. The integration interval is divided into
14: % mparts subintervals of equal length and integration
15: % over each part is performed with a Gauss formula
```

```
16: % making nquad function evaluations. Results are
17: % exact for polynomials of degree up to 2*nquad-1.
18: % ~~~~~~~~~~~~~~~~~~~~~~~~~~~~~~~~~~~~~~~~~~~~~~~~~~~~~
19: % func           - name of a function to be integrated
20: %                  having an argument list of the form
21: %                  func(x,p1,p2,...) where any auxiliary
22: %                  parameters p1,p2,.. are passed through
23: %                  variable varargin. Use [ ] for the
24: %                  function name if only the base points
25: %                  and weight factors are needed.
26: % xlow,xhigh     - integration limits
27: % nquad          - order of Gauss formula chosen
28: % mparts         - number of subintervals selected in
29: %                  the composite integration
30: % varargin       - variable length parameter used to
31: %                  pass additional arguments needed in
32: %                  the integrand func
33: % val            - numerical value of the integral
34: % bp,wf          - vectors containing base points and
35: %                  weight factors in the composite
36: %                  integral formula
37: %
38: % A typical calculation such as:
39: % Fun=inline('(sin(w*t).^2).*exp(c*t)','t','w','c');
40: % A=0; B=12; nquad=21; mparts=10; w=10; c=8;
41: % [value,pcterr]=integrate(Fun,A,B,nquad,mparts,w,c);
42: % gives value = 1.935685556078172e+040 which is
43: % accurate within an error of 1.9e-13 percent.
44: %
45: % User m functions called:  the function name passed
46: %                           in the argument list
47:
48: %----------------------------------------------
49:
50:  if isempty(nquad),  nquad=10; end
51:  if isempty(mparts), mparts=1; end
52:
53: % Compute base points and weight factors
54: % for the single interval [-1,1]. (Ref:
55: % 'Methods of Numerical Integration' by
56: % P. Davis and P. Rabinowitz, page 93)
57:
58: u=(1:nquad-1)./sqrt((2*(1:nquad-1)).^2-1);
59: [vc,bp]=eig(diag(u,-1)+diag(u,1));
60: [bp,k]=sort(diag(bp)); wf=2*vc(1,k)'.^2;
```

```
61:
62: % Modify the base points and weight factors
63: % to apply for a  composite interval
64: d=(xhigh-xlow)/mparts;  d1=d/2;
65: dbp=d1*bp(:); dwf=d1*wf(:);  dr=d*(1:mparts);
66: cbp=dbp(:,ones(1,mparts))+ ...
67: dr(ones(nquad,1),:)+(xlow-d1);
68: cwf=dwf(:,ones(1,mparts)); wf=cwf(:); bp=cbp(:);
69:
70: % Compute the integral
71: if isempty(func)
72:   val=[];
73: else
74:   f=feval(func,bp,varargin{:}); val=wf'*f(:);
75: end
```

Function gridview

```
1: function gridview(x,y,xlabl,ylabl,titl)
2: %
3: % gridview(x,y,xlabl,ylabl,titl)
4: % ~~~~~~~~~~~~~~~~~~~~~~~~~~~~~~
5: %
6: % This function views a surface from the top
7: % to show the coordinate lines of the surface.
8: % It is useful for illustrating how coordinate
9: % lines distort in a conformal transformation.
10: % Calling gridview with no arguments depicts the
11: % mapping of a polar coordinate grid map under
12: % a transformation of the form
13: % z=R*(zeta+m/zeta).
14: %
15: %  x,y          - real matrices defining a
16: %                 curvilinear coordinate system
17: %  xlabl,ylabl - labels for x and y axes
18: %  titl         - title for the graph
19: %
20: % User m functions called:  cubrange
21: %----------------------------------------------
22:
23: % close
24: if nargin<5
25:   xlabl='real axis'; ylabl='imaginary axis';
```

```
26:   titl='';
27: end
28:
29: % Default example using z=R*(zeta+m/zeta)
30: if nargin==0
31:   zeta=linspace(1,3,10)'* ...
32:        exp(i*linspace(0,2*pi,81));
33:   a=2; b=1; R=(a+b)/2; m=(a-b)/(a+b);
34:   z=R*(zeta+m./zeta); x=real(z); y=imag(z);
35:   titl=['Circular Annulus Mapped onto an ', ...
36:         'Elliptical Annulus'];
37: end
38:
39: range=cubrange([x(:),y(:)],1.1);
40:
41: % The data define a curve
42: if size(x,1)==1 | size(x,2)==1
43:   plot(x,y,'-k'); xlabel(xlabl); ylabel(ylabl);
44:   title(titl); axis('equal'); axis(range);
45:   grid on; figure(gcf);
46:   if nargin==0
47:     print -deps gridviewl;
48:   end
49: % The data define a surface
50: else
51:   plot(x,y,'k-',x',y','k-')
52:   xlabel(xlabl); ylabel(ylabl); title(titl);
53:   axis('equal'); axis(range); grid on;
54:   figure(gcf);
55:   if nargin==0
56:     print -deps gridview;
57:   end
58: end
59:
60: %===============================================
61:
62: function range=cubrange(xyz,ovrsiz)
63: %
64: % range=cubrange(xyz,ovrsiz)
65: % ~~~~~~~~~~~~~~~~~~~~~~~~~~
66: % This function determines limits for a square
67: % or cube shaped region for plotting data values
68: % in the columns of array xyz to an undistorted
69: % scale
70: %
```

```
71: % xyz    - a matrix of the form [x,y] or [x,y,z]
72: %          where x,y,z are vectors of coordinate
73: %          points
74: % ovrsiz - a scale factor for increasing the
75: %          window size. This parameter is set to
76: %          one if only one input is given.
77: %
78: % range  - a vector used by function axis to set
79: %          window limits to plot x,y,z points
80: %          undistorted. This vector has the form
81: %          [xmin,xmax,ymin,ymax] when xyz has
82: %          only two columns or the form
83: %          [xmin,xmax,ymin,ymax,zmin,zmax]
84: %          when xyz has three columns.
85: %
86: % User m functions called:  none
87: %----------------------------------------------
88:
89: if nargin==1, ovrsiz=1; end
90: pmin=min(xyz); pmax=max(xyz); pm=(pmin+pmax)/2;
91: pd=max(ovrsiz/2*(pmax-pmin));
92: if length(pmin)==2
93:   range=pm([1,1,2,2])+pd*[-1,1,-1,1];
94: else
95:   range=pm([1 1 2 2 3 3])+pd*[-1,1,-1,1,-1,1];
96: end
```

Function inputv

```
1: function varargout=inputv(prompt)
2: %
3: % [a1,a2,...,a_nargout]=inputv(prompt)
4: %~~~~~~~~~~~~~~~~~~~~~~~~~~~~~~~~~~~~~~~~~~~~~
5: %
6: % This function reads several values on one line.
7: % The items should be separated by commas or
8: % blanks.
9: %
10: % prompt              - A string preceding the
11: %                       data entry.  It is set
12: %                       to ' ? ' if no value of
13: %                       prompt is given.
14: % a1,a2,...,a_nargout - The output variables
```

```
15: %                             that are created. If
16: %                             not enough data values
17: %                             are given following the
18: %                             prompt, the remaining
19: %                             undefined values are
20: %                             set equal to NaN
21: %
22: % A typical function call is:
23: % [A,B,C,D]=inputv('Enter values of A,B,C,D: ')
24: %
25: %----------------------------------------------
26:
27: if nargin==0, prompt=' ? '; end
28: u=input(prompt,'s'); v=eval(['[',u,']']);
29: ni=length(v); no=nargout;
30: varargout=cell(1,no); k=min(ni,no);
31: for j=1:k, varargout{j}=v(j); end
32: if no>ni
33: for j=ni+1:no, varargout{j}=nan; end
34: end
```

Function lintrp

```
1: function y=lintrp(xd,yd,x)
2: %
3: % y=lintrp(xd,yd,x)
4: % ~~~~~~~~~~~~~~~~~
5: % This function performs piecewise linear
6: % interpolation through data values stored in
7: % xd, yd, where xd values are arranged in
8: % nondecreasing order. The function can handle
9: % discontinuous functions specified when some
10: % successive values in xd are equal. Then the
11: % repeated xd values are shifted by a small
12: % amount to remove the discontinuities.
13: % Interpolation for any points outside the range
14: % of xd is also performed by continuing the line
15: % segments through the outermost data pairs.
16: %
17: % xd,yd - vectors of interpolation data values
18: % x     - matrix of values where interpolated
19: %         values are required
20: %
```

```
21: % y      - matrix of interpolated values
22:
23: k=find(diff(xd)==0);
24: if length(k)~=0
25:   xd(k+1)=xd(k+1)+(xd(end)-xd(1))*1e3*eps;
26: end
27: y=interp1(xd,yd,x,'linear','extrap');
```

Function manyrts

```
1: function roots=manyrts(func,a,b,nsteps,...
2:                        maxrts,tol,varargin)
3: %
4: % roots=manyrts(func,a,b,nsteps,maxrts,tol,...
5: %                varargin)
6: % ~~~~~~~~~~~~~~~~~~~~~~~~~~~~~~~~~~~~~~~~~~~~
7: % This function attempts to find multiple roots
8: % of a function by searching an interval in steps
9: % of equal length and finding a root in each
10: % interval where a sign change occurs
11: % func     - name of a function of the form
12: %            func(x,p1,p2,...) where additional
13: %            parameters after the first are
14: %            passed through varargin
15: % a,b      - upper and lower limits of the
16: %            search interval
17: % nsteps   - number of intervals from a to b
18: %            which are checked to detect a
19: %            sign change
20: % maxrts   - maximum number of roots sought
21: %            within the search limits. The
22: %            search terminates when the number
23: %            of roots found equals maxrts.
24: % tol      - the root tolerance passed to
25: %            function fzero. A default value of
26: %            1e-10 is used if no value is given
27: % varargin - the cell variable provided to pass
28: %            multiple arguments to function func
29:
30: if nargin<6, tol=1e-10; end;
31: if nargin<5, maxrts=100; end
32: if isstruct(tol), options=tol;
33: else
```

```
34:   options=optimset('tolfun',tol,'tolx',tol);
35: end
36: x=linspace(a,b,nsteps); roots=[];
37: rtlast=-realmax;
38: for j=1:nsteps-1
39:   xj=x(j); xj1=x(j+1);
40:   fj=feval(func,xj,varargin{:});
41:   fj1=feval(func,xj1,varargin{:});
42:   if fj.*fj1<=0
43:      rt=fzero(func,[xj,xj1],...
44:                options,varargin{:});
45:      if  (rt-rtlast)>tol
46:         roots=[roots,rt]; rtlast=rt;
47:      end
48:   end
49:   if length(roots)==maxrts, break, end
50: end
```

Function membanim

```
1: function membanim(u,x,y,t)
2: %
3: % function membanim(u,x,y,t)
4: % ~~~~~~~~~~~~~~~~~~~~~~~~~
5: % This function animates the motion of a
6: % vibrating membrane
7: %
8: % u    array in which component u(i,j,k) is the
9: %      displacement for y(i),x(j),t(k)
10: % x,y  arrays of x and y coordinates
11: % t    vector of time values
12:
13: % Compute the plot range
14: if nargin==0;
15:   [u,x,y,t]=memrecwv(2,1,1,15.5,1.5,.5,5);
16: end
17: xmin=min(x(:)); xmax=max(x(:));
18: ymin=min(y(:)); ymax=max(y(:));
19: xmid=(xmin+xmax)/2; ymid=(ymin+ymax)/2;
20: d=max(xmax-xmin,ymax-ymin)/2; Nt=length(t);
21: range=[xmid-d,xmid+d,ymid-d,ymid+d,...
22:        3*min(u(:)),3*max(u(:))];
23:
```

```
24: while 1 % Show the animation repeatedly
25:   disp(' '), disp('Press return for animation')
26:   dumy=input('or enter 0 to stop > ? ','s');
27:   if ~isempty(dumy)
28:     disp(' '), disp('All done'), break
29:   end
30:
31:   % Plot positions for successive times
32:   for j=1:Nt
33:     surf(x,y,u(:,:,j)), axis(range)
34:     xlabel('x axis'), ylabel('y axis')
35:     zlabel('u axis'), titl=sprintf(...
36:     'MEMBRANE POSITION AT T=%5.2f',t(j));
37:     title(titl), colormap([1 1 1])
38:     colormap([127/255 1 212/255])
39:     % axis off
40:     drawnow, shg, pause(.1)
41:   end
42: end
```

Function plotmotn

```
 1: function plotmotn(x,y,titl,isave)
 2: %
 3: % plotmotn(x,y,titl,isave)
 4: % ~~~~~~~~~~~~~~~~~~~~~~~
 5: % This function plots the cable time
 6: % history described by coordinate values
 7: % stored in the rows of matrices x and y.
 8: %
 9: % x,y   - matrices having successive rows
10: %         which describe position
11: %         configurations for the cable
12: % titl  - a title shown on the plots
13: % isave - parameter controlling the form
14: %         of output. When isave is not input,
15: %         only one position at a time is shown
16: %         in rapid succession to animate the
17: %         motion. If isave is given a value,
18: %         then successive are all shown at
19: %         once to illustrate a kinematic
20: %         trace of the motion history.
21: %
```

```
22: % User m functions called:  none
23: %----------------------------------------
24:
25: % Set a square window to contain all
26: % possible positions
27: [nt,n]=size(x);
28: if nargin==4, save =1; else, save=0; end
29: xmin=min(x(:)); xmax=max(x(:));
30: ymin=min(y(:)); ymax=max(y(:));
31: w=max(xmax-xmin,ymax-ymin)/2;
32: xmd=(xmin+xmax)/2; ymd=(ymin+ymax)/2;
33: hold off; clf; axis('normal'); axis('equal');
34: range=[xmd-w,xmd+w,ymd-w,ymd+w];
35: title(titl)
36: xlabel('x axis'); ylabel('y axis')
37: if save==0
38:   for j=1:nt
39:     xj=x(j,:); yj=y(j,:);
40:     plot(xj,yj,'-k',xj,yj,'ok');
41:     axis(range), axis off
42:     title(titl)
43:     figure(gcf), drawnow, pause(.1)
44:   end
45:   pause(2)
46: else
47:   hold off; close
48:   for j=1:nt
49:     xj=x(j,:); yj=y(j,:);
50:     plot(xj,yj,'-k',xj,yj,'ok');
51:     axis(range), axis off, hold on
52:   end
53:   title(titl)
54:   figure(gcf), drawnow, hold off, pause(2)
55: end
56:
57: % Save plot history for subsequent printing
58: % print -deps plotmotn
```

Function polhedrn

```
1: function [v,rc,vrr,irr]=polhedrn(x,y,z,idface)
2: %
3: % [v,rc,vrr,irr]=polhedrn(x,y,z,idface)
```

```
4: % ~~~~~~~~~~~~~~~~~~~~~~~~~~~~~~~~~~~
5: %
6: % This function determines the volume,
7: % centroidal coordinates and inertial moments
8: % for an arbitrary polyhedron.
9: %
10: % x,y,z  - vectors containing the corner
11: %          indices of the polyhedron
12: % idface - a matrix in which row j defines the
13: %          corner indices of the j'th face.
14: %          Each face is traversed in a
15: %          counterclockwise sense relative to
16: %          the outward normal. The column
17: %          dimension equals the largest number
18: %          of indices needed to define a face.
19: %          Rows requiring fewer than the
20: %          maximum number of corner indices are
21: %          padded with zeros on the right.
22: %
23: % v      - the volume of the polyhedron
24: % rc     - the centroidal radius
25: % vrr    - the integral of R*R'*d(vol)
26: % irr    - the inertia tensor for a rigid body
27: %          of unit mass obtained from vrr as
28: %          eye(3,3)*sum(diag(vrr))-vrr
29: %
30: % User m functions called: pyramid
31: %----------------------------------------------
32:
33: r=[x(:),y(:),z(:)]; nf=size(idface,1);
34: v=0; vr=0; vrr=0;
35: for k=1:nf
36:   i=idface(k,:); i=i(find(i>0));
37:   [u,ur,urr]=pyramid(r(i,:));
38:   v=v+u; vr=vr+ur; vrr=vrr+urr;
39: end
40: rc=vr/v; irr=eye(3,3)*sum(diag(vrr))-vrr;
```

Function polyxy

```
1: function [area,xbar,ybar,axx,axy,ayy]=polyxy(x,y)
2: %
3: % [area,xbar,ybar,axx,axy,ayy]=polyxy(x,y)
```

```
4: % ~~~~~~~~~~~~~~~~~~~~~~~~~~~~~~~~~~~~~~~~~~~
5: %
6: % This function computes the area, centroidal
7: % coordinates, and inertial moments of an
8: % arbitrary polygon.
9: %
10: % x,y         - vectors containing the corner
11: %               coordinates. The boundary is
12: %               traversed in a counterclockwise
13: %               direction
14: %
15: % area        - the polygon area
16: % xbar,ybar - the centroidal coordinates
17: % axx         - integral of x^2*dxdy
18: % axy         - integral of xy*dxdy
19: % ayy         - integral of y^2*dxdy
20: %
21: % User m functions called: none
22: %----------------------------------------------
23:
24: n=1:length(x); n1=n+1;
25: x=[x(:);x(1)]; y=[y(:);y(1)];
26: a=(x(n).*y(n1)-y(n).*x(n1))';
27: area=sum(a)/2; a6=6*area;
28: xbar=a*(x(n)+x(n1))/a6; ybar=a*(y(n)+y(n1))/a6;
29: ayy=a*(y(n).^2+y(n).*y(n1)+y(n1).^2)/12;
30: axy=a*(x(n).*(2*y(n)+y(n1))+x(n1).* ...
31:     (2*y(n1)+y(n)))/24;
32: axx=a*(x(n).^2+x(n).*x(n1)+x(n1).^2)/12;
```

Function quadlsqrt

```
1: function v=quadlsqrt(fname,type,a,b,tol,trace,varargin)
2: %
3: % v=quadlsqrt(fname,type,a,b,tol,trace,varargin)
4: % ~~~~~~~~~~~~~~~~~~~~~~~~~~~~~~~~~~~~~~~~~~~~~~
5: %
6: % This function uses the MATLAB integrator quadl
7: % to evaluate integrals having square root type
8: % singularities at one or both ends of the
9: % integration interval a < x < b.
10: % The integrand has the form:
11: % func(x)/sqrt(x-a) if type==1.
```

```
12: % func(x)/sqrt(b-x) if type==2.
13: % func(x)/sqrt((x-a)*(b-x)) if type==3.
14: %
15: % func   - the handle for a function continuous
16: %          from x=a to x=b
17: % type   - 1 if the integrand is singular at x=a
18: %          2 if the integrand is singular at x=b
19: %          3 if the integrand is singular at both
20: %            x=a and x=b.
21: % a,b    - integration limits with b > a
22:
23: if nargin<6 | isempty(trace), trace=0; end
24: if nargin<5 | isempty(tol), tol=1e-8; end
25: if nargin<7
26:   varargin{1}=type; varargin{2}=[a,b];
27:   varargin{3}=fname;
28: else
29:   n=length(varargin); c=[a,b]; varargin{n+1}=type;
30:   varargin{n+2}=c; varargin{n+3}=fname;
31: end
32:
33: if type==1 | type==2
34:   v=2*quadl(@fshift,0,sqrt(b-a),...
35:     tol,trace,varargin{:});
36: else
37:   v=quadl(@fshift,0,pi,tol,trace,varargin{:});
38: end
39:
40: %==========================================
41:
42: function u=fshift(x,varargin)
43: % u=fshift(x,varargin)
44: % This function shifts arguments to produce
45: % a nonsingular integrand called by quadl
46: N=length(varargin); fname=varargin{N};
47: c=varargin{N-1}; type=varargin{N-2};
48: a=c(1); b=c(2); c1=(b+a)/2; c2=(b-a)/2;
49:
50: switch type
51:   case 1, t=a+x.^2; case 2, t=b-x.^2;
52:   case 3, t=c1+c2*cos(x);
53: end
54:
55: if N>3, u=feval(fname,t,varargin{1:N-3});
56: else, u=feval(fname,t); end
```

Function ratcof

```
1: function [a,b]=ratcof(xdata,ydata,ntop,nbot)
2: %
3: % [a,b]=ratcof(xdata,ydata,ntop,nbot)
4: % ~~~~~~~~~~~~~~~~~~~~~~~~~~~~~~~~~~~~~
5: %
6: % Determine a and b to approximate ydata as
7: % a rational function of the variable xdata.
8: % The function has the form:
9: %
10: %     y(x) = sum(1=>ntop) ( a(j)*x^(j-1) ) /
11: %          ( 1 + sum(1=>nbot) ( b(j)*x^(j)) )
12: %
13: % xdata,ydata - input data vectors (real or
14: %               complex)
15: % ntop,nbot   - number of series terms used in
16: %               the numerator and the
17: %               denominator.
18: %
19: %----------------------------------------------------
20:
21: ydata=ydata(:); xdata=xdata(:);
22: m=length(ydata);
23: if nargin==3, nbot=ntop; end;
24: x=ones(m,ntop+nbot); x(:,ntop+1)=-ydata.*xdata;
25: for i=2:ntop, x(:,i)=xdata.*x(:,i-1); end
26: for i=2:nbot
27:   x(:,i+ntop)=xdata.*x(:,i+ntop-1);
28: end
29: ab=pinv(x)*ydata; %ab=x\ydata;
30: a=ab(1:ntop); b=ab(ntop+1:ntop+nbot);
```

Function raterp

```
1: function y=raterp(a,b,x)
2: %
3: % y=raterp(a,b,x)
4: % ~~~~~~~~~~~~~~~
5: % This function interpolates using coefficients
6: % from function ratcof.
7: %
```

```
8: % a,b - polynomial coefficients from function
9: %         ratcof
10: % x   - argument at which function is evaluated
11: % y   - computed rational function values
12: %
13: %-----------------------------------------------
14:
15: a=flipud(a(:)); b=flipud(b(:));
16: y=polyval(a,x)./(1+x.*polyval(b,x));
```

Function smdsolve

```
1: function [x,v]=smdsolve(m,c,k,f1,f2,w,x0,v0,t)
2: %
3: % [x,v]=smdsolve(m,c,k,f1,f2,w,x0,v0,t)
4: % ~~~~~~~~~~~~~~~~~~~~~~~~~~~~~~~~~~~~~~~~~~
5: % This function solves the differential equation
6: % m*x''(t)+c*x'(t)+k*x(t)=f1*cos(w*t)+f2*sin(w*t)
7: % with x(0)=x0 and x'(0)=v0
8: %
9: % m,c,k   - mass, damping and stiffness coefficients
10: % f1,f2   - magnitudes of cosine and sine terms in
11: %           the forcing function
12: % w       - frequency of the forcing function
13: % t       - vector of times to evaluate the solution
14: % x,v     - computed position and velocity vectors
15:
16: ccrit=2*sqrt(m*k); wn=sqrt(k/m);
17:
18: % If the system is undamped and resonance will
19: % occur, add a little damping
20: if c==0 & w==wn; c=ccrit/1e6; end;
21:
22: % If damping is critical, modify the damping
23: % very slightly to avoid repeated roots
24: if c==ccrit; c=c*(1+1e-6); end
25:
26: % Forced response solution
27: a=(f1-i*f2)/(k-m*w^2+i*c*w);
28: X0=real(a); V0=real(i*w*a);
29: X=real(a*exp(i*w*t)); V=real(i*w*a*exp(i*w*t));
30:
31: % Homogeneous solution
```

```
32: r=sqrt(c^2-4*m*k);
33: s1=(-c+r)/(2*m); s2=(-c-r)/(2*m);
34: p=[1,1;s1,s2]\[x0-X0;v0-V0];
35:
36: % Total solution satisfying the initial conditions
37: x=X+real(p(1)*exp(s1*t)+p(2)*exp(s2*t));
38: v=V+real(p(1)*s1*exp(s1*t)+p(2)*s2*exp(s2*t));
```

Function splined

```
1: function val=splined(xd,yd,x,if2)
2: %
3: % val=splined(xd,yd,x,if2)
4: % ~~~~~~~~~~~~~~~~~~~~~~~~~
5: %
6: % This function evaluates the first or second
7: % derivative of the piecewise cubic
8: % interpolation curve defined by the intrinsic
9: % function spline provided in MATLAB.If fewer
10: % than four data points are input, then simple
11: % polynomial interpolation is employed
12: %
13: % xd,yd - data vectors determining the spline
14: %         curve produced by function spline
15: % x     - vector of values where the first or
16: %         the second derivative are desired
17: % if2   - a parameter which is input only if
18: %         y''(x) is required. Otherwise, y'(x)
19: %         is returned.
20: %
21: % val   - the first or second derivative values
22: %         for the spline
23: %
24: % User m functions called: none
25:
26: n=length(xd); [b,c]=unmkpp(spline(xd,yd));
27: if n>3 % Use a cubic spline
28:   if nargin==3, c=[3*c(:,1),2*c(:,2),c(:,3)];
29:   else, c=[6*c(:,1),2*c(:,2)]; end
30:   val=ppval(mkpp(b,c),x);
31: else % Use a simple polynomial
32:   c=polyder(polyfit(xd(:),yd(:),n-1));
33:   if nargin==4, c=polyder(c); end
```

```
34:   val=polyval(c,x);
35: end
```

Function splineg

```
 1: function [val,b,c]=splineg(xd,yd,x,deriv,endc,b,c)
 2: %
 3: % [val,b,c]=splineg(xd,yd,x,deriv,endc,b,c)
 4: % ~~~~~~~~~~~~~~~~~~~~~~~~~~~~~~~~~~~~~~~~~~~~
 5: %
 6: % For a cubic spline curve through data points
 7: % xd,yd, this function evaluates y(x), y'(x),
 8: % y''(x), or integral(y(x)*dx, xd(1) to x(j) )
 9: % for j=1:length(x).The coefficients needed to
10: % evaluate the spline are also computed.
11: %
12: % xd,yd    - data vectors defining the cubic
13: %            spline curve
14: % x        - vector of points where curve
15: %            properties are computed.
16: % deriv    - denoting the spline curve as y(x),
17: %            deriv=0 gives a vector for y(x)
18: %            deriv=1 gives a vector for y'(x)
19: %            deriv=2 gives a vector for y''(x)
20: %            deriv=3 gives a vector of values
21: %               for integral(y(z)*dz) from xd(1)
22: %               to x(j) for j=1:length(x)
23: % endc     - endc=1 makes y'''(x) continuous at
24: %            xd(2) and xd(end-1).
25: %            endc=[2,left_slope,right_slope]
26: %            imposes slope values at both ends.
27: %            endc=[3,left_slope] imposes the left
28: %            end slope and makes the discontinuity
29: %            of y''' at xd(end-1) small.
30: %            endc=[4,right_slope] imposes the right
31: %            end slope and makes the discontinuity
32: %            of y''' at xd(2) small.
33: % b,c        coefficients needed to perform the
34: %            spline interpolation. If these are not
35: %            given, function unmkpp is called to
36: %            generate them.
37: % val        values y(x),y'(x),y''(x) or
38: %            integral(y(z)dz, z=xd(1)..x) for
```

```
39: %             deriv=0,1,2, or 3, respectively.
40: %
41: % User m files called: splincof
42: % ----------------------------------------------
43: if nargin<5 | isempty(endc), endc=1; end
44: if nargin<7, [b,c]=splincof(xd,yd,endc); end
45: n=length(xd); [N,M]=size(c);
46:
47: switch deriv
48:
49: case 0 % Function value
50:   val=ppval(mkpp(b,c),x);
51:
52: case 1 % First derivative
53:   C=[3*c(:,1),2*c(:,2),c(:,3)];
54:   val=ppval(mkpp(b,C),x);
55:
56: case 2 % Second derivative
57:   C=[6*c(:,1),2*c(:,2)];
58:   val=ppval(mkpp(b,C),x);
59:
60: case 3 % Integral values from xd(1) to x
61:   k=M:-1:1;
62:   C=[c./k(ones(N,1),:),zeros(N,1)];
63:   dx=xd(2:n)-xd(1:n-1); s=zeros(n-2,1);
64:   for j=1:n-2, s(j)=polyval(C(j,:),dx(j)); end
65:   C(:,5)=[0;cumsum(s)]; val=ppval(mkpp(b,C),x);
66:
67: end
68:
69: %===============================================
70:
71: function [b,c]=splincof(xd,yd,endc)
72: %
73: % [b,c]=splincof(xd,yd,endc)
74: % ~~~~~~~~~~~~~~~~~~~~~~~~~~
75: % This function determines coefficients for
76: % cubic spline interpolation allowing four
77: % different types of end conditions.
78: % xd,yd - data vectors for the interpolation
79: % endc  - endc=1 makes y'''(x) continuous at
80: %         xd(2) and xd(end-1).
81: %         endc=[2,left_slope,right_slope]
82: %         imposes slope values at both ends.
83: %         endc=[3,left_slope] imposes the left
```

```
84: %          end slope and makes the discontinuity
85: %          of y''' at xd(end-1) small.
86: %          endc=[4,right_slope] imposes the right
87: %          end slope and makes the discontinuity
88: %          of y''' at xd(2) small.
89: %
90: if nargin<3, endc=1; end;
91: type=endc(1); xd=xd(:); yd=yd(:);
92:
93: switch type
94:
95: case 1
96:   % y'''(x) continuous at the xd(2) and xd(end-1)
97:   [b,c]=unmkpp(spline(xd,yd));
98:
99: case 2
100:   % Slope given at both ends
101:   [b,c]=unmkpp(spline(xd,[endc(2);yd;endc(3)]));
102:
103: case 3
104:   % Slope at left end given. Compute right end
105:   % slope.
106:   [b,c]=unmkpp(spline(xd,yd));
107:   c=[3*c(:,1),2*c(:,2),c(:,3)];
108:   sright=ppval(mkpp(b,c),xd(end));
109:   [b,c]=unmkpp(spline(xd,[endc(2);yd;sright]));
110:
111: case 4
112:   % Slope at right end known. Compute left end
113:   % slope.
114:   [b,c]=unmkpp(spline(xd,yd));
115:   c=[3*c(:,1),2*c(:,2),c(:,3)];
116:   sleft=ppval(mkpp(b,c),xd(1));
117:   [b,c]=unmkpp(spline(xd,[sleft;yd;endc(2)]));
118:
119: end
```

Function spterp

```
1: function [v,c]=spterp(xd,yd,id,x,endv,c)
2: % [v,c]=spterp(xd,yd,id,x,endv,c)
3:
4: % This function performs cubic spline interpo-
```

```
 5: % lation. Values of y(x),y'(x),y''(x) or the
 6: % integral(y(t)*dt, xd(1)..x) are obtained.
 7: % xd, yd - data vectors with xd arranged in
 8: %          ascending order.
 9: % id     - id equals 0,1,2,3 to compute y(x),
10: %          y'(x), integral(y(t)*dt,t=xd(1)..x),
11: %          respectively.
12: % v      - values of the function, first deriva-
13: %          tive, second derivative, or integral
14: %          from xd(1) to x
15: % c      - the coefficients defining the spline
16: %          curve.
17: % endv   - vector giving the end conditions in
18: %          one of the following five forms:
19: %          endv=1 or endv omitted makes
20: %            c(2) and c(n-1) zero
21: %          endv=[2,left_end_slope,...
22: %            right_end_slope] to impose slope
23: %            values at each end
24: %          endv=[3,left_end_slope] imposes the
25: %            left end slope value and makes
26: %            c(n-1) zero
27: %          endv=[4,right_end_slope] imposes the
28: %            right end slope value and makes
29: %            c(2) zero
30: %          endv=5 defines a periodic spline by
31: %            making y,y',y" match at both ends
32:
33: if nargin<5 | isempty(endv), endv=1; end
34: n=length(xd); sx=size(x); x=x(:); X=x-xd(1);
35:
36: if nargin<6, c=spcof(xd,yd,endv); end
37:
38: C=c(1:n); s1=c(n+1); m1=c(n+2); X=x-xd(1);
39:
40: if id==0       %  y(x)
41: v=yd(1)+s1*X+m1/2*X.*X+...
42:   powermat(x,xd,3)*C/6;
43: elseif id==1  % y'(x)
44:     v=s1+m1*X+powermat(x,xd,2)*C/2;
45: elseif id==2  % y''(x)
46: v=m1+powermat(x,xd,1)*C;
47: else           % integral(y(t)*dt, t=xd(1)..x)
48: v=yd(1)*X+s1/2*X.*X+m1/6*X.^3+...
49: powermat(x,xd,4)*C/24;
```

```
50: end
51: v=reshape(v,sx);
52:
53: %=============================================
54:
55: function c=spcof(x,y,endv)
56: % c=spcof(x,y,endv)
57: % This function determines spline interpolation
58: % coefficients consisting of the support
59: % reactions concatenated with y' and y'' at
60: % the left end.
61: % x,y  - data vectors of interplation points.
62: %        Denote n as the length of x.
63: % endv - vector of data for end conditions
64: %        described in function spterp.
65: %
66: % c    - a vector [c(1);...;c(n+2)] where the
67: %        first n components are support
68: %        reactions and the last two are
69: %        values of y'(x(1)) and y''(x(1)).
70:
71: if nargin<3, endv=1; end
72: x=x(:); y=y(:); n=length(x); u=x(2:n)-x(1);
73: a=zeros(n+2,n+2); a(1,1:n)=1;
74: a(2:n,:)=[powermat(x(2:n),x,3)/6,u,u.*u/2];
75: b=zeros(n+2,1); b(2:n)=y(2:n)-y(1);
76: if endv(1)==1     % Force, force condition
77:   a(n+1,2)=1; a(n+2,n-1)=1;
78: elseif endv(1)==2 % Slope, slope condition
79:   b(n+1)=endv(2); a(n+1,n+1)=1;
80:   b(n+2)=endv(3); a(n+2,:)=...
81: [((x(n)-x').^2)/2,1,x(n)-x(1)];
82: elseif endv(1)==3 % Slope, force condition
83:   b(n+1)=endv(2); a(n+1,n+1)=1; a(n+2,n-1)=1;
84: elseif endv(1)==4 % Force, slope condition
85:   a(n+1,2)=1; b(n+2)=endv(2);
86:   a(n+2,:)=[((x(n)-x').^2)/2,1,x(n)-x(1)];
87: elseif endv(1)==5
88:   a(n+1,1:n)=x(n)-x'; b(n)=0;
89:   a(n+2,1:n)=1/2*(x(n)-x').^2;
90:   a(n+2,n+2)=x(n)-x(1);
91: else
92:   error(...
93:   'Invalid value of endv in function spcof')
94: end
```

```
95: if endv(1)==1 & n<4, c=pinv(a)*b;
96: else, c=a\b; end
97:
98: %==================================================
99:
100: function a=powermat(x,X,p)
101: % a=powermat(x,X,p)
102: % This function evaluates various powers of a
103: % matrix used in cubic spline interpolation.
104: %
105: % x,X  - arbitrary vectors of length n and N
106: % a    - an n by M matrix of elements such that
107: %        a(i,j)=(x(i)>X(j))*abs(x(i)-X(j))^p
108: x=x(:); n=length(x); X=X(:)'; N=length(X);
109: a=x(:,ones(1,N))-X(ones(n,1),:); a=a.*(a>0);
110: switch p, case 0, a=sign(a); case 1, return;
111: case 2, a=a.*a; case 3; a=a.*a.*a;
112: case 4, a=a.*a; a=a.*a; otherwise, a=a.^p; end
```

Function srfv

```
1: function [v,rc,vrr]=srfv(x,y,z)
2: %
3: % [v,rc,vrr]=srfv(x,y,z)
4: % ~~~~~~~~~~~~~~~~~~~~~~~
5: %
6: % This function computes the volume, centroidal
7: % coordinates, and inertial tensor for a volume
8: % covered by surface coordinates contained in
9: % arrays x,y,z
10: %
11: % x,y,z    - matrices containing the coordinates
12: %            of a grid of points covering the
13: %            surface of the solid
14: % v        - volume of the solid
15: % rc       - centroidal coordinate vector of the
16: %            solid
17: % vrr      - inertial tensor for the solid with the
18: %            mass density taken as unity
19: %
20: % User functions called: scatripl proptet
21: %-----------------------------------------------
22:
```

```
23: % p=inline(...
24: %   'v*(eye(3)*(r(:)''*r(:))-r(:)*r(:)'')','v','r');
25:
26: %d=mean([x(:),y(:),z(:)]);
27: %x=x-d(1); y=y-d(2); z=z-d(3);
28:
29: [n,m]=size(x); i=1:n-1; I=i+1; j=1:m-1; J=j+1;
30: xij=x(i,j); yij=y(i,j); zij=z(i,j);
31: xIj=x(I,j); yIj=y(I,j); zIj=z(I,j);
32: xIJ=x(I,J); yIJ=y(I,J); zIJ=z(I,J);
33: xiJ=x(i,J); yiJ=y(i,J); ziJ=z(i,J);
34:
35: % Tetrahedron volumes
36: v1=scatripl(xij,yij,zij,xIj,yIj,zIj,xIJ,yIJ,zIJ);
37: v2=scatripl(xij,yij,zij,xIJ,yIJ,zIJ,xiJ,yiJ,ziJ);
38: v=sum(sum(v1+v2));
39:
40: % First moments of volume
41: X1=xij+xIj+xIJ; X2=xij+xIJ+xiJ;
42: Y1=yij+yIj+yIJ; Y2=yij+yIJ+yiJ;
43: Z1=zij+zIj+zIJ; Z2=zij+zIJ+ziJ;
44: vx=sum(sum(v1.*X1+v2.*X2));
45: vy=sum(sum(v1.*Y1+v2.*Y2));
46: vz=sum(sum(v1.*Z1+v2.*Z2));
47:
48: % Second moments of volume
49: vrr=proptet(v1,xij,yij,zij,xIj,yIj,zIj,...
50:     xIJ,yIJ,zIJ,X1,Y1,Z1)+...
51:     proptet(v2,xij,yij,zij,xIJ,yIJ,zIJ,...
52:     xiJ,yiJ,ziJ,X2,Y2,Z2);
53: rc=[vx,vy,vz]/v/4; vs=sign(v);
54: v=abs(v)/6; vrr=vs*vrr/120;
55: vrr=[vrr([1 4 5]), vrr([4 2 6]), vrr([5 6 3])]';
56: vrr=eye(3,3)*sum(diag(vrr))-vrr;
57:
58: %vrr=vrr-p(v,rc)+p(v,rc+d); rc=rc+d;
```

Function strdynrk

```
1: function [t,x,v]=strdynrk(t,x0,v0,m,c,k,functim)
2: % [t,x,v]=strdynrk(t,x0,v0,m,c,k,functim)
3: % This function uses ode45 to solve the matrix
4: % differential equation: M*X"+C*X'+K*X=F(t)
```

```
5: % t       - vector of solution times
6: % x0,v0   - initial position and velocity vectors
7: % m,c,k   - mass, damping and stiffness matrices
8: % functim - character name for the driving force
9: % x,v     - arrays containing solution values for
10: %           position and velocity
11: %
12: % A typical call to strdynrk function is:
13: % m=eye(3,3); k=[2,-1,0;-1,2,-1;0,-1,2];
14: % c=.05*k; x0=zeros(3,1); v0=zeros(3,1);
15: % t=linspace(0,10,101);
16: % [t,x,v]=strdynrk(t,x0,v0,m,c,k,'func');
17:
18: global Mi C K F n n1 n2
19: Mi=inv(m); C=c; K=k; F=functim;
20: n=size(m,1); n1=1:n; n2=n+1:2*n;
21: [t,z]=ode45(@sde,t,[x0(:);v0(:)]);
22: x=z(:,n1); v=z(:,n2);
23:
24: %===============================
25:
26: function zp=sde(t,z)
27: % zp=sde(t,z)
28: global Mi C K F n n1 n2
29: zp=[z(n2); Mi*(feval(F,t)-C*z(n2)-K*z(n1))];
30:
31: %===============================
32:
33: function f=func(t)
34: % f=func(t)
35: % This is an example forcing function for
36: % function strdynrk in the case of three
37: % degrees of freedom.
38: f=[-1;0;2]*sin(1.413*t);
```

Function surf2surf

```
1: function [d,r,R]=surf2surf(x,y,z,X,Y,Z,n)
2: % [d,r,R]=surf2surf(x,y,z,X,Y,Z,n)
3: % ~~~~~~~~~~~~~~~~~~~~~~~~~~~~~~~~~~~
4: % This function determines the closest points on two
5: % surfaces and the distance between these points. It
6: % is similar to function srf2srf except that large
```

```
 7: % arrays can be processed.
 8: %
 9: % x,y,z  -  arrays of points on the first surface
10: % X,Y,Z  -  arrays of points on the second surface
11: % d      -  the minimum distance between the surfaces
12: % r,R    -  vectors containing the coordinates of the
13: %           nearest points on the first and the
14: %           second surface
15: % n      -  length of subvectors used to process the
16: %           data arrays. Sending vectors of length
17: %           n to srf2srf and taking the best of the
18: %           subresults allows processing of large
19: %           arrays of data points
20: %
21: % User m functions used: srf2srf
22:
23: if nargin<7, n=500; end
24: N=prod(size(x)); M=prod(size(X)); d=realmax;
25: kN=max(1,floor(N/n)); kM=max(1,floor(M/n));
26: for i=1:kN
27:   i1=1+(i-1)*n; i2=min(i1+n,N); i12=i1:i2;
28:   xi=x(i12); yi=y(i12); zi=z(i12);
29:   for j=1:kM
30:     j1=1+(j-1)*n; j2=min(j1+n,M); j12=j1:j2;
31:     [dij,rij,Rij]=srf2srf(...
32:                   xi,yi,zi,X(j12),Y(j12),Z(j12));
33:     if dij<d, d=dij; r=rij; R=Rij; end
34:   end
35: end
36:
37: %==============================================
38:
39: function [d,r,R]=srf2srf(x,y,z,X,Y,Z)
40: % [d,r,R]=srf2srf(x,y,z,X,Y,Z)
41: % ~~~~~~~~~~~~~~~~~~~~~~~~~~~~
42: % This function determines the closest points on two
43: % surfaces and the distance between these points.
44: % x,y,z  -  arrays of points on the first surface
45: % X,Y,Z  -  arrays of points on the second surface
46: % d      -  the minimum distance between the surfaces
47: % r,R    -  vectors containing the coordinates of the
48: %           nearest points on the first and the
49: %           second surface
50:
51: x=x(:); y=y(:); z=z(:); n=length(x); v=ones(n,1);
```

```
52: X=X(:)'; Y=Y(:)'; Z=Z(:)'; N=length(X); h=ones(1,N);
53: d2=(x(:,h)-X(v,:)).^2; d2=d2+(y(:,h)-Y(v,:)).^2;
54: d2=d2+(z(:,h)-Z(v,:)).^2;
55: [u,i]=min(d2); [d,j]=min(u); i=i(j); d=sqrt(d);
56: r=[x(i);y(i);z(i)]; R=[X(j);Y(j);Z(j)];
```

Function surfmany

```
1: function surfmany(varargin)
2: %function surfmany(x1,y1,z1,x2,y2,z2,...
3: %                x3,y3,z3,..,xn,yn,zn)
4: % This function plots any number of surfaces
5: % on the same set of axes without shape
6: % distortion. When no input is given, then a
7: % six-legged solid composed of spheres and
8: % cylinders is shown.
9: %
10: % User m functions called: none
11: %----------------------------------------------
12:
13: if nargin==0
14:   % Default data for a six-legged solid
15:   n=10; rs=.25; d=7; rs=2; rc=.75;
16:   [xs,ys,zs]=sphere;  [xc,yc,zc]=cylinder;
17:   xs=rs*xs; ys=rs*ys; zs=rs*zs;
18:   xc=rc*xc; yc=rc*yc; zc=2*d*zc-d;
19:   x1=xs; y1=ys; z1=zs;
20:   x2=zs+d; y2=ys; z2=xs;
21:   x3=zs-d; y3=ys; z3=xs;
22:   x4=xs; y4=zs-d; z4=ys;
23:   x5=xs; y5=zs+d; z5=ys;
24:   x6=xs; y6=ys; z6=zs+d;
25:   x7=xs; y7=ys; z7=zs-d;
26:   x8=xc; y8=yc; z8=zc;
27:   x9=zc; y9=xc; z9=yc;
28:   x10=yc; y10=zc; z10=xc;
29:   varargin={x1,y1,z1,x2,y2,z2,x3,y3,z3,...
30:        x4,y4,z4,x5,y5,z5,x6,y6,z6,x7,y7,z7,...
31:        x8,y8,z8,x9,y9,z9,x10,y10,z10};
32: end
33:
34: % Find the data range
35: n=length(varargin);
```

```
36: r=realmax*[1,-1,1,-1,1,-1];
37: s=inline('min([a;b])','a','b');
38: b=inline('max([a;b])','a','b');
39: for k=1:3:n
40: x=varargin{k}; y=varargin{k+1};
41: z=varargin{k+2};
42: x=x(:); y=y(:); z=z(:);
43: r(1)=s(r(1),x); r(2)=b(r(2),x);
44: r(3)=s(r(3),y); r(4)=b(r(4),y);
45: r(5)=s(r(5),z); r(6)=b(r(6),z);
46: end
47:
48: % Plot each surface
49: hold off, newplot
50: for k=1:3:n
51: x=varargin{k}; y=varargin{k+1};
52: z=varargin{k+2};
53: surf(x,y,z); axis(r), hold on
54: end
55:
56: % Set axes and display the combined plot
57: axis equal, axis(r), grid on
58: xlabel('x axis'), ylabel('y axis')
59: zlabel('z axis')
60: title('SEVERAL SURFACES COMBINED')
61: % colormap([127/255 1 212/255]); % aquamarine
62: colormap([1 1 1]);, figure(gcf), hold off
```

Function volrevol

```
1: function [v,rg,Irr,X,Y,Z,aprop,xd,zd,kn]=...
2:                    volrev(xd,zd,kn,th,nth,noplot)
3: %
4: % [v,rg,Irr,X,Y,Z,aprop,xd,zd,kn]=...
5: %                  volrev(xd,zd,kn,th,nth,noplot)
6: %~~~~~~~~~~~~~~~~~~~~~~~~~~~~~~~~~~~~~~~~
7:
8: % This function computes geometrical properties
9: % for a volume of revolution resulting when a
10: % closed curve in the (x,z) plane is rotated,
11: % through given angular limits, about the z axis.
12: % The cross section of the volume is defined by
13: % a spline curve passed through data points
```

```
14: % (xd,zd) in the same manner as was done in
15: % function areaprop for plane areas.
16:
17: % xd,zd - data vectors defining the spline
18: %         interpolated boundary, which is
19: %         traversed in a counterclockwise
20: %         direction
21: % kn    - indices of any points where slope
22: %         discontinuity is allowed to turn
23: %         sharp corners
24: % p     - vector of volume properties containing
25: %         [v, xcg, ycg, zcg, vxx, vyy, vzz,...
26: %         vxy, vyz, vzx] where v is the volume,
27: %         (xcg,ycg,zcg) are coordinates of the
28: %         centroid, and the remaining properties
29: %         are volume integrals of the following
30: %         integrand:
31: %         [x.^, y.^2, z.^2, xy, yz, zx]*dxdyxz
32: % X,Y,Z - data arrays containing points on the
33: %         surface of revolution. Plotting these
34: %         points shows the solid volume with
35: %         the ends left open. Function fill3
36: %         is used to plot the surface with ends
37: %         closed
38: % aprop - a vector containing properties of the
39: %         area in the (x,z) plane that was used
40: %         to generate the volume. aprop=[area,...
41: %         xcentroidal, ycentroidal, axx, axz, azz].
42:
43: % User m functions called: rotasurf, gcquad,
44: %         curve2d, anglefun, splined
45: %-----------------------------------------------
46: if nargin==0
47:   t1=-pi:pi/6:0; t2=0:pi/6:pi;
48:   Zd=[0,exp(i*t1),1/2+i+exp(i*t2)/2,0,-1];
49:   xd=real(Zd)+4; zd=imag(Zd);
50:   kn=[1,2,8,9,15,16];
51:   th=[-pi/2,pi]; nth=31;
52: end
53:
54: % Plot a surface of revolution based on the
55: % input data points
56: if nargin==6
57:   [X,Y,Z]=rotasurf(xd,zd,th,nth,1);
58: else
```

```
59:     [X,Y,Z]=rotasurf(xd,zd,th,nth); pause
60: end
61:
62: % Obtain base points and weight factors for the
63: % composite Gauss formula of order seven used in
64: % the numerical integration
65: nd=length(xd); nseg=nd-1;
66: [dum,bp,wf]=gcquad([],1,nd,7,nseg);
67:
68: % Evaluate complex points and derivative values
69: % on the spline curve which is rotated to form
70: % the volume of revolution
71: [u,uplot,up]=curve2d(xd,zd,kn,bp);
72: % plot(real(uplot),imag(uplot)), axis equal,shg
73: u=u(:); up=up(:); n=length(bp);
74: x=real(u); dx=real(up); z=imag(u);
75: dz=imag(up); da=x.*dz-z.*dx;
76:
77: % Evaluate line integrals for area properties
78: p=[ones(n,1), x, z, x.^2, x.*z, z.^2, x.^3,...
79:            (x.^2).*z, x.*(z.^2)].*repmat(da,1,9);
80: p=(wf(:)'*p)./[2 3 3 4 4 4 5 5 5];
81:
82: % Scale area properties by multipliers involving
83: % the rotation angle for the volume
84: f=anglefun(th(2))-anglefun(th(1));
85: v=f(1)*p(2); rg=f([2 3 1]).*p([4 4 5])/v;
86: vrr=[f([4 5 2]); f([5 6 3]); f([2 3 1])].*...
87:     [p([7 7 8]); p([7 7 8]); p([8 8 9])];
88: Irr=eye(3)*sum(diag(vrr))-vrr;
89: aprop=[p(1),p(2:3)/p(1),p(4:6)];
```

References

[1] M. Abramowitz and I.A. Stegun. *Handbook of Mathematical Functions with Formulas, Graphs, and Mathematical Tables*. National Bureau of Standards, Applied Math. Series #55. Dover Publications, 1965.

[2] J. H. Ahlberg, E. N. Nilson, and J. L. Walsh. *The Theory of Splines and Their Applications*. Mathematics in Science and Engineering, Volume 38. Academic Press, 1967.

[3] J. Albrecht, L. Collatz, W. Velte, and W. Wunderlich, editors. *Numerical Treatment of Eigenvalue Problems*, volume 4. Birkhauser Verlag, 1987.

[4] E. Anderson, Z. Bai, C. Bischof, J. Demmel, J. Dongarra, J. Du Croz, A. Greenbaum, S. Hammarling, A. McKenney, S. Ostrouchov, and D. Sorensen. *LAPACK User's Guide*. SIAM, Philadelphia, 1992.

[5] F. Arbabi and F. Li. Macroelements for variable-section beams. *Computers & Structures*, 37(4):553–559, 1990.

[6] B. A. Barsky. *Computer Graphics and Geometric Modeling Using Beta-splines*. Computer Science Workbench. Springer-Verlag, 1988.

[7] K. J. Bathe. *Finite Element Procedures in Engineering Analysis*. Prentice-Hall, 1982.

[8] E. Becker, G. Carey, and J. Oden. *Finite Elements, An Introduction*. Prentice-Hall, 1981.

[9] F. Beer and R. Johnston, Jr. *Mechanics of Materials*. McGraw-Hill, second edition, 1992.

[10] K. S. Betts. Math packages multiply. *CIME Mechanical Engineering*, pages 32–38, August 1990.

[11] K.E. Brenan, S.L. Campbell, and L.R. Petzold. *Numerical Solution of Initial-Value Problems in Differential-Algebraic Equations*. Elsevier Science Publishers, 1989.

[12] R. Brent. *Algorithms for Minimization Without Derivatives*. Prentice-Hall, 1973.

[13] P. Brown, G. Byrne, and A. Hindmarsh. VODE: A variable coefficient ODE solver. *SIAM J. Sci. Stat. Comp.*, 10:1038–1051, 1989.

[14] G. Carey and J. Oden. *Finite Elements, Computational Aspects*. Prentice-Hall, 1984.

[15] B. Carnahan, H.A. Luther, and J. O. Wilkes. *Applied Numerical Methods*. John Wiley & Sons, 1964.

[16] F. E. Cellier and C. M. Rimvall. Matrix environments for continuous system modeling and simulation. *Simulation*, 52(4):141–149, 1989.

[17] B. Char, K. Geddes, G. Gonnet, and S. Watt. *MAPLE User's Guide*, chapter First Leaves: A Tutorial Introduction to MAPLE. Watcom Publications Ltd., Waterloo, Ontario, 1985.

[18] R. V. Churchhill, J. W. Brown, and R. F. Verhey. *Complex Variables and Applications*. McGraw-Hill, 1974.

[19] Column Research Committee of Japan. *Handbook of Structural Stability*. Corona Publishing Company, Tokyo, 1971.

[20] S. D. Conte and C. de Boor. *Elementary Numerical Analysis: An Algorithmic Approach*. McGraw-Hill, third edition, 1980.

[21] J. W. Cooley and J. W. Tukey. An algorithm for the machine calculation of complex fourier series. *Math. Comp.*, 19:297–301, 1965.

[22] R. Courant and D. Hilbert. *Methods of Mathematical Physics*. Interscience Publishers, 1953.

[23] R. R. Craig Jr. *Structural Dynamics*. John Wiley & Sons, 1988.

[24] J. K. Cullum and R. A. Willoughby, editors. *Large Scale Eigenvalue Problems*, chapter High Performance Computers and Algorithms From Linear Algebra, pages 15–36. Elsevier Science Publishers, 1986. by J. J. Dongarra and D. C. Sorensen.

[25] J. K. Cullum and R. A. Willoughby, editors. *Large Scale Eigenvalue Problems*, chapter Eigenvalue Problems and Algorithms in Structural Engineering, pages 81–93. Elsevier Science Publishers, 1986. by R. G. Grimes, J. G. Lewis, and H. D. Simon.

[26] P. J. Davis and P. Rabinowitz. *Methods of Numerical Integration*. Computer Science and Applied Mathematics. Academic Press, Inc., second edition, 1984.

[27] C. de Boor. *A Practical Guide to Splines*, volume 27 of *Applied Mathematical Sciences*. Springer-Verlag, 1978.

[28] J. Dennis and R. Schnabel. *Numerical Methods for Unconstrained Optimization and Nonlinear Equations*. Prentice-Hall, 1983.

[29] J. Dongarra, E. Anderson, Z. Bai, A. Greenbaum, A. McKenney, J. Du Croz, S. Hammerling, J. Demmel, C. Bischof, and D. Sorensen. LAPACK: A

portable linear algebra library for high performance computers. In *Supercomputing 1990*. IEEE Computer Society Press, 1990.

[30] J. Dongarra, J. Du Croz, I. Duff, and S. Hammarling. A set of level 3 basic linear algebra subprograms. Technical report, Argonne National Laboratory, Argonne, Illinois, August 1988.

[31] J. Dongarra, P. Mayes, and G. R. di Brozolo. The IBM RISC System/6000 and linear algebra operations. Technical Report CS-90-122, University of Tennessee Computer Science Department, Knoxville, Tennessee, December 1990.

[32] J. J. Dongarra, J. Du Croz, I. Duff, and S. Hammarling. A set of level 3 basic linear algebra subprograms. *ACM Transactions on Mathematical Software*, December 1989.

[33] J. J. Dongarra, J. Du Croz, S. Hammarling, and R. Hanson. An extended set of fortran basic linear algebra subprograms. *ACM Transactions on Mathematical Software*, 14(1):1–32, 1988.

[34] J.J. Dongarra, J.R. Bunch, C.B. Moler, and G.W. Stewart. *LINPACK User's Guide*. SIAM, Philadelphia, 1979.

[35] T. Driscoll. Algorithm 756: A MATLAB toolbox for Schwarz-Christoffel mapping. *ACM Transactions on Mathematical Software*, 22(2), June 1996.

[36] A. C. Eberhardt and G. H. Williard. Calculating precise cross-sectional properties for complex geometries. *Computers in Mechanical Engineering*, Sept./Oct. 1987.

[37] W. Flugge. *Handbook of Engineering Mechanics*. McGraw-Hill, 1962.

[38] G. Forsythe and C. B. Moler. *Computer Solution of Linear Algebraic Systems*. Prentice-Hall, 1967.

[39] G. Forsythe and W. Wasow. *Finite Difference Methods for Partial Differential Equations*. John Wiley & Sons, 1960.

[40] G. E. Forsythe, M. A. Malcolm, and C. B. Moler. *Computer Methods for Mathematical Computations*. Prentice-Hall, 1977.

[41] R. L. Fox. *Optimization Methods for Engineering Design*. Addison-Wesley Publishing Company, 1971.

[42] B. S. Garbow, J. M. Boyle, J. Dongarra, and C. B. Moler. *Matrix Eigensystem Routines — EISPACK Guide Extension*, volume 51 of *Lecture Notes in Computer Science*. Springer-Verlag, 1977.

[43] C. W. Gear. *Numerical Initial Value Problems in Ordinary Differential Equations*. Prentice-Hall, 1971.

[44] J. M. Gere and S. P. Timoshenko. *Mechanics of Materials*. Wadsworth, Inc., second edition, 1984.

[45] J. Gleick. *Chaos: Making a New Science*. Viking, 1987.

[46] G.H. Golub and J. M. Ortega. *Scientific Computing and Differential Equations: An Introduction to Numerical Methods*. Academic Press, Inc., 1992.

[47] G.H. Golub and C.F. Van Loan. *Matrix Computations*. Johns Hopkins University Press, second edition, 1989.

[48] D. Greenwood. *Principles of Dynamics*. Prentice-Hall, 1988.

[49] R. Grimes and H. Simon. New software for large dense symmetric generalized eigenvalue problems using secondary storage. *Journal of Computational Physics*, 77:270–276, July 1988.

[50] R. Grimes and H. Simon. Solution of large, dense symmetric generalized eigenvalue problems using secondary storage. *ACM Transactions on Mathematical Software*, 14(3):241–256, September 1988.

[51] P. Henrici. *Discrete Variable Methods in Ordinary Differential Equations*. John Wiley & Sons, 1962.

[52] P. Henrici. *Applied Complex Analysis*, volume 3. John Wiley & Sons, 1986.

[53] E. Horowitz and S. Sohni. *Fundamentals of Computer Algorithms*. Computer Science Press, 1978.

[54] T. J. Hughes. *The Finite Element Method — Linear Static and Dynamic Finite Element Analysis*. Prentice-Hall, 1987.

[55] J. L. Humar. *Dynamics of Structures*. Prentice-Hall, 1990.

[56] L.V. Kantorovich and V.I. Krylov. *Approximate Methods of Higher Analysis*. Interscience Publishers, 1958.

[57] W. Kerner. Large-scale complex eigenvalue problems. *Journal of Computational Physics*, 85(1):1–85, 1989.

[58] H. Kober. *Dictionary of Conformal Transformations*. Dover Publications, 1957.

[59] E. Kreyszig. *Advanced Engineering Mathematics*. John Wiley & Sons, Inc., 1972.

[60] C. Lanczos. *Applied Analysis*. Prentice-Hall, 1956.

[61] L. Lapidus and J. Seinfeld. *Numerical Solution of Ordinary Differential Equations*. Academic Press, 1971.

[62] C. Lawson and R. Hanson. *Solving Least Squares Problems*. Prentice-Hall, 1974.

[63] C. Lawson, R. Hanson, D. Kincaid, and F. Krogh. Basic linear algebra subprograms for fortran usage. *ACM Transactions on Mathematical Software*, 5:308–325, 1979.

[64] Y. T. Lee and A. A. G. Requicha. Algorithms for computing the volume and other integral properties of solids, i. known methods and open issues. *Communications of the ACM*, 25(9), 1982.

[65] I. Levit. A new numerical procedure for symmetric eigenvalue problems. *Computers & Structures*, 18(6):977–988, 1984.

[66] J. A. Liggett. Exact formulae for areas, volumes and moments of polygons and polyhedra. *Communications in Applied Numerical Methods*, 4, 1988.

[67] J. Marin. Computing columns, footings and gates through moments of area. *Computers & Structures*, 18(2), 1984.

[68] L. Meirovitch. *Analytical Methods in Vibrations*. Macmillan, 1967.

[69] L. Meirovitch. *Computational Methods in Structural Dynamics*. Sijthoff & Noordhoff, 1980.

[70] C. Moler and G. Stewart. An algorithm for generalized matrix eigenvalue problems. *SIAM Journal of Numerical Analysis*, 10(2):241–256, April 1973.

[71] C.B. Moler and C.F. Van Loan. Nineteen dubious ways to compute the exponential of a matrix. *SIAM Review*, 20:801–836, 1979.

[72] N.I. Muskhelishvili. *Some Basic Problems of the Mathematical Theory of Elasticity*. P. Noordhoff, Groninger, Holland, 4th edition, 1972.

[73] N.I. Muskhelishvili. *Singular Integral Equations*. P. Noordhoff, Groninger, Holland, 2nd edition, 1973.

[74] N.W. McLachlan. *Theory and Application of Mathieu Functions*. Dover Publications, 1973.

[75] Z. Nehari. *Conformal Mapping*. McGraw-Hill, 1952.

[76] D. T. Nguyen and J. S. Arora. An algorithm for solution of large eigenvalue problems. *Computers & Structures*, 24(4):645–650, 1986.

[77] J. M. Ortega. *Matrix Theory: A Second Course*. Plenum Press, 1987.

[78] J. M. Ortega and W. C. Rheinboldt. *Iterative Solution of Nonlinear Equations in Several Variables*. Academic Press, 1970.

[79] B. Parlett. *The Symmetric Eigenvalue Problem*. Prentice-Hall, 1980.

[80] M. Paz. *Structural Dynamics: Theory & Computation*. Van Nostrand Reinhold Company, 1985.

[81] R. Piessens, E. de Doncker-Kapenga, C.W. Uberhuber, and D.K. Kahaner. *QUADPACK: A Subroutine Package for Automatic Integration*, volume 1 of *Computational Mathematics*. Springer-Verlag, 1983.

[82] P. Prenter. *Splines and Variational Methods*. John Wiley & Sons, 1975.

[83] W. H. Press, B. P. Flannery, S. A. Teukolsky, and W. T. Vetterling. *Numerical Recipes: The Art of Scientific Computing*. Cambridge University Press, 1986.

[84] J. R. Rice. *The Approximation of Functions, Volumes 1 and 2*. Addison-Wesley, 1964.

[85] R. J. Roark and W. C. Young. *Formulas for Stress and Strain*. McGraw-Hill, 1975.

[86] Scientific Computing Associates, Inc., New Haven, CT. *CLAM User's Guide*, 1989.

[87] N. S. Sehmi. *Large Order Structural Analysis Techniques*. John Wiley & Sons, New York, 1989.

[88] L. Shampine and M. Gordon. *Computer Solutions of Ordinary Differential Equations: The Initial Value Problem*. W. H. Freeman, 1976.

[89] B. T. Smith, J. M. Boyle, J. Dongarra, B. S. Garbow, Y. Ikebe, V. C. Klema, and C. Moler. *Matrix Eigensystem Routines — EISPACK Guide*, volume 6 of *Lecture Notes in Computer Science*. Springer-Verlag, 1976.

[90] I. S. Sokolnikoff. *Mathematical Theory of Elasticity*. McGraw-Hill, 1946.

[91] M. R. Spiegel. *Theory and Problems of Vector Analysis*. Schaum's Outline Series. McGraw-Hill, 1959.

[92] M. R. Spiegel. *Theory and Problems of Complex Variables*. Schaum's Outline Series. McGraw-Hill, 1967.

[93] R. Stepleman, editor. *Scientific Computing*, chapter ODEPACK, A Systemized Collection of ODE Solvers. North Holland, 1983. by A. Hindmarsh.

[94] G. W. Stewart. *Introduction to Matrix Computations*. Academic Press, 1973.

[95] G. Strang. *Introduction to Applied Mathematics*. Cambridge Press, 1986.

[96] G. Strang. *Linear Algebra and Its Applications*. Harcourt Brace Jovanovich, 1988.

[97] The MathWorks Inc. *MATLAB User's Guide*. The MathWorks, Inc., South Natick, MA, 1991.

[98] The MathWorks Inc. *The Spline Toolbox for Use With MATLAB*. The MathWorks, Inc., South Natick, MA, 1992.

[99] The MathWorks Inc. *The Student Edition of MATLAB For MSDOS Personal Computers*. The MATLAB Curriculum Series. Prentice-Hall, Englewood Cliffs, NJ, 1992.

[100] S. Timoshenko. *Engineering Mechanics*. McGraw-Hill Book Company, fourth edition, 1956.

[101] S. Timoshenko and D. H. Young. *Advanced Dynamics*. McGraw-Hill Book Company, 1948.

[102] L. H. Turcotte and H. B. Wilson. *Computer Applications in Mechanics of Materials Using MATLAB*. Prentice-Hall, 1998.

[103] C. Van Loan. A survey of matrix computations. Technical Report CTC90TR26, Cornell Theory Center, Ithaca, New York, October 1990.

[104] G. A. Watson, editor. *Lecture Notes in Mathematics*, volume 506. Springer-Verlag, 1975. *An Overview of Software Development for Special Functions* by W. J. Cody.

[105] R. Weinstock. *Calculus of Variations: With Applications to Physics and Engineering*. Dover Publications, 1974.

[106] D. W. White and J. F. Abel. Bibliography on finite elements and supercomputing. *Communications in Applied Numerical Methods*, 4:279–294, 1988.

[107] J. H. Wilkinson. *Rounding Errors in Algebraic Processes*. Prentice-Hall, 1963.

[108] J. H. Wilkinson. *The Algebraic Eigenvalue Problem*. Oxford University Press, 1965.

[109] J. H. Wilkinson and C. Reinsch. *Handbook for Automatic Computation, Volume II: Linear Algebra*. Springer-Verlag, 1971.

[110] H. B. Wilson. *A Method of Conformal Mapping and the Determination of Stresses in Solid-Propellant Rocket Grains*. PhD thesis, Dept. of Theoretical and Applied Mechanics, University of Illinois, Urbana, IL, February 1963.

[111] H. B. Wilson and G. S. Chang. Line integral computation of geometrical properties of plane faces and polyhedra. In *1991 ASME International Computers in Engineering Conference and Exposition*, Santa Clara, CA, August 1991.

[112] H. B. Wilson and K. Deb. Inertial properties of tapered cylinders and partial volumes of revolution. *Computer Aided Design*, 21(7), September 1989.

[113] H. B. Wilson and K. Deb. Evaluation of high order single step integrators for structural response calculation. *Journal of Sound and Vibration*, 141(1):55–70, 1991.

[114] H. B. Wilson and D. S. Farrior. Computation of geometrical and inertial properties for general areas and volumes of revolution. *Computer Aided Design*, 8(8), 1976.

[115] H. B. Wilson and D. S. Farrior. Stress analysis of variable cross section indeterminate beams using repeated integration. *International Journal of Numerical Methods in Engineering*, 14, 1979.

[116] H. B. Wilson and S. Gupta. Beam frequencies from finite element and finite difference analysis compared using MATLAB. *Sound and Vibration*, 26(8), 1992.

[117] H. B. Wilson and J. L. Hill. Volume properties and surface load effects on three dimensional bodies. Technical Report BER Report No. 266-241, Department of Engineering Mechanics, University of Alabama, Tuscaloosa, Alabama, 1980. U.S. Army Engineer Waterways Experiment Station, Vicksburg, MS, 1980.

[118] S. Wolfram. *A System for Doing Mathematics by Computer*. Addison-Wesley, 1988.

[119] C. R. Wylie. *Advanced Engineering Mathematics*. McGraw-Hill, 1966.

[120] D. Young and R. Gregory. *A Survey of Numerical Mathematics, Volume 1 and 2*. Chelsea Publishing Co., 1990.

Index

C

D

E

F

G

K

L

M

N

O

P

Q

R

S

T

U

V

W

X

Y

Z